Organometallic Compounds

Organometallic Compounds

Synthesis, Reactions, and Applications

Edited by
Dakeshwar Kumar Verma
Jeenat Aslam

WILEY-VCH

Editors

Dr. Dakeshwar Kumar Verma
Govt. Digvijay Autonomous PG College
Department of Chemistry
Rajnandgaon, Chhattisgarh
491441 Chhattisgarh
India

Dr. Jeenat Aslam
Taibah University
Department of Chemistry
Yanbu
30799
Saudi Arabia

Cover Image: © Stampf/pixabay

Library of Congress Card No.: applied for

British Library Cataloguing-in-Publication Data
A catalogue record for this book is available from the British Library.

Bibliographic information published by the Deutsche Nationalbibliothek
The Deutsche Nationalbibliothek lists this publication in the Deutsche Nationalbibliografie; detailed bibliographic data are available on the Internet at <http://dnb.d-nb.de>.

Print ISBN: 978-3-527-35178-7
ePDF ISBN: 978-3-527-84092-2
ePub ISBN: 978-3-527-84093-9
oBook ISBN: 978-3-527-84094-6

Typesetting Straive, Chennai, India
Printing and Binding CPI Group (UK) Ltd, Croydon, CR0 4YY

C9783527351787_130624

Contents

Preface

The present edited book titled *Organometallic Compounds: Synthesis, Reactions, and Applications* discusses the basics, current trends, challenges, and future prospects, showing the value and scope of organometallic compounds toward current applications. This edited book will be carefully written to present a modern account of traditional methods and the latest details of recent advances in organometallic chemistry. The edited book will contain a traditional and logical approach in detail about introduction, nomenclature, synthesis methods, current applications, metallocenes, natural organometallic compounds, emerging role of organometallic compounds, F-block-based organometallic compounds, toxicity and applications of computational modeling for organometallic compounds. The book will be of significant interest to students of chemistry, pharmacy, biochemistry, and chemical engineering at the advanced undergraduate, graduate, and postgraduate levels, as well as academic and industry researchers who wish to familiarize themselves with the concepts and applications of organometallic chemistry. A book to wrap the developments in detailed synthetic reaction mechanisms and industrial applications is long overdue, and the present one will be a milestone in the field.

This will be the only book that provides chronological advancements of organometallic compounds (both synthetic and natural), their synthesis mechanisms, and recent industrial applications in the fields of material science, engineering, and science. The book will also serve as a valuable source for new learners about fundamentals, basics, reactions, catalytic mechanisms, and modern applications such as carbon dioxide fixation, reduction, gas adsorption and gas purification, drug delivery, renewable energy, and waste water treatment. It serves as a valuable reference for scientists, organochemists, biochemists, pharmacists, and engineers who are searching information on organometallic compounds, their current applications, toxicity, and computational modeling.

To confine the comprehensive description of organometallic compounds and to propose a rational and expressive design of the topic and a concentrated up-to-date reference, the book is divided into many chapters. Topics covered in Chapters 1–4 are fundamental aspects, nomenclature, classification properties, and synthesis methods of organometallic compounds. Chapter 5 explains the metal carbonyls' synthesis, properties, and structure. Chapter 6 covers metal-carbon multiple-bonded compounds. Chapter 7 describes the metallocenes' synthesis, properties, and structure. Chapter 8 explains σ-complexes, p-complexes, and

η_n-C_nR_n carbocyclic polyenes-based organometallic compounds. Chapter 9 covers the organometallic complexes of the lanthanoids and actinoids. Chapter 10 describes bio-organometallic chemistry. Chapter 11 explains the important reactions of organometallic compounds. Chapter 12 shows the characterization techniques of organometallic compounds. Chapter 13 discusses the organometallic compounds based on important reagents. Chapter 14 covers homogeneous and heterogeneous catalysis by organometallic complexes. Chapter 15 explains the cluster compounds boranes, heteroboranes, and metallaboranes. Chapter 16 covers the applications of organometallic compounds for carbon dioxide fixation, reduction, gas adsorption, and gas purification. Chapter 17 shows the emerging role of organometallic compounds for drug delivery, renewable energy, and waste water treatment. Chapter 18 explains the toxicity of organometallic compounds. Chapter 19 describes the computational approaches for some important organometallic catalysis reactions.

This book aims to deliver the recent evidences from fundamentals and synthesis to applications on organometallic compounds. The book will be intended for a very broad audience working in the fields of organic synthesis, environmental science and engineering, nanotechnology, energy, chemistry, etc. This book will be a valuable reference source for libraries in universities and industrial institutions, government and independent institutes, individual research groups, and scientists working in the field. Overall, this will be a valuable reference for government and non-government agencies, research scholars of the field, teachers and research supervisors, policy makers, organometallic-compounds-related industries, chemists and chemical engineers working in both R&D and academia who want to learn more on fundamental aspects of organometallic chemistry. The book will be a valuable source and guidebook for science (B. Sc. and M. Sc.), environmental science, pharmaceuticals, biomedical engineering, and engineering (B. Tech. and M. Tech.) students to learn the basics of recent evidences on the proposed title.

The editors and contributors of all chapters are well-known researchers, scientists, and experts from academia and industry.

On behalf of John Wiley & Sons, Inc., we thank all contributors for their exceptional and whole-hearted contribution. Invaluable thanks to Dr. Sakeena Quraishi (Associate Commissioning Editor), Miss Katherine Wong (Senior Managing Editor), and the Editorial Team at John Wiley & Sons, Inc. for their wholehearted support and help during this project. In the end, all appreciation to John Wiley & Sons, Inc. for publishing the book.

02 September 2022

Dakeshwar Kumar Verma
Govt. Digvijay Autonomous
Postgraduate College
Rajnandgaon, Chhattisgarh,
491441, India

Jeenat Aslam
Department of Chemistry,
College of Science, Taibah University
Yanbu-30799, Al-Madina, Saudi Arabia

About the Editors

Dakeshwar Kumar Verma, PhD, is an Assistant Professor of Chemistry at Govt. Digvijay Autonomous Postgraduate College, Rajnandgaon, Chhattisgarh, India. His research is mainly focused on the preparation and designing of organic compounds for various applications and green chemistry. Dr. Verma is the author of more than 60 research papers, review articles, and book chapters in peer-reviewed international journals of ACS, RSC, Wiley, Elsevier, Springer, Taylor & Francis, etc. He has also worked as an editor/co-editor/author on various books published by Elsevier, Wiley Science, and De Gruyter. He has more than 870 citations with an H-index of 16 and an i-10 index of 21. Recently, two full-time Ph.D. research scholars have been working under his guidance. Dr. Verma received a Council of Scientific and Industrial Research Junior Research Fellowship award in 2013. He also availed the MHRD National Fellowship during his Ph.D. in 2013.

Jeenat Aslam, PhD, is currently working as an Associate Professor at the Department of Chemistry, College of Science, Taibah University, Yanbu, Al-Madina, Saudi Arabia. She obtained her PhD in Surface Science/Chemistry at the Aligarh Muslim University, Aligarh, India. Her research is mainly focused on materials and corrosion, nanotechnology, and surface chemistry. Dr. Jeenat has published several research and review articles in peer-reviewed international journals of ACS, Wiley, Elsevier, Springer, Taylor & Francis, Bentham Science, etc. She has authored/edited many books and has contributed to twenty-seven book chapters.

1

Organometallic Compounds: The Fundamental Aspects

Geetha B. Markandeya and Srinivasa Budagumpi

Jain University, Centre for Nano and Material Sciences, Jain Global Campus, 45 km, NH – 209, Jakkasandra Post, Kanakapura Taluk, Ramanagaram 562112, Bangalore, Karnataka, India

1.1 Introduction

1.1.1 Organometallic Chemistry

The branch of chemistry deals with the study of molecules having a metal–carbon bond, in which a compound is said to be an organometallic compound when the metal–carbon bond in a molecule should be completely or partially covalent. Depending upon the elements in the periodic table, organometallic chemistry is mainly classified into main groups metal, transition metal, lanthanide, and actinide-based organometallics.

1.1.2 Organometallic Compounds

The organic compounds contain at least one metal–carbon bond in which metal is directly attached to the carbon atom in which there should be a bonding interaction (covalent, ionic, localized/delocalized) between the metal and the carbon atom are defined as organometallic compounds. The metals may be alkaline metals, alkaline earth metals, and metalloids (boron, silicon, arsenic, germanium, tellurium, and selenium). The bond present between the metal atom and the carbon is likely covalent in nature.

Examples: Organocadmium compounds, organoboron compounds, organozinc compounds, organomagnesium compounds, organolithium compounds, organolead compounds, and organotin compounds.

1.1.3 Structure of Organometallic Compound

The nature of the metal–carbon bond varies from ionic to covalent. The organometallic compounds have some effect on the nature of metal–carbon and these compounds have both organic and metal portions in which the metallic

Organometallic Compounds: Synthesis, Reactions, and Applications, First Edition.
Edited by Dakeshwar Kumar Verma and Jeenat Aslam.

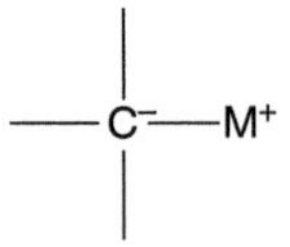

Figure 1.1 General representation of organometallic compound.

portion has greater importance. The electropositive nature of metal will be the ionic nature of the metal–carbon bond.

In organometallic compounds, the carbon is bonded to an electropositive atom, which has a negative charge, whereas the metal has a slightly positive charge; hence, the organometallic compounds having the organic part behave as nucleophilic or basic as shown in Figure 1.1.

1.2 Milestones in Organometallic Compounds

In 1760, the first organometallic compound of the main group (Eq. (1.1)), cadet fuming liquid was discovered. In the Paris military pharmacy, a cadet discovered a fuming liquid while working on ink while preparing cobalt salt from cobalt minerals consisting of arsenic, which is called cacodyl(malodorous)oxide [1, 2].

$$As_2O_3 + 4CH_3COOK \longrightarrow [(CH_3)_2As]_2O \qquad (1.1)$$

1.2.1 Equation (1.1): Synthesis of First Organometallic Compound

In 1827, W. C. Zeise, a Danish pharmacist, discovered the first organometallic compound with a transition metal called Zeise's, $K[PtCl_3C_2H_4]H_2O$. When K_2PtCl_4 was refluxed in ethanol, it resulted in the formation of Zeise's salt (Eq. (1.2)). It was characterized as the first organometallic olefin complex. Herein, they have used platinum because the Nobel metal complexes are stable toward air and moisture [3, 4].

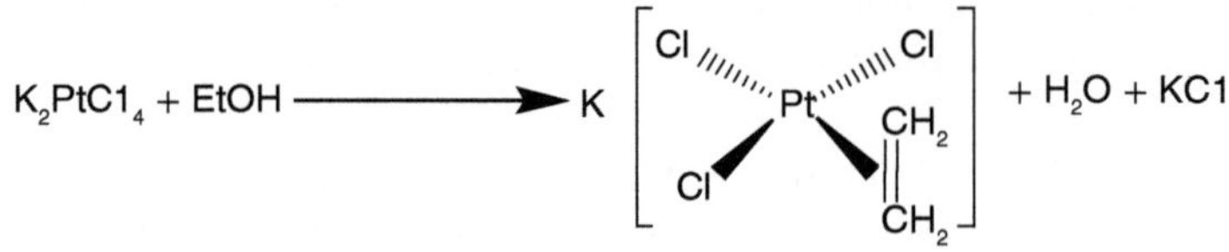

(1.2)

1.2.2 Equation (1.2): Preparation of Zeise's Salt

In 1849, Edward Frankland prepared diethylzinc (a pyrophoric liquid), while trying to prepare an ethyl radical, he ended up with ethylzinc iodide (solid) and diethylzinc (liquid) [5, 6]. Whereas in 1852, he used sodium amalgam and methyl halide to prepare dimethyl mercury. Furthermore, many people have used R_2Hg and R_2Zn following an alkyl transfer reaction to prepare main group elements of organometallic compounds.

In the same year 1852, Schweizer and Lowig used an alloy of Na/Pb to prepare tetraethyllead (Eqs. (1.3) and (1.4)). In 1863, organochlorosilane was prepared by Friedel and Craft using alkylzinc as a reagent (Eq. (1.5)).

$$3C_2H_5I + 3Zn \longrightarrow (C_2H_5)_2Zn + C_2H_5ZnI + ZnI_2 \tag{1.3}$$

$$2CH_3X + 2Na/Hg \longrightarrow (CH_3)_2Hg + 2NaH \tag{1.4}$$

$$SiCl_4 + m/2\,ZnR_2 \longrightarrow RmSiCl_{4-m} + m/2\,ZnCl_2 \tag{1.5}$$

1.2.3 Equations (1.3)–(1.5): Preparation of Organochlorosilane Compound

In 1868, $[PtCl_2(CO)]_2$ the first metal carbonyl compound (Eq. (1.6)) was prepared by Schutzenberger [7].

$$PtCl_2 + CO \longrightarrow [PtCl_2(CO)]_2 + \text{cis } PtCl_2(CO)_2 \tag{1.6}$$

1.2.4 Equation (1.6): Synthesis of First Metal Carbonyl Compound

In 1890, Ludwig Mond prepared the first binary metal carbonyl $Ni(CO)_4$ (Eq. (1.7)), which is used for the refining of nickel [8–10].

$$Ni + 4CO \xrightarrow{1\text{ bar},25^\circ C} Ni(CO)_4 \tag{1.7}$$

1.2.5 Equation (1.7): Synthesis of First Binary Metal Carbonyl Complex

In 1899, the Philippe Barbier, Grignard's teacher introduced Barbier reaction (Eq. (1.8)). It was a one pot reaction carried out in the presence of water. Compared to Grignard reaction this is less versatile.

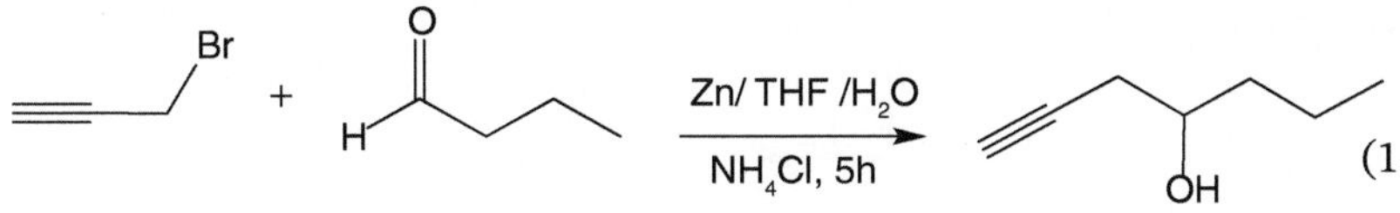

(1.8)

1.2.6 Equation (1.8): Barbier Reaction

Whereas in 1900, in RMgX, Zn was replaced by Mg in Barbier by Grignard and called as Grignard reagent (Eq. (1.9)). This reagent is more versatile than the Barbier reaction and has more applications compared to organozinc reagents [11].

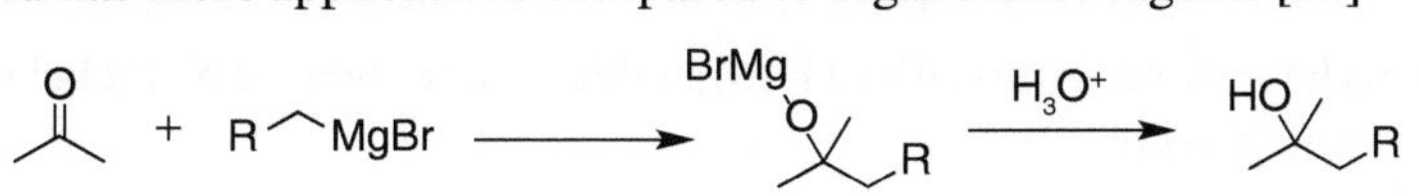

(1.9)

1.2.7 Equation (1.9): Synthesis of Organic Compound Using a Grignard Reagent

In 1912, P. Sabatier and V. Grignard were awarded the Nobel Prize for Grignard reagent, and they followed Sabatier's method for the hydrogenation by using metal powders.

In 1917, the first alkyllithium derivatives (Eqs. (1.10) and (1.11)) were prepared by Wilhelm Schlenk, and he followed the transalkylation of organomercury compounds. Whereas the synthetic strategies of lithium derivatives replace the Grignard reagent as the primary anionic intermediate (1.10).

$$2Li + R_2Hg \longrightarrow 2RLi + Hg \quad (1.10)$$

$$2EtLi + Me_2Hg \longrightarrow 2MeLi + Et_2Hg \quad (1.11)$$

1.2.8 Equations (1.10) and (1.11): Synthesis of Alkyllithium Compound

In 1921, Thomas Midgley introduced tetraethyllead as an additive in gasoline while working on GE motors [12].

In 1930, K. Ziegler prepared organolithium compounds (Eqs. (1.12) and (1.13)) by a simple synthetic procedure and, furthermore, fine-tuned by Gilman. Later on, this compound was widely used.

$$2Li + PhCH_2OMe \longrightarrow PhCH_2Li + MeOLi \quad (1.12)$$

$$RX + 2Li \longrightarrow RLi + LiX \quad (1.13)$$

1.2.9 Equations (1.12) and (1.13): Synthesis of Organolithium Compound

In 1938, hydroformylation (Eq. (1.14)) was discovered by Otto Roelen, for the first time in homogeneous catalysis an organometallic compound has been used [13, 14].

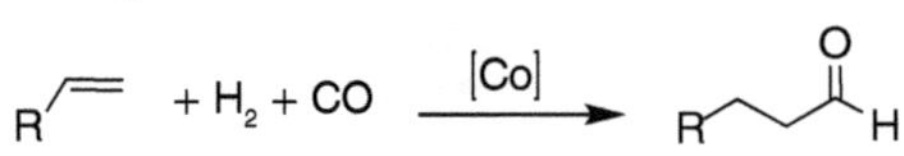

(1.14)

1.2.10 Equation (1.14): Hydroformylation Reaction

In 1943, direct synthesis of organochlorosilane (Eq. (1.15)) was discovered by E G Rochow, further it initiated for large scale production of silicones [15].

$$2CH_3Cl + Si(Cu) \longrightarrow (CH_3)_2SiCl_2 \quad (1.15)$$

1.2.11 Equation (1.15): Synthesis of Organochlorosilane Compound

In 1948, acetylene trimerization (Eq. (1.16)), which was catalyzed by nickel was discovered by W. Reppe [16].

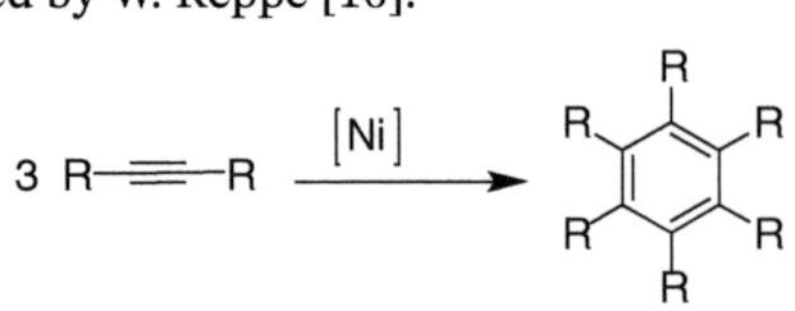

(1.16)

1.2.12 Equation (1.16): Trimerization of Acetylene

In 1951, the sigma bonded structure of ferrocene (Eq. (1.17)) was suggested and independently prepared by two groups. Fischer, Woodward, and Wilkinson proposed the sandwich structure of ferrocene [17].

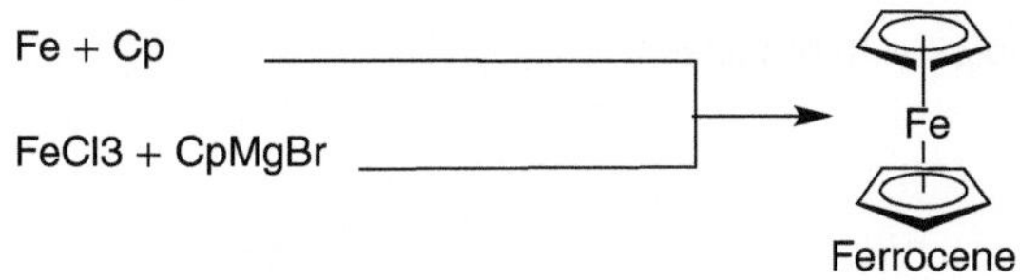

(1.17)

1.2.13 Equation (1.17): Synthesis of Ferrocene

In 1955, W. Hafner and E. O. Fischer followed rational synthesis to prepare bis (benzene)chromium (Figure 1.2), even though in 1919 same reaction was carried by F. Hein with $CrCl_3$ and PhMgBr [18, 19].

In 1955, G Natta and K Ziegler used a mixed metal catalyst to develop olefin polymerization at lower pressure.

In 1959, the stabilization of cyclobutadiene was done by complexation in $[(C_4Me_4)NiCl_2]_2$.

In 1961, Vaska's complex was discovered, which binds reversibly to O_2 trance $IrCl(CO)(PPh_3)_2$ [20].

In 1963, Ziegler and Natta both were awarded Nobel prize for Ziegler Natta catalyst.

In 1964, the first carbene complex with metal W (tungsten) (Figure 1.3) by E. O. Fischer [21].

In 1965, Coffey and Wilkinson, for hydrogenation of alkenes, they have used $(PPh_3)_3RhCl$ as a homogeneous catalyst [22].

In 1968, asymmetric catalysis (Eq. (1.18)) was discovered by William S. Knowles, the achiral substrates can be converted into chiral products with the help of complexes consisting of ligands that are chiral with high enantiomers [23].

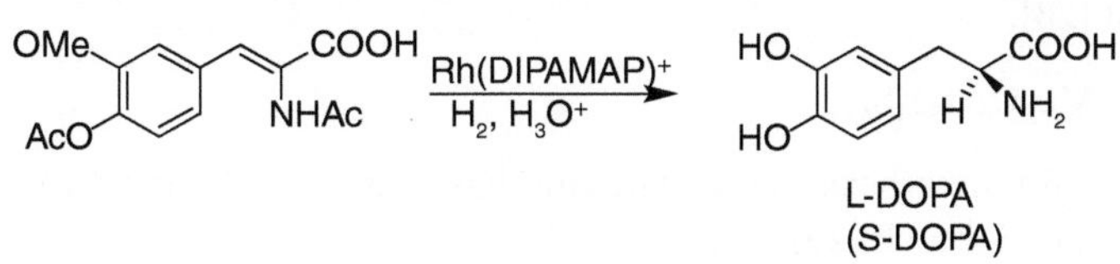

(1.18)

1.2.14 Equation (1.18): Asymmetric Catalysis Reaction

In 1972, T Mizorki and R. F. Heck discovered the substitution reaction of aryl halides with vinylic hydrogen atom a palladium based catalysis reaction [24].

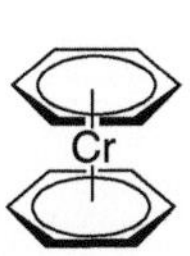

Figure 1.2 Bis(benzene)chromium compound.

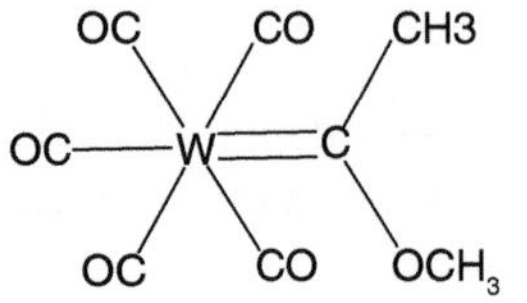

Figure 1.3 First carbene complex with tungsten.

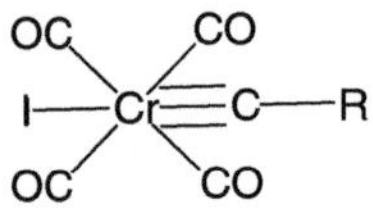

Figure 1.4 First chromium carbene complex.

In 1973, the first Chromium carbene complex (Figure 1.4) was prepared by E. O. Fischer. In the same year G. Wilkinson and E. O. Fischer were awarded Nobel Prize for working on metal sandwich compound [25].

In 1976, M. F. Lappert prepared first double bonded compound tin–tin using dimetallenes first main group element [26].

$$[(Me_3Si)_2CH]_2Sn = Sn[CH(SiMe_3)_2]_2$$

In 1979, the first paper on palladium catalyzed Suzuki coupling reaction (Eq. (1.19)) of aryl boronic acids was published by Suzuki and Miyaura [27].

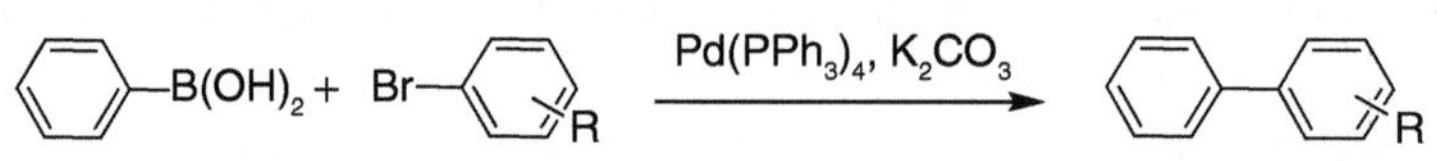

(1.19)

1.2.15 Equation (1.19): Palladium Catalyzed Suzuki Coupling Reaction

In 1980, the zirconocene-based catalyst (Figure 1.5a,b) was prepared by Walter Kaminsky only for syndiotactic and isotactic polypropylene.

In 1981, the compound having Si=Si (silicon–silicon double bond) was prepared by Robert West [28].

$$(1,3,5 - Me_3C_6H_2)_4Si_2$$

In 1990, for olefin metathesis, a molybdenum-based catalyst (Figure 1.6) was discovered by Richard Schrock [29].

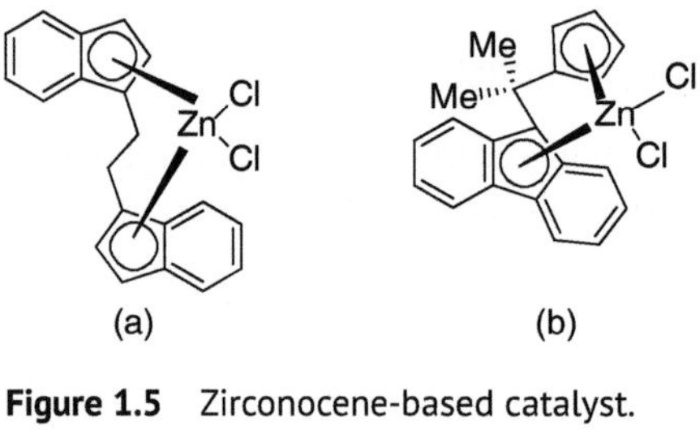

Figure 1.5 Zirconocene-based catalyst.

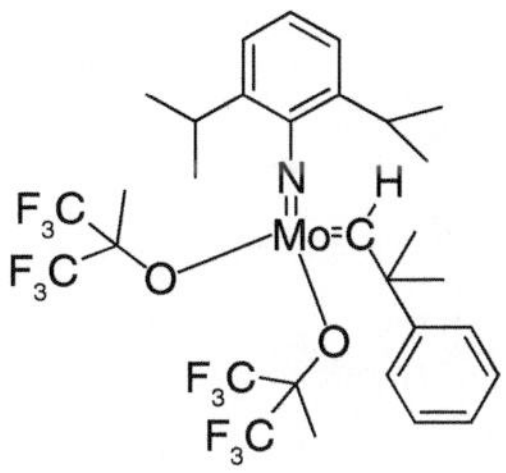

Figure 1.6 Molybdenum-based catalyst.

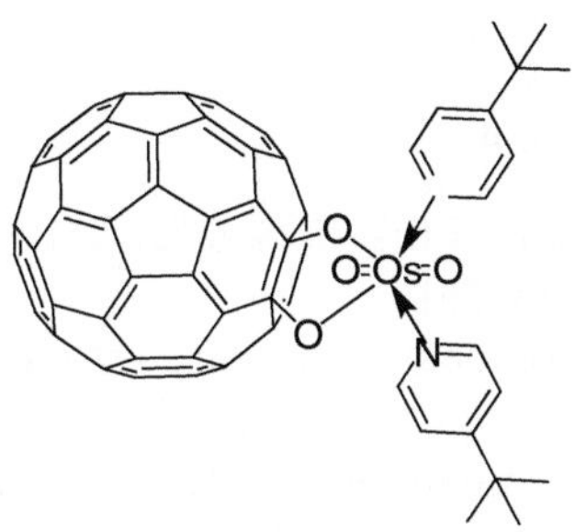

Figure 1.7 Fullerene-based organometallic compound.

In 1991, the fullerene-based organometallic compound (Figure 1.7) was made by J. M. Hawkins, which is also the derivative of structurally characterized fullerene $C_{60}(OsO_4)(4\text{-}t\text{-}BuPy)_2$. Furthermore, the organometallic compounds with η^2-bonding directly with metal–carbon bonds were discovered [30].

In 1995, the olefin metathesis catalyst, also called first-generation Grubbs catalyst (Figure 1.8), was prepared by Robert Grubbs [31].

In 1997, the organometallic compound having C atom as a ligand was showed by C C Cummins $[(R_2N)_3MoC]^-$. In the same year G. M. Robinson prepared the salt of sodium $Na_2[ArGaGaAr]$ and postulated [32, 33].

$$Ga \equiv Ga$$

In 2001, for asymmetric hydrogenation, W. S. Knowles, K. B. Sharpless, and R. Noyori received Nobel Prize.

In 2002, the molecular hybrids of fullerene and ferrocene together called bucky ferrocene (Eq. (1.20)) was prepared by E. Nakamura by treating $[FeCp(CO)_2]_2$ with $C_{70}HMe_3$ or $C_{60}HMe_5$ [34].

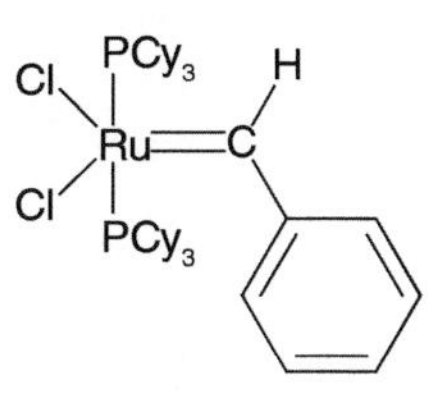

Figure 1.8 First generation Grubbs catalyst.

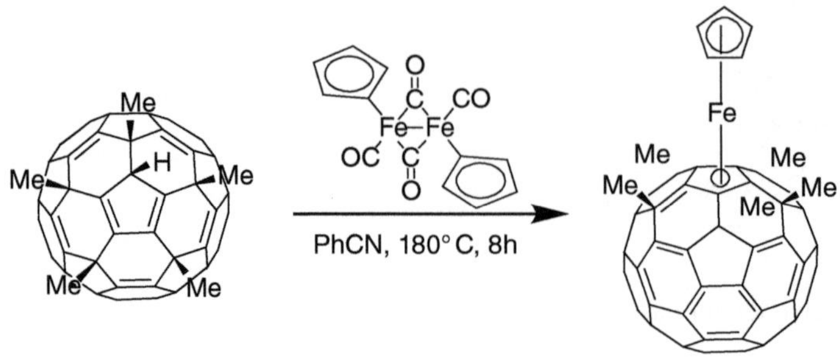

(1.20)

1.2.16 Equation (1.20): Synthesis of Bucky Ferrocene

In 2004, the preparation and structural characterization of first Si–Si (Figure 1.9, Si—Si triple bonded) molecule was done by A. Sekiguchi [35].

In the same year 2004, the first zinc organometallic compound $Zn_2(Cp)_2$ was prepared by E Carmona [36].

In 2005, for olefin metathesis R. R. Schrock, Y. Chauvin, and R. H. Grubbs, received Nobel Prize. In the same year, the first stable organometallic compound with quintuple bond was prepared by P. Power [37].

In 2006, hexaferrocenylbenzene (Figure 1.10) was prepared by P. C. Vollhardt [38].

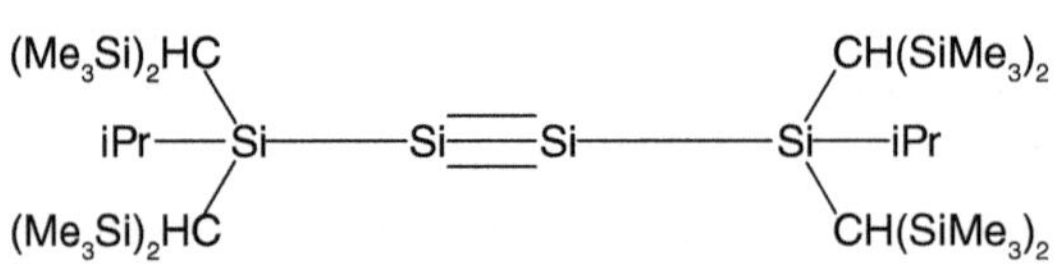

Figure 1.9 First Si compound.

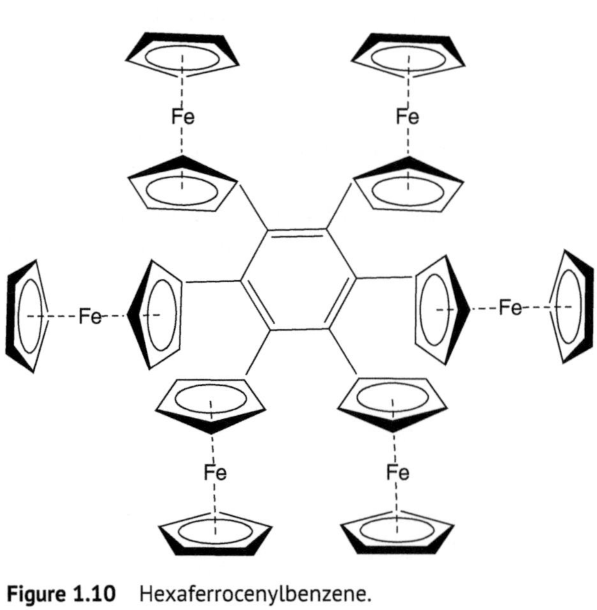

Figure 1.10 Hexaferrocenylbenzene.

1.3 Stability of Organometallic Compounds

The oxidation of the organometallic compounds will plays a vital role in the stability of the compounds, whereas the organometallic compounds are thermally unstable due to the negative free energy of carbon dioxide, metal oxide, and water production. In addition to that, at lower room temperatures, the organometallic compounds are unstable to oxidation due to the presence of empty orbitals of metal or due to non-bonding electrons [39].

The hydrolysis of organometallic compounds will also affect the stability of compounds

1. Hydrolysis involves the reaction of nucleophilic attack of water.
2. The rate of hydrolysis is determined by the polarity of the metal–carbon bond, the higher the polarity of the metal–carbon bond, the faster will be the rate of hydrolysis.

1.4 Properties of Organometallic Compounds

1. Most organometallic compounds with aromatic and hydrocarbon groups are solid.
2. The metal–carbon bond is covalent.
3. Organometallic compounds with metals such as lithium and sodium are volatile and burn spontaneously.
4. Organometallic compounds having electropositive metals behave as reducing agents.
5. Organometallic compounds are toxic to human beings in most cases.
6. Organometallic compounds have low melting point.
7. Organometallic compounds are soluble in ether, whereas insoluble in water.
8. Organometallic compounds are very reactive compounds.
9. The electronegativity of carbon in organometallic compound is 2.5 and for metals the electronegativity is below 2.0.
10. Many organometallic compounds occur in solid state, particularly the compounds having hydrocarbons and aromatic groups. Few of organometallic compounds are liquids and gases.

1.5 Basic Concepts in Organometallic Compounds

(1) 18-electron rule
(2) Π – back bonding or back donation
(3) Hapticity

1.5.1 18-Electron Rule

The structure and bonding present in the organometallic compound are studied based on the 18-electron rule and this is the combination of the ability of the ligand for pi acceptor or sigma donor and the back bonding and bonding nature of the ligand. In addition to that, the stability of the metal complexes is predicted and characterized by the 18-electron rule. A total of 18 electrons are present in the valence shell of MT, whereas a total of 10 electrons in the d orbital (2 each from five d orbitals), 6 electrons from the p orbital (2 each from three p orbitals) and 2 electrons from 1 s orbital. The combination of these atomic orbitals (5d+3p+1s) results in 9 molecular orbitals. These 9 molecular orbitals are metal–ligand non-bonding or bonding orbitals, and there might be a few higher energy antibonding molecular orbitals. These 9 molecular orbitals are filled by electrons originating from ligand or metal [39].

1.5.1.1 Statement of 18 Electron Rule

Thermodynamically stable transition metal organometallic compounds are formed when the sum of the metal d electrons and the electrons conventionally considered as being supplied by the surrounding ligands equals 18. In this way, the metal attains the electronic configuration of the next higher Nobel gas in general *n*s2 (*n*–1)d10 *n*p6. It is also called the EAN (effective atomic number) rule or inert gas rule [39].

The presence or absence of a metal–metal bond and the stability of the complex can be determined by counting the number of electrons surrounded in the outer shell of the metal atom in the complex.

Two methods were used for counting the electrons

1. Neutral atom counting method–it is difficult for organometallic compounds because it doesn't require oxidation state assignment
2. Oxidation state/ionic counting method in this method a change in the oxidation state of the metal is required.

In 1921, Irwing Langmuir formulated the 18-electron rule and derived the equation.

$$V_c = s - e$$

where

V_c = number of shared electrons in a metal complex or compound.
s = number of electrons for the completion of the valence shell.
e = number of valence electrons in an isolated atom.

In organic chemistry, the compounds have to obey the octet rule, the value of s is 8.

$$V_c = 8 - e$$

For Example, NH_3, 3 = 8–5

In organometallic complexes, such as transition metal carbonyls, the value of s is 18.

$$V_c = 18 - \mathrm{e}$$

Example, $Ni(CO)_4$, 8 = 18–10, $Cr(CO)_6$, 12 = 18–6

1.5.1.2 Examples

(1) $TiCl_4$
Electronic configuration: $3s^2 3p^6 3d^2 4s^2$
By neutral counting method: for Ti 4 electron, for Cl 1 electron:

$$TiCl_4 : 4 + 4(1) = 8 \text{ valence electrons.}$$

By oxidation/ionic counting method: for Ti^{4+} 0 electron, for Cl 2 electron

$$TiCl_4 : 0 + 4(2) = 8 \text{ valence electrons.}$$

$TiCl_4$ is having only 8 electrons, but it should be 18 electrons to satisfy 18- electron rule. $TiCl_4$ will act as good Lewis acid and it reacts with alcohol, water and amines.

(2) $Fe(CO)_5$
Electronic configuration: $3s^2 3p^6 3d^6 4s^2$
By neutral counting method: for Fe 8 electron, for CO 2 electron:

$$Fe(CO)_5 : 8 + 2(5) = 8 + 10 = 18 \text{ valence electrons.}$$

By oxidation/ionic counting method: in this case all the fragments are neutral in nature.
This is a stable compound with 18-electron complex.

(3) $Fe(C_5H_5)_2$ or $FeCp_2$
Electronic configuration: $3s^2 3p^6 3d^6 4s^2$
By neutral counting method: for Fe 8 electron, for C_5H_5 2 electron:

$$Fe(C_5H_5)_2 : 8 + 2(5) = 8 + 10 = 18 \text{ valence electrons.}$$

By oxidation/ionic counting method: for Fe^{2+} 6 electron, for C_5H_5 2(6) = 12 electron

$$Fe(C_5H_5)_2 : 6 + 12 = 18 \text{ electrons}$$

This is a stable compound with 18-electron complexes.

(4) $[CpFe(CO)_2]_2$
Neutral counting method

$$Cp\ 5 + Fe\ 8 + 2CO\ 4 + Fe{-}Fe\ 1 = 18$$

Electrons were counted for only one iron center since the other iron will contribute one electron.

(5) $CpFe(CH_3)(CO)_2$
Neutral counting method

$$Cp\ 5 + Fe\ 8 + CH_3 1 + 2\ CO\ 4 = 18$$

(6) $Fe_2(CO)_9$
Neutral counting method

$$3CO\ 6 + 3 - CO\ 3 + Fe\ 8 + Fe{-}Fe\ 1 = 18$$

1.5.2 Π –Back Bonding or Back Donation

When electrons move from one atomic orbital to another anti–bonding orbital π^* of a ligand or atom in the process relieves the metal that has an excess negative charge.

Example: Zeise's salt, $Ni(CO)_4$.

1.5.3 Hapticity η^x

The term hapticity is defined as total number of donor atoms of a ligand that are coordinated (attached) to the central metal atom (it is also defined as how a contiguous group of atoms of a ligand, which are coordinated to the central metal atom).

Hapticity is denoted by a Greek word η^x.

Where, x – number of donor atoms attached to the metal.

1.6 Hapticity of Ligands

Ligands are an atom or group of atoms, ions, or molecules that are capable of donating a pair of electrons to the metal atom and are called ligands.

The hapticity of ligands ranges from 1 to 8 that is monohapto ligands to octohapto ligands.

(1) Monohapto ligands η^1: these are the ligands joined with one atom and it has the capacity to donate one electron (1 electron donor ligands).
Examples: Alkyl groups($-CH_3$) ($M-CH_3$), Aryl($-C_6H_5$), Alkenyl($M-CR{=}CR_2$) (where R_2-alkyl)

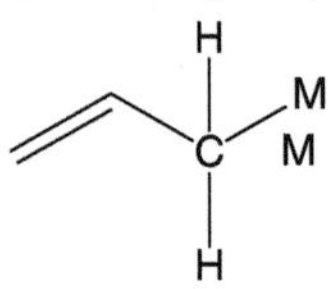

$\eta^x = \eta^1$ number of donor atoms.

(2) Dihapto ligands η^2: Two ligands atoms are directly bonded to metal (2 electron donor ligands).
Examples: Alkenes ($=CH_2=CH_2=$), Zeise's salt($K[PtCl_3C_2H_4]H_2O$)

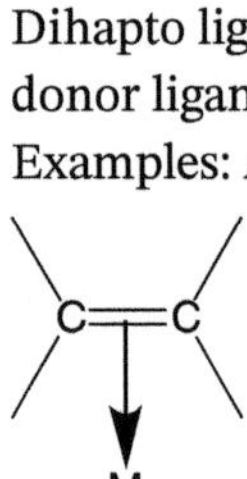

(3) Trihapto ligands η^3: three atoms of the ligands are directly bonded to metal atom (three-electron donor). There are two types of complexes, π-allyl complexes and σ-allyl complexes.

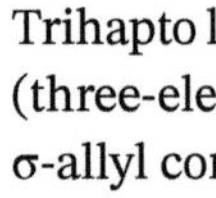

$\eta^x = \eta^3$ number of donor atoms.

π-allyl complexes: when π-bond will be delocalized around all the three carbon atoms time to time. The π-allyl complexes are 3 electron donor complexes.

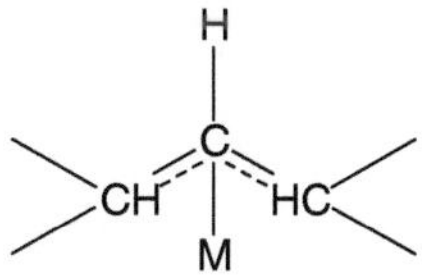

σ-allyl complexes: when metal is directly attached to one carbon atom of the ligand.

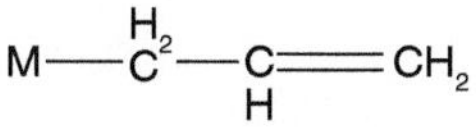

(4) Tetrahapto ligands η^4: these are the ligands joined with four atoms and it has the capacity to donate 4 electrons.
Example: Cyclobutadiene it has delocalized electron around all the four–carbon atom.

(5) Pentahapto ligands η^5: these are the ligands joined with five atoms and it has the capacity to donate 5 electrons.
Example: Ferrocene bis[η^5-cyclopentadienyl]iron

(6) Hexahapto ligands η^6: these are the ligands joined with six atoms and it has the capacity to donate 6 electrons.
Example: benzene

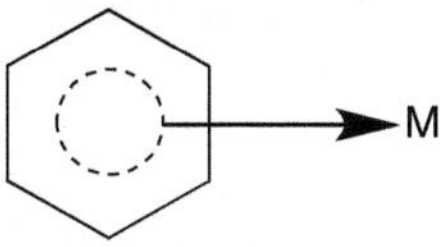

(7) Heptahapto ligands η^7: these are the ligands joined with seven atoms and it has the capacity to donate 7 electrons.

(8) Octahapto ligands η^8: these are the ligands joined with eight atoms and it has the capacity to donate 8 electrons.
Example: Uranocene-bis(η^8-1,3,5,7-cyclooctatetraene)uranium.

1.7 Change in Hapticity

During the course of the reaction the hapticity of the ligand changes (Schemes 1.1 and 1.2).

The η^5-Cp changes to η^3-Cp, by making room on the metal where one molecule of CO is removed and ligand L is donating extra two more electrons.

- The hapticity will changes the electron count.
- Hapticity will change for dienes, indenyl and it will change for cyclopentadienyl for some time.

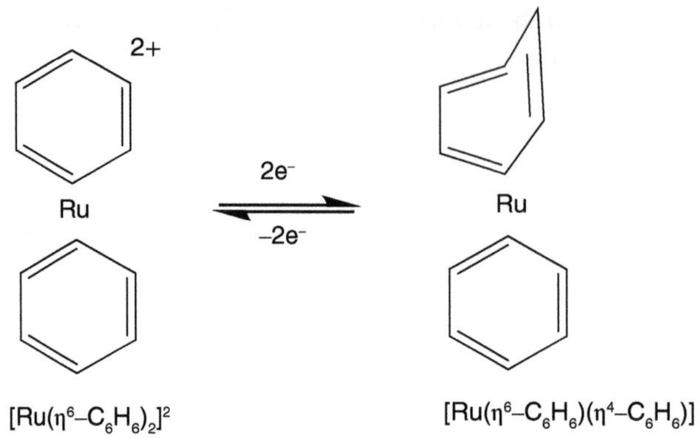

Scheme 1.1 Change in hapticity in Ru compound.

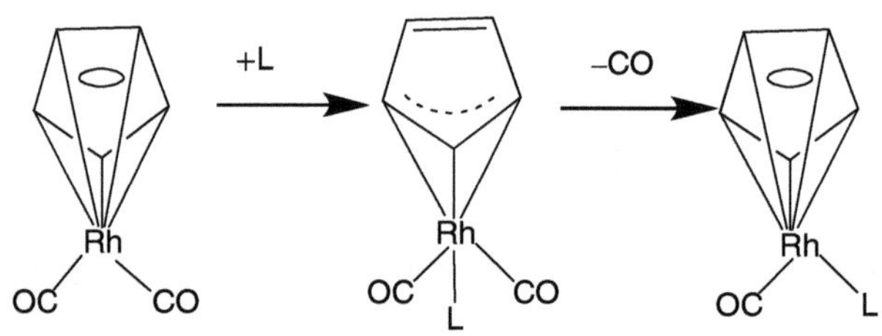

Scheme 1.2 Change in hapticity in Rh compound.

- Hapticity does not change the oxidation state.
- It makes a path for a system "giving room" for other reactions or it will avoid electron counts above 18 electron.

1.8 Hapticity Verses Denticity

Table 1.1 Differences between hapticity verses denticity[40].

Hapticity	Denticity
• Hapticity refers to how a contiguous group of atoms of a ligand is coordinated to the central metal atom • It is denoted by η. • The contiguous atoms are involved in the coordination process. • Ferrocene where the iron(Fe) is sandwiched between two hydrocarbon rings.	• Denticity refers to the number of donor atoms in the same ligand that binds to the central metal atom in the coordination complex. • It is denoted by k. • The donor atoms will attach to the central metal atom. • EDTA coordinated to central metal atom through six atoms two nitrogen and the other four oxygen.

1.9 Counting of Electrons and Finding out Metal–Metal Bonds

Metal carbonyl clusters are the compounds that contain metal in the lower oxidation state, these compounds also consist of metal–metal bonds. The larger and structurally complex clusters come under this category. The total number of metal–metal bonds and the number of bonds present in one metal to the other metal in a complex cluster can be determined by following the procedure and this holds good for the complexes with nuclearity ≤ 4. The electronic structure of the complex can be understood by the bond between the metal atoms, whereas each metal atom has to attain 18 electronic configurations [39].

1.9.1 Calculating the Number of Metal–Metal Bonds

1. First, find out the total number of valence electrons in the molecule and consider it as X.
2. Let us consider an equation, Y = (*n*X8)-X, where "*n*" is the number of metals present in the complex.
   i) Y/2 is the number of metal–metal bonds present in the complex.
   ii) X/2 is the number of electrons per metal. If the number of electrons is equal to 18, that shows there is no metal–metal is present. If the number of electrons is equal to 17, it shows one metal–metal bond is present. If the number of electrons is equal to 16, two metal–metal bonds are present.

1.9.2 Writing the Probable Structure of Compound

1. The first step is to write the metal core(center)
2. The ligands and metal carbonyls can be arranged to the metal core in which the metal core has to satisfy the 18-electron rule, additionally the carbonyl group has to occupy either a bridged or terminal position.

1.9.3 How to Draw the Probable Structure of Ni(η^1-C_3H_5) (η^3-C_3H_5)

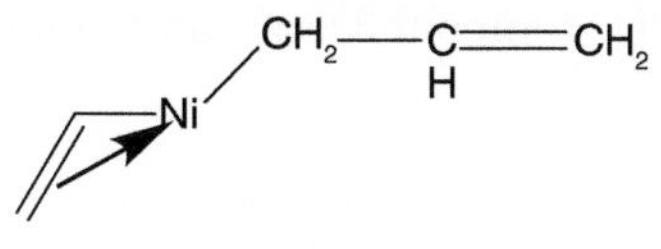

1.9.4 How to Draw the Probable Structure of (μ-CO)-[η^5-CpRh]$_3$(CO)

Total valence electron = 4 + [(6 + 8)X3] + 2 = 48
The number of metal–metal bond = (3X8)–48 = 6/2 = 3
Number of bonds per metal center = 2 (48/3 = 16)

1 and 2 are the plausible structure of (μ-CO)-[η^5-CpRh]$_3$(CO). Among 1 and 2 structures, 1 is proper, whereas, in structure 2, the electron count on metal 3Rh is found to be 17, 18, and 19e$^-$, but it should be 16.

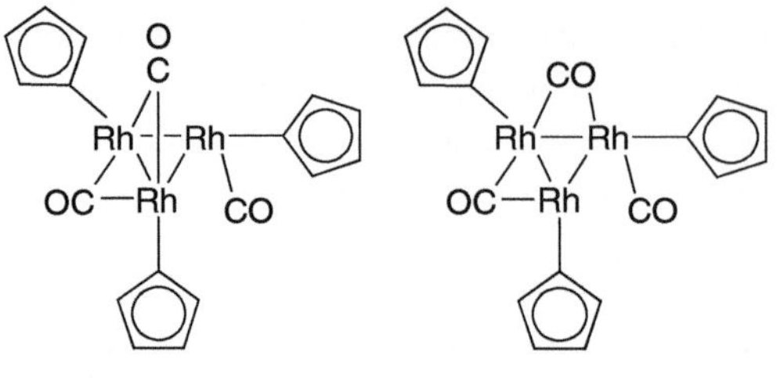

1 2

The 18e$^-$ rule is not applicable for all the f-block elements, there are exceptions to transition metal organometallics. The 18e$^-$ rule for organo–transition metal compounds can be explained by various physical and chemical properties.

From 18e$^-$ rule, one can predict the total number of ligands coordinated to a particular metal and also the reactivity/stability of the complex. This method holds good for the metals having low valency and also with the small ligands that are strong sigma donors and π–acceptors. The ligands are small enough that allow metal that coordinates saturated and that give more ligand field splitting value. 18e$^-$ are required to fill the dπ orbitals similar to the octet rule. There are some exceptions to 18e$^-$ rule also few of them are mentioned below.

1. Metals having d^8 electrons: the metals having d^8 electrons have the tendency to form a square planar complex with 16e$^-$ this tendency is strong for the groups 9, 10, and 11 except for higher energy orbitals $d_{x2\text{-}y2}$. The d^8 metals have a square planar complex with 16e$^-$ have a completely filled orbital.
2. Metals having d^0 electrons: the d^0 complexes formed by high valency have a less electron count than 18. Also, the complexes having sterically demanding bulky ligands as substitution force the complexes to have less than 18 electrons.
3. When the size is considered the organometallic clusters with moderate to big size fails to obey 18e$^-$ rule while bonding. Therefore, the stability and bonding of organometallic clusters are considered on the electron counting rule, not on the 18e$^-$ rule.
4. This rule is not applicable for organometallic compounds of lanthanides, actinides, and main group metals.

1.10 Metals of Organometallic Compounds

As we discussed above organometallic compounds are the substances having at least one metal–carbon bond, where the carbon is directly attached to carbon atom, in which the carbon atom is found to be a part of organic group. Organometallic compounds play an important role as catalysts and also act as intermediates. One such example is ferrocene, where the iron(Fe) is sandwiched between two hydrocarbon rings.

The metals of organometallic compounds are discussed as transition metals or main group compounds. It includes s-block elements (group 1 and 2), p-block elements (group 13–15) transition metals of d and f block elements (group 3–12) [39].

1.10.1 Organometallic Compounds of Transition Metals

- (n-1)d, ns and np orbitals are the valence orbitals for transition metals.
- When (n-1)d, ns and np orbitals occupy partially, it gives the electron-donating and electron-accepting property for transition metals.
- It forms different multiple bonds between metal–metal with or without an bridging ligand.
- It has the capacity to vary the coordination number.

1.10.2 The Bonding and Structure in Different Metal complexes [37]

(1) Alkene complexes
(2) Allyl complexes
(3) Carbonyl complexes
(4) Metallocene
(5) Dihydrogen complexes
(6) Transition metal carbene complexes

1.10.2.1 Alkene Complexes

- In alkene complexes, the bond between the alkene and transition metal will activate the ligand, either to nucleophilic or electrophilic attack, and that will be decided by the nature and charge of metal center.
- If σ-donation is greater to metal center, π-back donation is also greater.
- If the electron density of alkene complex is greater due to back donation of π* orbital, the reduction in C=C bond order will be greater.
- The stability of alkene complexes that are substituted is less when compared to ethylene complexes.

  Example: Zeise's salt.

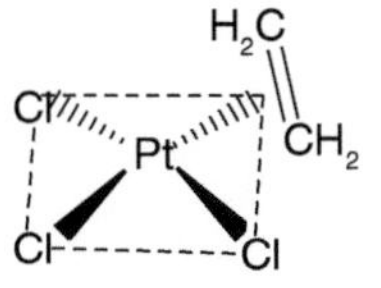

1.10.2.2 Allyl Complexes

The ligands in allyl complexes are ambidentate, which can bind to both trihapto and monohapto form (Scheme 1.3).

1.10.2.3 Carbonyl Complexes

These complexes have (carbon monoxide) CO as coordinating ligand. These complexes have synergistic bonding. Greater the sigma donation given by carbonyl complexes, stronger will be the π-back donation.

Examples: $Ni(CO)_4$, $Co(CO)_4$

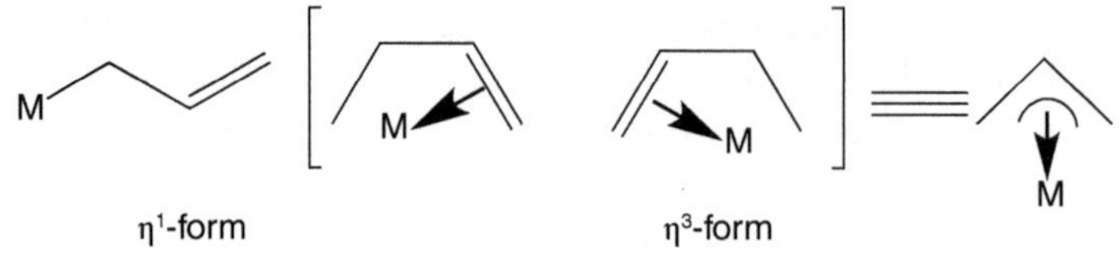

Scheme 1.3 Hapticity in allyl complexes.

1.10.2.4 Metallocenes

The compounds have two cyclopentadienyls (Cp) anions that bind to a metal with an oxidation state of two (II). They are also known as "sandwich compounds". Metallocene derivatives with two Cp ring like Cp_2TiCl_2 are known as "bent metallocene". The complex with one Cp ring is known as a "half sandwich," which has the geometry of three-legged piano stool (Figure 1.11).

1.10.2.5 Dihydrogen Complexes

The transition metal complexes having molecular hydrogen that donates two electrons to central metal atom (Figure 1.12).

1.10.2.6 Transition Metal Carbene Complex

The organometallic compounds with transition metals bear a divalent carbene carbon as a ligand. Carbenes are specific species that contain a neutral carbon atom with valence electrons in which two electrons are nonbonding in nature and they are highly reactive neutral species with a divalent carbon atom. The two nonbonding electrons of the carbene carbon can exist in two states singlet and triplet. In the singlet state, the two unshared electrons are paired in the same σ and π-orbitals, whereas in the triplet state two unshared electrons in two different orbitals (Figure 1.13).

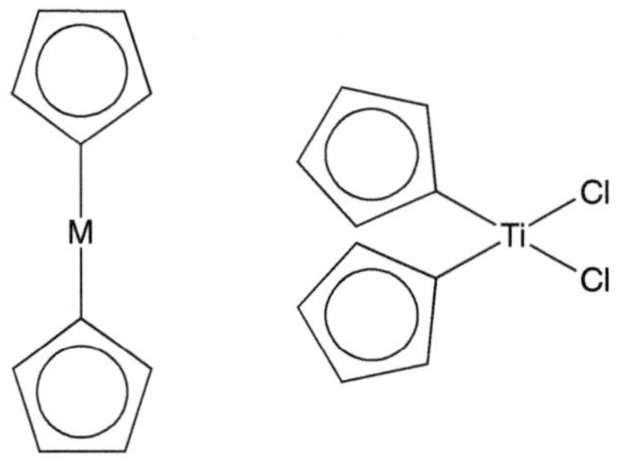

Figure 1.11 Metallocene compounds.

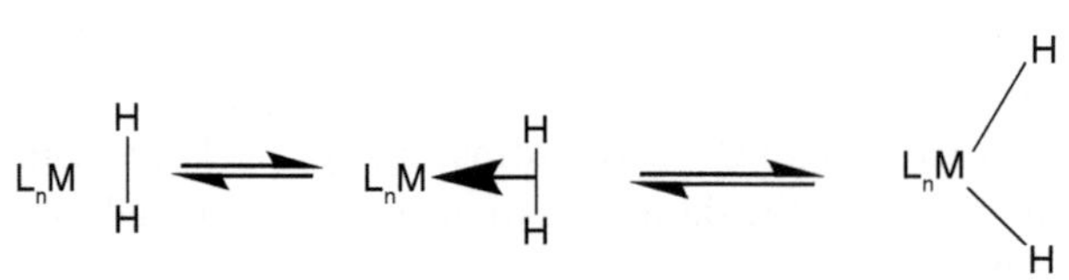

Figure 1.12 Dihydrogen complexes.

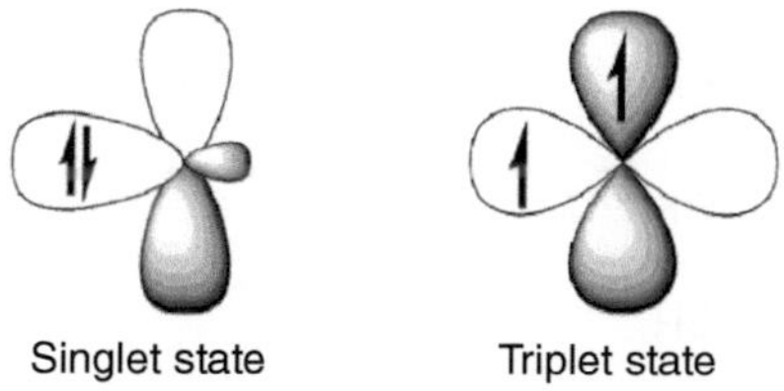

Figure 1.13 Singlet and triplet state carbene complex.

1.11 Importance of Organometallic Compounds

Most organometallic compounds are used for synthesis purposes because bond polarity made them highly reactive.

(1) The Grignard reagent (R-Mg-X) organomagnesium halides are used in organic synthesis for organoboron and organolithium compounds.
(2) Alkylaluminium compounds are used for organic synthetic purposes.
(3) When an organometallic compound is used with titanium salts can act as a catalyst for the polymerization of ethylene and propylene that is unsaturated hydrocarbons.
(4) The organometallic compounds of tin, mercury, and lead are used for commercial purposes.
For example, organotin compounds are used as pesticides, pharmaceuticals, and stabilizers for polyvinylchloride (PVC) and also, they are used as fire retardants.
(5) Metal carbonyls are used as catalysts in most reactions such as in the petrochemical industry.

1.11.1 Types of Organometallic Compounds

Depending upon the nature of the metal–carbon bond, these organometallic compounds are classified into four types.

1. Ionic bonded compounds
2. σ–σ bonded compounds
3. π–π bonded compounds
4. Multicenter bonded compounds
5. Ionic bonded compounds: The elements of lanthanides, actinides, alkaline metals, and alkaline earth metals will form ionic bonded organometallic compounds. These are colorless solid salts that are insoluble in nonpolar solvents and upon hydrolysis liberate hydrocarbons which also act as electrolytes.
6. σ–σ bonded compounds: The metal atom and carbon atom will form σ/covalent. Generally, the metals/nonmetals with less/weak electropositive atoms will form this type of compound. The pair of electrons are shared by a metal atom and carbon atom to form a σ–σ bond.

Examples:

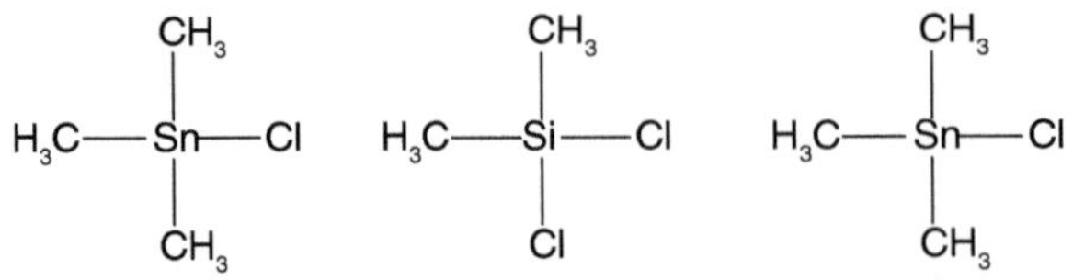

7. π–π bonded compounds: The compounds containing metals with an alkene, alkyne, benzene, and aromatic ring compound will together form π–π bonded organometallic compounds. These compounds have metal and ligands that form a π–π bond.
   Examples: dibenzene and ferrocene chromium, Zeise's salt, aromatic, acetylene, and ethylene compounds.
8. Multicenter bonded compounds: These are the compounds where the electron-deficient species are coordinated to metals like Be, Li, and Al.
   Examples: MeLiMeLi and $Al_2Me_6Al_2Me$

1.11.2 Uses of Organometallic Compounds

Generally, organometallic compounds are used as reagents and catalysts, such as homogeneous and heterogeneous catalysts, for most synthetic reactions.

1. Organometallic compounds as homogeneous and heterogeneous catalysts: The organometallic compounds with a transition metal such as Rh i.e. $[RhCl(PPh_3)_3]$ act as catalysts for the hydrogenation of alkenes. Similarly, organometallic compounds such as trialkyl aluminum with titanium tetrachloride or trichloride are used as heterogeneous catalysts for the polymerization of alkanes.
2. Organometallic compounds for metal purification: During the process of metal purification initially, the impure metals are converted into carbonyls, further they should be decomposed to form pure metals.
3. Organometallic compounds for the synthesis of organic compounds: Organometallic compounds, such as organolithium and organomagnesium compounds, were used to prepare for organic synthesis.
4. Organometallic compounds can be used as reagents.
5. Organometallic compounds can be used as additives.

Agriculture and medicine: Organometallic compounds were also used in the field of agriculture and medicine. When seeds were treated with ethyl mercury chloride to prevent infection in plants. The organoarsenic compounds were used in the treatment of syphilis, in modern surgery, silicone rubbers were used.

1.12 Conclusions

The compounds containing metal–carbon bonds are meant to be organometallic compounds in the current chapter the fundamental aspects of organometallic compounds were discussed and they are used for different applications such as the synthesis of medicinal products, organic compounds, and agricultural products, they were used as a catalyst both heterogeneous and homogeneous, used for the purification of metals. In addition to that they are used as additives and reagents.

References

1 Seyferth, D. (2001). Cadet's fuming arsenical liquid and the cacodyl compounds of Bunsen. *Organometallics* 20 (8): 1488–1498.
2 Thayer, J.S. (1966). Cadet's fuming liquid: an historical survey. *J. Chem. Educ.* 43 (11): 594–595.
3 Hunt, L.B. (1984). The first organometallic compounds. *Platinum Met. Rev.* 28 (2): 76–83.
4 Thayer, J.S. (1969). Historical origins of organometallic chemistry – part 1 Zeise's salt. *J. Chem. Educ.* 46 (7): 442–443.
5 Thayer, J.S. (1969). Historical origins of organometallic chemistry. Part II, Edward Frankland and diethylzinc. *J. Chem. Educ.* 46 (11): 764.
6 Seyferth, D. (2001). Zinc alkyls, Edward Frankland, and the beginnings of main-group organometallic chemistry. *Organometallics* 20 (14): 2940–2955.
7 Davis, T.L. (1929). Paul Schutzenberger. *J. Chem. Educ.* 6 (9): 1403.
8 Morris, P. (1989). The legacy of Ludwig Mond. *Endeavour* 13 (1): 34–40.
9 Abel, E. (1990). Ludwig Mond – father of metal carbonyls – and so much more (7 March 1839–11 December 1909). *J. Organomet. Chem.* 383 (1–3): 11–20.
10 Herrmann, W.A. (1990). 100 years of metal carbonyls: a serendipitous chemical discovery of major scientific and industrial impact. *J. Organomet. Chem.* 383 (1–3): 21–44.
11 Rheinholdt, H. (1950). Fifty years of the Grignard reaction. *J. Chem. Educ.* 27 (9): 476.
12 Seyferth, D. (2003). The rise and fall of tetraethyllead. 2. *Organometallics* 22 (25): 5154–5178.
13 Cornils, B., Herrmann, W.A., and Rasch, M. (1994). Otto Roelen, pioneer in industrial homogeneous catalysis. *Angew. Chem., Int. Ed. Engl.* 33 (21): 2144–2163.
14 Herrmann, W.A. and Cornils, B. (1997). Organometallic homogeneous catalysis – Quo vadis? *Angew. Chem., Int. Ed. Engl.* 36 (10): 1048–1067.
15 Seyferth, D. (2001). Dimethyldichlorosilane and the direct synthesis of methylchlorosilanes. The key to the silicones industry. *Organometallics* 20 (24): 4978–4992.
16 Reppe, W., v. Kutepow, N., and Magin, A. (1969). Cyclization of acetylenic compounds. *Angew. Chem., Int. Ed. Engl.* 8 (10): 727–733.
17 Wilkinson, G., Rosenblum, M., Whiting, M.C., and Woodward, R.B. (1952). The structure of iron bis-cyclopentadienyl. *J. Am. Chem. Soc.* 74 (8): 2125–2126.
18 Seyferth, D. (2002). Bis(benzene)chromium. 2. Its discovery by E. O. Fischer and W. Hafner and subsequent work by the research groups of E. O. Fischer, H. H. Zeiss, F. Hein, C. Elschenbroich, and others. *Organometallics* 21 (14): 2800–2820.

19 Seyferth, D. (2002). Bis(benzene)chromium. 1. Franz Hein at the University of Leipzig and Harold Zeiss and Minoru Tsutsui at Yale. *Organometallics* 21 (8): 1520–1530.

20 Vaska, L. and DiLuzio, J.W. (1961). Carbonyl and hydrido-carbonyl complexes of iridium by reaction with alcohols. Hydrido complexes by reaction with acid. *J. Am. Chem. Soc.* 83 (12): 2784–2785.

21 Fischer, E.O. and Maasböl, A. (1964). Zur frage eines wolfram-carbonyl-carben-komplexes. *Angew. Chem.* 76 (14): 645–645.

22 Osborn, J.A., Jardine, F.H., Young, J.F., and Wilkinson, G. (1966). The preparation and properties of tris(triphenylphosphine)halogenorhodium(I) and some reactions thereof including catalytic homogeneous hydrogenation of olefins and acetylenes and their derivatives. *J. Chem. Soc. A* 1: 1711.

23 Knowles, W.S. and Sabacky, M.J. (1968). Catalytic asymmetric hydrogenation employing a soluble, optically active, rhodium complex. *Chem. Commun.* 22: 1445–1446.

24 Dieck, H.A. and Heck, R.F. (1975). A palladium-catalyzed conjugated diene synthesis from vinylic halides and olefinic compounds. *J. Org. Chem.* 40 (8): 1083–1090.

25 Fischer, E.O., Kreis, G., Kreiter, C.G. et al. (1973). *trans*-Halogeno[alkyl(aryl) carbyne]tetracarbonyl complexes of chromium, molybdenum, and tungsten – A new class of compounds having a transition metal-carbon triple Bond. *Angew. Chem., Int. Ed. Engl.* 12 (7): 564–565.

26 Goldberg, D.E., Harris, D.H., Lappert, M.F., and Thomas, K.M. (1976). A new synthesis of divalent group 4B alkyls M[CH(SiMe3)2]2 (M = Ge or Sn), and the crystal and molecular and molecular strcuture of the tin compound. *J. Chem. Soc., Chem. Commun.* 227 (7): 261–262.

27 Miyaura, N., Yamada, K., and Suzuki, A. (1979). A new stereospecific cross-coupling by the palladium-catalyzed reaction of 1-alkenylboranes with 1-alkenyl or 1-alkynyl halides. *Tetrahedron Lett.* 20 (36): 3437–3440.

28 West, R., Fink, M.J., and Michl, J. (1981). Tetramesityldisilene, a stable compound containing a silicon-silicon double Bond. *Science (80-.).* 214 (4527): 1343–1344.

29 Schrock, R.R., Murdzek, J.S., Bazan, G.C. et al. (1990). Synthesis of molybdenum imido alkylidene complexes and some reactions involving acyclic olefins. *J. Am. Chem. Soc.* 112 (10): 3875–3886.

30 Hawkins, J.M., Meyer, A., Lewis, T.A. et al. (1991). Crystal Structure of Osmylated C60: Confirmation of the Soccer Ball Framework. *Science (80-.).* 252 (5003): 312–313.

31 Schwab, P., France, M.B., Ziller, J.W., and Grubbs, R.H. (1995). A series of well-defined metathesis catalysts–synthesis of [$RuCl_2(CHR')(PR_3)_2$] and its reactions. *Angew. Chem. Int. Ed. Engl.* 34 (18): 2039–2041.

32 Peters, J.C., Odom, A.L., and Cummins, C.C. (1997). A terminal molybdenum carbide prepared by methylidyne deprotonation. *Chem. Commun.* 20: 1995–1996.

33 Grunenberg, J. and Goldberg, N. (2000). How strong is the gallium≡gallium triple bond? Theoretical compliance matrices as a probe for intrinsic bond strengths. *J. Am. Chem. Soc.* 122 (25): 6045–6047.

34 Sawamura, M., Kuninobu, Y., Toganoh, M. et al. (2002). Hybrid of ferrocene and fullerene. *J. Am. Chem. Soc.* 124 (32): 9354–9355.

35 Sekiguchi, A., Kinjo, R., and Ichinohe, M. (2004, 5691). A Stable Compound Containing a Silicon-Silicon Triple Bond. *Science (80-.).* 305: 1755–1757.

36 Resa, I., Carmona, E., Gutierrez-Puebla, E., and Monge, A. (2004). Decamethyldizincocene, a Stable Compound of Zn(I) with a Zn-Zn Bond. *Science (80-.).* 305 (5687): 1136–1138.

37 Nguyen, T., Sutton, A.D., Brynda, M. et al. (2005). Synthesis of a Stable Compound with Fivefold Bonding Between Two Chromium(I) Centers. *Science (80-.).* 310 (5749): 844–847.

38 Yu, Y., Bond, A.D., Leonard, P.W. et al. (2006). Hexaferrocenylbenzene. *Chem. Commun.* 24: 2572–2574.

39 Elias, A.J. and Gupta, B.D. (2010). *Basic Organometallic Chemistry*. University Press (India) Pvt, Ltd.

40 Difference between hapticity and denticity. Chemistry LibreTexts, Libretexts, 23 Feb. 2019.

2

Nomenclature of Organometallic Compounds

Raj K. Rathore[1], Sadhucharan Mallick[2], and Bhavabhuti Prasad[2]

[1]*Atal Bihari Vajpayee Vishwavidyalaya, Government E.V. Post Graduate College Korba, Department of Chemistry, Rajgamar Road, Korba 495677, Chhattisgarh, India*
[2]*Indira Gandhi National Tribal University (Central University), Department of Chemistry, Amarkantak 484887, India*

2.1 Introduction

Organometallic compounds are organic molecules, ions, or radicals with at least one metal–to–carbon bond. Metal is a main group element, transition, actinide, or lanthanide in an organometallic compound. Some metalloids like silicon, boron, arsenic, germanium, and tellurium are also included in this type of compound. The organic ligand could be an alkyl, aryl, or aryl group, or an alkene, alkyne, or aryne. In organometallic compounds, metal has lower electropositivity compared to a carbon atom. Organometallic chemistry covers a versatile and diverse area of chemistry that bridges inorganic and conventional organic chemistry.

By their very nature, organometallic compound names should integrate organic and coordination chemistry rules. In general, however, these two different naming systems developed independently. The goal of this section is to develop an organometallic nomenclature system that, whereas primarily based on the additive system of coordination nomenclature (Nomenclature of Inorganic Chemistry, Recommendations 1990), also incorporates, to the extent possible, the rules for naming organic groups and substituents (A Guide to IUPAC Nomenclature of Organic Compounds, Recommendations 1993). Furthermore, new criteria have been proposed that clearly define the peculiar bonding modes of organometallic compounds [1–4].

Organometallic Compounds: Synthesis, Reactions, and Applications, First Edition.
Edited by Dakeshwar Kumar Verma and Jeenat Aslam.

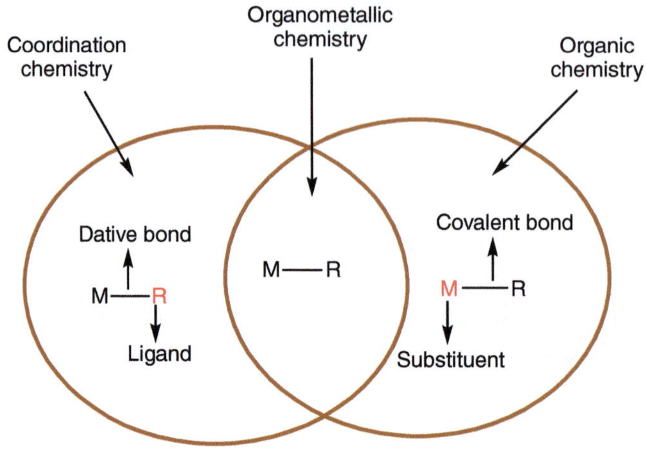

2.2 Aim of the Nomenclature

A collection of guidelines used to produce systematic names for chemical substances is known as a chemical nomenclature. The International Union of Pure and Applied Chemistry designed and produced the most widely used nomenclature (IUPAC).The main objective of chemical nomenclature is to identify a chemical species using written or spoken words. Additionally, each compound should only have one name; however, alternate names may occasionally be approved.

Nomenclature is based on the user's needs; there is no one appropriate nomenclature. Instead, a variety of nomenclatures are suitable for various contexts [5, 6].

2.3 Type of Nomenclature System

Three types of nomenclature are used for various types of chemical entities, reflecting the evolution of inorganic compound naming over time.

2.3.1 Binary Nomenclature

The term "binary" comes from the fact that it is most commonly used to describe simple salts with only a cation and an anion, although it can also be used for more complex compositions. This empirical rule for naming is extensively used for salt-like ionic species. The components have to be in a definite order, and a modification of the element name is sometimes required e.g. iodide, bromide, chloride, telluride, etc. It is beneficial and common when a material's composition is specified, but information on the specific structure is unknown or not mandatory. This nomenclature system has also been extended to the more basic forms of organometallic species that can be considered derivatives of inorganic salts or molecules, and it is most often used in commercial product identification [7, 8].

1. Diethylaluminum bromide Et_2AlBr
2. Ethylmagnesium bromide EtMgBr

2.3.2 Substitutive Nomenclature

This type of nomenclature originated from an organic compound's nomenclature; it has been extended to name certain organometallic compounds of some main-group elements of groups 13, 14, 15, and 16 (which closely resemble organic compounds both in bonding and properties) [7–9].

For examples

(i) $C_6H_5AsH_2$ Dihydridophenylarsenic
(ii) C_2H_5BeH Ethylhydridoberyllium
(iii) $(C_6H_5)_4Pb$ Tetraphenylplumbane
(iv) $(C_6H_5)_2SnH_2$ Diphenylstannane

The basis of this system is the concept of a parent hydride (an alkane in organic nomenclature), for example, silane stands for SiH_4, whose hydrogen atoms have been partially or completely replaced by organic groups/substituents.

(i) methylalumane AlH_2Me
(ii) triethylalumane $AlEt_3$
(iii) triethenylstilbane $Sb(CH{=}CH_2)_3$
(iv) chlorotrimethylsilane $Si(CH_3)_3Cl$
(v) dicyclohexylborane $B(C_6H_{11})_2H$
(vi) diphenylselane $Se(C_6H_5)_2$
(vii) triethylarsane $As(C_2H_5)_3$

Organometallic compounds with double bonds between the elements of the main group may also be mentioned as alkenes [7–9].

For example:

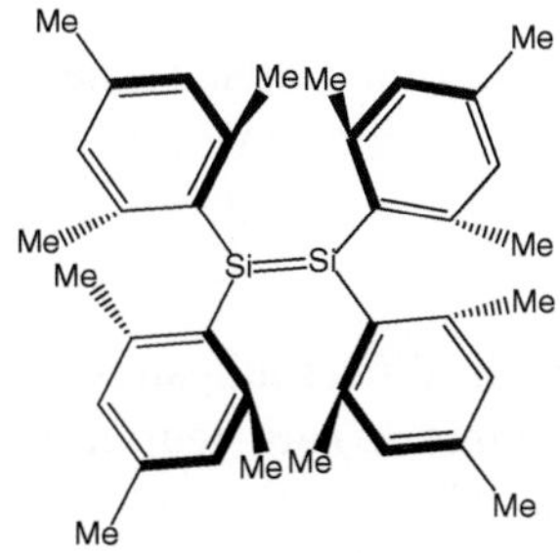

1,1,2,2-Tetramesityldisilene

2.3.3 Additive Nomenclature or Coordination nomenclature

This nomenclature system is based on the addition principle that addition reactions produce the coordination compounds. As shown in the example below, the coordination entity is named after the central metal ion.

Example

$[Co(NH_3)_6]^{3+}$ hexaamminecobalt(III) ion

The above compound is named by adding the ligand names to a central atom. The same naming system is also followed for the organometallic compounds.
Example

$[Ni(CO)_4]$	tetracarbonylnickel(0)
$[Co_2(CO)_8]$	octacarbonyldicobalt(0)
$[Fe_2(CO)_9]$	Nonacarbonyldiiron(0)
$[Mn_2(CO)_{10}]$	Decacarbonyldimanganese(0)

This additive nomenclature extends to more complicated structures involving binuclear, trinuclear, or polynuclear compounds formed by central atoms.

The central core of organometallic compound nomenclature is based on the various rules adopted for coordination nomenclature with specific additional rules; a basic summary of the general definitions and regulations of coordination nomenclature used for naming organometallic compounds is discussed in Section 2.5 Regulations concerning the nomenclature of transition element organometallic compounds [4, 8].

2.4 Concepts and Conventions

2.4.1 Oxidation Number

A central atom is defined as the charge that a central atom would have if all ligands and their shared pair of electrons were removed. A Roman numeral usually represents it [3, 7, 10].

The oxidation state on a central atom can be calculated as follows considering an example $[Os(C_2H_5)\,(NH_3)_5]Cl$

i. Separate the counter ions (ions outside the square bracket) in the given complex.

   $[Os(C_2H_5)(NH_3)_5]^+ \quad Cl^-$

ii. Find out the charges that the different ligands in the complex possess.
   C_2H_5 (ethyl) and NH_3 (ammonia) are neutral ligands; they do not possess any charge, so their charge is "0."
iii. Assume the charge on the central atom to be x.
   The charge on the complex equals the sum of the charges on the central atom and all the ligands, i.e.

x	+	0	+	5X0	=	+1
Os		C_2H_5		$5NH_3$		$[Os(C_2H_5)(NH_3)_5]^+$

Therefore, $x = +1$
So, the central atom Os (x) charge is calculated as +1.

The oxidation number does not indicate electron distribution. Sometimes, when ligands addition cannot be determined to be an association of Lewis acid or Lewis base or as an oxidative addition, the formalism does not give acceptable central atom oxidation numbers. In such cases, the net charge of the coordination entity is preferred in most nomenclature practices.

In organometallic compounds, the oxidation numbers cannot be assigned unambiguously to many organometallic compounds, so often, no formal oxidation numbers are assigned to the central atoms in the nomenclature [3, 7, 10].

2.4.2 Coordination Number

The number of sigma bonds between the central atom and the ligands in an organometallic compound is called the coordination number.

For example, in the $[Ni(CO)_4]$ complex, the coordination number of Ni in the above complex is 4. However, this definition does not apply to several organometallic compounds because the interaction between the central atom and two or more ligand atoms takes place via a combination of sigma (σ), pi (π), and delta (δ) bonds [3, 10, 11].

2.4.3 Chelation

When the bonding between a ligand (except monodentate) and the central atom results in a ring-like structure of the organometallic complex, the compound is termed a chelate, and the phenomenon is called chelation. This concept is strictly limited to organometallic compounds in which the ligand bonding atoms are connected to the central atom via a sigma bond. For example, propane–1,3-diyl makes chelate complexes with metal (M) [8, 11, 12].

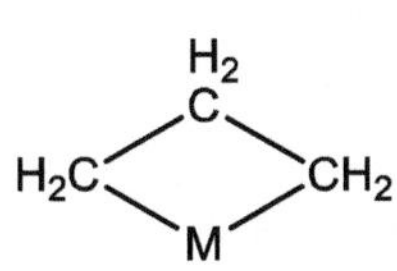

Propane–1,3-diyl

2.4.4 Ligands

A ligand is a species/ion/group that acts as a Lewis base and donates electron pairs to the Lewis acid (metal ion) to form a coordination or organometallic complex. It generally forms a σ-bond (in actuality, a coordinate bond) with the donation of an electron pair. In an organometallic complex, the formation of π- or/and δ bond with the central atom along with the σ-bond may occur. Based on the interaction of the ligand with the central atom, it may be classified into strong and weak field ligands; on the number of donating atoms per ligand, it may be termed as monodentate, bidentate, polydentate, etc. If a ligand possesses two different coordinating atoms (only a single atom coordinates at a time), such ligand is called an ambidentate ligand [3, 4, 7, 8, 13, 14].

Table 2.1 Few examples of negatively charged ligands.

S. No.	Ligand	Alternative ligand name	Systematic ligand name
1	$(C_6H_5S)^-$	(benzenethiolato)	(phenylsulfido)
2	$[(CH_3)_2P]^-$	(dimethylphosphanido)	(dimethylphosphanyl)
3	$(CH_3CO_2)^-$	(ethanoato)	(acetato)
4	$(C_2H_5O)^-$	(ethanolato)	ethoxido
5	$(HCO_2)^-$	(methanoato)	(formato)
6	$(CH_3O)^-$	(methanolato)	methoxido
7	$(CH_3P)^{2-}$	(methylphosphanediido)	(methylphosphanediyl)
8	$(CH_3PH)^-$	(methylphosphanido)	(methylphosphino)
9	$[(CH_3)_2N]^-$	(N-methylmethanaminato)	(dimethylamido)
10	$(C_6H_5O)^-$	(phenolato)	phenoxido
11	$(NO_2)^-$	[dioxonitrato(1−)-kN]	nitrito-N,nitro
12	-ONO	[dioxonitrato(1−)-kO]	nitrito-O
13	$(CH_3N)^{2-}$	[methanaminato(2-)]	(methylimido)
14	$(NO_3)^-$	[trioxonitrato(1−)]	nitrato
15	$(NH)^{2-}$	azanediido	imido
16	$(NH_2)^-$	azanido	amido
17	Br^-	bromido	Bromo
18	Cl^-	chlorido	chloro
19	CN^-	cyanido	cyano
20	OH^-	hydroxido	hydroxo
21	I^-	iodido	iodo
22	N^{3-}	nitrido	—
23	O^{2-}	oxido	oxo
24	P^{3-}	phosphido	—
25	S^{2-}	sulfido	thio

A wide variety of organic and inorganic ligands are tabulated below, along with their systematic and alternative names used in the nomenclature of the complexes (Tables 2.1–2.6).

2.4.5 Specifying Connectivity – The Kappa (κ) Convention

When a ligand contains more than one donor atom, the point of attachment to the metal atom needs to be specified. For this, the kappa (κ) convention is used. The ligating carbon atoms are often satisfactorily described in the ligand name in organometallic nomenclature. When bridging multiple metal centers in a polynuclear complex; however, kappa notation is necessary to show the attachment of heteroatoms and identify the exact locations of attachment of a single ligand [7, 12]. The κC^1 specification, for example, is used for clarification but is not

Table 2.2 Few examples of neutral ligands.

S. No.	Ligand	Alternative Ligand Name	Systematic Ligand Name
1	NH_3	Azane	Ammine
2	CO_2	Carbondioxide	—
3	CS	Carbonmonosulfide	Thiocarbonyl
4	CO	Carbonmonoxide	Carbonyl
5	$(CH_3)_2PH$	Dimethylphosphane	Dimethylphosphine
6	N_2	Dinitrogen	—
7	CH_3NH_2	Methanamine	Methylamine
8	CH_3PH_2	Methylphosphane	Methylphosphine
9	$(CH_3)_3N$	*N,N*-dimethylmethanamine	Trimethylamine
10	NO	Nitrogenmonoxide	Nitrosyl
11	$(CH_3)_2NH$	*N*-methylmethanamine	Dimethylamine
12	H_2O	Oxidane	Aqua
13	PH_3	Phosphane	Phosphine
14	$(CH_3)_3P$	Trimethylphosphane	Trimethylphosphine

strictly required because the name phenyl implies the bonding from carbon atom number 1. Tetra carbonyl [2-(2-phenyldiazen-1-yl-κN^2)phenyl-κC^1]manganese is an example.

Example

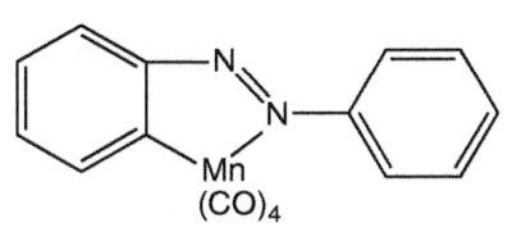

Tetra carbonyl [2-(2-phenyldiazen-1-yl-κN^2)phenyl-κC^1]manganese

2.4.6 Bridging Ligands – The Mu (μ) Convention

Bridging ligands are ligands in which one or two individual donor atoms each have at least two pairs of electrons and share these electron pairs with two metal ions or atoms simultaneously. A bridging complex can form when an organic ligand coordinates with two or more metal atoms. The Greek letter μ denotes a bridging bond formation mode.

When a metal atom is coupled to a bridging ligand, it is represented by the μ_n, where $n \geq 2$.

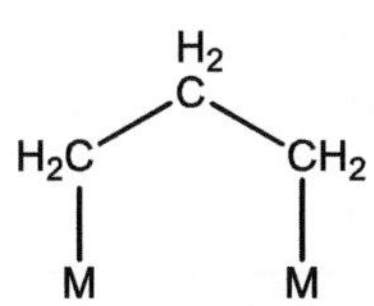

μ-Propane-1,3-diyl
(bridging)

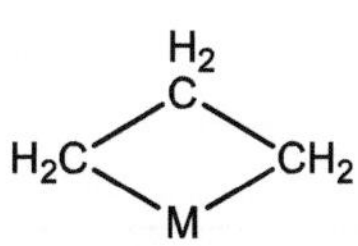

Propane-1,3-diyl
(chelating)

Table 2.3 Few common organic ligands and their names.

S. No.	Ligands	Systematic substitutive name	Systematic additive name
1	—C_6H_5	—	Benzenido
2	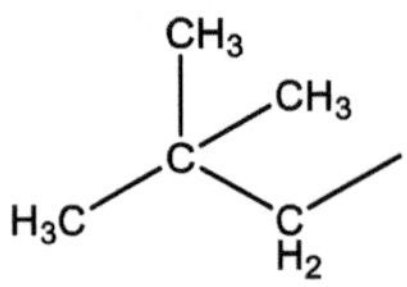	2,2-dimethylpropyl	2,2-dimethylpropanido
3	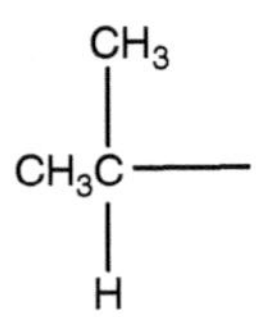	butan-2-yl	butan-2-ido
4	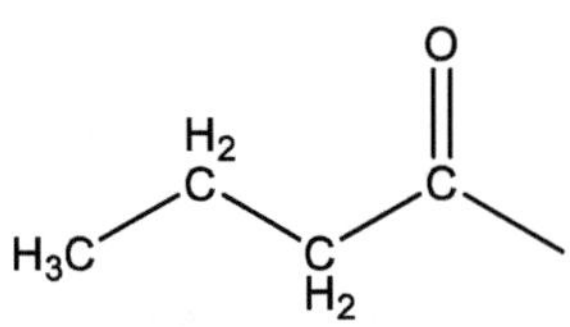	butanoyl	1-oxobutanido
5	$-CH_2CH_2CH_2CH_3$	butyl	butan-1-ido
6	—C_5H_5	cuclopentadienyl	cyclopentadienido
7	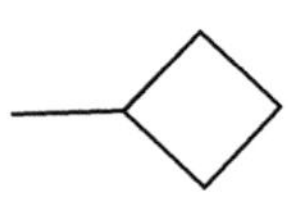	cyclobutyl	cyclobutanido
8	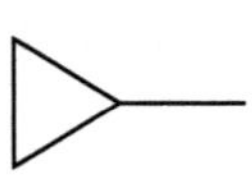	Cyclopropyl	Cyclopropanido
9	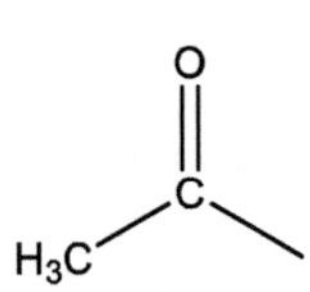	ethanoyl	1-oxoethanido
10	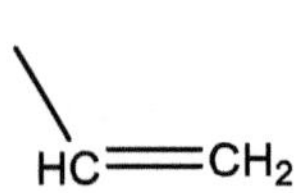	ethynyl	thenido
11	—GeH_3	germyl	germanido
12	$-CH_3$	methyl	methanido

(Continued)

Table 2.3 (Continued)

S. No.	Ligands	Systematic substitutive name	Systematic additive name
13	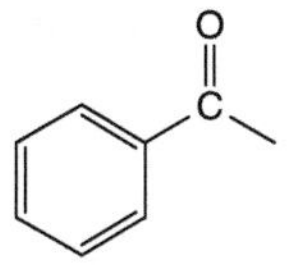	Phenylcarbonyl	Oxophenylmethanido
14	$CH_2{=}CHCH_2$-	Prop-2-en-1-yl	Prop-2-en-1-ido
15	$-CH(CH_3)_2$	propan-2-yl	propan-2-ido
16	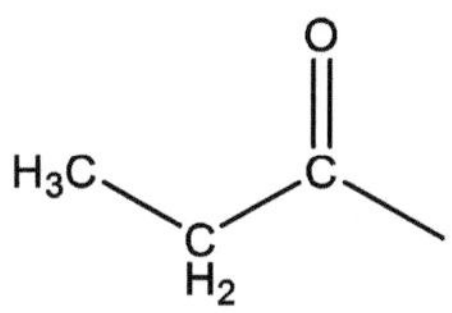	propanoyl	1-oxopropanido
17	$-CH_2CH_2CH_3$	Propyl	Propan-1-ido

In the chelating complex, there is also more than one metal–ligand bond, but here the same metal atom coordinates with a different atom of the same ligand. As a result, the mu (μ) convention cannot be used for the chelating complex [4, 8, 13, 15].

2.4.7 Hapticity – The Eta (η) Convention

Hapticity measures how many nearby atoms in the ligand are coordinated to the metal atom. The number of neighboring ligating atoms needed to form a bond with the metal is denoted by the eta (η) convention. The numeral of adjacent atoms in the ligand coordinated to the metal atom is represented with a correct superscript number to the eta (η), e.g. η^2 (eta two or dihapto), η^3 (eta three or trihapto), etc. The nomenclature prefixes the ligand with the sign "η." The particular way that unsaturated hydrocarbons attach to metals through their "π-electrons" necessitates the usage of this convention.

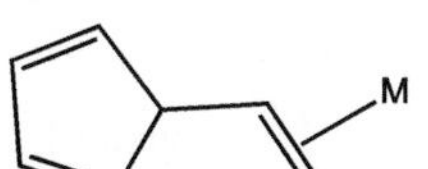

Cyclopenta-2,4-dien-1-yl-η^2-ethene

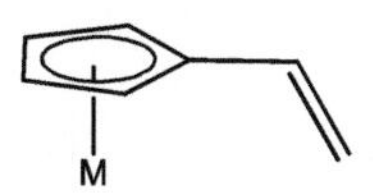

Vinyl-η^5-cyclopentadienyl

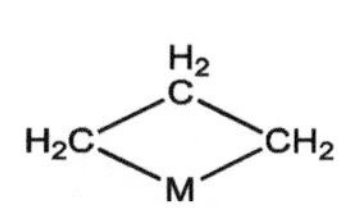

Propane-1,3-diyl (chelating)

In a chelate, the same ligand coordinates through two different atoms to a single metal atom, but the atoms of ligands that are coordinating are not adjacent (contagious). So the eta convention should not be used for chelates [4, 8, 9, 15].

Table 2.4 Organic ligand their systematic substitutive and systematic additive name.

S. No.	Ligand formula	Systematic substitutive name	Systematic additive name
1	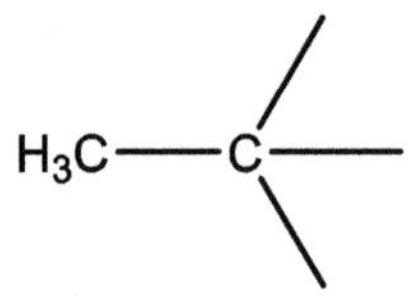	μ_3-ethane-1,1,1-triyl	μ_3-ethane-1,1,1-triido
2	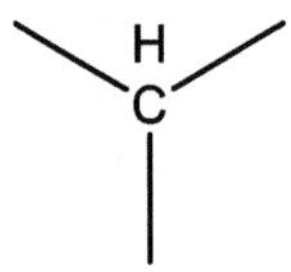	μ_3-methanetriyl	μ_3-methanetriido
3	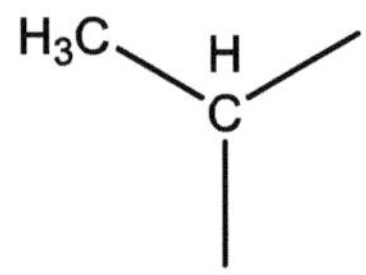	μ-ethane-1,1-diyl	μ-ethane-1,1-diido
4	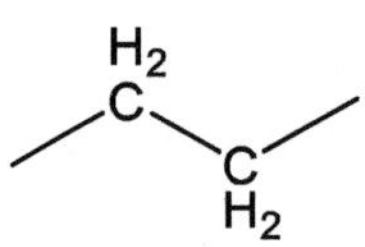	μ-ethane-1,2-diyl	μ-ethane-1,2-diido
5	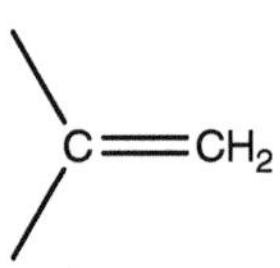	μ-ethene-1,1-diyl	μ-ethene-1,1-diido
6	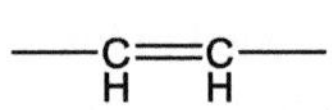	μ-ethene-1,2-diyl	μ-ethene-1,2-diido
7		μ-ethyne-1,2-diyl	μ-ethyne-1,2-diido
8	$-CH_{2}-$	μ-methanediyl	μ-methanediido
9	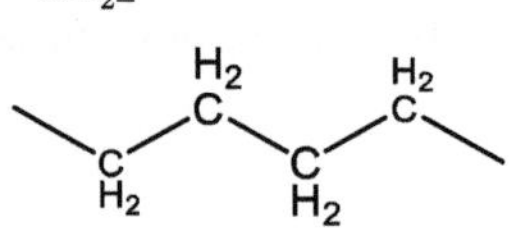	butane-1,4-diyl	butane-1,4-diido
10	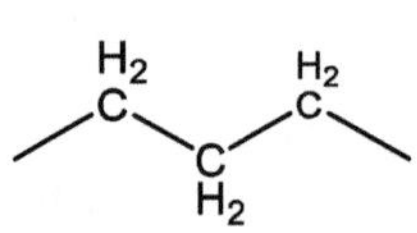	propane-1,3-diyl	propane-1,3-diido

Table 2.5 Ligands that coordinate during the formation of metal–carbon multiple bonds.

S. No.	Ligand	Systematic name
1	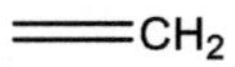	Methylidene
2	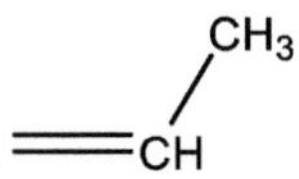	Ethylidene
3	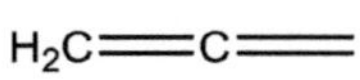	Ethenylidene
4	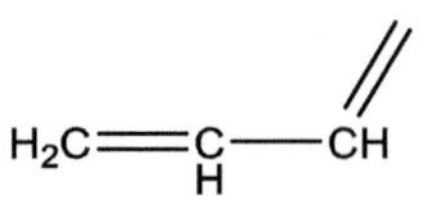	Prop-2-enylidene
5	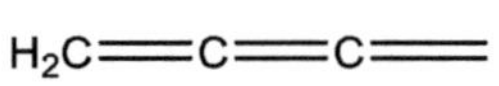	Prop-1,2-dienylidene
6	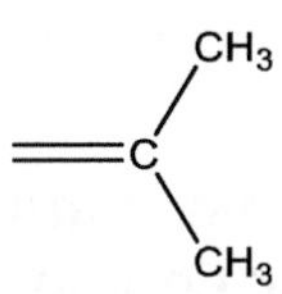	Propan-2-ylidene
7	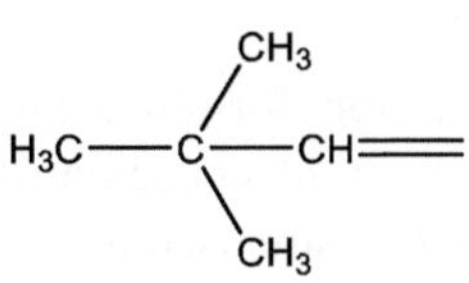	2,2-Dimethylpropylidene
8	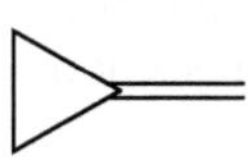	Cyclopropylidene
9	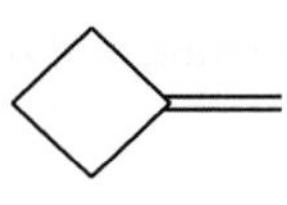	Cyclobutylidene
10	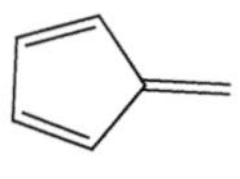	Cyclopenta-2,4-dienylidene
11	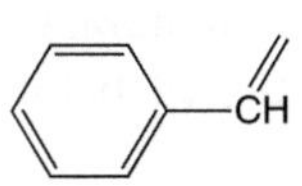	Phenylmethylidene
12	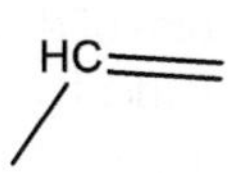	Methanylylidene
13		Methylidyne

(Continued)

Table 2.5 (Continued)

S. No.	Ligand	Systematic name
14	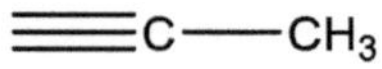	Ethylidyne
15	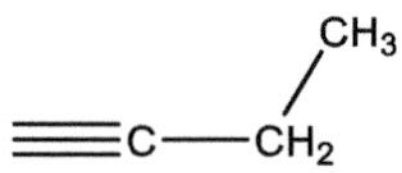	Propylidyne
16	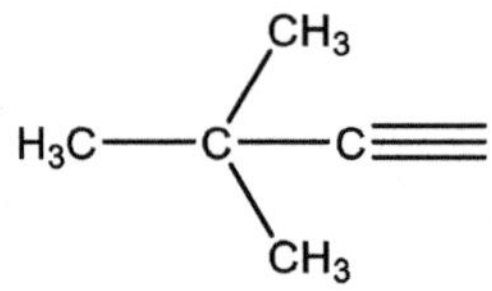	2,2-Dimethylpropylidyne
18	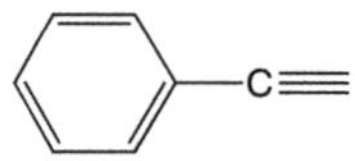	Phenylmethylidyne

2.5 Regulations Concerning the Nomenclature of Transition Element Organometallic Compounds

According to the 2005 International Union for Pure and Applied Chemistry nomenclature guidelines, systematic names of coordination entities formed by adhering to additive nomenclature's concepts are used to designate organometallic compounds. Specific additional rules are also incorporated for the nomenclature of organometallic compounds, which are described in sections 2.5 (XII). The rules for naming organometallic compounds are as follows:

(I) The compound's name starts with a small letter, and the full name is written as one word without any gap [7, 10].
Example

[BeEtH] ethylhydridoberyllium

(II) *Sequences of ligands and central atoms with in names*:
- (a) The names of the ligands are mentioned before those of the center atom(s),
- (b) There should be no space between the names of the coordination entities.
- (c) The names of the ligands are given alphabetically (the number of ligands indicated by multiplicative prefixes is not considered when establishing that order). There are two different multiplicative prefixes that can be used to specify the number of each form of ligand in the coordination entity's name.
- (d) Prefixes like di, tri, etc., are applied to simple ligands. No enclosing marks are necessary.

Table 2.6 Names of organic ligands with unsaturated groups and molecules.

S. No.	Ligands	Systematic substitutive name	Systematic additive name
1		η^3-propenyl	η^3-propenido
2		η^3-(*Z*)-butenyl	η^3-(*Z*)-butenido
3	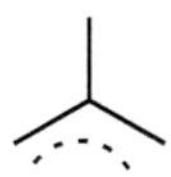	η^3-2-methylpropenyl	η^3-2-methylpropenido
4		η^4-2-methylidene-propane-1,3-diyl	η^4-2-methylidene-propane-1,3-diido
5	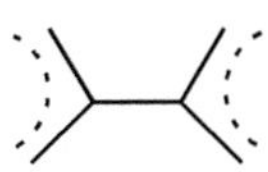	η^6-2,3-dimethylidene-butane-1,4-diyl	η^6-2,3-dimethylidene-butane-1,4-dido
6		η^5-(*Z*,*Z*)-pentadienyl	η^5-(*Z*,*Z*)-pentadienido
7		η^5-cyclopentadienyl	η^5-cyclopentadienido
8	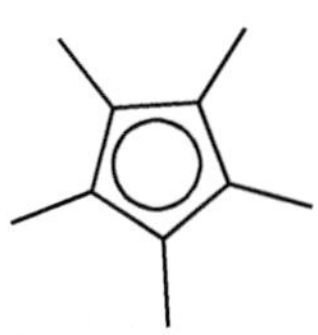	Pentamethyl-η^5-cyclopentadienyl	Pentamethyl-η^5-cyclopentadienido
9	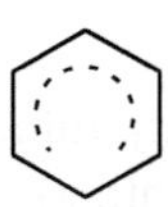	η^5-cyclopentadienyl	η^5-cyclopentadienido
10		η^7-cycloheptatrienyl	η^7-cycloheptatrienido
11	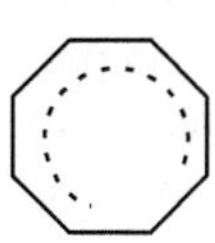	η^7-cyclooctatrienyl	η^7-cyclooctatrienido

(Continued)

Table 2.6 (Continued)

S. No.	Ligands	Systematic substitutive name	Systematic additive name
12	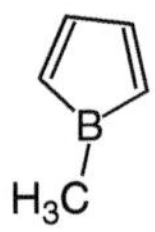	1-methyl-η^5-borole	
13	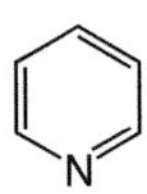	η^5-pyrrolyl	η^5-azacyclo-pentadienido
14	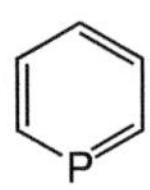	η^5-phospholyl	η^5-phosphacyclo-pentadienido
15	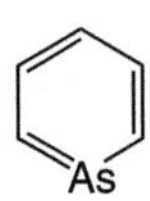	η^5-arsolyl	η^5-arsacyclo-pentadienido
16	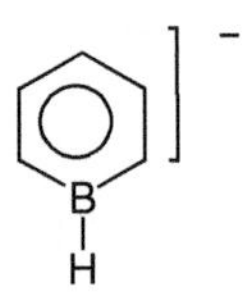	η^6-boratabenzene	η^6-borinin-1-uido
17	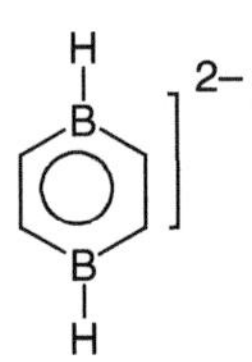	η^6-1,4-diboratabenzene	η^6-1,4-diborinine-1,4-diuido

(e) Prefixes like bis, tris, tetrakis, etc., are used to simplify things, especially with complex ligands. There must be parenthesis around the multiplicand [7, 10, 13]. For example, $[V(dipy)_3]$ it has named tris(dipyridine)vanadium(I) (Table 2.7).

(III) *Representing ligands in names*:

(a) The "o" suffix is added to the end of the ligand if it is anionic and either organic or inorganic in nature. For instance, when "e" is present in an anionic ligand like ide, ate, or ite, "o" is present instead, creating ido, ato, and ito, respectively. The classifications for carboxylates, thiolates, alcoholates, etc., are the same. Halide also transforms into halido. For instance, fluoride, bromide, and coordinated ligands like cyanide transform into cyanido. Except for those of molecular hydrogen, hydrogen is always considered as an anion in its complexes. Hydrogen coordinates to all elements, including boron, and is known as "hydrido."

Table 2.7 Prefixes for complex and simple ligands.

No. of ligands	Prefixes for complex ligands	Prefixes for simple ligand
2	bis	di
3	tris	tri
4	tetrakis	tetra
5	pentakis	penta
6	hexakis	hexa
7	heptakis	hepta
8	octakis	octa
9	nonakis	nona
10	decakis	deca

(b) Names of neutral and cationic ligands, including organic ligands, are used without modification, even if they have the ends "ite," "ide," or "ate."

(c) Enclosing marks are required whenever it is essential to avoid confusion, such as when naming neutral or cationic ligands, inorganic anionic ligands with multiplicative prefixes (such as triphosphato), or compositional names (such as carbon disulfide), substituted organic ligands (even when there is no ambiguity in their use). Common ligand names like aqua, ammine, ethyl, methyl, carbonyl, and nitrosyl do not need enclosing marks.

(d) In the rule of nomenclature, abbreviations are not used for compound naming [7, 13, 14].

Examples

$[Fe(C_4H_6)(CO)_3]$	(butadiene)tricarbonyliron
$K[PtCl_3(C_2H_4)]$	potassium trichkiri(ethane)platinate

(IV) *Representation of charge numbers and oxidation state*:

It might be challenging to apply the idea of oxidation number or state to organometallic compounds. This is particularly true when it is unclear whether complexation by a ligand should be viewed as an oxidative addition, a Lewis-acid or Lewis-base interaction, or both. Therefore, formal oxidation numbers won't be ascribed to the central atoms of the organometallic complexes because, for naming purposes, only the net charge on the coordination entity matters. However, the following techniques can be used to help describe the composition of a compound if the oxidation number is indicated:

(a) The central atom's oxidation number may be denoted by a Roman numeral written in parenthesis to the central atom (such as the ending "ate, if needed)" provided the oxidation state can be delineated.

The number is sometimes preceded by a negative sign. The oxidation number zero is denoted by the Arabic zero.

(b) Alternately, it is possible to state the charge against a coordination entity. The total charge is stated in parenthesis, with the number written in Arabic before the charge symbol. Without using space, it repeats the name of the central atom (including, if applicable, the ending "ate."

(c) By utilizing multiplicative prefixes, the ratios of ionic units in a coordination complex can be determined [7, 14].

Examples

$[Ni(CO)_4]$	tertranickelcobalt(0)
$Li[CuMe_2]$	lithium dimethylcuprate(1−)

(V) *The complex containing bridging ligands*

(a) Bridging ligands are listed alphabetically along with the other ligands.

(b) When a bridging ligand and matching nonbridging ligands are both present, then the bridging ligand is cited first when it appears in names.

(c) Multiple ligands should be stated to increase complexity; for example, the μ_2 bridging ligand should appear before the μ_3 bridging ligand [7, 14].

Example

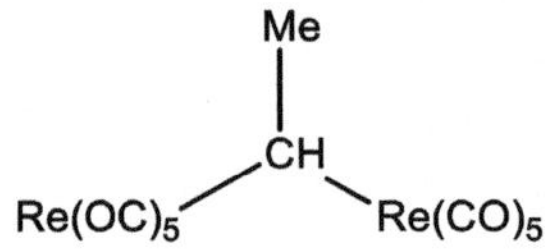

(μ–ethane-1,1-diyl)bis(pentacarbonylrhenium)

(VI) *The complex contains multiple bonding*:

Ligands that form metal–carbon double or triple bonds end with ylidene for a double bond and ylidyne for a triple bond, with the prefix names taken from the parent hydrides. These suffixes can be used in two ways. The parent hydride name's "ane" ending is replaced by the suffix "ylidene" or "ylidyne." The atom with the free valencies is considered to conclude the chain if the parent hydride is a chain. In every instance, this atom has the locant "1" (omitted from the name). Only saturated acyclic and monocyclic hydrocarbon substituent groups and the mononuclear parent hydrides of germanium, tin, silicon, and lead are treated using this procedure. Please take note that the suffix "ylene" can only be used in conjunction with μ to signify bridging $-C_6H_4-$(phenylene) or $-CH_2-$ (methylene).

The suffix "ylidene" or "ylidyne" replaces the ending "ane" of the parent hydride name. If the parent hydride is a chain, the atom with the free valencies is understood to terminate the chain. This atom has, in all cases, the locant "1" (which is omitted from the name). This method is used only for saturated acyclic and monocyclic hydrocarbon substituent groups and for the mononuclear parent hydrides of silicon, germanium, tin, and lead. Note that the suffix "ylene" should only be used in conjunction with μ to designate bridging $-CH_2-$ (methylene) or $-C_6H_4-$ (phenylene).

In a simpler method, the suffix "ylidene" or "ylidyne" is added to the name of the parent hydride, replacing any terminal "e" that may be present. The atom also with free valence is assigned a number that is as low as is consistent with the established numbering of the parental hydride [7, 14].
Example

EtCH=	propylidene
$Me_2C=$	propan-2-ylidene

It must be mentioned that the parent chain with the longest chain of carbon atoms is chosen when numbering a ligand with multiple attachment sites before providing the lowest locant to the atom with the free valence. The numbering's direction in a metallacycle is selected to provide feasible side chains or substituents with the lowest locants. Once more, heterocyclic and polycyclic systems are subject to particular numbering methods. If a ligand creates single or multiple metal–carbon bonds, the order of ends is as follows: "yl," "ylidene," "ylidyne." To provide the smallest set of locants for the free valencies, method (b) should then be employed. If -yl, -ylidene, side chains, or substituents are present, -yl yields a lower number than -ylidene following that side chain or a substituent [7, 14].
Example

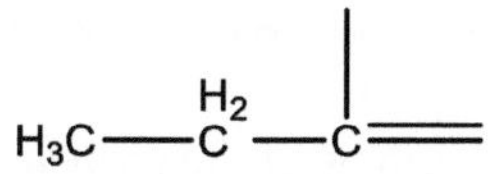

Propan-1-yl-1-ylidene

(VII) Nomenclature for kappa (κ) convention:
The κ convention is used to coordinate those flexidentate ligands having more than one donor atom and coordinating with metal with one particular atom at a specific time. Similarly to thiocyanato-kN for SCN ligands bonded with metal atoms via the N atom, thioyanato-S for SCN ligands bonded with metal atoms via the S atom. When a nitrogen atom is a donor in nitrito ligand, it is represented as Nitrito-κN while oxygen is a donor atom; it is designated as Nitrito-κO in this convention. An appropriate superscript numeral may be written to the "κ" sign to show the number of identical bonds that attach a binding atom of the ligand to the central atom(s); nonequivalent ligating atoms of the ligand should each be represented by an italics element symbol accompanied by "κ."

One or even more superscript primes may also be given to the element symbol to easily differentiate between various donor atoms of the same element under specific circumstances. If not, the identity of the ligating atom is clearly defined using a right superscript numeral that corresponds to the usual numbering of the atoms in the ligand. After the part of the ligand name that designates the particular function, substituent group, ring, or chain in which the ligating atom is located, these symbols are used to denote those elements [7, 12, 14].

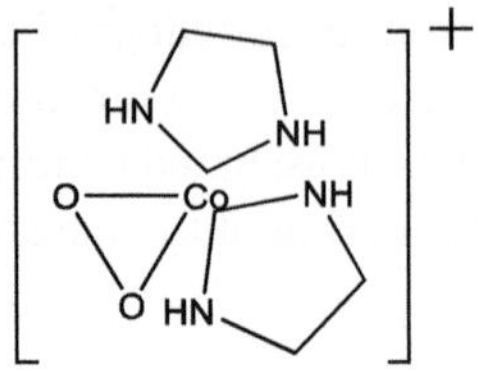

Bis(ethane-1,2-diamine-κ2N)-(η2-peroxido)-cobalt(III)

(VIII) If more than one contiguous (adjacent) ligand coordinates to the metal-ligand, then the eta (η) convention must be used before the name of such ligand. A right-superscript number to the eta (η) is used in this convention to indicate the number of neighboring atoms in the ligand coordinated to the metal atom [4, 7–9, 15].
Example

Bis(η6-benzene)chromium (η7-cycloheptatrienl)(η5-cyclopentadieny)vanadium

(IX) The italicized element symbols of the corresponding metal atoms, separated by a "em" dash and contained in parentheses (M–M), is inserted after the list of central atom names and before the ionic charge to signify metal–metal bonding. The element symbols are arranged in the same order as the names' central atoms. An Arabic numeral should appear before the first element symbol in the parenthesis if the complex has more than one metal–metal link [7, 14].
Example

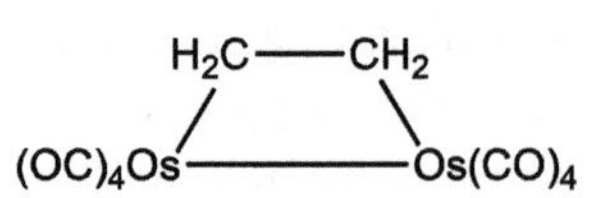

μ–ethane-1,2-diyl)bis(tetracarbonylosmium) (*Os–Os*)

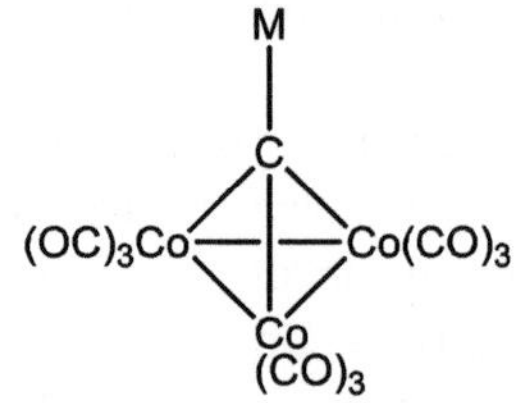

(μ3–ethane-1,1,1-triyl)-triangulo-tris(tricarbonycobalt) (3 *Co–Co*)

(X) Organometallic compounds of group 13 to group 16 element nomenclatures are based on the substitutive system.

(a) Each substituent that takes the place of a hydrogen atom in the parental hydride changes the compound's name by adding a prefix. Examples are given below in the table:

S. No.	Structure of the compound	Name of the compound
1	AlH_2Me	Methylalumane
2	$AlEt_3$	Triethylalumane
3	$Sb(CHCH_2)_3$	triethenylstibane, ortrivinylstibane

(b) The name of the parent hydride carrying the highest ranking of these characteristic groups is modified by the suffix in the presence of one or more characteristic groups that can be expressed using one or more suffixes ($-NH_2$, –OH, –COOH, etc.), and other substituents are then denoted by their prefixes. The name of the group13 to group 16 element parent hydride in question is altered if it is serving as a substituent by changing the final "ane" to "anyl" (or "yl" for the group 14 elements), "anediyl," etc. [14]. Examples are given below in the table:

S. No.	Structure of the compound	Name of the compound
1	$(EtO)_3GeCH_2CH_2COOMe$	Methyl 3-(triethoxygermyl)propanoate
2	$H_2As(CH_2)_4SO_2Cl$	4-Arsanylbutane-1-sulfonyl chloride
3	$SiMe_3NH_2$	Trimethylsilaneamine

(c) Sometimes it might be essential or advantageous to consider a parent hydride in which a few (four or more) of a hydrocarbon's skeleton carbon atoms have been swapped out for main group elements. The heteroatoms in this form of skeletal replacement are identified by the replacement nomenclature's "a" word and are preceded by the proper locant(s) [14]. Examples are given below in the table:

S. No.	Structure of the compound	Name of the compound
1	$MeSiH_2CH_2CH_2SiH_2CH_2CH_2SiH_2CH_2CH_2SiH_2Me$	2,5,8,11-Tetrasiladodecane
2	$MeSiH_2OP(H)OCH_2Me$	3,5-Dioxa-4-phospha-2-silaheptane

(XI) *Priority order of central atoms in polynuclear organometallic compounds*:
The name's foundation must be decided upon when an organometallic complex comprises two or more distinct metal atoms.
The potential central atoms that could be associated with one of the given

(i) Additive nomenclature for the element belongs to groups 1–12
(ii) Substitutive nomenclature for the element belongs to groups 13–16

(a) **Central atoms belongs to s and d groups elements only**
If any or all of the prospective central atoms fall into class I, the compound's name is given using an additive system of nomenclature.

(b) **Central atoms belongs to s and d groups elements and groups 13–16**
If at least one prospective central atom belongs to class I and one or more others belong to class II, the compound is identified additively with the metal atom(s) of class I serving as the central atom(s). The complex's remaining atoms are referred to as ligands [14]. Examples are given below in the table:

S. No.	Structure of the compound	Name of the compound
1	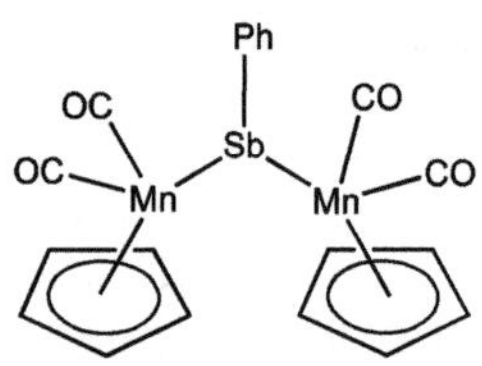	(phenylstibanediyl)bis[dicarbonyl (η^5-cyclopentadienyl)manganese]
2	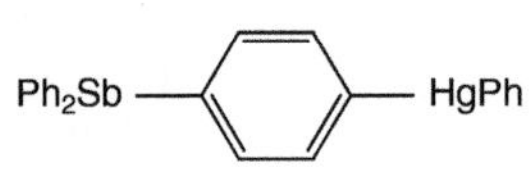	4-(diphenylstibanyl)phenyl] (phenyl)mercury
3	$[Li(GePh_3)]$	(triphenylgermyl)lithium

(c) **Central atoms belongs to groups 13–16 only**
If the central atom of an organometallic compound belongs to class (II), their nomenclature is followed by substitutively [14]. Based on the following elemental order (">" denoting "picked before"), the parent hydride is selected:
N > P > As > Sb > Bi > Si > Ge > Sn > Pb > B > Al > Ga > In > Tl > S > Se > Te > C
When the hydride of lead and arsenic are both present in a chemical compound, then AsH_3 has chosen parent hydride rather than PbH_4. The lead atom would then appear in the name as a prefixed substituent, frequently with its own substituent groups [14].

Examples

S. No.	Structure of the compound	Name of the compound
1	$Et_3PbCH_2CH_2CH_2BiPh_2$	diphenyl[3-(triethylplumbyl)propyl] bismuthane
2	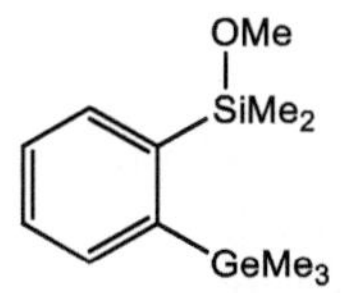	Methoxydimentyl[2-(trimethylgermyl) phenyl]silane
3	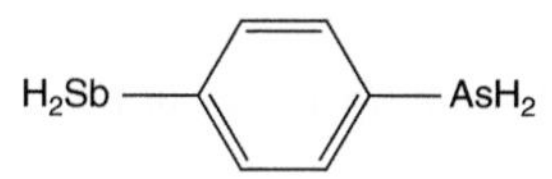	(4-stilbanylphenyl)arsane

(XII) Nomenclature of Metallocene

Metallocene derivatives can be named using either prefix nomenclature or the typical organic suffix (functional) nomenclature. The names of the groups of metallocene substituents end in "yl," "ocenediyl," "ocenetriyl," etc.

Examples

S. No.	Structure of the compound	Name of the compound
1	$[Cr(\eta^5\text{-}C_5H_5)_2]$	Chromocene
2	$[Co(\eta^5\text{-}C_5H_5)_2]$	Cobaltocene
3	$[Os(\eta^5\text{-}C_5H_5)_2]$	Osmocene
4	$[Fe(\eta^5\text{-}C_5H_5)_2]$	Ferrocene

Not all transitional components, however, fall under the metallocene terminology. For instance, there exist at least two isomers of the empirical formula $C_{10}H_{10}Ti$, but neither of them should be called "titanocene" because neither has the normal sandwich structure like that of ferrocene. Similarly, "manganocene" is an incorrect term for $[Mn(\eta^5\text{-}C_5H_5)_2]$, which in its solid state has a chain structure rather than separate sandwich entities. Decamethylrhenocene, $[Re(\eta^5\text{-}C_5Me_5)_2]$, and decamethyl manganocene, $[Mn(\eta^5\text{-}C_5Me_5)_2]$, both have a typical sandwich structure. The traditional bis(η^5-cyclopentadienyl) sandwich structure of the ferrocene type decreases with increasing atomic number.

Since the cyclopentadienyl rings are essentially parallel and the metal is in the d-block, the name ending "ocene" should only be used for discrete molecules of the form bis(η^5-cyclopentadienyl) metal (and ring-substituted analogs) [i.e. the terminology does not apply to compounds of the s- or p-block elements like $Ba(C_5H_5)_2$ or $Sn(C_5H_5)_2$

Although it should be noted that in this context, the ending "ium" does not have the same significance it does in substitutive nomenclature, i.e. the addition of a hydron to a neutral parent molecule, the oxidized species that have been referred to as metallocenium (n+) salts [7, 11, 14].

References

1 Shriver, D.F. *Organometallic Compound*. Encyclopedia Britannica https://www.britannica.com/science/organometallic-compound.

2 Gupta, B.D. (2011). *Basic Organometallic Chemistry: Concepts, Syntheses and Applications*. Universities Press.

3 Crabtree, R.H. (2009). *The Organometallic Chemistry of the Transition Metals*. Wiley.

4 Hartshorn, R.M., Hellwich, K.-H., Yerin, A. et al. (2015). Brief guide to the nomenclature of inorganic chemistry. *Pure Appl. Chem.* 87 (9–10): 1039–1049.

5 Libretexts Chemistry. 3.5: Importance of nomenclature [Internet]. Chemistry LibreTexts. Libretexts; 2021 [cited 2022 Aug 15]. Available from: https://chem.libretexts.org/Bookshelves/General_Chemistry/Map%3A_General_Chemistry_(Petrucci_et_al.)/03%3A_Chemical_Compounds/3.5%3A_Importance_of_Nomenclature.

6 Rigaudy, J. and Klesney, S.P. (1979). *Nomenclature of Organic Chemistry, 1979 edition (The Blue Book 79)*. Oxford: Pergamon Press.

7 Salzer, A. (1999). Nomenclature of organometallic compounds of the transition elements. *Pure Appl. Chem.* 71 (8): 1557–1585.

8 Leigh, G.J. (ed.) (2011). *Principles of Chemical Nomenclature – A Guide to IUPAC Recommendations, 2011 Edition*. Cambridge: Royal Society of Chemistry. ISBN: 978-1-84973-007-5.

9 Mehrotra, R.C. (2007). *Organometallic Chemistry*. New Age International.

10 Malik, W.U., Tuli, G.D., and Madan, R.D. (1998). *Selected Topics in Inorganic Chemistry*. S. Chand Publishing.

11 Huheey, J.E., Keiter, E.A., and Keiter, R.L. (1993). *Inorganic Chemistry: Principles of Structure and Reactivity*. New York, NY: Harper Collins College Publishers.

12 Douglas, B.E., McDaniel, D.H., and Alexander, J.J. (1983). *Concepts and Models of Inorganic Chemistry*, 2e. Wiley.

13 Miessler, G.L. and Tarr, D.A. (2004). *Inorganic Chemistry*. Upper Saddle River, NJ: Pearson Education.

14 Connely, N.G., Damhus, T., Hartshorn, R.M., and Hutton, A.T. (2005). *Nomenclature of Inorganic Chemistry, IUPAC Recommendations 2005*. Cambridge, UK: RSC Publishing.

15 Electron, C. and Model, C. (2014). Making sense of organometallic complexes. *Organomet. Chem. Transit. Met.* 40–68.

3

Classification of Organometallic Compounds

Mukesh K. Tyagi[1], Gokul R. Nishad[2], Dakeshwar Kumar Verma[2], Lei Guo[3], and Elyor Berdimurodov[4]

[1]*Hemchand Yadav University, Atal Bihari Vajpai, Govt. College, Department of Chemistry, Pandatarai, Kabirdham 491559, India*
[2]*Hemchand Yadav University, Government Digvijay Autonomous Postgraduate College, Department of Chemistry, Rajnandgaon 491441, India*
[3]*Tongren University, School of Materials and Chemical Engineering, Tongren 554300, China*
[4]*National University of Uzbekistan, Faculty of Chemistry, Tashkent 100034, Uzbekistan*

3.1 Introduction

Organometallic compounds are compounds in which at least one bond is found between metal and carbon, or in other words, those organic compounds which have direct linkage with one or more metal atoms of carbon. Alkaline metals and alkaline earth metals mediated organometallic compounds are common and very useful. Various metalloids such as selenium, silicon, and boron have been synthesized from organometallic compounds with various classes of organic compounds, but the most important organometallic compound has been synthesized from transition metals. Transition metal-mediated synthesized organometallic compounds are widely used and have numerous applications. Their applications the include synthesis of catalysts, drugs, paints, metal organic framework (MOF), polymers, composites, adsorbents, supercapacitors, electrode materials, etc. Similarly, lanthanides and actinides metal-based organometallic materials have also come into existence and play a very vital role in different fields.

Overall, it can be said that organometallic mainly deals with the transformations of organic compounds by reacting metals of different groups like transition metals, s-block and p-block elements, lanthanides and actinides series. Organometallic chemistry in inorganic chemistry (EVM) is a part of coordination chemistry, which specifically studies M—C bonds, whereas coordination chemistry studies both metal elements and M—C bonds.

The discovery of Zeise's salts in the nineteenth century brought a revolution, which was mainly complex and containing platinum. The salt, which was discovered in 1827, is represented by $[PtCl_3(\eta_2\text{-}C_2H_4)]$ [1] (Scheme 3.1).

Organometallic Compounds: Synthesis, Reactions, and Applications, First Edition.
Edited by Dakeshwar Kumar Verma and Jeenat Aslam.

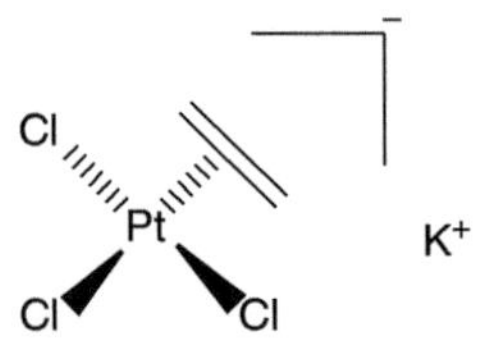

Scheme 3.1

Similarly, in the twentieth century, French scientist Victor Grignard gave a new direction in the field of chemistry by discovering Grignard reagents in the field. They synthesized RMgX by reacting magnesium by oxidative addition of RX, whose key function was to alkylate carbonyl derivatives (Scheme 3.2) [2].

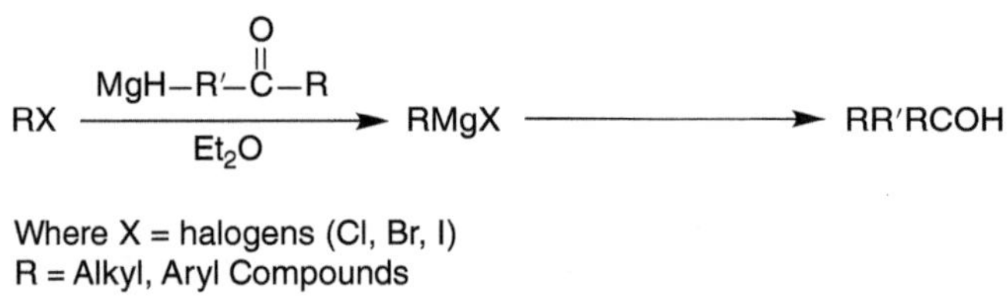

Scheme 3.2

3.2 Classification of Organometallic Compound

3.2.1 Sigma-Bonded Organometallic Compound

A sigma-bonded organometallic compound is a two-centered, two-electron (2c–2e) bond, in which the organic compound is bonded to the metal by a common covalent sigma bond. In sigma-bonded organometallic compounds, mainly metals having low electronegativity, form sigma bonds with nonmetallic elements, such as carbon atoms. Transition elements, alkali earth metals, and some main group elements form sigma-bonded organometallic compounds and are generally covalent in nature [3, 4].

For example, $W(CH_3)_6$, $Mn(CO)_5$, $Ta(CH_3)_5$, $Ph(CH_3)_4$, $Al(CH_3)_3$, $Si(CH_3)_4$, $B(CH_3)_3$.

Figure 3.1 shows the classification of organometallic compounds.

Also, shapes of organometallic compounds depend on the hybridization of central metal ions.

Hybridization	Shape	Examples
sp	Linear	$Be(CH_3)_2$
sp^2	Triangular planner	$Al(CH_3)_3$, $B(CH_3)_3$
sp^3	Tetrahedral	$Pb(CH_3)_4$, $Si(CH_3)_4$
sp^3d	Trigonal bipyramidal	$Ta(CH_3)_5$
sp^3d^2	Octahedral	$W(CH_3)_6$, $Mn(CO)_5$

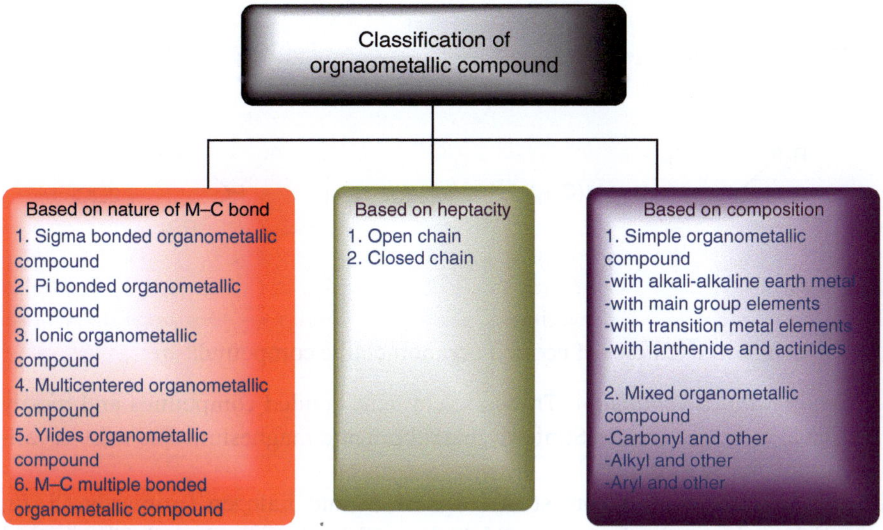

Figure 3.1 Classification of organometallic compounds.

Transition metals and carbon-headedness are present in the coordinately sigma-bonded organometallic compound in the form of terminal metal–allyl complexes such as $[W(CH_3)_6]$. Similarly, tetrahedral compounds also form complexes in which they form various bridging complexes as shown in Scheme 3.3.

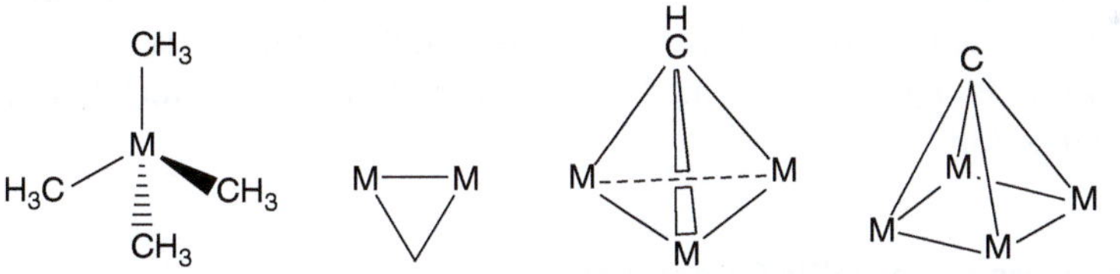

Scheme 3.3

Metallocycles: In this regard, metallocycles are a class of organometallic compound in which a metal ion is the ring atom (Scheme 3.4).

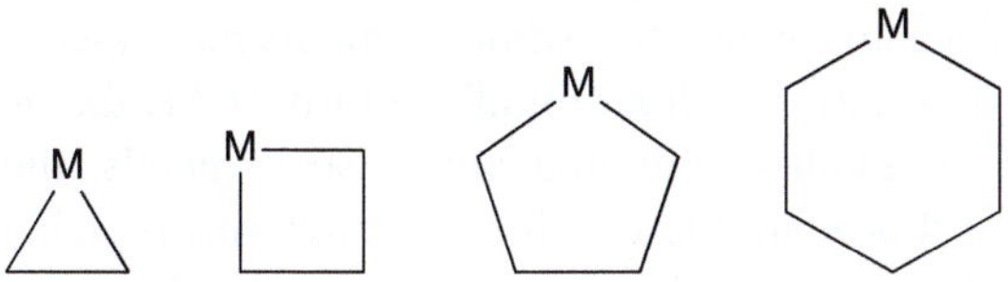

Scheme 3.4

These metallocycle organometallic compounds can be synthesized as given in Scheme 3.5.

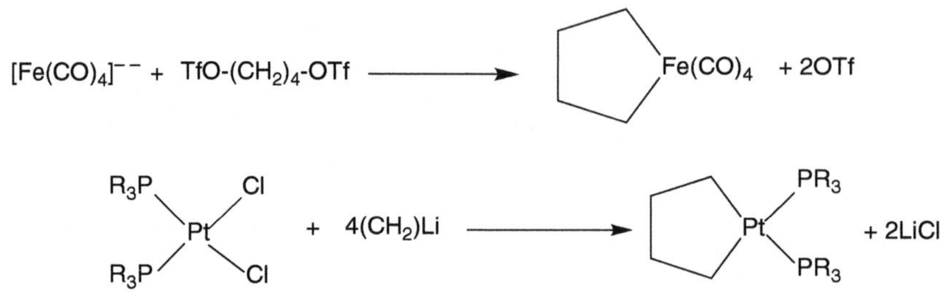

Scheme 3.5

Some common examples of bonded organometallic compounds are:

i. *Group I and II compound*: These are covalent bonded compounds and mostly exist as tetrameric units. Similarly, berkelium and magnesium also exist in various polymeric forms.
ii. *Grignard reagents*: It consist of magnesium and halogen (primarily Cl, Br, and I) and can covalent most aldehydes and ketones to alcohol. The general structure of Grignard reagents is RMgX, in which R = alkyl or aryl and X = Cl, Br, and I.
iii. *Trimethyl aluminum and Tetra alkyl tri organometallic compound*: $Al_2(CH_3)_6$ (trimetyhyl aluminum) exist in the dimer form in which alkyl group linked as bridge compound between the metals. $Al_2(CH_3)_6$ formed 2-Al and 1 atom multicenter bond.
    Similarly, tetramethyl tin $Sn(CH_3)_4$, tetra ethyl tin, $Sn(Et)_4$, and tetra methyl arsine ($As(CH_3)_4$) are important classes of organometallic compounds.
iv. *Diethylzinc ($Zn(C_2H_5)_2$)*: This zinc mediate compound is provide nucleophile form organic reaction.

3.2.2 π-Bonded Organometallic Compounds

Generally, unsaturated compounds such as alkenes, alkynes, and other carbon-containing multiple bonds are electron-rich. When a metal with a vacant orbital or high oxidation state approaching the electron rich multiple bonded moiety, then they form a stable bond by approaches the electron rich center. Metal cation when it interacts with multiple carbon centers. It gets attached to all the carbon atoms linked by multiple bonds. Since all d-block transition elements have vacant d-orbital's, they readily interact with electron-rich π orbital's to form stable dπ–pπ bonds. Unlike other metals such as alkali metals and alkali earth metals, the main group element lacks a vacant d-orbital, due to which it rarely forms stable bonds with unsaturated compounds. In this context, the credit for the discovery of the most stable and complex ferrocene goes to the scientist who discovered it in 1951. In ferrocene, there is a stable bond between iron and cyclopentadiene ions, and in this the center metal ion is equally attached to all the five carbon atoms through their nuclei, which are mainly covalent in nature [5, 6] (Scheme 3.6).

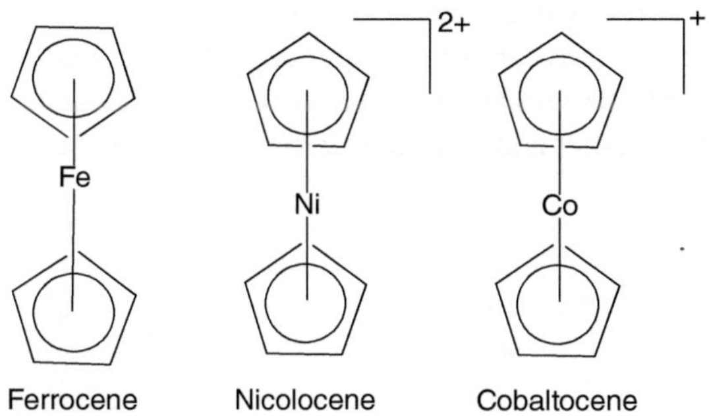

Scheme 3.6

Similarly, after a few decades, sandwich compounds such as nicolocene and cobaltocene were also synthesized (Scheme 3.6). A classic example of a pi-bonded organometallic compound is usually given as a metal with multiple covalent complexes (Scheme 3.7).

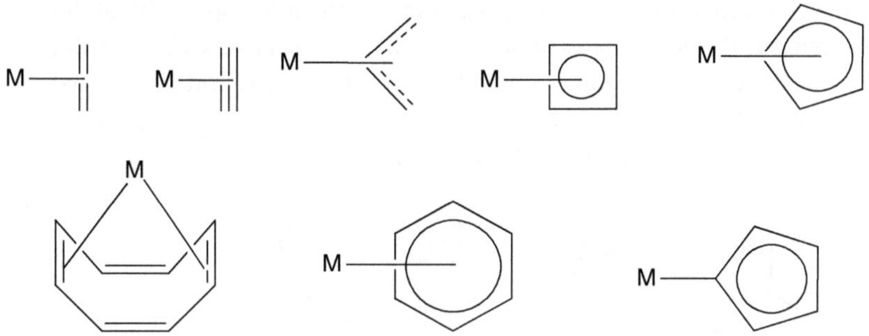

Scheme 3.7

3.2.3 Ionic Bonded Organometallic Compounds

Bond formation in ionic bonded organometallic complexes occurs mainly by electrostatic bond formation between a metal cation and carbon having an electronegativity value lower than that of carbon. Earth metal cations are highly electropositive in character, with the shared electron being more localized toward the carbon atom. As a result, the metal ion becomes slightly positively charged and the carbon negatively charged. The alkali metals, alkaline earth metals, lanthanides, and actinide metals are slightly less electronegative and mostly form ionic bonds [7].

Exceptions: alkali and alkali earth metals, such as lithium, beryllium, and magnesium, which are relatively smaller in size and higher electronegative, resulting in mostly covalent bonds and being less reactive than other related organometallic compounds [8].

Examples: Sodium cyclopentadenide ($C_5H_5^-Na^+$), phenyl sodium ($Na^+C_6H_5$) and methyl potassium ($CH_3^-K^+$) are well-known examples of organometallic compounds in this class. Their organic compounds are colorless, soluble in polar solvents and solids such as salts, and very reactive toward water and air [8].

Generally the stability and reactivity of this class of compound depend on the organic anion. Extended conjugated alkenes containing organometallic compounds are more stable as compared to simple alkenes because in previous care, delocalization has taken place from one and two other ends, hence reactivity will be decreased (Scheme 3.8).

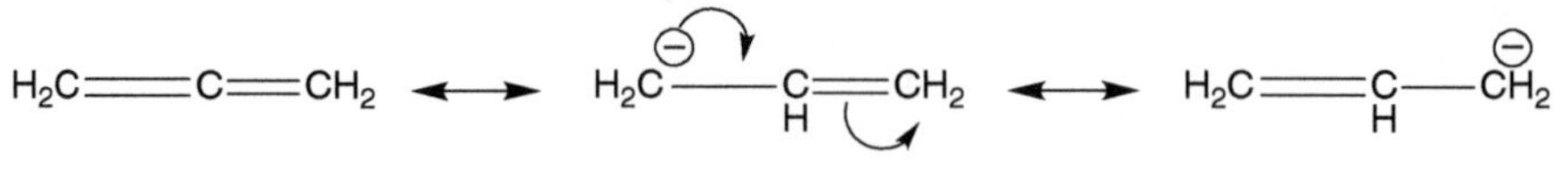

Scheme 3.8

Because sp hybridized carbon atoms easily form carbanian due to the 50%–80% sp character of the carbon bond, electronegativity of alkyne carbon has been slightly increased and can easily remove attached proton by attracting shared paired electrons.

Similarly, C_6H_5Na and $C_5H_5CH_2Na$ are both states of organometallic compounds. Compared to both, $C_5H_5CH_2Na$ is more stable because the distributed negative charge is on all carbon atoms in the benzene ring and the nucleophilicity of the CH_2 group is decreased. While in the case of $C_6H_5^-Na^+$ the negatively charged delocalized are the sp^2 hybridized carbon atoms. These are sigma type and hence unstable. Overall, due to the above effect, C_6H_5Na is more reactive toward electrophiles as a compound to $C_5H_5CH^-Na^+$ (Scheme 3.9).

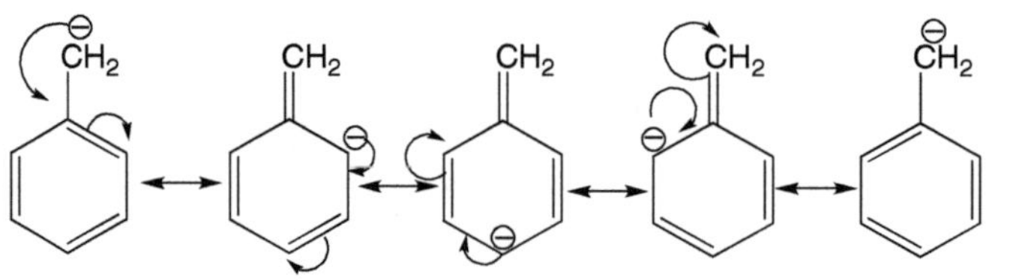

Scheme 3.9

Another example includes the nucleophillic comparison between $C_5H_5^-M^+$ and $C_5H_{11}^-M^+$. In this case, C_5H_5 is more stable as it follows the Huckel rule $(4n+2\pi)$ e^- system and has a total 6π electrons, whereas $C_6H_4^-$ doesn't follow the Huckel rule and the negative charge localized only on one carbon atom (Scheme 3.10). Delocalization on $C_6H_5^-$ ion:

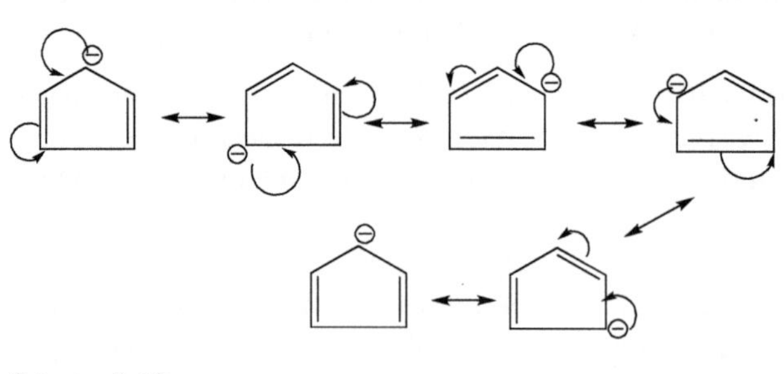

Scheme 3.10

3.2.4 Multicentered Bonded Organometallic Compounds

Elements such as Al, B, Mg, Be, and Li exhibited a higher tendency to make multicentered bonds with alkyls and aryl groups. In this case, alkyls and aryls acted as bridges between two HOMO-centered or hetero-centered metal ions. There, bridge compounds can be considered as intermediates between sigma-bonded and ionic organometallic compounds [9, 10].

Well-known examples of this class are dimeric trimethyl aluminum ($Al_2(CH_3)_6$), $[Be(CH_3)_2]_n$, polymeric dimethyl–beryllium, dimeric tryphenyl aluminum $[Al_2(C_6H_5)_6]$, etc. (Scheme 3.11).

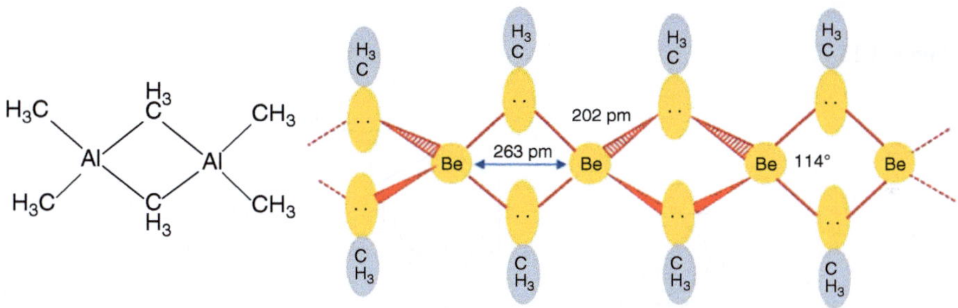

Scheme 3.11

5. Ylide: In this case, multiple bond, especially double bonded organometallic compounds, were extensively studied. Here the center metal ions bond with the carbon atom of the ligand. Transitional elements, such as Fe, W, Re, Mo, and Cr, primarily form double bonds with carbon atoms of ligands.

3.2.4.1 Based on Heptacity (η1 to η8):

In organometallic chemistry, heptacity can be considered as the coordination of an unsaturated or multiple bond containing ligand (aliphatic or aryl) through a continuous and uninterrupted chain of atoms. Usually word heptacity has been described from Greece letter η form instand ita 4 demonstrated the coordination though, four contagious carbon atoms. In addition here are the examples of heptacity [11] (Scheme 3.12).

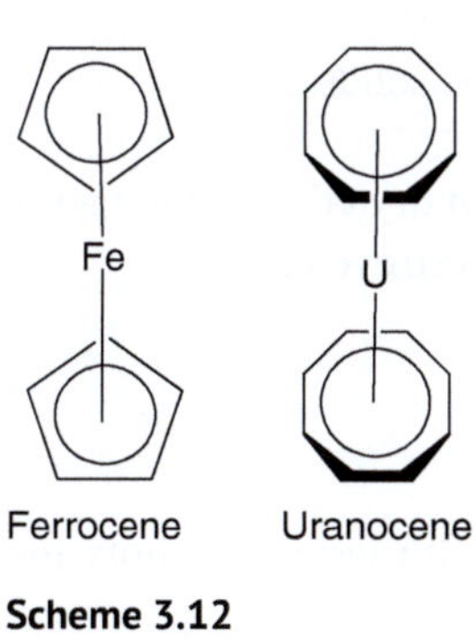

Scheme 3.12

Various and noteworthy contribution of heptacity in various forms:

- Side wise bonding of ligands containing multiple bond: Example – Zeise's Salt (Scheme 3.13)
- 18-electron rule applicable compound with heptacity (Scheme 3.14).

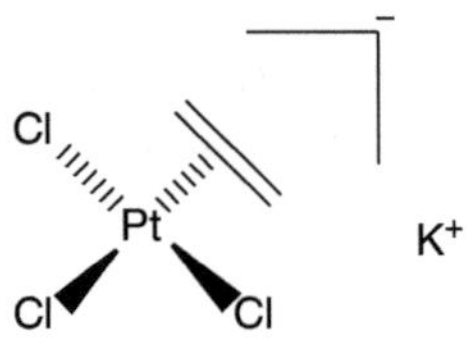

Scheme 3.13

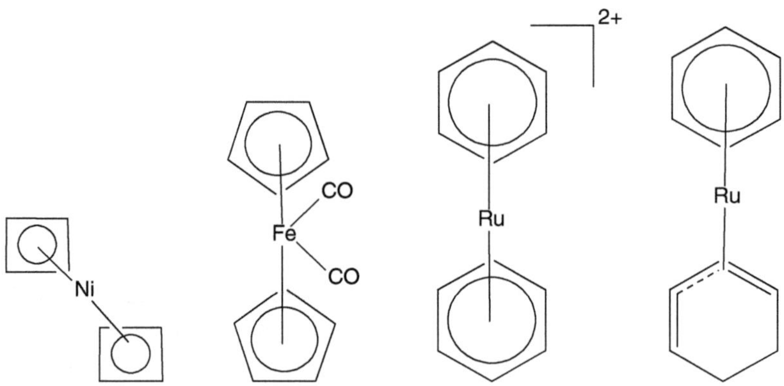

Scheme 3.14

Additionally, Table 3.1 demonstrated the electron donation by unsaturated compound various heptacity.

3.3 Grignard Reagent (G.R.)

3.3.1 Physical Properties

- Grignard reagents are colorless and nonvolatile crystalline solids.
- Grignard reagents are strong nucleophile.
- Grignard reagent should be kept in a dry environment (in dry ether). If moisture or water comes into contact with Grignard, Grignard is destroyed [12].

3.3.2 Chemical Properties

3.3.2.1 Alkanes

Alkanes can be obtained when Grignard reagent treated with water, alcohols and amines (Scheme 3.15) [13].

Table 3.1 Electron donation by unsaturated compound various heptacity.

S. No.	Name and structure of ligands	Structure of compound	Number of electron which contributed
1	Allyl ligands: η1, η3	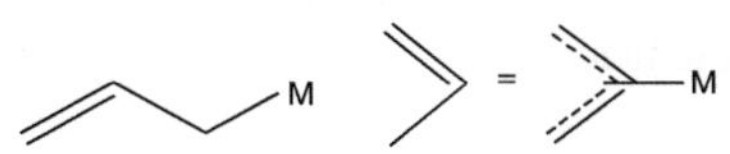	1 and 3
2	Alkene/Ethane : η2, η2, η4	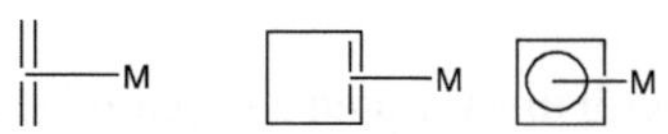	2, 2, and 4
3	Cyclopentadienyl: η1, η3, η5	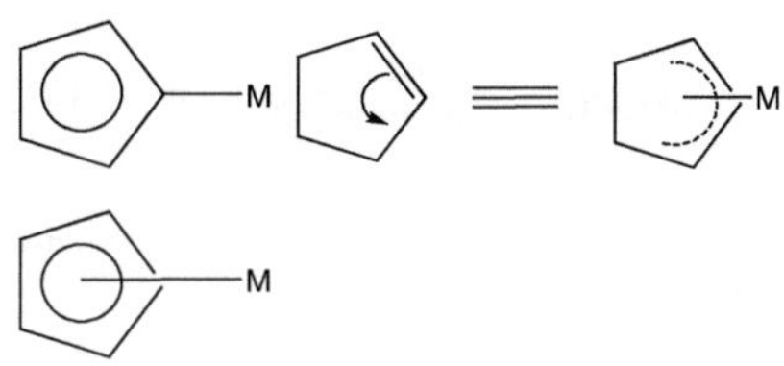	1, 3, and 5
4	Cyclohexane: η1, η2, η3, η4, η6	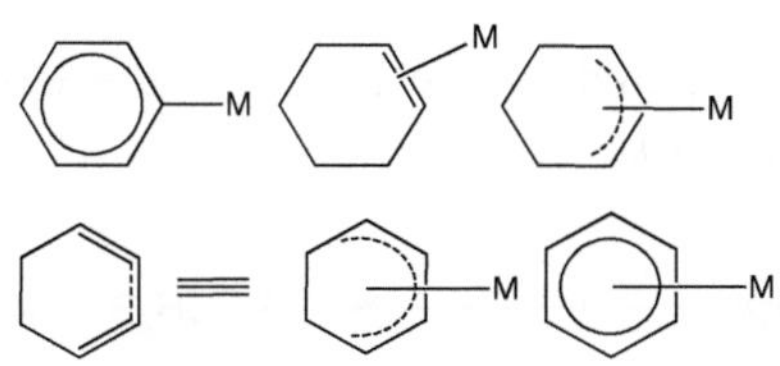	1, 2, 3, 4, and 5
5	Polycyclic compound : η7, η8	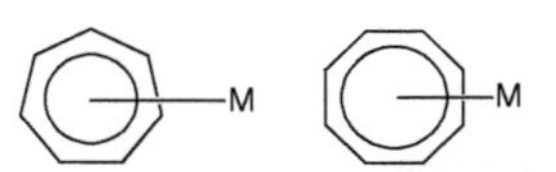	7 and 8

Source: Adapted from Soto and Salcedo [11].

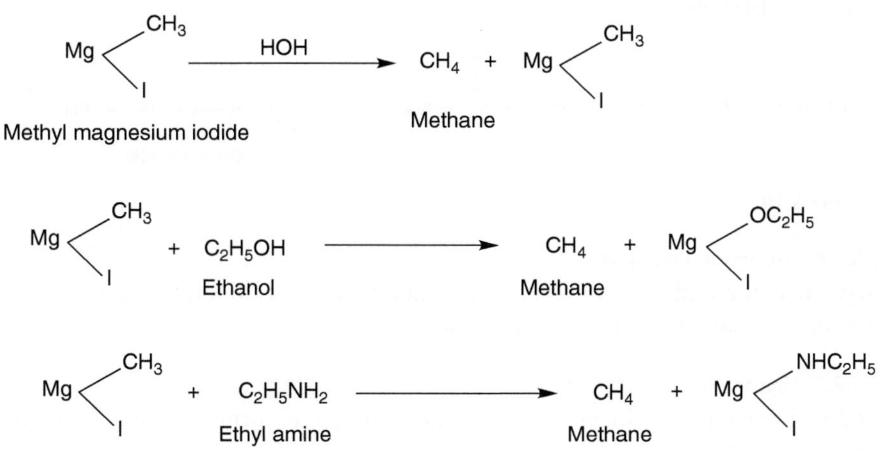

Scheme 3.15

3.3.2.2 Alkenes

These are obtained by the reaction of unsaturated halides on the Grignard reagent (Scheme 3.16).

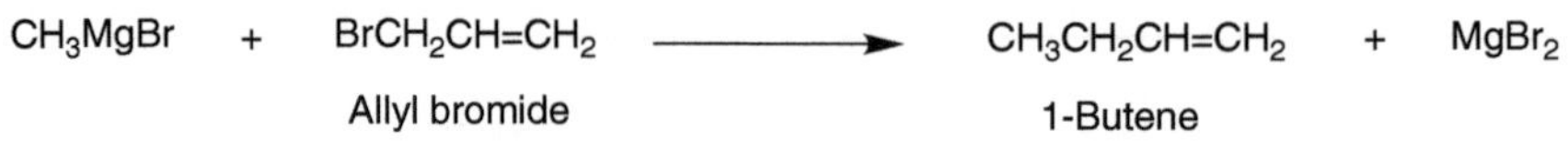

Scheme 3.16

3.3.2.3 Alkynes

Grignard reagents react with unsaturated compound which possess active hydrogen (Scheme 3.17).

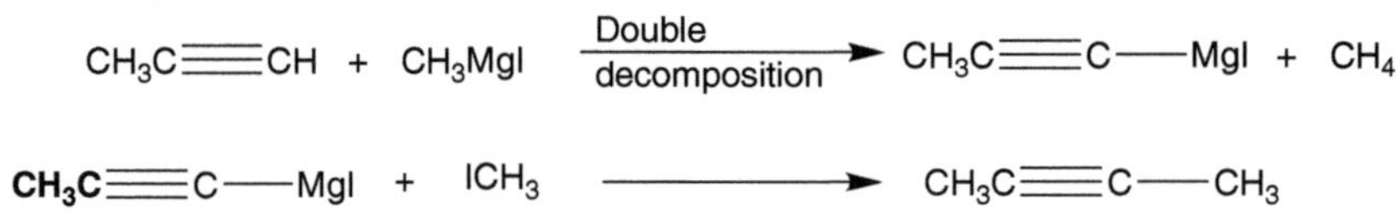

Scheme 3.17

3.3.2.4 Ethers

Higher ethers are prepared by the action of the Grignard reagent on a monocholoether (Scheme 3.18).

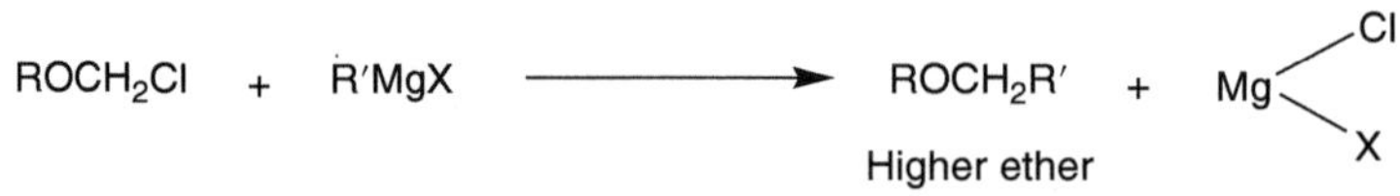

Scheme 3.18

3.3.2.5 Reaction with carbon dioxide

Grignard reagents when treated with solid carbon dioxide form nucleophilic addition products, which on treatment with dilute acid yield a monocarboxylic acid (Scheme 3.19) [14].

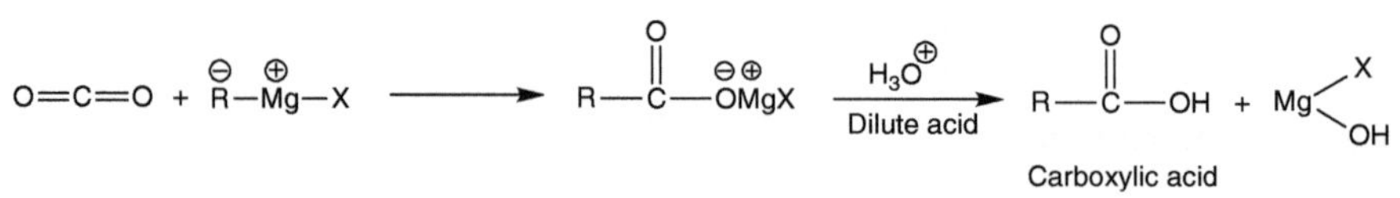

Scheme 3.19

3.3.2.6 Insertion Reaction

Oxygen, sulfur, and halogen react with Grignard reagents to form alcohols, thiols, and organic halides, respectively (Scheme 3.20) [14].

3.3.2.7 Synthesis of Silicones

Grignard reagent and silicone tetrachloride combine to form silicones, which are long-chain anhydrides. They are employed in the production of electrical insulators, silicone rubber, and premium lubricants (Scheme 3.21).

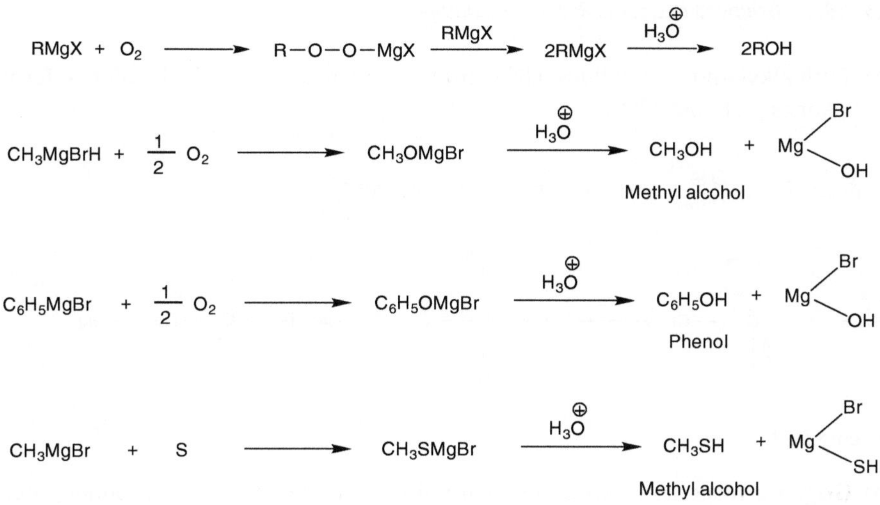

Scheme 3.20

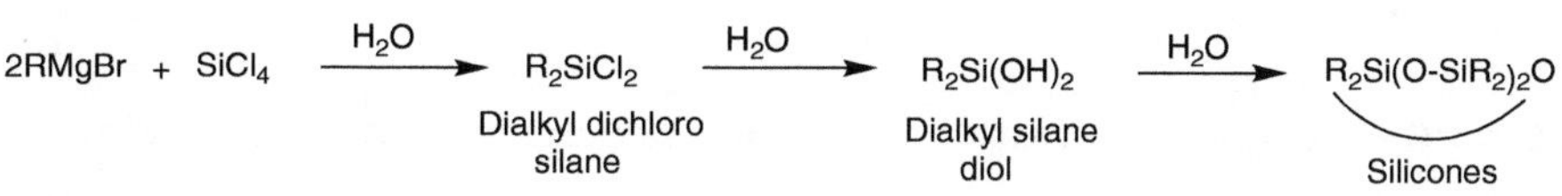

Scheme 3.21

3.3.2.8 Nucleophilic Substitution

3.3.2.8.1 Grignard Reagent React with Active Hydrogen (Scheme 3.22) [15].

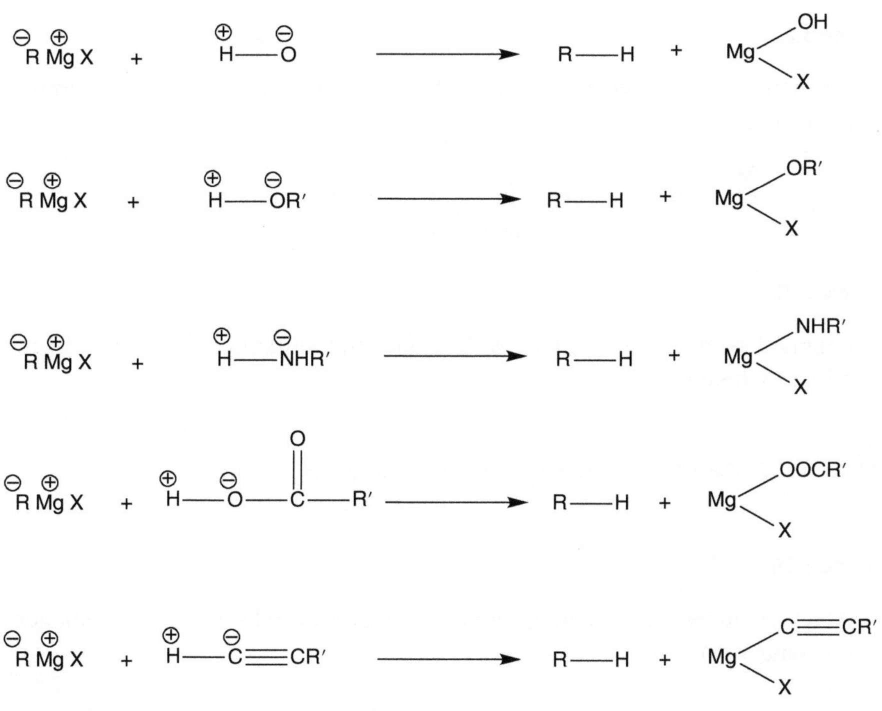

Scheme 3.22

3.3.2.8.2 Grignard Reagents React with Halogen

(a) With alkyl and acyl halides- Grignard reagents reacts with alkyl halide to form alkanes (Scheme 3.23).

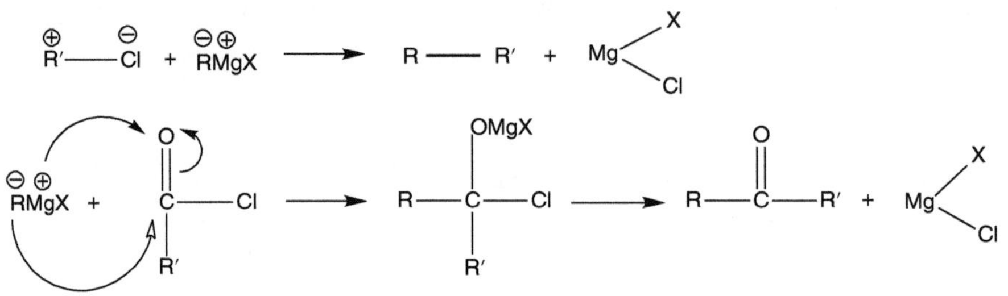

Scheme 3.23

(b) Grignard reagents when treated with allyl bromide and benzyl bromide, they produce 1-alkene and higher alkyl benzene (Scheme 3.24).

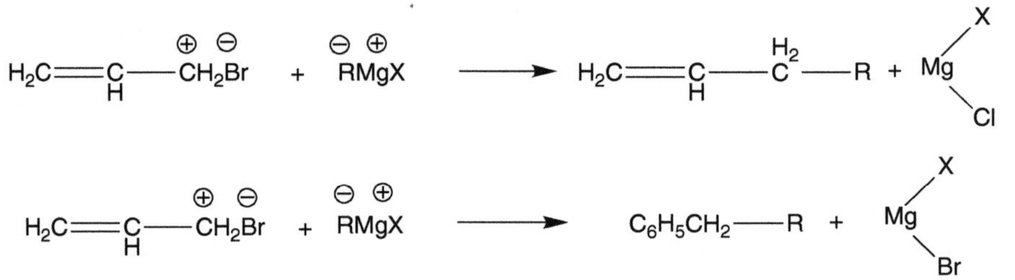

Scheme 3.24

(c) Grignard reagents when treated with chloramine it gives primary amines (Scheme 3.25):

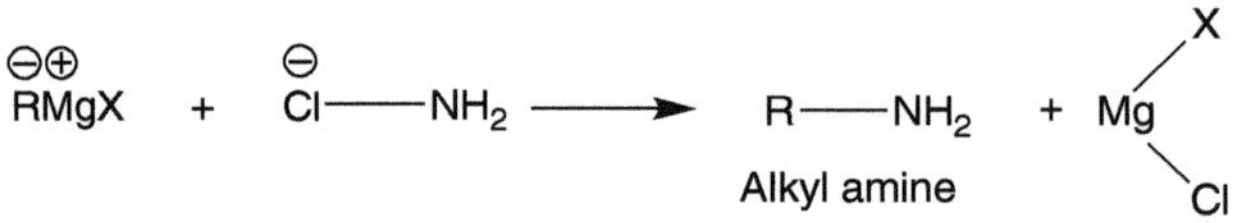

Scheme 3.25

(d) Grignard reagents on reaction with monochlorodimethyl ether gives higher ethers (Scheme 3.26).

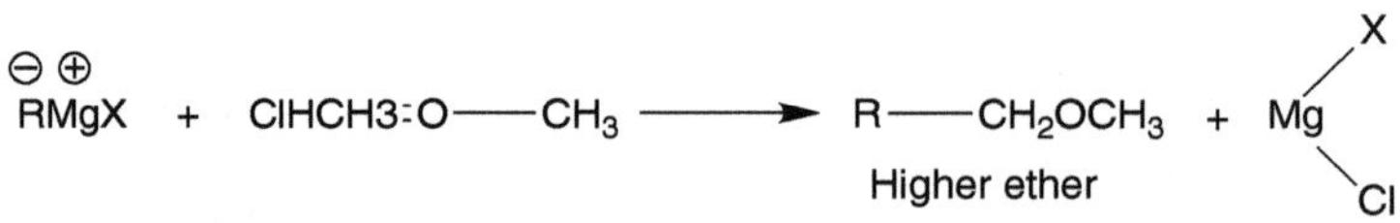

Scheme 3.26

(e) Alkyl cyanides are obtained with cyanogenchloride or even cyanogens (Scheme 3.27)

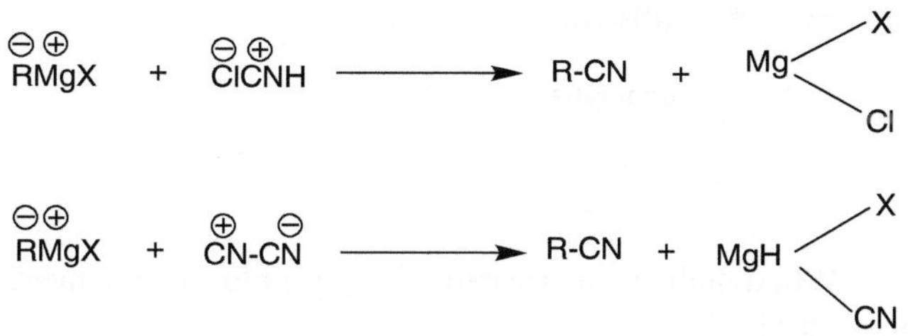

Scheme 3.27

(f) *With halide of certain metals and nonmetal*: Organometallic and organo nonmetallic compound are formed (Scheme 3.28).

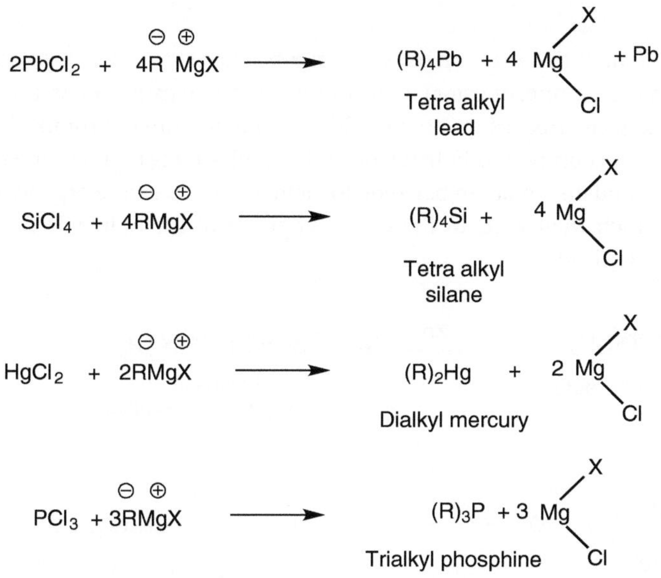

Scheme 3.28

3.4 Organozinc Compounds

3.4.1 Physical Properties

Dialkyl zincs are colorless volatile liquids with unpleasant smells and low boiling points [$(CH_3)_2Zn$, 319 K; $(C_2H_5)_2Zn$, 391 K]. They are spontaneously inflammable in the air [16].

3.4.2 Chemical Properties

Dialkyl zinc has been used in various synthetic reactions.

(i) *Preparation of hydrocarbon containing quaternary carbon atom*: Dialkyl zinc reacts with tertiary butyl chloride to give neopatane (Scheme 3.29) [16].

$$(CH_3)_3CCl + (CH_3)_2Zn \longrightarrow (CH_3)_4C + CH_3ZnCl$$

Tert. Butyl chloride → Neopentane

Scheme 3.29

(ii) *Preparation of ketones*: When dialkyl zinc treated with acyl chloride it converts into ketones (Scheme 3.30) [17].

$$R'COCl + R_2Zn \longrightarrow R'COR + RZnCl$$

Acyl chloride → Ketone

Scheme 3.30

(iii) *Reformatsky Reaction*: In this reaction between an α-bromo ester and a carbonyl compound (aldehyde, ketone, or ester) in the presence of zinc using benzene as the solvent to yield β-hydroxy ester. Zinc is added to the mixture of bromoacid ester and the carbonyl compound in benzene and the mixture is then warmed. The α-bromo ester and zinc react in benzene to yield an intermediate organozinc compound, which then adds to the carbonyl group of the aldehyde and ketone (Scheme 3.31) [18].

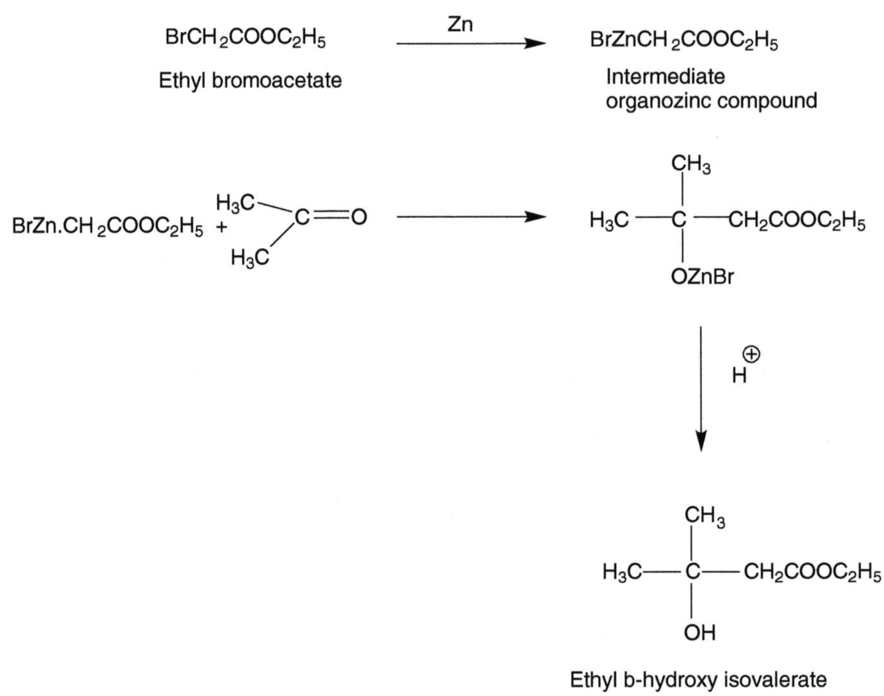

Scheme 3.31

3.5 Organolithium Compounds

Alkyl lithium is either liquid or low-melting solids. They are insoluble in water but soluble in nonpolar hydrocarbons as solvents. Organolithium compounds resemble Grignard reagents, but they are more reactive. The greater reactivity of organolithium compounds over Grignard reagents is due to greater ionic character of the C—Li bond in comparison to the C—Mg bond [19].

3.5.1 Reaction Resembling Grignard Reagents

(i) *Preparation of hydrocarbons*: Organolithium reagents react with active hydrogen-containing compounds (Scheme 3.32) [19].

$$\overset{-\delta}{R}—\overset{+\delta}{Li} + \overset{\oplus\ \ominus}{HOH} \longrightarrow R—H + LiOH$$

$$R—Li + R'OH \longrightarrow R—H + LiOR'$$

$$R—Li + BrCH_3 \longrightarrow R—CH_3 + LiBr$$

Scheme 3.32

(ii) *Preparation of alcohols*: Formaldehyde gives primary alcohol; other aldehydes give secondary alcohol while ketones give tertiary alcohol (Scheme 3.33).

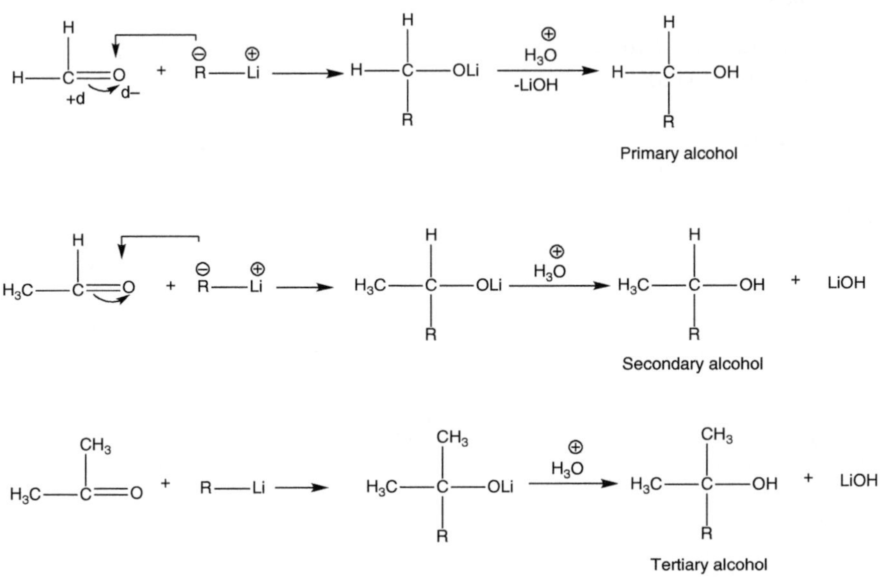

Scheme 3.33

(iii) *Preparation of higher ethers*: When treated with lower ether chloride it gives higher ether (Scheme 3.34).

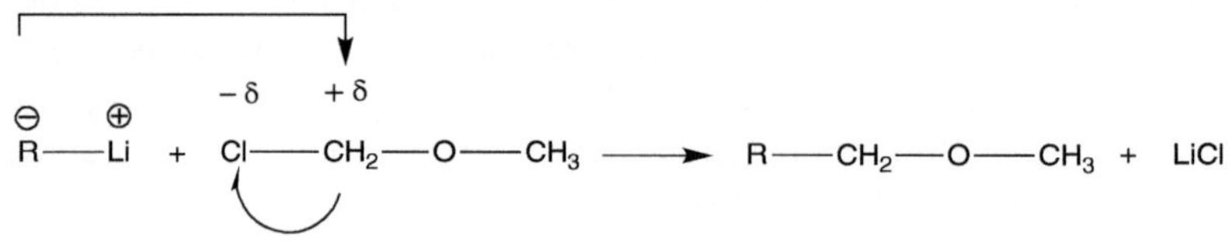

Scheme 3.34

(iv) *Preparation of aldehydes*: Organolithium compounds on reaction with ethyl orthoformate or hydrogen cyanide give aldehyde (Scheme 3.35).

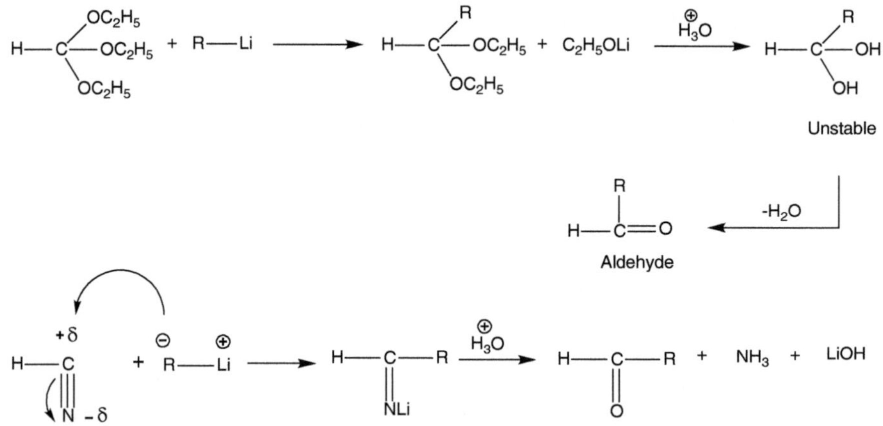

Scheme 3.35

(v) *Preparation of ketones*: Ethyl ortho ester and alkyl cyanide give ketone (Scheme 3.36) [20].

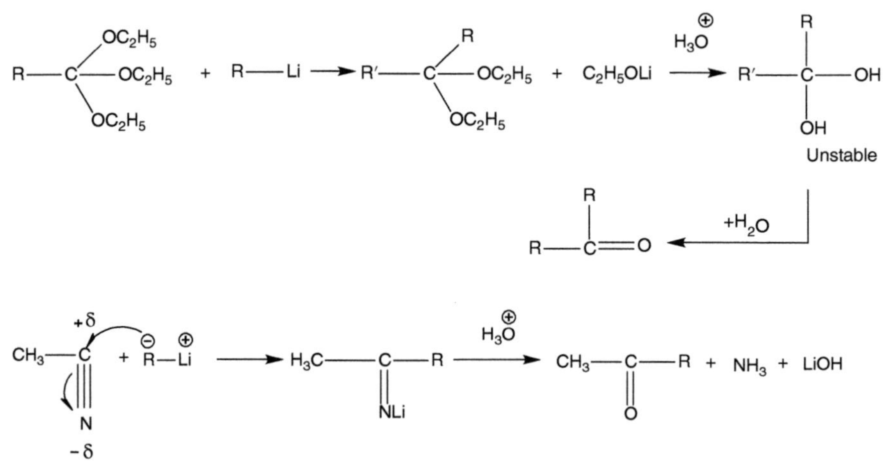

Scheme 3.36

(vi) *Preparation of other metallic compounds*: When treated with other metal halides it gives organometallic compounds (Scheme 3.37).

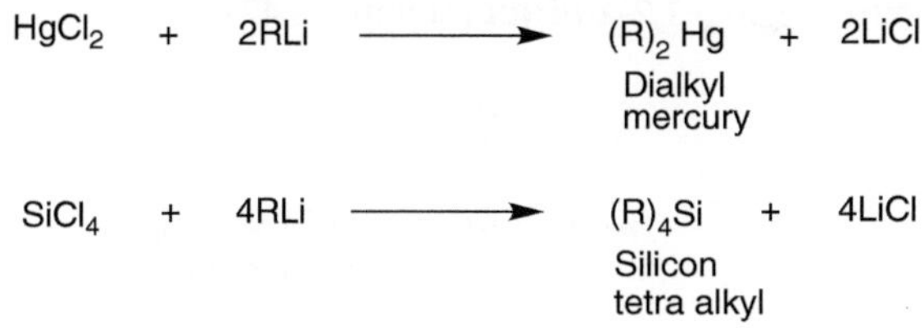

Scheme 3.37

(vii) *Preparations of amines*: It gives alkyl amines when react with amine chloride (Scheme 3.38).

Scheme 3.38

(viii) Preparations of higher alkenes from allyl halide: When it reacts with allyl bromide, it gives higher alkenes (Scheme 3.39).

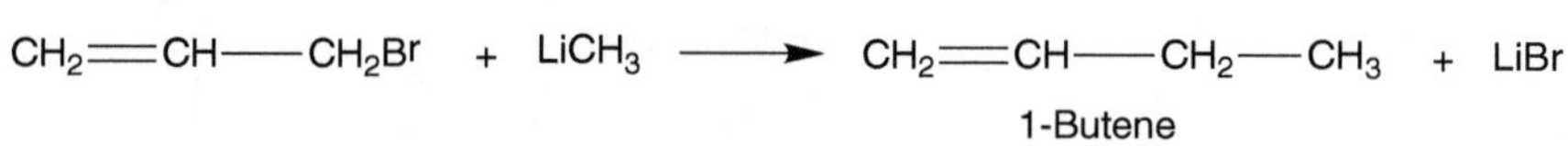

Scheme 3.39

(ix) *Preparations of alkyl cyanides*: When they react with halide cyanides (Scheme 3.40).

Scheme 3.40

(x) *Preparations of thio-alcohols*: First alkyl lithium reacts with sulfur then the formed product reacts with water to give thio-alcohols (Scheme 3.41).

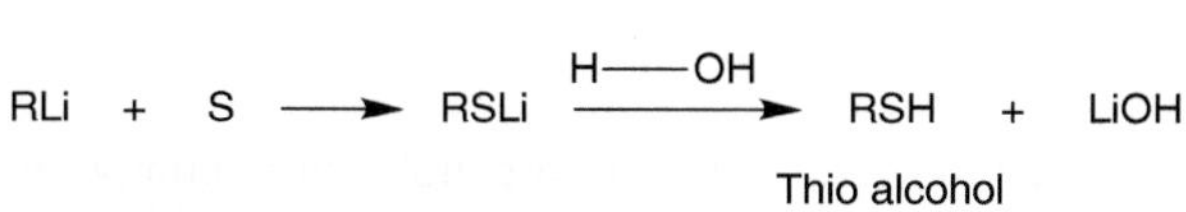

Scheme 3.41

3.5.2 Reactions Different from Grignard Reagents

(i) Grignard reagent with α, β-unsaturated ketones gives a **1, 4-addition** product while organolithium compounds give a **1,2-adduct** (Scheme 3.42).

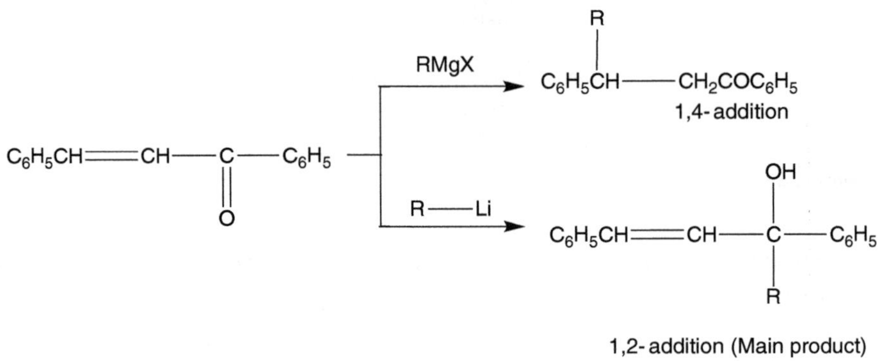

Scheme 3.42

(ii) *Reaction with carbon dioxide*: Organolithium compounds gives ketone (Scheme 3.43).

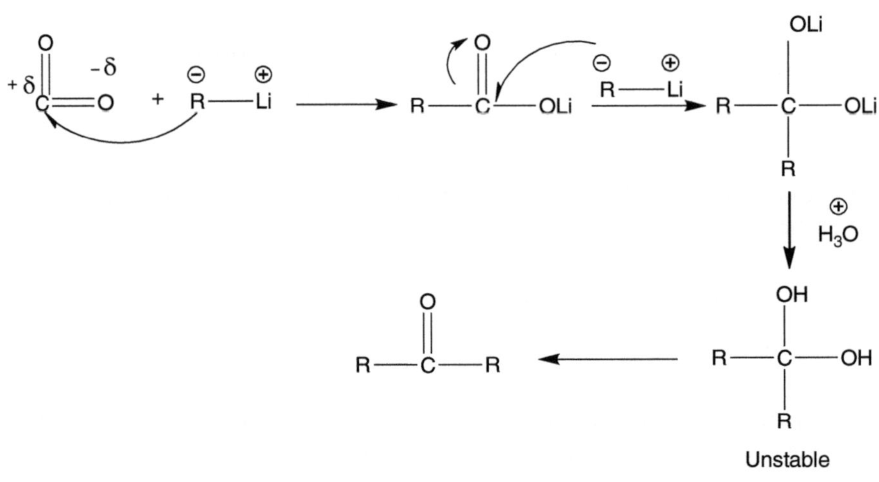

Scheme 3.43

Acid also gives a ketone when reacted with an organolithium compound (Scheme 3.44).

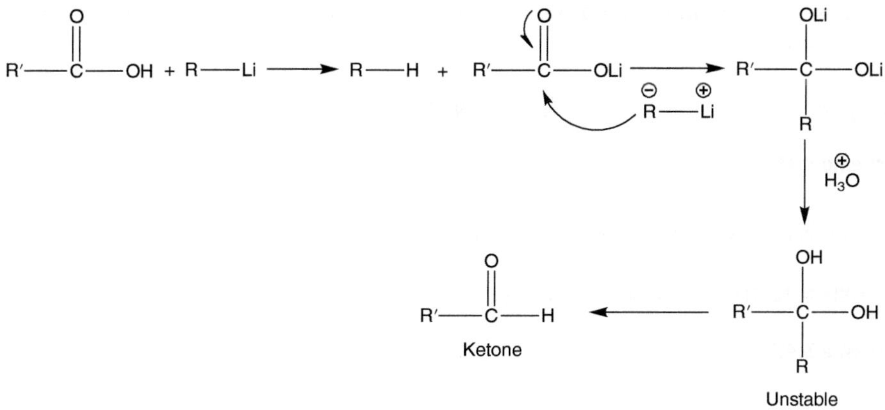

Scheme 3.44

3.6 Organosulfur Compounds

3.6.1 Physical Properties

- Methanethiol is a gas while higher homologues are volatile liquids having an extremely unpleasant odor. The boiling points of methanethiol is 0^0 and that of ethanethiol is 37^0. Their disagreeable odor decreases with an increase in the molecular weight.
- The boiling points of thiols are much lower than corresponding alcohols, which is due to weaker intermolecular hydrogen bonding in thiols than alcohols. Due to lesser electronegativity of sulfur the character of sulfur hydrogen bond is much less ionic.
- Weak hydrogen bonding makes them less soluble in water as compress to alcohols. However they readily soluble in organic solvents such as ether and alcohol [21].

3.6.2 Chemical Properties

a. *Reaction with alkali metals*: Thiols react with alkali metals like sodium, potassium, etc., to from mercaptides with the evolution of hydrogen (Scheme 3.45) [21].

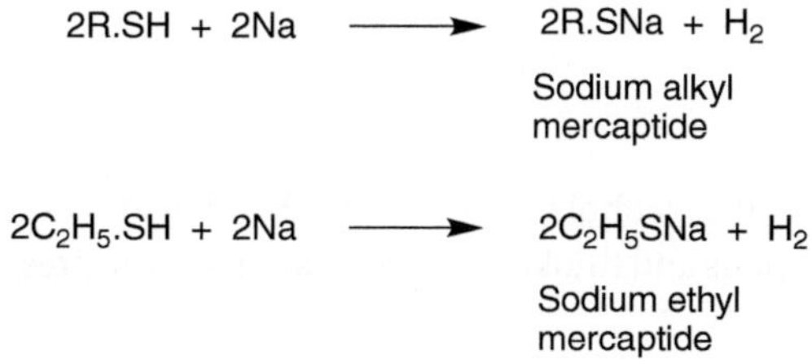

Scheme 3.45

These mercaptides could be decomposed with water to give back thiols (Scheme 3.46).

$$C_2H_5SNa + H_2O \longrightarrow C_2H_5SH + NaOH$$

Scheme 3.46

Thiols also react with alkalis to form mercaptides (Scheme 3.47).

$$C_2H_5SH + NaOH \longrightarrow C_2H_5SNa + H_2O$$

Scheme 3.47

b. *Reaction with acid and acid derivatives*: Like alcohol, thiols react with acid, acid chlorides, and acid anhydrides (Scheme 3.48).

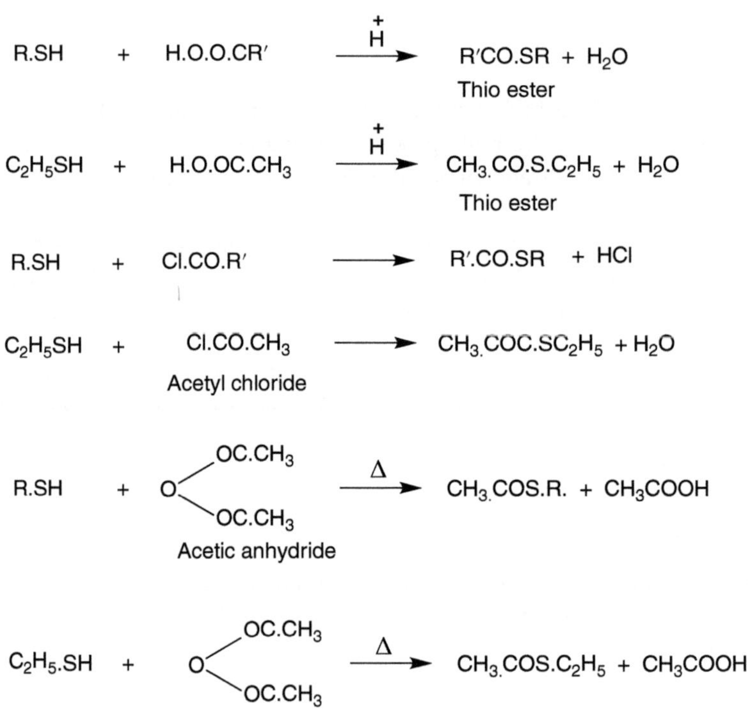

Scheme 3.48

c. *Reaction with aldehyde and ketones*: Like alcohols, they react with aldehydes and ketones to form thioacetals or mercaptals and thioketals or mercaptals in the presence of hydrochloric acid (Scheme 3.49).

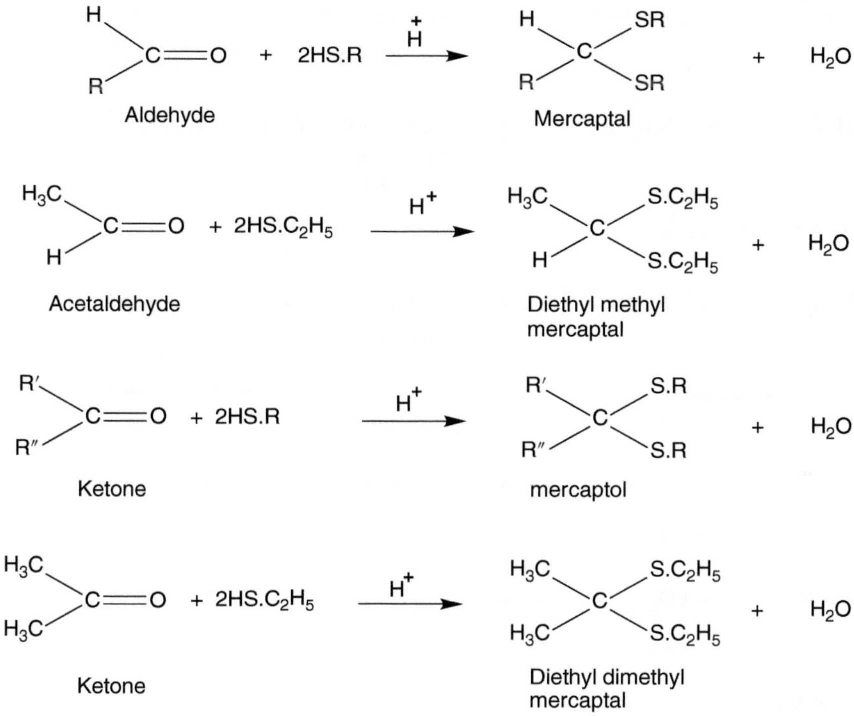

Scheme 3.49

3.6.3 Properties Different from Alcohols

a. *Reaction with alkyls*: It gives sodium alkyl mercaptide (Scheme 3.50).

$$R.SH + NaOH \longrightarrow R.S.Na + H_2O$$

$$C_2H_5.SH + NaOH \longrightarrow C_2H_5.S.Na + H_2O$$

Sodium ethyl mercaptide

Scheme 3.50

b. *Reaction with metallic salts and metallic oxides*: It gives different mercaptide products (Scheme 3.51 and 3.52).

$$2R.SH + HgO \longrightarrow (R.S)_2Hg + H_2O$$

Mercury mercaptide

$$2C_2H_5.SH + HgO \longrightarrow (C_2H_5.S)_2Hg + H_2O$$

Mercury ehtyl mercaptide

Scheme 3.51

$$2R.SH + (CH_3COO)_2Pb \longrightarrow \underset{\text{Lead mercaptide}}{(RS)_2Pb} + 2CH_3COOH$$

$$2C_2H_5.SH + (CH_3COO)_2Pb \longrightarrow \underset{\text{Lead ehtyl mercaptide}}{(C_2H_5S)_2Pb} + 2CH_3COOH$$

Figure 3.51 (*Continued*)

$$2R.SH + 2CuCl_2 \longrightarrow \underset{\text{Disulphide}}{R.S.S.R} + Cu_2Cl_2 + 2HCl$$

$$2C_2H_5.SH + 2CuCl_2 \longrightarrow \underset{\text{Diethyl sulphide}}{C_2H_5.S.S.C_2H_5} + Cu_2Cl_2 + 2HCl$$

$$2R.SH + \underset{\text{Sodium plumbite}}{Na_2PbO_2} + S \longrightarrow R.S.S.R + 2NaOH + PbS$$

$$2C_2H_5.SH + \underset{\text{Sodium plumbite}}{Na_2PbO_2} + S \longrightarrow C_2H_5.S.S.C_2H_5 + 2NaOH + PbS$$

Scheme 3.52

c. *Oxidation*: Thiols differ significantly from alcohols in their behavior toward oxidizing agents
   (i) With mild oxidizing agents like air, hydrogen peroxide, cupric chloride, iodine, sodium hypochloride, etc., thiols are oxidized to disulfides (Scheme 3.53).

$$RSH + H_2O_2 \longrightarrow R—S—S—R + 2H_2O$$

$$C_2H_5SH + H_2O_2 \longrightarrow C_2H_5—S—S—C_2H_5 + 2H_2O$$

Scheme 3.53

   (ii) With strong oxidizing agents like nitric acid and $KMnO_4$ they are oxidized to sulphonic acids (Scheme 3.54).

$$RSH + 3[O] \xrightarrow{HNO_3} \underset{\text{Alkyl sulphonic acid}}{RSO_3H}$$

Scheme 3.54

3.7 Conclusion

Organometallic compounds and their classification are mainly focused on the chemistry of s-block elements, p-block elements, transition elements, and lanthanide and actinide series with carbon-containing moieties. Here M—C bond is mainly formed, which is covalent in nature, through sometimes it behaves as an ionic character. Due to this polarity, these organometallic compounds play a vital role in various reactions.

References

1 Thayer, J.S. (1969). Historical origins of organometallic chemistry. Part I, Zeise's salt. *J. Chem. Educ.* 46 (7): 442.

2 Gilman, H. and Schulze, F. (1925). A qualitative color test for the Grignard reagent. *J. Am. Chem. Soc.* 47 (7): 2002–2005.

3 Craig, P.J. and Jenkins, R. (2004). Organometallic compounds in the environment: an overview. In: *Organic Metal and Metalloid Species in the Environment; Analysis, Distribution, Processes and Toxicological Evaluation* (ed. A.V. Hirner and H. Emons), 1–15. Springer.

4 Mudi, S., Usman, M., and Ibrahim, S. (2015). Clinical and industrial application of organometallic compounds and complexes: a review. *Am. J. Chem. Appl.* 2 (6): 151–158.

5 Swope, W.C. and Schaefer, H.F. III, (1977). Model studies of π-bonded organometallic systems Mn-C_2H_2 and Mn-C_2H_4. *Mol. Phys.* 34 (4): 1037–1048.

6 Rayón, V.M. and Frenking, G. (2003). Bis (benzene) chromium is a δ-bonded molecule and ferrocene is a π-bonded molecule. *Organometallics* 22 (16): 3304–3308.

7 Gutmann, V. (1976). Solvent effects on the reactivities of organometallic compounds. *Coord. Chem. Rev.* 18 (2): 225–255.

8 Crabtree, R.H. (2009). *The Organometallic Chemistry of the Transition Metals.* Wiley.

9 Tang, J. and Zhao, L. (2020). Polynuclear organometallic clusters: synthesis, structure, and reactivity studies. *Chem. Commun.* 56 (13): 1915–1925.

10 Oliver, J.P. (1977). Rearrangements of organoaluminum compounds and their group III analogs. In: *Advances in Organometallic Chemistry*, vol. 16 (ed. F.G.A. Stone and R. West), 111–130. Elsevier.

11 Soto, D. and Salcedo, R. (2012). Coordination modes and different hapticities for fullerene organometallic complexes. *Molecules* 17 (6): 7151–7168.

12 Seyferth, D. (2009). *The Grignard Reagents*, 1598–1605. ACS Publications.

13 Tamura, M. and Kochi, J. (1971). Iron catalysis in the reaction of grignard reagents with alkyl halides. *J. Organomet. Chem.* 31 (3): 289–309.

14 Ashby, E. (1967). Grignard reagents. Compositions and mechanisms of reaction. *Q. Rev. Chem. Soc.* 21 (2): 259–285.

15 Ashby, E. (1980). A detailed description of the mechanism of reaction of Grignard reagents with ketones. *Pure Appl. Chem.* 52 (3): 545–569.

16 Knochel, P., Millot, N., Rodriguez, A.L., and Tucker, C.E. (2004). Preparation and applications of functionalized organozinc compounds. *Org. React.* 58: 417–759.

17 Furukawa, J. and Kawabata, N. (1974). Organozinc compounds in synthesis. In: *Advances in Organometallic Chemistry*, vol. 12 (ed. F.G.A. Stone and R. West), 83–134. Elsevier.

18 Fürstner, A. (1989). Recent advancements in the Reformatsky reaction. *Synthesis* 1989 (08): 571–590.

19 Wakefield, B.J. (2013). *The Chemistry of Organolithium Compounds*. Elsevier.

20 Gilman, H. and Cartledge, F.K. (1964). The analysis of organolithium compounds. *J. Organomet. Chem.* 2 (6): 447–454.

21 Block, E. (2013). *Reactions of Organosulfur Compounds: Organic Chemistry: A Series of Monographs*, vol. 37. Academic Press.

4

Synthesis Methods of Organometallic Compounds

Omar Dagdag[1], Rajesh Haldhar[2], Seong-Cheol Kim[2], Elyor Berdimurodov[3], Ekemini D. Akpan[1], and Eno E. Ebenso[1]

[1] *University of South Africa, Centre for Materials Science, College of Science, Engineering and Technology, Johannesburg 1710, South Africa*
[2] *Yeungnam University, School of Chemical Engineering, Gyeongsan, 38541, South Korea*
[3] *National University of Uzbekistan, Faculty of Chemistry, Tashkent, 100034, Uzbekistan*

4.1 Introduction

Organic synthesis of organometallic chemistry is widely used in synthetic organic chemistry [1, 2]. Polyorganic synthesis defines the process of formation of organometallic compounds [3]. Organic cursive chemistry is one of the major areas of research in organic, inorganic, biochemistry, and catalytic chemistry [4, 5]. Indeed, organometallic reagents are used to synthesize many of the commercial compounds used in the pharmaceutical, petrochemical, and polymer production sectors [6, 7].

A metal atom is bound to carbon in an organometallic molecule [8]. For example, compounds with metal–nitrogen, metal–oxygen, or metal–phosphorus bonds are called coordination complexes, but they are often referred to as organometallic [9]. Metalloids like boron or silicon may be included, as well as transition metals group 3 to 12, alkali metals group 2, alkali metals group 1, and main group 13 to 15 elements [10]. Organometallic compounds can be formed into a large number of elements, and the research field in organometallic synthesis procedures is immense. Researchers are also experimenting with the use of organometallic reagents to synthesize more complex and tailor-made compounds by triggering specific bonds and/or catalytic reactions [11].

There are many other methods to synthesize organometallic compounds. One of the main methods of synthesizing organometallic compounds is to cause a reaction between a pure metal and specific organic molecules. This method is used in particular to synthesize two types of organometallic reagents among the most frequently used: organolithiums and organomagnesium, which are formed by reacting the metal with an alkyl or aryl halide.

Organometallic Compounds: Synthesis, Reactions, and Applications, First Edition.
Edited by Dakeshwar Kumar Verma and Jeenat Aslam.

In recent years, efforts have been made to develop synthesis trends in organometallic compounds. This book chapter reviews recent advances in electrochemical methods and the nucleoside method in the field of synthetic organic metal chemistry.

4.2 Synthesis Methods of Organometallic Compounds

4.2.1 Electrochemical Methods for the Synthesis of Organometallic Compounds

4.2.1.1 Synthesis of Cyano Cu(I) Complexes in the Electrochemical Cell

In this research work [12], the Cu(I) complex with cyano was synthesized by using an electrochemical cell (Figure 4.1), in which Cu is the anode and Pt is the cathode electrode. The Cu(I) complex with cyano was synthesized in the Bu_4NBF_4 and malononitrile solutions. At the cathode, the malononitrile $CNCH_2$ radicals were formed from the malononitrile molecules on the surface of the Pt electrode. At the anode, the copper (I) ions were extracted from the Cu electrodes. Finally, the Cu(I) reacted with the malononitrile $CNCH_2$ radicals and Bu_4NBF_4 molecules to form the cyano Cu(I) complex.

4.2.1.2 Synthesis of an Organorhenium Cyclopentadienyl Complex in the Electrochemical Cell

Utilizing an electrochemical cell (Figure 4.2) [14] with electrodes made of Pt as the anode and carbon mesh as the cathode, the [ReClCp(H)(dppe)] complex was created. The Re(III) ions are formed on the surface anode electrode while free radicals are formed on the cathode. In this complex [ReClCp(H)(dppe)], the Cp is

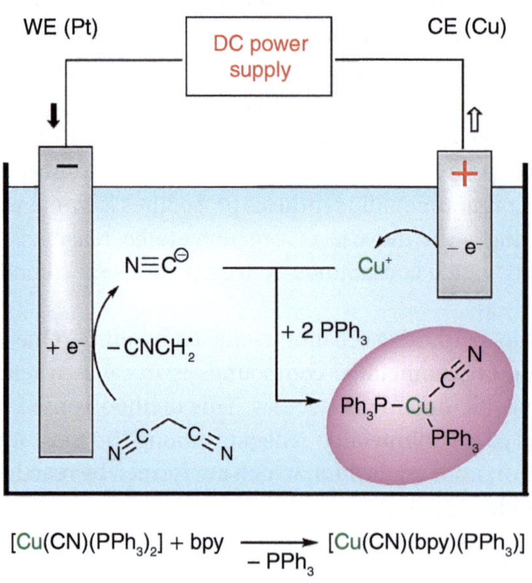

Figure 4.1 Synthesis of Cu(I) complex with cyano by using an electrochemical cell. Source: Reproduced with permission from [13]/Elsevier.

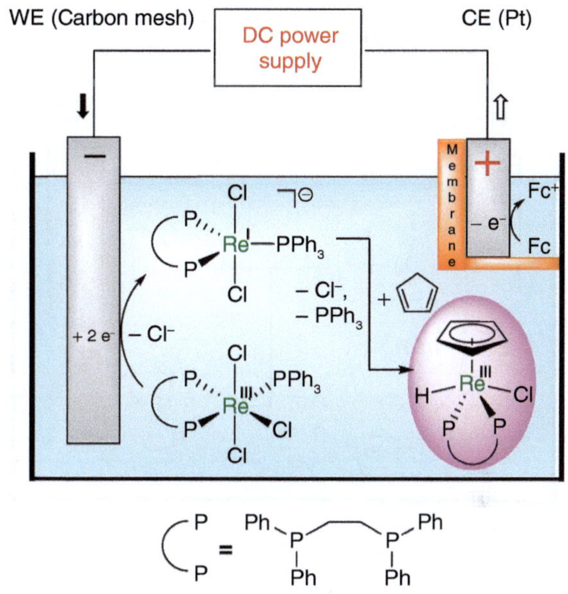

Figure 4.2 Syntheses of [ReClC$_p$(H)(dppe)] complex by using electrochemical cell. Source: Reprinted with permission from [13], © 2021 Elsevier Publications.

cyclopentadienyl and dppe is 1,2-bis(diphenylphosphine)ethane. It was found that the reaction yield in this electrochemical cell is 68.7%.

4.2.1.3 Synthesis of N-heterocyclic Carbene Complexes in the Electrochemical Cell

The electrochemical approach can be used to create N-heterocyclic carbene complexes of Fe, Ni, Cu, and Au [15]. Liu et al. explained the first model of this technique [16], using an unshared electrochemical cell. Imidazole salts are used as a carbon source and thin films as a source of metal ions. In the first phase of the electrolyte pair, the imidazolium cations pass through the grooves of a single electrode to the PC cathode, whereas the N-heterocyclic carbene compound passes through the hydrogen radical (Figure 4.3).

4.2.1.4 Synthesis of Organocopper (I) π-Complexes in the Electrochemical Cell

Hordiychuk et al. describe the synthesis of the π-organic-copper complex with various heterocyclic allyl derivatives in the electrochemical cell (Figure 4.4) [17, 18]. In this electrochemical reaction, the Ag, Zn, and Cu electrodes (anode and cathodes) were performed as ionization booms.

4.2.1.5 Synthesis of Organonickel σ-Complexes in the Electrochemical Cell

The following study used an electrochemical cell (Figure 4.5) in which Ni, Al, Zn, or Mg metals serve as the counter electrode (CE, cathode) and Ni or Pt serves as the working electrode to create the organonickel -complexes [NiBr(Aryl)(bpy)] (WE, anode). The Ni(II) ions are extracted from the counter electrode. This is due to the oxidation potential of this electrode being lower than the WE. In comparison, the Br anions are formed on the WE. Finally, the Ni(II) ions easily reacted with the Br anions and aromatic compounds at high reaction yields [19, 20].

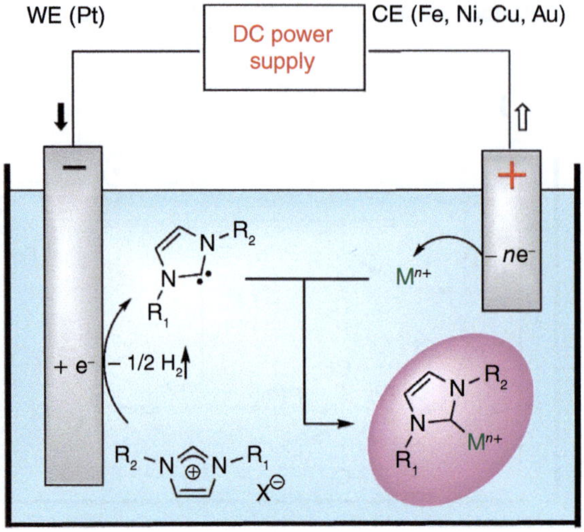

Figure 4.3 Syntheses of N-heterocyclic carbine complexes by using electrochemical cell. Source: Reprinted with permission from [13], © 2021 Elsevier Publications.

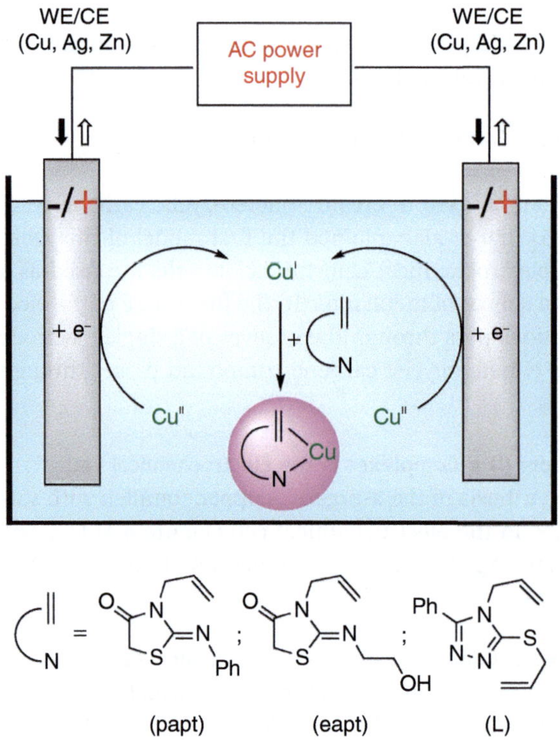

Figure 4.4 Synthetics of organocopper π-complexes complexes by using electrochemical cell. Source: Reprinted with permission from [13], © 2021 Elsevier Publications.

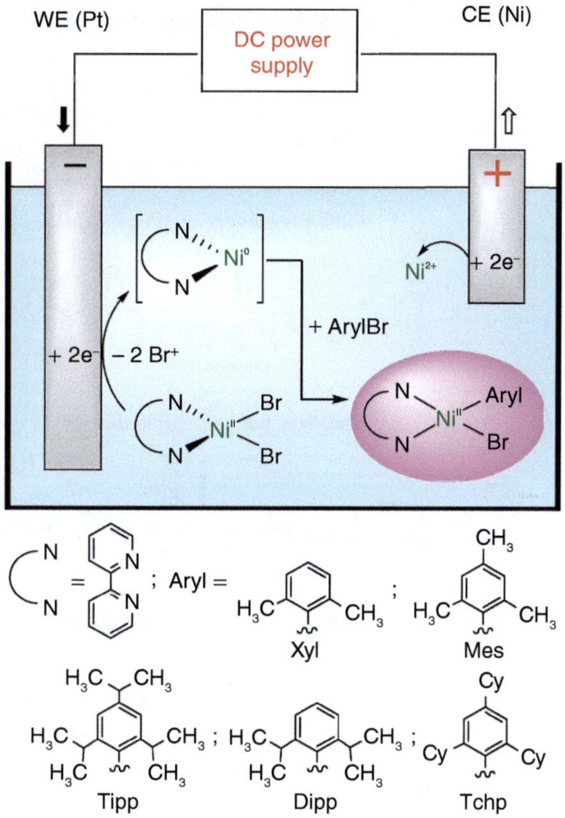

Figure 4.5 Syntheses of organonickel σ-complexes [NiBr(Aryl)(bpy)] complex by using electrochemical cell. Source: Reprinted with permission from [13], © 2021 Elsevier Publications.

4.2.2 Synthesis of Organic Compounds in the Electrochemical Cell by Metal organic Catalysts

4.2.2.1 The Synthesis of Organic Compounds in the Electrochemical Cell by the Ni-Organic Catalyze

In recent research, catalysts based on Ni-based organic compounds are used to synthesize some organic compounds in an electrochemical cell. Figure 4.6 indicates the synthesis reactions of organic compounds in the electrochemical cell by the Ni-organic catalyses. These reactions are also called electrochemical coupling reactions. In this synthesis, the Ni electrode was used as an anode or cathode. The Ni-based organic compounds are used as catalytic agents in organic syntheses.

4.2.2.2 The Synthesis of Organic Compounds in the Electrochemical Cell by the Pd-Organic Catalyses

Huang and Lai developed Zn-catalyzed electrochemical alkalinity by reacting alkyl halides with alkyl halides in an aqueous solution (Figure 4.7) [21]. Various Pd-organic catalysts and electrodes tested in this study were obtained from $Pd(OAc)_2$/Cu(OAc). The H_2O system with a Pt cathode and Zn anode is the most efficient of these processes. Based on mechanistic studies, a Barbie–Negishi fusion

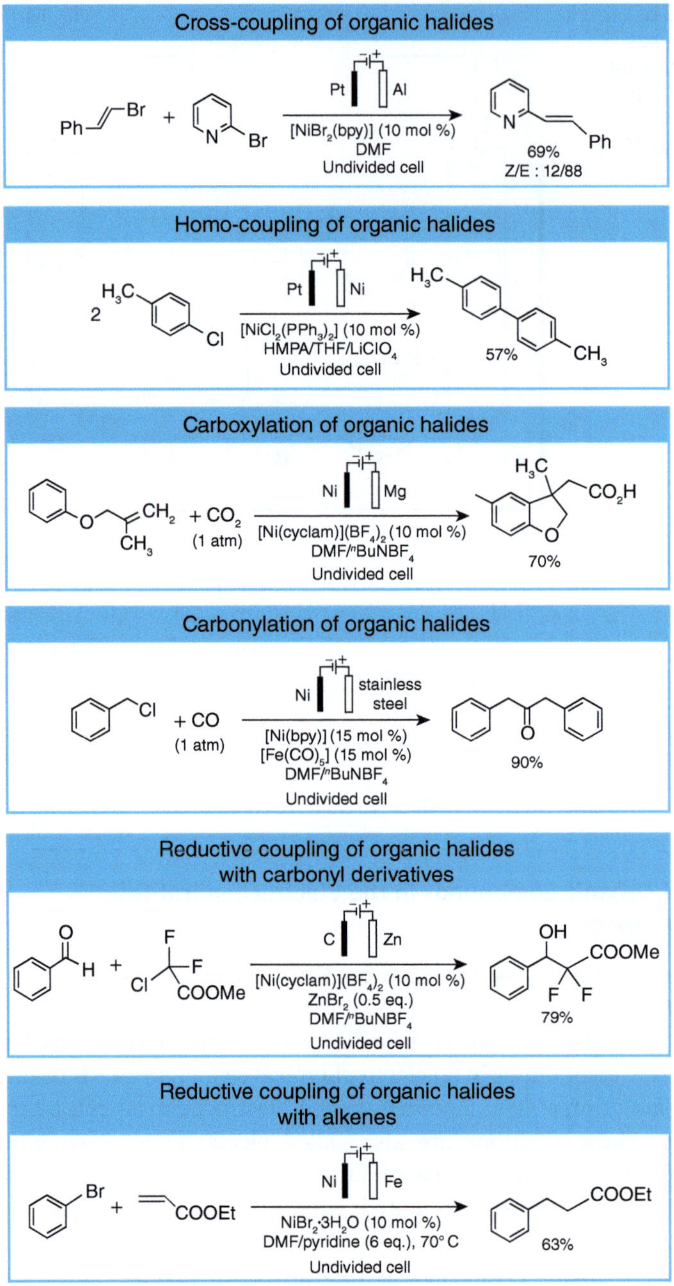

Figure 4.6 Synthesis reactions of organic compounds in the electrochemical cell are catalyzed by the Ni-organic. Source: Reprinted with permission from [13], © 2021 Elsevier Publications.

Figure 4.7 Synthesis reactions of organic compounds in the electrochemical cell by the Pd-organic catalyst. Source: Reprinted with permission from [13], © 2021 Elsevier Publications.

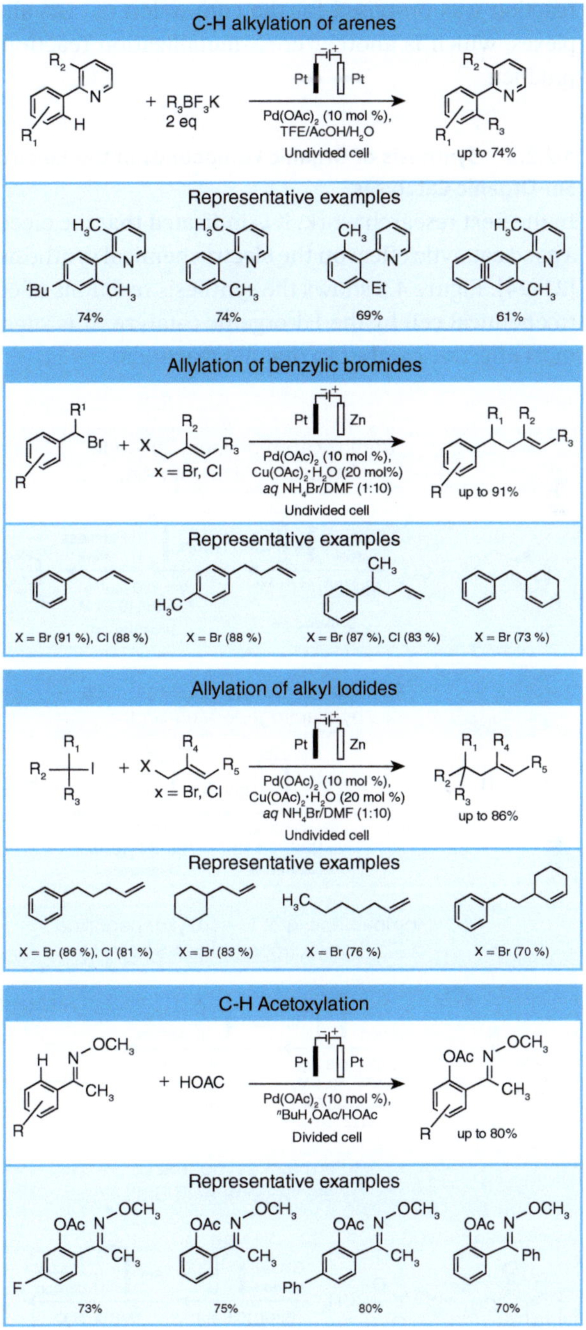

reaction was proposed for the conversion of zinc alkyl reactors to π-allyl-Pd complexes, which is another cross-metallization reaction leading to the final coupling product.

4.2.2.3 Synthesis of Organic Compounds in the Electrochemical Cell by the Sm-Organic Catalyses

In the next research work, it is indicated that the electrolytic salts of samarium have a good catalytic effect on the electrochemical synthesis of various organic substances [22–24]. Figure 4.8 shows the synthesis reactions of organic compounds in the electrochemical cell by the Pd-organic catalyze. It is suggested that the SmI_2 salt is the most effective catalyst in organic reactions.

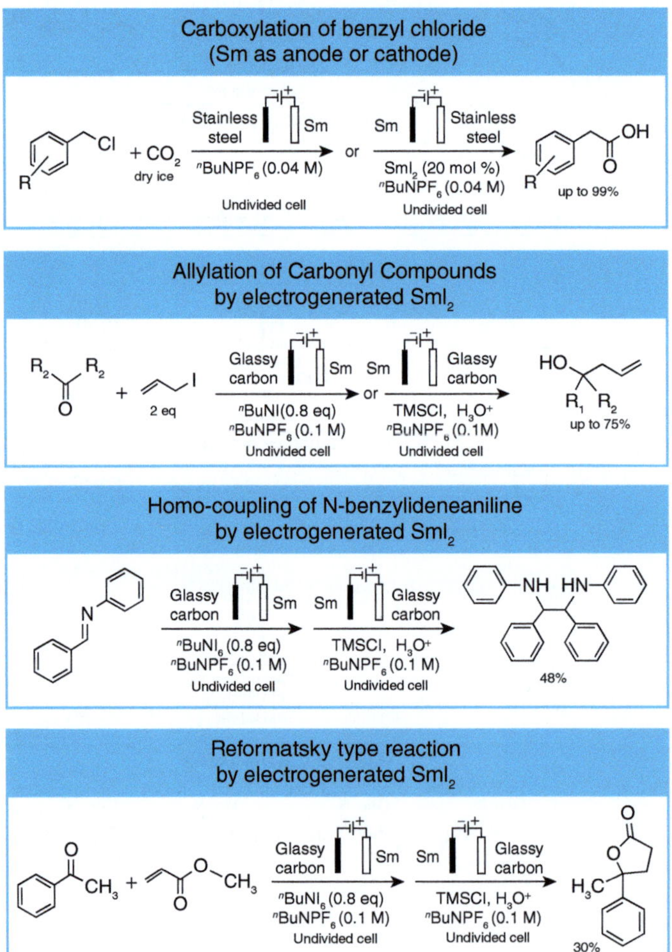

Figure 4.8 The synthesis reactions of organic compounds in the electrochemical cell by the Sm-organic catalyses. Source: Reprinted with permission from [13], © 2021 Elsevier Publications.

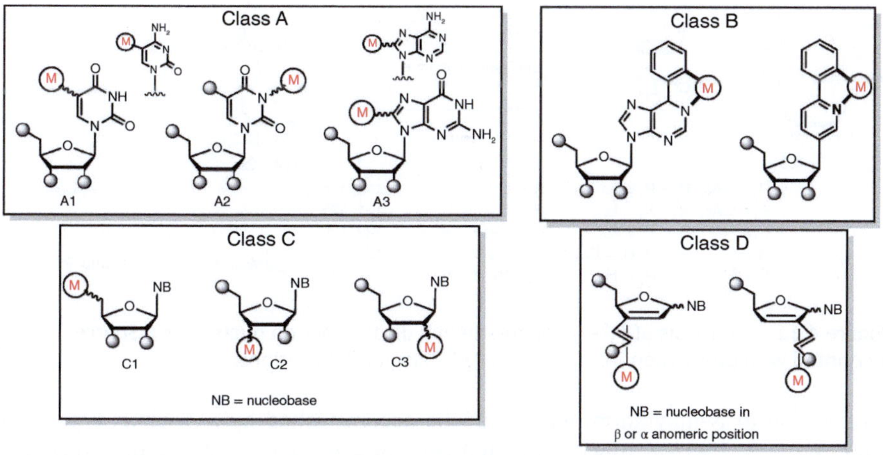

Figure 4.9 Representations into the 4 (A, B, C, and D) categorized synthesis of organometallic nucleosides. Source: Reprinted with permission from [25], © 2021 Elsevier Publications.

4.2.3 Synthesis of Organometallic Nucleosides

The synthesis of organometallic nucleosidesis 4 (A, B, C, and D) are categorized. Their representations are shown in Figure 4.9. They differ from each other according to the molecular structure, aromatic rings, functional groups, and hetero compounds.

4.2.3.1 A Category: Main Compounds

4.2.3.2 A1 Subcategory: Main Compounds

The 5-mercury and 5-mercury cytidine derivatives are the main A1 compounds. For example, Figure 4.10 indicates the 5-chloromercuri-2′-deoxycytidine syntheses. In order to create this molecule, an electrophilic substitution was used to replace the hydrogen atom in the C-5-pyrimidine nucleotide's electron-rich region. The presence of mercury can change the electronic balance of aromatic rings, as a result, the chemical and physical properties are also changed.

4.2.3.2.1 A2 Subcategory: Main Compounds

Hoogsteen and Watson–Crick base pairs in nucleic acids require the N-3 hydrogen found in thymine and uracil bases. Because it is acidic in nature, it makes pyrimidine

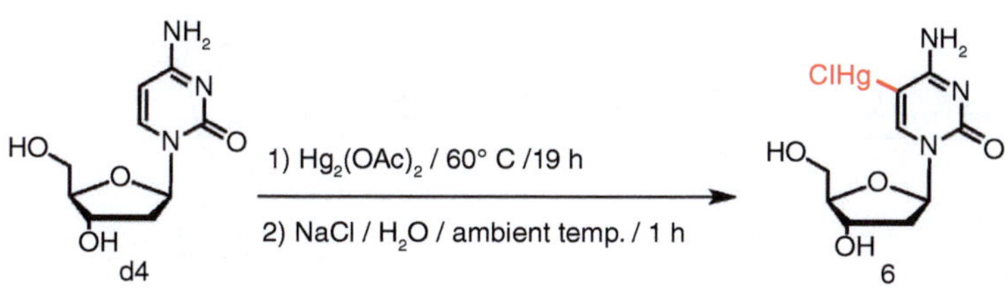

Figure 4.10 Synthesis of 5-chloromercuri-2′-deoxycytidine. Source: Reprinted with permission from [25], © 2021 Elsevier Publications.

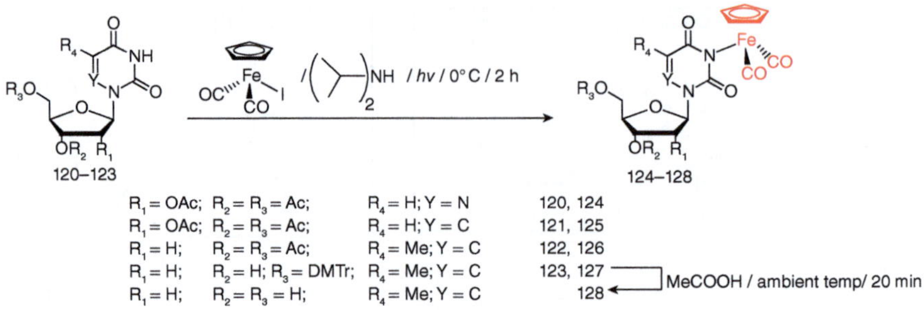

Figure 4.11 Synthesis of (η^5-cyclopentadienyl) dicarbonyliron nucleosides. Source: Reprinted with permission from [25], © 2021 Elsevier Publications.

susceptible to N-alkylation, which can be accomplished by a Michael-type addition reaction [26] or substituting halogen processes [27]. Alkylation reactions with N-3 have also been extensively investigated as a preparation method for the organometallic derivation of pyrimidine nuclei [28]. This successful method involves a photochemical reaction between (η5-cyclopentadienyl) dicarbonyl iodide and 6-azauridine, uridine, and thymidine, which are sufficiently protected to form diisopropylamine iron derivatives, as shown in Figure 4.11.

4.2.3.2.2 A3 Subcategory: Main Compounds

The Pd(II) guanosine derivatives are A3 compounds. Figures 4.12 and 4.13 show the synthesis of guanosine and adenosine derivatives. In these syntheses, the nucleosides were synthesized through the C-8 position modification. In this process, the Br was exchanged with the Pt complex. Finally, the ionic liquids (salts) are formed. These ionic liquids are good water-soluble and then they are separated by recrystallization.

4.2.3.3 B Category: Main Compounds

Sierra, Esteruelas, Gómez Gallego, et al. have investigated the chemistry of class B1 nucleosides [25]. It is described as an N-regulated cyclic reaction of 6-phenylpurine nucleotides and nucleotides with pentamethylcyclopentadienylolide dimers and rhodium dichloride (Figure 4.14) [29]. In this reaction, the CH band is activated, leading to the formation of the Rh and Ir bands. It is first described as the nucleotide/nucleotide cyclometallate.

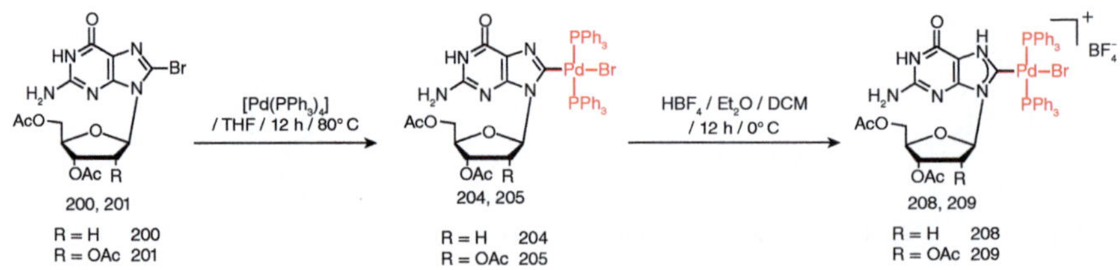

Figure 4.12 Synthesis of guanosine derivatives. Source: Reprinted with permission from [25], © 2021 Elsevier Publications.

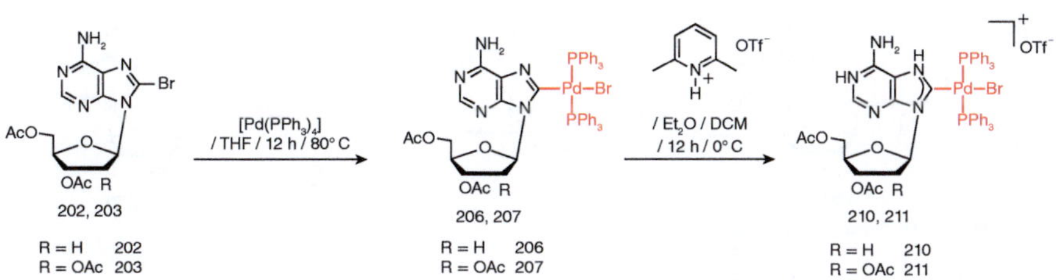

Figure 4.13 Synthesis of adenosine derivatives. Source: Reprinted with permission from [25], © 2021 Elsevier Publications.

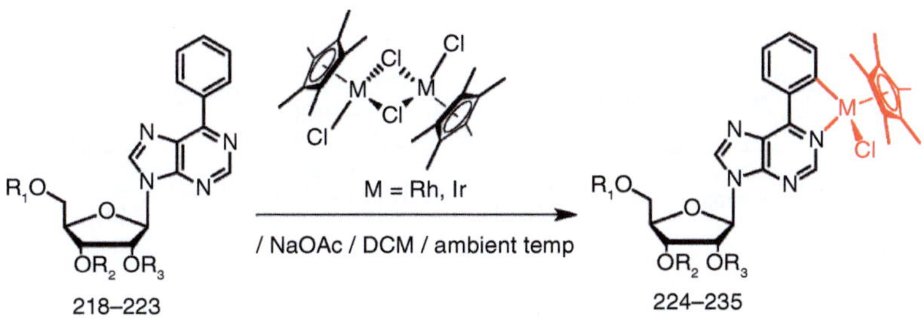

	Rh	Ir
$R_1 = R_2 = R_3 = Ac$	218, 224, 230	
$R_1 = H$; $R_2 = R_3$ = isopropylidene	219, 225, 231	
$R_1 = R_2$; $R_3 = H$	220, 226, 232	
$R_1 = H$; $R_2 = H$; $OR_3 = H$	221, 227, 233	
$R_1 = PO(OMe)_2$; $= R_2 = R_3 = H$	222, 228, 234	
$R_1 = PO(OMe)_2$; $= R_2 = R_3$ = isopropylidene	223, 229, 235	

Figure 4.14 Synthesis of cyclometallated nucleosides. Source: Reprinted with permission from [25], © 2021 Elsevier Publications.

4.2.3.4 C Category: Main Compounds

4.2.3.5 C1 Subcategory: Main Compounds

The 5′-carboxamide nucleosides of rhenium and technetium containing iminodiacetic acid as part of iron chelating agents has been investigated [30]. The compounds were treated with a solution of ebac-$[ReBr_3(CO)_3]_2$-methanol and water to produce rhenium derivatives, as shown in Figure 4.15.

4.2.3.5.1 C2 Subcategory: Main Compounds

The pentaharbon-tungsten nucleosides are C1 compounds, as described by Bergström et al. [31]. These were obtained by synthesizing oligonucleotides in which phosphoramidites reacted with methanol-acetonitrile and 1H-tetrazole (Figure 4.16). The intermediates reacted with pentacarbonyl to form products (Figure 4.16).

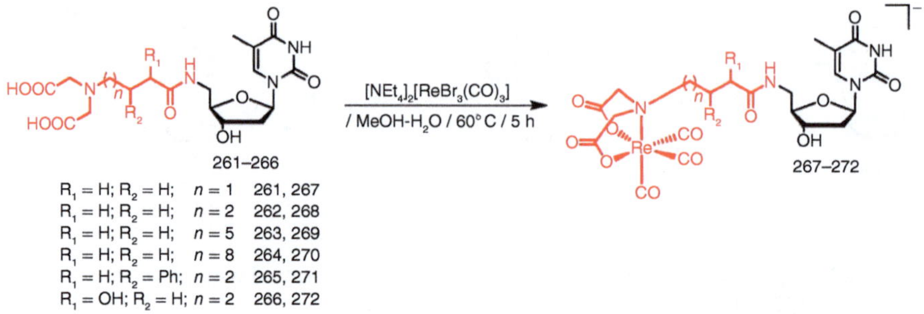

Figure 4.15 Synthesis of nucleosides. Source: Reprinted with permission from [25], © 2021 Elsevier Publications.

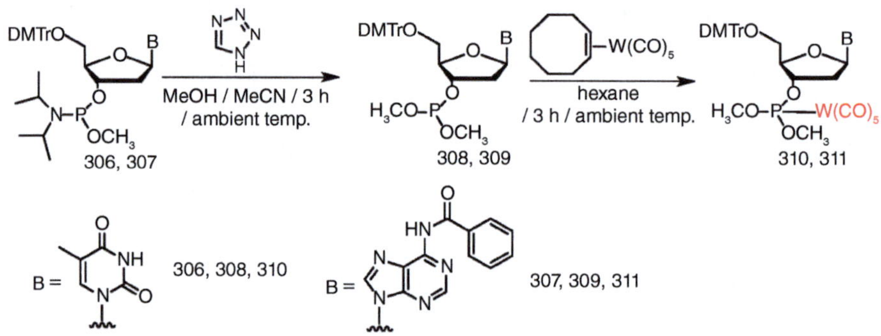

Figure 4.16 Synthesis of compounds 310 and 311. Source: Reprinted with permission from [25], © 2021 Elsevier Publications.

4.2.3.5.2 C3 Subcategory: Main Compounds

Zubieta et al. introduced the C3 compounds and studied their main synthesis characteristics [32]. The synthesis was done with the reaction between the 2′-amino-2′-deoxyuridine and dipyridine or dinolinylcarboxylic acid to form ligands (Figure 4.17). The resulting ligands bind to $[Re(CO)_3Br_3]^{2-}$ to form nucleosides (Figure 4.17).

4.2.3.6 D Categories: Main Compounds

Class D organometallic nuclides are the last group of compounds. Due to their altered glucose content, these compounds are similar to nucleoside xenonucleic acid [33] and also involved in the unstable 4-cyclohexa-2,4-dienyl chlorofleobase/nucleoside complex [25]. Schmalz et al. described some types of D categories [34, 35]. There are two main problems with the binding of this molecule: (i) the pure water needed to bind the three-carbon metal parts and the 4-butadiene ligand, and (ii) the adhesion of the nitrogen-containing veins. To solve the first problem, it is used the simulation plan shown in Figures 4.18 and 4.19 applied.

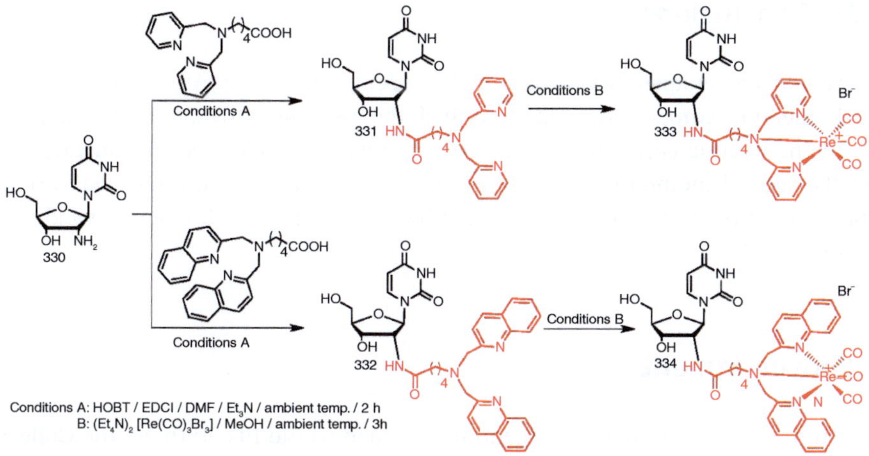

Figure 4.17 Synthesis of nucleosides 333 and 334. Source: Reprinted with permission from [25], © 2021 Elsevier Publications.

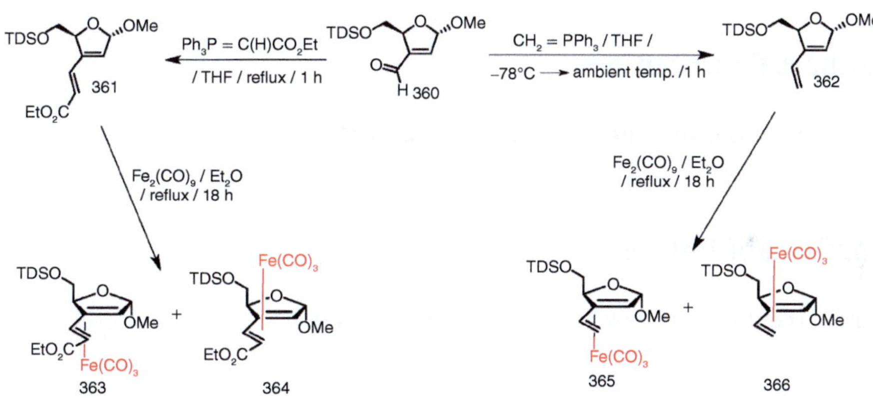

Figure 4.18 Synthesis of η^4-butadiene-tricarbonyliron compounds 363–366. Source: Reprinted with permission from [25], © 2021 Elsevier Publications.

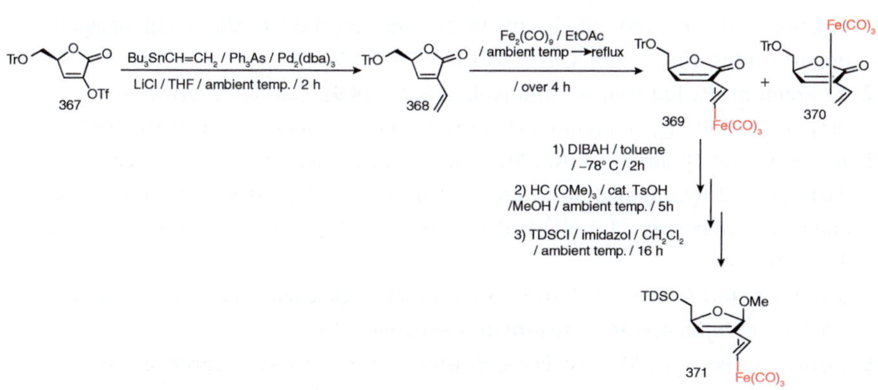

Figure 4.19 Synthesis of η^4-butadiene-tricarbonyliron compound 371. Reprinted with permission from [25], © 2021 Elsevier Publications.

4.3 Conclusions

In this current chapter, we reviewed the recent trends in synthesis methods of organometallic compounds and described the two types of synthesis methods for organometallic compounds. The first method is the electrochemical method, and the second method is nucleosides. It is suggested that these methods would replace traditional methods in the future because of their high reaction yield and greenness.

Acknowledgment

The authors Omar Dagdag and Ekemini D. Akpan would like to thank the College of Science, Engineering, and Technology, University of South Africa for the postdoctoral research fellowships.

Authors Contributions

All the authors contributed equally to the chapter.

Conflicts of Interest

Authors confirm that they have no conflict of interest.

References

1 García-Álvarez, J., Hevia, E., and Capriati, V. (2015). Reactivity of polar organometallic compounds in unconventional reaction media: challenges and opportunities. *Eur. J. Org. Chem.* 2015: 6779–6799.

2 de Fremont, P., Marion, N., and Nolan, S.P. (2009). Carbenes: synthesis, properties, and organometallic chemistry. *Coord. Chem. Rev.* 253: 862–892.

3 Jia, H.-P. and Quadrelli, E.A. (2014). Mechanistic aspects of dinitrogen cleavage and hydrogenation to produce ammonia in catalysis and organometallic chemistry: relevance of metal hydride bonds and dihydrogen. *Chem. Soc. Rev.* 43: 547–564.

4 Swart, M. and Costas, M. (2015). *Spin States in Biochemistry and Inorganic Chemistry: Influence on Structure and Reactivity*. Wiley.

5 Ibáñez, S., Poyatos, M., and Peris, E. (2020). N-heterocyclic carbenes: a door open to supramolecular organometallic chemistry. *Acc. Chem. Res.* 53: 1401–1413.

6 Coulembier, O., Degée, P., Hedrick, J.L., and Dubois, P. (2006). From controlled ring-opening polymerization to biodegradable aliphatic polyester: especially poly (β-malic acid) derivatives. *Prog. Polym. Sci.* 31: 723–747.

7 Sadjadi, S., Heravi, M.M., Malmir, M., and Masoumi, B. (2018). HPA decorated halloysite nanoclay: an efficient catalyst for the green synthesis of spirooxindole derivatives. *Appl. Organomet. Chem.* 32: e4113.

8 Haiduc, I. and Zuckerman, J.J. (2011). *Basic Organometallic Chemistry: Containing Comprehensive Bibliography*, I–XVI. Berlin, New York: De Gruyter https://doi.org/10.1515/9783110847888.fm.

9 Fehlner, T.P. (2013). *Inorganometallic Chemistry*. Springer Science & Business Media.

10 Beck, S. (2019). *Organometallic Chemistry*. Scientific e-Resources.

11 Das, T., Mohar, M., and Bag, A. (2021). Simple and cost-efficient chlorination of electron deficient aromatics to provide templates for organogelation and fluoride sensing. *Colloid Interface Sci. Commun.* 45: 100534.

12 Kamte, M.F., Wagner, C., and Schaefer, W. (2004). Direct electrosynthesis of cyano copper (I) complexes: X-ray crystal structure of (cyano-C) (2, 2′-bipyridine-N, N′)(triphenylphosphane) copper (I). *J. Coord. Chem.* 57: 55–60.

13 Gafurov, Z.N., Kantyukov, A.O., Kagilev, A.A. et al. (2021). Electrochemical methods for synthesis and in situ generation of organometallic compounds. *Coord. Chem. Rev.* 442: 213986.

14 Huang, T., Rodriguez, T.M., Gruninger, C.T. et al. (2020). Electrosynthetic route to cyclopentadienyl rhenium hydride complexes enabled by electrochemical investigations of their redox-induced formation. *Organometallics* 39: 1730–1743.

15 Gafurov, Z., Kantyukov, A., Kagilev, A. et al. (2021). Electrochemical generation of pyrazolyl-pyridyl N-heterocyclic carbene complexes of nickel. *Russ. J. Electrochem.* 57: 134–140.

16 Liu, B., Zhang, Y., Xu, D., and Chen, W. (2011). Facile synthesis of metal N-heterocyclic carbene complexes. *Chem. Commun.* 47: 2883–2885.

17 Hordiichuk, R., Kinzhybalo, V.V., Goreshnik, E.A. et al. (2017). Influence of apical ligands on Cu–(CC) interaction in copper (I) halides (Cl−, Br−, I−) π-complexes with an 1, 2, 4-triazole allyl-derivative: syntheses, crystal structures and NMR spectroscopy. *J. Organomet. Chem.* 838: 1–8.

18 Hordiichuk, R., Slyvka, Y.I., Kinzhybalo, V.V. et al. (2019). Construction of heterometallic and mixed-valence copper (I/II) chloride π-complexes with 1, 2, 4-triazole allyl-derivative. *Inorg. Chim. Acta* 495: 119012.

19 Sakhapov, I., Gafurov, Z., Babaev, V. et al. (2015). Electrochemical properties and reactivity of organonickel sigma-complex [NiBr (Mes)(bpy)](Mes = 2, 4, 6-trimethylphenyl, bpy = 2, 2′-bipyridine). *Russ. J. Electrochem.* 51: 1061–1068.

20 Gafurov, Z.N., Sinyashin, O.G., and Yakhvarov, D.G. (2017). Electrochemical methods for synthesis of organoelement compounds and functional materials. *Pure Appl. Chem.* 89: 1089–1103.

21 Lai, Y.-L. and Huang, J.-M. (2017). Palladium-catalyzed electrochemical allylic alkylation between alkyl and allylic halides in aqueous solution. *Org. Lett.* 19: 2022–2025.

22 Bazzi, S., Schulz, E., and Mellah, M. (2019). Electrogenerated Sm (II)-catalyzed CO_2 activation for carboxylation of benzyl halides. *Org. Lett.* 21: 10033–10037.

23 Sun, L., Sahloul, K., and Mellah, M. (2013). Use of electrochemistry to provide efficient SmI_2 catalytic system for coupling reactions. *ACS Catal.* 3: 2568–2573.

24 Zhang, Y.-F. and Mellah, M. (2017). Convenient electrocatalytic synthesis of azobenzenes from nitroaromatic derivatives using SmI_2. *ACS Catal.* 7: 8480–8486.

25 Kowalski, K. (2021). Organometallic nucleosides–synthesis, transformations, and applications. *Coord. Chem. Rev.* 432: 213705.

26 Mag, M. and Engels, J.W. (1988). Synthesis and structure assignments of amide protected nucleosides and their use as phosphoramidites in deoxyoligonucleotide synthesis. *Nucleic Acids Res.* 16: 3525–3543.

27 Spork, A.P., Wiegmann, D., Granitzka, M. et al. (2011). Stereoselective synthesis of uridine-derived nucleosyl amino acids. *J. Org. Chem.* 76: 10083–10098.

28 Zakrzewski, J., Tosik, A., and Bukowska-Strzyżewska, M. (1995). Photochemical reaction of (η5-cyclopentadienyl) dicarbonyliron iodide with 1-substituted uracils in the presence of diisopropylamine: crystal structure of the (η5-C5H5Fe $(CO)_2$ complex of deprotonated 5-fluoro-1-(tetrahydro-2-furyl) uracil (Ftorafur)). *J. Organomet. Chem.* 495: 83–90.

29 Martín-Ortíz, M., Gómez-Gallego, M., Ramirez de Arellano, C., and Sierra, M.A. (2012). The selective synthesis of metallanucleosides and metallanucleotides: a new tool for the functionalization of nucleic acids. *Chemistry* 18: 12603–12608.

30 Schibli, R., Netter, M., Scapozza, L. et al. (2003). First organometallic inhibitors for human thymidine kinase: synthesis and in vitro evaluation of rhenium (I)-and technetium (I)-tricarbonyl complexes of thymidine. *J. Organomet. Chem.* 668: 67–74.

31 Dalla Riva Toma, J.M. and Bergstrom, D.E. (1994). Transition metal labeling of oligodeoxyribonucleotides: synthesis and characterization of (pentacarbonyl) tungsten (0) nucleoside phosphites. *J. Org. Chem.* 59: 2418–2422.

32 Wei, L., Babich, J., Eckelman, W.C., and Zubieta, J. (2005). Rhenium tricarbonyl core complexes of thymidine and uridine derivatives. *Inorg. Chem.* 44: 2198–2209.

33 Chaput, J.C. and Herdewijn, P. (2019). What is XNA? *Angew. Chem. Int. Ed.* 58: 11570–11572.

34 Hirschhäuser, C., Velcicky, J., Schlawe, D. et al. (2013). Nucleoside analogues with a 1,3-diene–Fe $(CO)_3$ substructure: stereoselective synthesis, configurational assignment, and apoptosis-inducing activity. *Chemistry* 19: 13017–13029.

35 Schmalz, H.-G., Heßler, E., Bats, J.W., and Dürner, G. (1994). An approach to chiral η4-butadiene-Fe $(CO)_3$ complexes via diastereoselective complexation of nonracemic 2-alkoxy-4-vinyl-2, 5-dihydrofuran derivatives. *Tetrahedron Lett.* 35: 4543–4546.

5

Metal Carbonyls: Synthesis, Properties, and Structure

Ramesh Rengan

Bharathidasan University, Centre for Organometallic Chemistry, School of Chemistry, Palkalaiperur, 620024, Tiruchirappalli, India

5.1 Introduction

Metal carbonyls are organometallic complexes of transition metals with carbon monoxideligand (CO). The synthesis, structures, properties, and applications of metal carbonyl complexes have been the subject of intensive study for many years. CO is one of the most significant π-acceptor ligands and is widely studied in organometallic chemistry. Because of the C≡O multiple bonds, it is an unsaturated ligand and can act as a spectator or an actor ligand. Though it is an extremely weak Lewis base, it forms numerous complexes with almost all of the transition metals. It also forms complexes with some alkaline earth metals and p-block elements. A CO ligand can either bind to a single metal or act as a bridge between two or more metals. The number of CO ligands coordinated to metal generally follows the 18-electron rule. The rule is followed by almost 99% of metal carbonyls. Justus von Liebig initiated experiments on the reaction of CO with metals in 1834, marking the beginning of the era of metal carbonyls. Conversely, it was later revealed that the substance he stated to be potassium carbonyl was not in fact a metal carbonyl [1]. After Schutzenberger and Mond published the synthesis of $[PtCl_2(CO)_2]$ and $[PtCl_2(CO)]_2$ in 1868 and $[PtCl_2(CO)]_2$ in 1890 [2] respectively, Hieber produced a number of metal-carbonyl compounds [3]. Metal carbonyls play an important role in many catalytic reactions as well as in industrially important organic synthesis.

5.2 Classification of Metal Carbonyls [4]

5.2.1 Classification Based on Coordinated Ligands

Metal carbonyl complexes can be divided into two types based on the coordinated ligands ashomoleptic carbonyls and heteroleptic carbonyls.

Organometallic Compounds: Synthesis, Reactions, and Applications, First Edition.
Edited by Dakeshwar Kumar Verma and Jeenat Aslam.

5.2.1.1 Homoleptic Carbonyls

Homoleptic carbonyl complexes are those in which a metal is coordinated to only CO ligands such as $Ni(CO)_4$, $Fe(CO)_5$, $Cr(CO)_6$, $Co_2(CO)_8$, $Mn_2(CO)_{10}$, $Fe_3(CO)_{12}$, $Ir_4(CO)_{12}$, etc. (Figure 5.1).

5.2.1.2 Heteroleptic Carbonyls

Heteroleptic carbonyl complexes are those in which the metal is coordinated to carbon monoxide as well as other ligands including triphenylphosphine, nitrosyl, thiocarbonyls, isocyanides, etc., such as $(CO)_3Ni(PPh_3)$, $(CO)_3Mo(PF_3)_3$, $(CO)_3Cr(NO)_2$, $(CO)_4W$(1,2-ethylenediamine), etc. (Figure 5.2).

5.2.2 Classification Based on Number of Metals and the Constitution of Carbonyls

Metal carbonyl complexes can be divided into two categories: mononuclear and polynuclear carbonyl complexes, based on the number of metals and the constitution of carbonyls.

5.2.2.1 Mononuclear Carbonyl Complexes

There is only one metallic atom in these carbonyls, and there is no bridging CO ligand. For example, $Ru(CO)_5$,$W(CO)_6$, $V(CO)_6$, $Mo(CO)_6$, and so on (Figure 5.3).

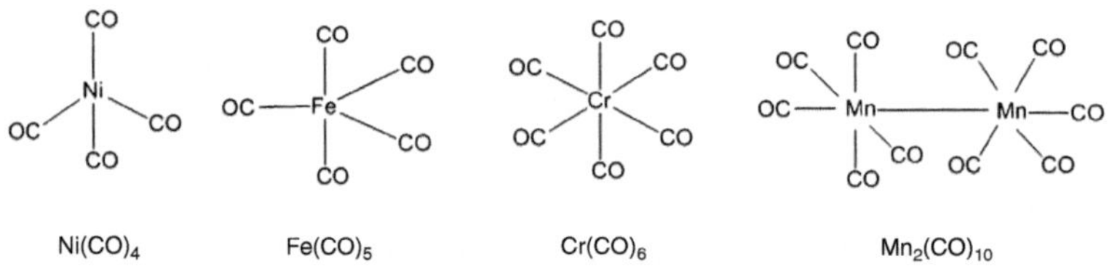

Figure 5.1 Structures of some homoleptic carbonyl complexes.

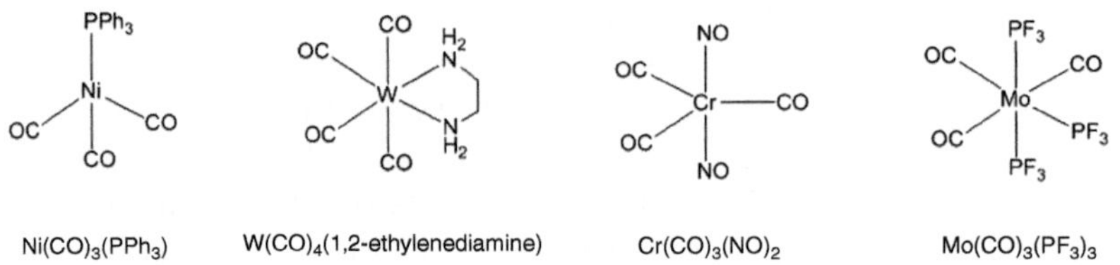

Figure 5.2 Structures of some heteroleptic carbonyl complexes.

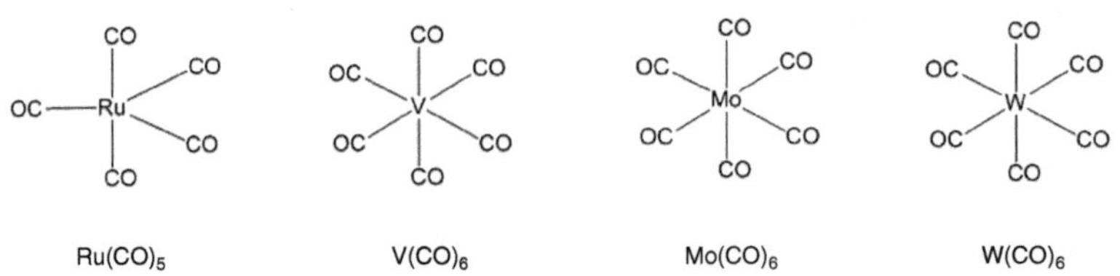

Figure 5.3 Structures of some mononuclear carbonyl complexes

5.2.2.2 Polynuclear Carbonyl Complexes

Polynuclear carbonyls are composed of two or more metal atoms. These are categorized as follows:

5.2.2.2.1 Homonuclear Carbonyl Complexes

These contain metal atoms of only one element.

For example, $Fe_2(CO)_9$, $Mn_2(CO)_{10}$, $Co_2(CO)_8$, $Fe_3(CO)_{12}$, $Co_4(CO)_{12}$, $Rh_4(CO)_{12}$, $Ir_4(CO)_{12}$, and so on (Figure 5.4).

5.2.2.2.2 Heteronuclear Carbonyls

These carbonyls contain metals of different elements. For example, $MnCo(CO)_{10}$ and $MnRe(CO)_{10}$ (Figure 5.5).

The polynuclear metal carbonyl complexes are also classified as follows:

Nonbridged metal carbonyl complexes and bridged metal carbonyl complexes.

5.2.2.2.3 Nonbridged Metal Carbonyl Complexes

These carbonyls contain terminal CO ligands and M–M bonds.

For example, $Tc_2(CO)_{10}$, $Re_2(CO)_{10}$, $Ir_4(CO)_{12}$, and so on (Figure 5.6).

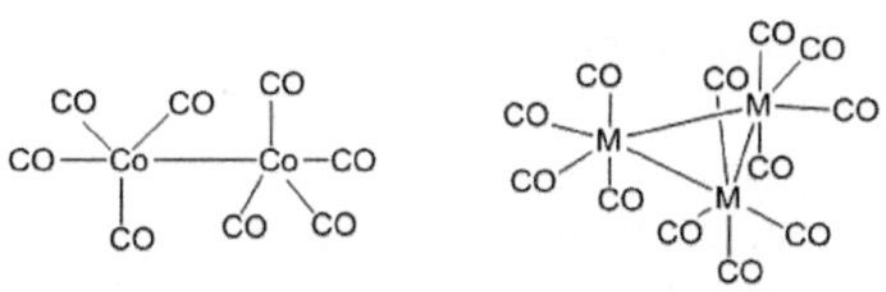

$CO_2(CO)_8$ $M_3(CO)_{12}$, M = Ru, Os

Figure 5.4 Structures of some homonuclear carbonyl complexes.

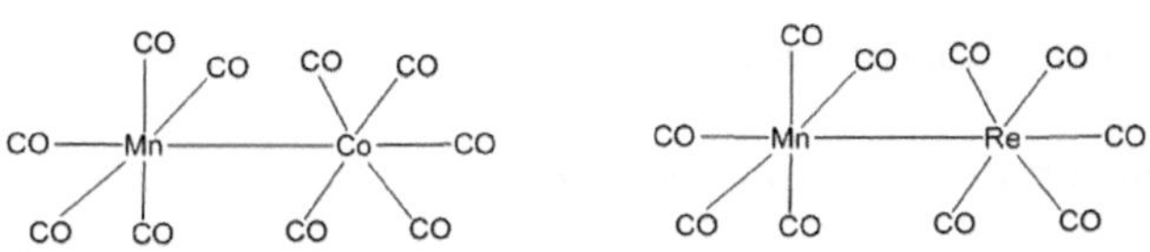

$MnCo(CO)_{10}$ $MnRe(CO)_{10}$

Figure 5.5 Structures of some heteronuclear carbonyl complexes.

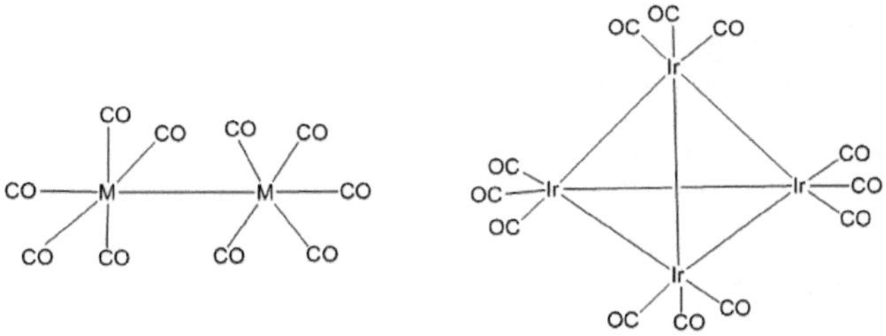

$M_2(CO)_{10}$, M = Tc, Re $Ir_4(CO)_{12}$

Figure 5.6 Structures of some nonbridged metal carbonyl complexes.

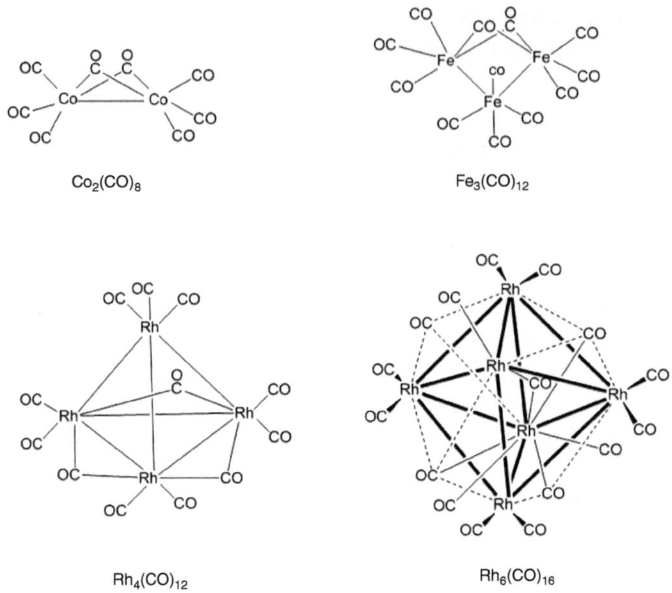

Figure 5.7 Structures of some bridged metal carbonyl complexes.

5.2.2.2.4 Bridged Metal Carbonyl Complexes

These carbonyl complexes comprise terminal as well as bridged CO ligands and M–M bonds. For example, $Co_2(CO)_8$ (in solid-state), $Fe_3(CO)_{12}$, $Rh_4(CO)_{12}$, $Rh_6(CO)_{16}$, and so on (Figure 5.7).

5.3 Synthesis of Metal Carbonyls

5.3.1 Direct Reaction of Metal with Carbon Monoxide

This method requires metals in a reduced oxidation state because only π-basic metals can bind with CO [5]

$$\mathrm{Ni} + 4\,\mathrm{CO} \xrightarrow[\text{1 atm}]{30\,^\circ\mathrm{C}} \underset{\text{Nickel tetracarbonyl}}{\mathrm{Ni(CO)_4}}$$

$$\mathrm{Fe} + 5\,\mathrm{CO} \xrightarrow[\text{200 atm}]{200\,^\circ\mathrm{C}} \underset{\text{Iron pentacarbonyl}}{\mathrm{Fe(CO)_5}}$$

$$2\,\mathrm{Co} + 8\,\mathrm{CO} \xrightarrow[\text{35 atm}]{150\,^\circ\mathrm{C}} \underset{\text{Dicobalt octacarbony}}{\mathrm{Co_2(CO)_8}}$$

5.3.2 Reductive Carbonylation

This technique is most commonly used to produce binary carbonyl complexes for the majority of the transition metals. In the reaction of carbon monoxide with metal salts in which metal is in higher oxidation state, reducing agents including sodium, aluminum, zinc, trialkylaluminum, Grignard reagents, and H_2 have been employed. The reducing agent initially reduces the metal to a low oxidized state before CO bonds to it to produce metal carbonyl compounds [6, 7].

$$\mathrm{CrCl_3 + Al + 6\,CO \xrightarrow[\text{AlCl}_3,\text{C}_6\text{H}_6]{140\,^\circ\text{C}} \underset{\text{Chromium hexacarbonyl}}{Cr(CO)_6} + AlCl_3}$$

$$\mathrm{MoCl_5 + 5Na + 6\,CO \xrightarrow[\text{62 bar}]{\text{Diglyme}} \underset{\text{Molybdenum hexacarbonyl}}{Mo(CO)_6} + 5\,NaCl}$$

In the process of reducing metal oxides, CO acts as a reducing agent and transforms into carbon dioxide.

$$\mathrm{2\,CoO + 10\,CO \xrightarrow[\text{1900 atm}]{80\,^\circ\text{C}} \underset{\text{Dicobalt octacarbonyl}}{Co_2(CO)_8} + 2\,CO_2}$$

$$\mathrm{Fe_2O_3 + 13\,CO \xrightarrow[\text{2000 atm}]{225\,^\circ\text{C}} \underset{\text{Iron pentacarbony}}{2\,Fe(CO)_5} + 3\,CO_2}$$

5.3.3 Photolysis and Thermolysis

Bimetallic and polymetallic carbonyls are generated from the thermal and photochemical reactions of lower carbonyl complexes.

$$\mathrm{2\,Os(CO)_5 \xrightarrow{\Delta} \underset{\text{Triosmium dodecacarbonyl}}{Os_3(CO)_{12}} + 3\,CO}$$

$$\mathrm{2[Co_2(CO)_8] \xrightarrow[70\,^\circ\text{C}]{\Delta} \underset{\text{Tetracobalt dodecacarbonyl}}{Co_4(CO)_{12}} + 4\,CO}$$

$$\mathrm{3[Fe(CO)_5] \xrightarrow{\Delta} \underset{\text{Triiron dodecacarbony}}{Fe_3(CO)_{12}} + 3\,CO}$$

$$\mathrm{2[Fe(CO)_5] \xrightarrow{h\nu} \underset{\text{Diiron nonacarbonyl}}{Fe_3(CO)_9} + CO}$$

5.3.4 Abstraction of CO from a Reactive Organic Carbonyl Compounds

This process abstracts CO from organic carbonyl compounds [8]

$$\underset{\text{Wilkinson's catalyst}}{\mathrm{[RhCl(PPh_3)_3]}} + \mathrm{RCHO} \xrightarrow{\text{Oxidative addition}} \mathrm{\{XRhCl(COR)(PPh_3)_3\}}$$

$$\xrightarrow{\text{Retro migratory insertion}} \mathrm{\{XRhCl(CO)R(PPh_3)_2\}}$$

$$\xrightarrow{\text{Reductive elimination}} \mathrm{RX} + \underset{\text{Bis(triphenylphosphine)rhodium(I) carbonyl chloride}}{\mathrm{RhCl(CO)(PPh_3)_2}}$$

5.4 Properties of Metal Carbonyls

5.4.1 Physical Properties

Metallic carbonyl complexes are mostly liquids or volatile solids in nature. Mononuclear carbonyl complexes of nickel, iron, ruthenium, and osmium, such as $Ni(CO)_4$, $Fe(CO)_5$, Ru(CO), and $Os(CO)_5$, are liquids at ambient temperature and pressure, whereas other carbonyl complexes are solids. The physical properties of some transition metal carbonyl complexes are summarized in Table 5.1 [9–11]. The majority of mononuclear carbonyl complexes are colorless or only barely colored, whereas polynuclear carbonyl complexes are colored. The increase in the number of metal ions increases the intensity of the color of polynuclear carbonyl complexes. For example, $Fe(CO)_5$ is a colorless liquid, $Fe_2(CO)_9$ is a yellow solid, and $Fe_3(CO)_{12}$ is a deep green solid. Electron transfer among metal orbitals gives intense color to polynuclear carbonyl complexes. Metal carbonyl complexes are soluble in organic solvents. The mononuclear carbonyl complexes are volatile. Because of their low melting points and low thermal stability, they are toxic when compared to an analogs metal and CO ligand. These compounds can harm the brain, lungs, kidneys, and liver. $Ni(CO)_4$ shows the highest inhalation poisoning. Long-term exposure to these compounds is carcinogenic. They exhibit toxicity through inhalation, skin contact, inhalation, or intake in part due to their capability to carbonylate hemoglobin to produce carboxyhemoglobin, which inhibits oxygen binding [9]. With the exception of $V(CO)_6$, all the metal carbonyl complexes are diamagnetic. Mononuclear carbonyls are generated by metals with even atomic numbers. As a result, all electrons in metal atoms are paired. The unpaired electrons in binuclear metal carbonyl complexes, which are produced by metals having odd atomic numbers, are used to form M—M bonds. The solid carbonyl complexes sublime under reduced pressure with degradation to some extent. At lower temperatures, the majority of metal carbonyl complexes undergo melting. Metal carbonyls have a high thermodynamic instability and oxidize even in the aerobic atmosphere at different rates. Binuclear cobalt and iron carbonyl complexes oxidize at ambient temperature while chromium and molybdenum hexacarbonyl complexes oxidize on heating.

5.4.2 Chemical Properties

Metal carbonyl complexes undergo a broad spectrum of reactions [4, 12]. All of the reactions are highly reliant on the polarization of the CO ligand on binding and thus change in the net charge. Some reactions are aided by the electrophilic nature of carbonyl carbon, while others are aided by the nucleophilic nature of oxygen atom.

5.4.2.1 Ligand Substitution Reactions

Different mono- and bidentate ligands can be employed to substitute the CO ligand from the metal carbonyls through thermal and photochemical processes. CO can be partially or completely substituted by monodentate ligands such as phosphine

Table 5.1 Physical properties of some important metal carbonyl complexes.

Group	5	6	7	8	9	Group
3d	**$V(CO)_6$** • Black-blue solid • Volatile • Thermally unstable. • Soluble in organic solvents and insoluble in water. • It decomposes at 323 K. • Paramagnetic • Octahedral structure	**$Cr(CO)_6$** • Colorless crystalline solid; • Stable at room temperature. • Soluble in organic solvents and insoluble in water. • Sublimes readily. • It decomposes at 483 K. • Diamagnetic • Octahedral structure	**$Mn_2(CO)_{10}$** • Yellow crystalline solid. • Soluble in organic solvents and insoluble in water. • Sublimes at 60 °C. • Melting point 427 K. • Diamagnetic • Octahedral structure	**$Fe(CO)_5$** • Yellow-orange liquid • Soluble in organic solvents and insoluble in water • Melting point 252 K • Boiling point 376 K • Trigonalbipyramidal structure **$Fe_2(CO)_9$** • Golden yellow crystals. • Non-volatile. • Insoluble in all common solvents • Decomposes at 373 K. • Octahedral structure **$Fe_3(CO)_{12}$** • Deep green solid • Soluble in non- polar organic solvents • Sublimes under reduced pressure. • It decomposes slowly in air • Melting point 438 K.	**$Co_2(CO)_8$** • Red-orange-crystalline solid • Air-sensitive, • Melting point 325 K **$Co_4(CO)_{12}$** • Black crystalline compound • Insoluble in water • Air-sensitive, • Decomposes at 60 °C • Molecular symmetry is C_3v **$Co_6(CO)_{16}$** • Black solid • Slowly dec. in air • Td symmetry • Octahedral cluster	• 3d

(Continued)

Table 5.1 (Continued)

Group	5	6	7	8	9	Group
4d		**$Mo(CO)_6$** • Vivid, white, translucent crystals • Soluble in benzene, paraffin oil slightly soluble in ether • Sublimes in vacuo • Melting point 423 K • Boiling point 429 K • Octahedral geometry	**$Te_2(CO)_{10}$** • White solid; • Slowly dec. in air; • Melting point 433 K	**$Ru(CO)_5$** • Colorless liquid; • Melting point 257-290 K; • Dec. in air at 298 K to $Ru_3(CO)_{12}$ + CO **$Ru_3(CO)_{12}$** • Orange solid; • Soluble in nonpolar organic solvents • Sublimes in vacuo • Melting point 427 K • Molecular shape D_3h cluster	**$Rh_4(CO)_{12}$** • Red solid • Soluble in chlorocarbons, toluene, tetrahydrofuran • >403 K dec to $Rh_6(CO)_{16}$ **$Rh_6(CO)_{16}$** • Purple-brown crystals • Slightly soluble in dichloromethane and chloroform • Dec. >573 K • Melting point 508 K	**4d**
5d		**$W(CO)_6$** • White solid; • Sparingly in THF • Dec.443 K • Sublimes in vacuo • Octahedral geometry	**$Re_2(CO)_{10}$** • White solid; • Melting point 443 K • Pair of square pyramidal $Re(CO)_5$	**$Os(CO)_5$** • Colorless volatile liquid • Melting point 275 K **$Os_3(CO)_{12}$** • Yellow solid slightly in organic solvents • Melting point 497 K • Cluster has D_3h symmetry	**$Ir_4(CO)_{12}$** • Canary-yellow crystals • Air-stable species • Poorly soluble in organic solvents • Soluble in chlorocarbons,toluene, THF • Melting point 443 K • Tetrahedral cluster **$Ir_6(CO)_{16}$** • Red solid	**5d**

(PR3), isocyanide (CNR), cyanide (CN^-), and ethers.

$$Mn_2(CO)_{10} + PR_3 \longrightarrow 2\,Mn(CO)_4(PR_3) + 2\,CO$$

$$Fe(CO)_5 + 2\,CNR \longrightarrow Fe(CO)_3(CNR)_2 + 2\,CO$$

$$2\,Fe_2(CO)_{12} + 3\,py \longrightarrow Fe_3(CO)_9(py)_3 + 3\,Fe(CO)_5$$

$$Ni(CO)_4 + 4\,CNR \longrightarrow Ni(CNR)_4 + 4\,CO$$

o-phenylene-bis(dimethyl arsine) (diars) and *o*-phenanthroline (*o*-phen) are bidentate ligands that can CO ligands in multiples of two.

$$Cr(CO)_6 + 2\,diars \longrightarrow Cr(CO)_2(diars)_2 + 4\,CO$$

$$Ni(CO)_4 + o\text{-phen} \longrightarrow Ni(CO)_2(o\text{-phen})_2 + 2\,CO$$

$$Mo(CO)_6 + diars \longrightarrow Mo(CO)_4(diars) + 2\,CO$$

5.4.2.2 Reaction with Sodium Metal

Metal carbonyls can be reduced using sodium metal and its amalgam.

$$Cr(CO)_6 + 2\,Na \longrightarrow Na_2[Cr(CO)_5] + CO$$

$$Mn_2(CO)_{10} + 2\,Na \longrightarrow 2\,Na[Mn(CO)_5] + CO$$

The zero oxidation states of Cr and Mn metals are reduced to Cr^{2-} and Mn^-, respectively.

5.4.2.3 Reaction with Sodium Hydroxide

When sodium hydroxide reacts with metal carbonyl complexes, the OH^- ion attacks the carbonyl group as a nucleophile, forming a metal carboxylic acid complex. When the formed metal carboxylic acid complex reacts with NaOH, it releases carbon dioxide and a hydrido anion. When the formed anion is protonated, iron tetracarbonyl hydride is formed and the reaction is termed as the Heiber base reaction.

$$Fe(CO)_5 + NaOH \longrightarrow Na[Fe(CO)_4COOH]$$

$$Na[Fe(CO)_4COOH] + NaOH \longrightarrow Na[HFe(CO)_4] + NaHCO_3$$

$$Na[HFe(CO)_4] + H^+ \longrightarrow (H)_2Fe(CO)_4 + Na^+$$

5.4.2.4 Reaction with Halogens

The reaction of metal carbonyl complexes with halogens furnishes metal carbonyl halides

$$Fe(CO)_5 + X_2 \longrightarrow Fe(CO)_4X_2 + CO$$

$$Mo(CO)_6 + Cl_2 \longrightarrow Mo(CO)_4Cl_2 + 2\,CO$$

In polynuclear carbonyl complexes, halogens cleave M—M bonds.

$$Mn_2(CO)_{10} + X_2 \longrightarrow 2\,Mn(CO)_5X$$

Certain carbonyl complexes decompose when they react with halogens.

$$Co_2(CO)_8 + 2\,X_2 \longrightarrow 2\,CoX_2 + 8\,CO$$
$$Ni(CO)_4 + Br_2 \longrightarrow NiBr_2 + 4\,CO$$

5.4.2.5 Reaction with Hydrogen

Metal carbonyl hydrides are formed when some carbonyl complexes undergo a reduction reaction with hydrogen.

$$Mn_2(CO)_{10} + H_2 \xrightarrow{200\ \text{atm}} 2[Mn(CO)_5H]$$
$$Co_2(CO)_8 + H_2 \xrightarrow{165\ ^\circ\text{C},\ 200\ \text{atm}} 2[Co(CO)_4H]$$

Despite the fact that these compounds are termed as hydrides, they act as proton donors. The $[Mn(CO)_5H]$ and$[Co(CO)_4H]$ are neutral hydrides and act as acids as follows.

$$[Mn(CO)_5H] \longrightarrow [Mn(CO)_5]^- + H^+$$
$$[Co(CO)_4H] \longrightarrow [Co(CO)_4]^- + H^+$$

$[HFe(CO)_4]^-$is a true anion ichydride that acts as a reducing agent for halide derivatives.

$$RX + [HFe(CO)_4]^- \longrightarrow RH + [XFe(CO)_4]^-$$

5.4.2.6 Reaction with Nitricoxide (NO)

Several metal carbonyl complexes interact with NO to form carbonyl nitrosyl complexes.

$$Fe(CO)_5 + 2\,NO \xrightarrow{95\ ^\circ\text{C}} Fe(CO)_2(NO)_2 + 3CO$$
$$Co_2(CO)_8 + 2\,NO \xrightarrow{40\ ^\circ\text{C}} 2\,Co(CO)_3(NO) + 2\,CO$$

In the reaction between $Fe(CO)_5$ and NO, three carbon monoxide ligands are replaced by two NO molecules. NO ligand is a three-electron donor, whereas CO ligand is a two-electron donor, hence this is electronically equivalent.

5.4.2.7 Disproportionation

Some of the metal carbonyl complexes react with coordinating ligands like ammonia and encounter disproportionation.

$$Co_2(CO)_8 + 6\,NH_3 \longrightarrow [Co(NH_3)_6]\ [Co(CO)_4]_2$$
$$2\,Fe(CO)_5 + 6NH_3 \longrightarrow [Fe(NH_3)_6]^{2+}\ [Fe(CO)_4]^{2-} + 6CO$$

For metal carbonyl complexes, there are different types of chemical properties that only a few can be illustrated. The scheme depicts the comprehensive chemistry of $Fe(CO)_5$and $Mo(CO)_6$that each carbonyl goes through (Figure 5.8a,b)[12].

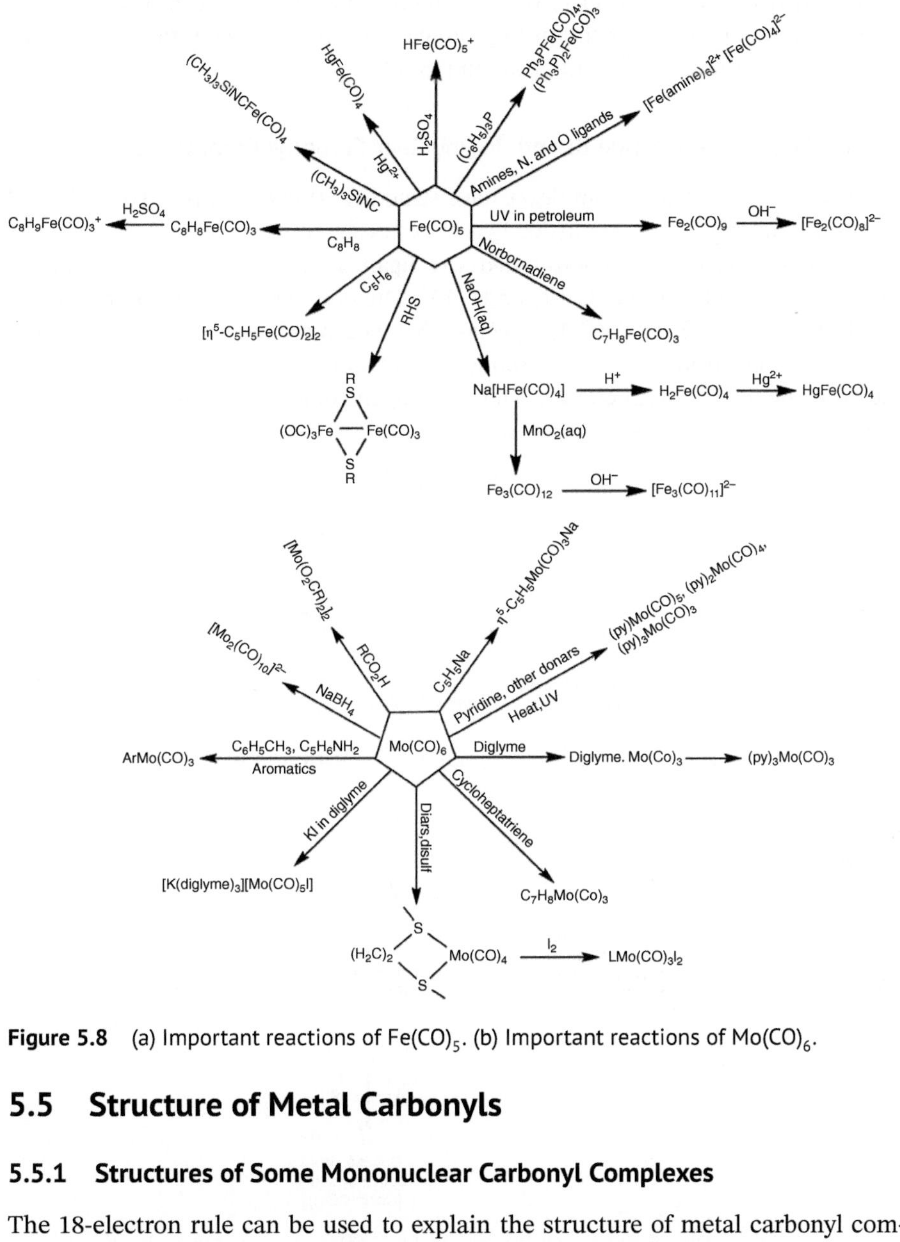

Figure 5.8 (a) Important reactions of $Fe(CO)_5$. (b) Important reactions of $Mo(CO)_6$.

5.5 Structure of Metal Carbonyls

5.5.1 Structures of Some Mononuclear Carbonyl Complexes

The 18-electron rule can be used to explain the structure of metal carbonyl complexes. The requisite and symmetrical configurations of the M-CO σ-bonding pairs establish the stoichiometry and frameworks of mononuclear carbonyl complexes, for example, $Ni(CO)_4$ is tetrahedral, $Fe(CO)_5$ is trigonalbipyramidal and $Cr(CO)_6$ is octahedral. $V(CO)_6$ is a simple paramagnetic carbonyl complex with octahedral geometry. It is a 17-valent electron system. To attain the 18-electron count, it is required to undergo dimerization to produce a seven-coordinate fragment, which is sterically an unfavorable configuration [13, 14].

5.5.2 Structures of Some Bi and Polynuclear Carbonyl Complexes

Binuclear metal carbonyl complexes have structures that encompass either M—M bonds or bridging carbon monoxide ligands, or even both. $Mn_2(CO)_{10}$, for example, contains only terminally coordinated CO ligands, whereas $Co_2(CO)_8$ contains both bridging and terminal CO ligands [15]. The odd electrons are shared by two 17-electron $Mn(CO)_5$ or $Co(CO)_4$ species, giving each metal a total of 18 electrons. In terms of electron counting, carbonyl ligands are generally considered as two electron donors. It can function as a terminal ligand or a bridge among two or three metal centers in a variety of ways (Figure 5.9).

1. Terminal carbonyls: M–C–O angles near 180°.

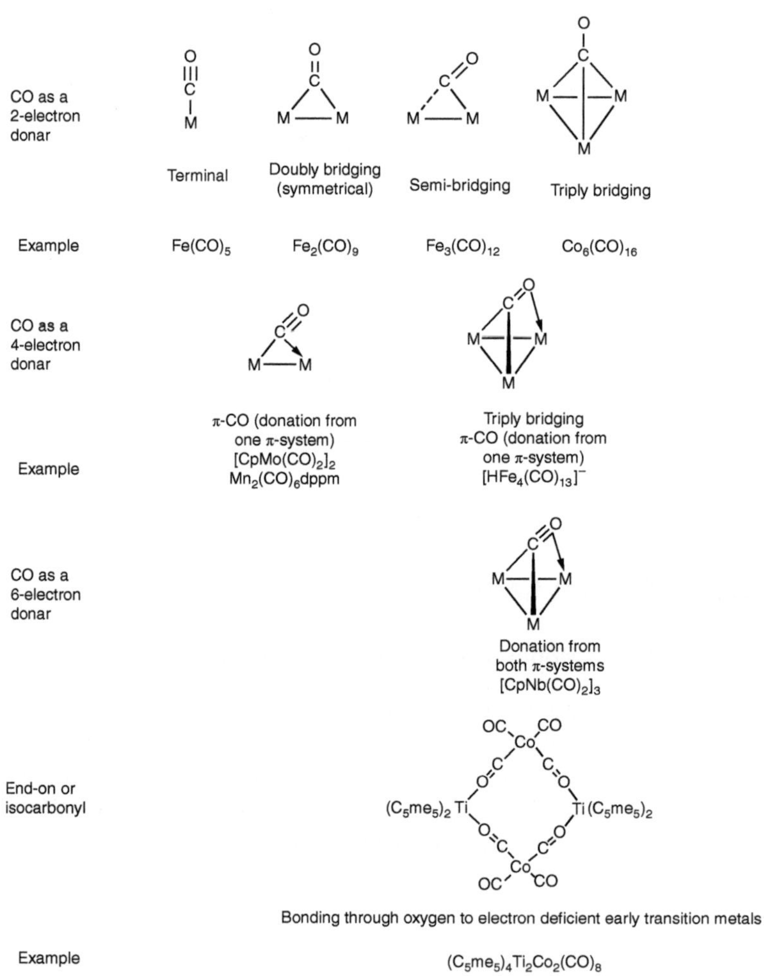

Figure 5.9 Various modes of bonding of CO.

2. Bridging carbonyl
   - Doubly bridging μ_2-CO
     It is the most common type of bridging carbonyl, μ_2- bridging two metals with M–C–M angles are usually 77–90^0.
   - Triply/Quadruply bridging μ_3-CO
     This type of bridging can be seen in cluster-type complexes.
   - Semi-bridging carbonyl
     It is a terminal carbonyl that has a secondary interaction with another metal through its carbon.
   - Four electron donor carbonyl or π –CO
     It is a terminal carbonyl on one metal that is bound to another metal in much the same way as an olefin (a σ donor and a π acceptor)
   - End-on or isocarbonyl
     It is a terminal carbonyl on one metal which serves through oxygen, as a Lewis base to another metal, M=C=O→M

5.6 Bonding in Metal Carbonyls

CO is regarded as an excellent sigma (σ) donor and pi(π) acceptor ligand. The molecular orbital (MO) diagram of CO must be studied to understand the bonding in metal carbonyls. The MO representation for carbon monoxide is given in Figure 5.10. The CO ligand contains ten electrons. The energy order of the molecular orbitals and the filling up of the electrons in CO can be represented as follows:

$$(\sigma_{1s}{}^{b})^2(\sigma_{1s}{}^{*})^2(\sigma_{2s}{}^{b})^2(\pi_{2px}{}^{b}=\pi_{2Py}{}^{b})^4(\sigma_{2pz}{}^{b})^2(\sigma_{2s}{}^{*})^2(\pi_{2px}{}^{*}=\pi_{2py}{}^{*})^0(\sigma_{2pz}{}^{*})^0$$

The 2s orbital of carbon and the 2s orbital of oxygen are mixed to produce bonding molecular orbital σ_{2s} and antibondingmolecular orbital $\sigma_{2s}{}^{*}$ which possess σ-symmetry about the internuclear CO axis (z-axis). Similarly, σ_{2pz}(bonding) and $\sigma_{2pz}{}^{*}$ (antibonding) molecular orbitals are formed by the overlap of $2p_z$ orbitals of carbon and oxygen directed along this axis. The $2p_y$ orbitals, however, overlap to form orbitals of π-symmetry, ($\pi_{2Py}{}^{b}$ and $\pi_{2py}{}^{*}$) about the z axis. The orthogonal $2p_x$ orbitals act similarly, resulting in doubly degenerate molecular orbitals $\pi_{2Px}{}^{b}$ and $\pi_{2px}{}^{*}$.

5.6.1 Formation of Mixed Atomic Orbitals

CO belongs to the $C_{\infty h}$ point group. All the orbitals possessing σ-symmetry can combine, despite they sourced from s or p orbitals, as they all correspond to the same symmetry class(σ^+). Due to the small difference in energy between 2s and 2p orbitals for carbon, the 2s and $2p_z$ orbitals on carbon combine to form two sp_z mixed atomic orbitals. The high-energy sp_z orbital of carbon is close to its original $2p_z$ orbital and thus has significant participation from it, whereas its low energy sp_z orbital is close to its unmixed 2s orbital and thus has significant participation from it. In the same way, two mixed sp_z orbitals are produced from the mixing of 2s and $2p_z$ orbitals of oxygen, perhaps not to the extent as the orbitals of carbon, since the difference in energy between the 2s and $2p_z$ orbitals of oxygen is greater than that the same atomic

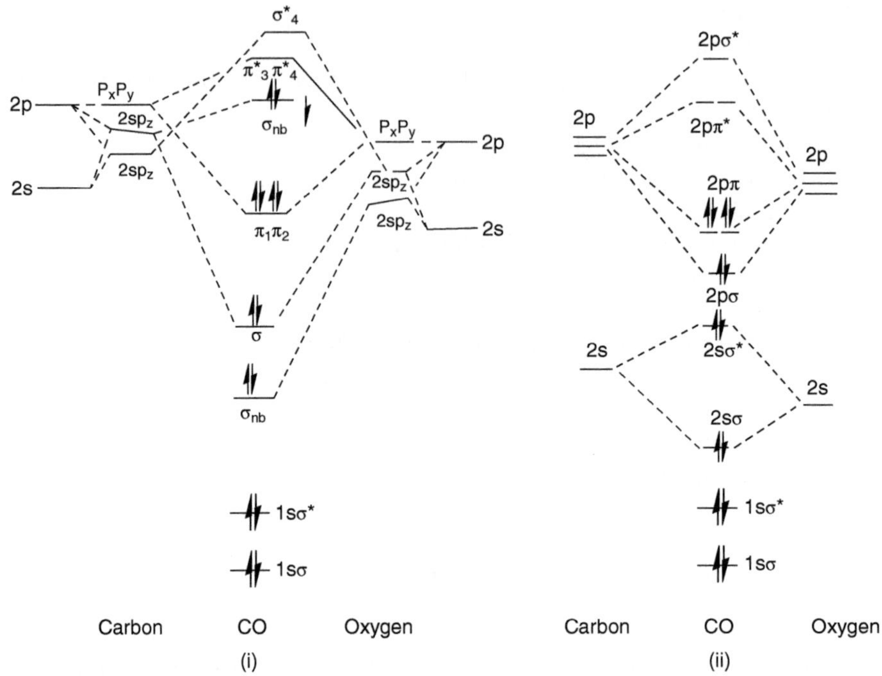

Figure 5.10 Molecular orbital diagram of CO (i) s and p mixing (ii) s and p mixing neglected.

orbitals of carbon. The higher energy sp_z orbital of oxygen is close to its unmixed $2p_z$ orbital and contributes significantly, whereas the low energy sp_z orbital is closer to its original 2s orbital and makes a significant contribution.

Since oxygen is more electronegative than carbon, its orbitals have lower energies than those of carbon. The low energy mixed sp_z orbital of carbon has equivalent energy to the high-energy mixed sp_z orbital of oxygen. The linear combination of these two mixed orbitals of the same symmetry and equivalent energy produces σ-bonding and antibonding molecular orbitals. The mixed sp_z orbital on carbon and the low energy sp_z orbital on oxygen possesses a larger difference in energy. Despite having the same symmetry, these orbitals do not overlap and thus remain almost nonbondings and are confined to oxygen and carbon atoms.

Though the high-energy sp_z orbital of carbon is nonbonding, it possesses some antibonding nature. Furthermore, it is closer to its p_z orbital and possesses the characteristics of its $2p_z$ orbital. Since the nonbonding sp_z orbital of oxygen has closer energy to its 2s orbital, it largely has the s-character. Both the nonbonding orbitals are populated by lone pairs of carbon and oxygen and are pointed away from the CO bond. The degenerate π-symmetric $2p_x$, $2p_y$ orbitals of oxygen combine to form two degenerate π-bonding and two π*-antibonding orbitals of the same energy (Figure 5.11).

The high-energy atomic orbital of carbon is highly involved in furnishing anti-bonding molecular orbitals, whereas the low-energy atomic orbital of oxygen has

Figure 5.11 (i) π-bonding MO (ii) π*-antibonding MO.

more impact on the bonding molecular orbitals. As a result, the bonding molecular orbitals of CO have the character of electro negative oxygen's orbitals, whereas the antibonding molecular orbitals exhibit the nature of the orbitals of carbon because of orbital conservation.

The carbon and oxygen atoms both have doubly filled and singly filled sp_z orbitals. A σ-bond and a π-bond, respectively, are produced when the sp_z and p_x orbitals of both atoms are overlapped (Figure 5.12i). As a result, the p_y orbital on carbon is left unoccupied, whereas the p_y orbital on oxygen is double inhabited. The donation of a lone pair of electrons from the oxygen's p_y orbital to the carbon's open p_y orbital results in the formation of yet another π-bond (Figure 5.12ii). This dative bond formation results in C^-—O^+ polarization. However, the greater electronegativity of oxygen causes the partial C^+—O^- polarization of the three bonding orbitals, which cancels exactly the C^-—O^+ polarization of the C≡O ligand.

As discussed previously, oxygen is more electronegative than carbon, and its orbitals have lower energies than those of carbon. Hence, the highest occupied molecular orbital (HOMO) is an σ-orbital confined to carbon, and it is not surprising that the metal binds to carbon and not oxygen. Furthermore, the CO ligand has two degenerate π* orbitals as the lowest unoccupied molecular orbitals (LUMO), which are also centered at carbon.

The bonding in metal carbonyls has usually been discussed in terms of two different interactions [16].

Figure 5.12 Bonding interactions in CO.

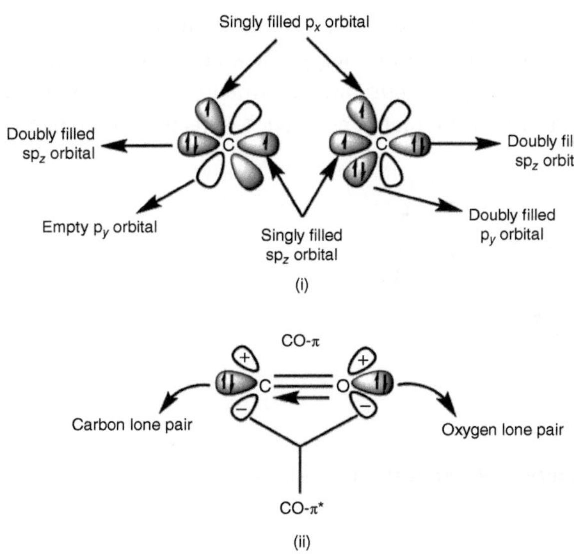

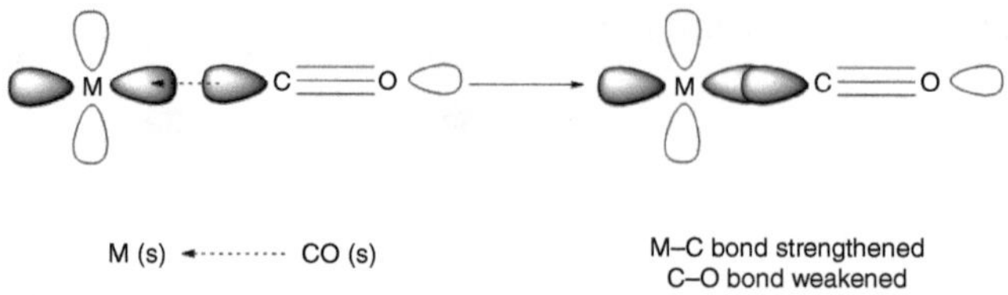

Figure 5.13 Formation of M—C σ-bond.

(i) Donation of a lone pair of electrons from the HOMO of the CO molecule to the metal's LUMO of appropriate symmetry (directed along the axis) to produce the $(d_\sigma)M \leftarrow C\equiv O(\sigma)\sigma$-bond.
(ii) Transfer of electrons from the HOMO of metal, the filled $(d_\pi)M$ to the LUMO, the vacant antibonding π* orbital of CO. Since electron density flows from the d-orbitals of metal to π* orbital of CO (Figure 5.13), this bonding is termed as back bonding, and the CO ligand is referred to as a π–acceptor orπ-acid (Figure 5.14).

The formation of this σ-bond removes the electron density from the carbon of CO and the M—C bond is strengthened. The formation of the π -bond enhances the electron density at both carbon and oxygen, since the π* of CO possesses both the characteristics of C and O. As a result, carbon turns more positive in bonding while oxygen gets more negative (M—C bond is strengthened and C—O bond is weakened). This leads to polarization of the CO upon binding. The CO ligand was chemically activated by the metal-induced polarization. It renders carbon highly susceptible to nucleophilic attack and oxygen more susceptible to electrophilic attack. The presence of the other coordinated ligands to metal and the overall charge of the complex will modify the polarization [17].

Although the occupation of the energetically high-lying π*orbital of CO may destabilize the system, it must be remembered that such 2π* levels have comparable in energy to 3d, 4d, or 5d orbitals of metal, and thus back-bonding results in a significant overall gainin energy. Back bonding becomes ineffective and CO is removed when the d-orbitals have contracted (i.e. lowered in energy) by placing positive charges on the metal, as in Ni^{2+}. This is demonstrated by the disproportionation of

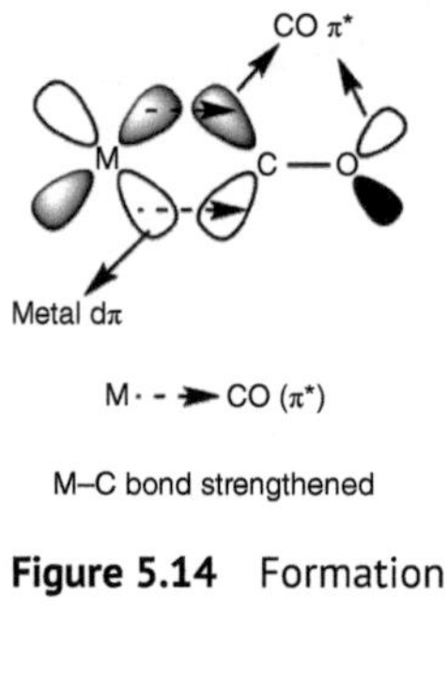

Figure 5.14 Formation of π–back bonding.

$Co_2(CO)_8$ in polar solvents: CO is still bound to Co^{-1}, but not Co^{2+}:

$$Co_2(CO)_8 \xrightarrow{\text{pyridine}} 2[CO(py)_4]^{2+}\left[Co(CO)_4{}^-\right]_2 + 8\,CO$$

5.7 Synergistic Effect

Both the σ-bond and π-bond reinforce one another. The σ-bond formation increases the electron density on the metal, making the CO ligand more positive, δ^+ (removal of the electron density from the carbon of CO). The excess electron density on metal as well as the positive charge on carbonyl ligand both boosts the π-accepting capability of CO. The greater the metal's electron density and the greater the δ^+character of carbonyl ligand facilitate the back donation from metal to ligand, which in turn boosts the flow of electrons from ligand to metal. As a result of this two-way electron transfer, the M—L bond is strengthened. The extent of the strength of the two-way bonding is greater than the sum of the individual M←L σ-bonding and M→L π-bonding. This type of concerted reinforcement of σ- and π-bonding is termed as synergistic effect.

As a result of the back donation, both the M—C bond strength and bond order increases. Furthermore, the back-bonding provokes the occupancy of π* of CO ligand, the C—O bond order decreases, making the C—O bond weakened and elongated. Valence bond resonance hybrids also result in equivalent explanations and demonstrate that the M—C bond order is greater than 1 and the C—O bond order is less than three (Figure 5.15) as follows.

When metal is at a higher oxidation state, it shows poor π -back bonding. Then, the M—C bond order is nearer to one and that of C—O is near three (M—C≡O). There will be greater π back-bonding between the metal and carbonyl ligand when the metal is electron-rich (at a lower oxidation state), and both the bond orders of M—C and C—O are nearer to two (M=C=O) [18].

5.8 Conclusion

Metal carbonyl complexes have been the topic of extensive research for many years because they play a significant role in a variety of catalytic processes as well as industrially relevant chemical synthesis. This chapter delves into the chemistry of metal carbonyl complexes by embodying their different synthetic methods, classifications, and physical, and chemical properties. The common reactions of metal carbonyls, such as replacement reactions, metal-centered oxidation,

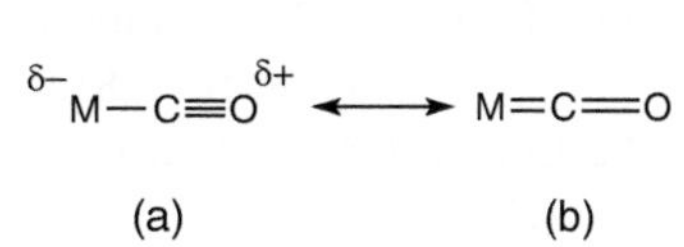

Figure 5.15 Valence bond resonance structures for M—CO bond order.

or reduction reactions, and disproportionation reactions, have been described. Furthermore, the constitution and binding of mono, bi, and poly metallic carbonyls have been discussed. In addition, the sigma donation and pi back bonding in metal carbonyl have been clearly explained by the molecular orbital theory and synergic metal-carbonyl orbitals overlap.

Further Reading

The concepts covered in this chapter, "Metal Carbonyls: Synthesis, Properties, and Structure," are a portion of a graduate program. The majority of textbooks with titles relevant to organometallic chemistry can serve as a resource for additional reading. There are also various web resources available for additional learning. Some examples are provided below.

1. https://en.wikipedia.org/wiki/Metal_carbonyl
2. https://onlinelibrary.wiley.com/doi/abs/10.1002/0471238961.0301180223010714.a01.pub2
3. https://onlinelibrary.wiley.com/doi/abs/10.1002/0471227617.eoc137
4. https://www.sciencedirect.com/topics/chemistry/metal-carbonyl
5. http://nptel.ac.in/courses/104106064/lectures.pdf

References

1 Trout, W.E. (1937). The metal carbonyls. I. History; II. Preparation. *J. Chem. Educ.* 14 (10): 453.

2 Schützenberger, P. (1868). Sur Quelques Réactions donnant lieu à la Formation de l'Oxychlorure de Carbone. *Compt. Rendus* 66: 747–748.

3 Mond, L., Langer, C., and Quincke, F. (1890). L.–Action of carbon monoxide on nickel. *J. Chem. Soc. Trans.* 57: 749–753.

4 Bhatt, V.D. (2014). *Selected Topics in Co-ordination Chemistry*. National Programme on Technology Enhanced Learning (NPTEL).

5 Heiber, W. and Anorg, Z. (1940). *Allgem. Chem.* 245: 295.

6 Ellis, J.E. (1990). *Adv. Organomet. Chem.* 31: 1.

7 Heiber, W., Fischer, E.O. et al. (1952). *Z. Anorg. Allgem. Chem.* 269: 308.

8 Baird, M.C., Osborn, J.A., and Wilkinson, G. (1966). *Chem. Commun.* 129.

9 Elschenbroich, C. (2016). *Organometallics*. Wiley.

10 Holleman, A.F. (2019). *Lehrbuch der Anorganischen Chemie*. Walter de Gruyter GmbH & Co KG.

11 McFarlane, W., Wilkinson, G., and Hübel, W. (1966). Triiron dodecacarbonyl. *Inorg. Synth.* 8: 181–183.

12 Cotton, F.A., Wilkinson, G., and Gaus, P.L. (1995). *Basic Inorganic Chemistry*. Wiley.

13 Abel, E.W. and Stone, F.G.A. (1969). The chemistry of transition-metal carbonyls: structural considerations. *Q. Rev. Chem. Soc.* 23 (3): 325–371.

14 Frenking, G., Fernández, I., Holzmann, N. et al. (2021). Metal–CO bonding in mononuclear transition metal carbonyl complexes. *JACS Au* 1 (5): 623–645.

15 Ponec, R., Lendvay, G., and Chaves, J. (2008). Structure and bonding in binuclear metal carbonyls from the analysis of domain averaged Fermi holes. I. $Fe_2(CO)_9$ and $Co_2(CO)_8$. *J. Comput. Chem.* 29 (9): 1387–1398.

16 Crabtree, R.H. (2009). *The Organometallic Chemistry of the Transition Metals.* Wiley.

17 Bochmann, M. (1994). *Organometallics: Complexes with Transition Metal-Carbon [Sigma]-Bonds*, vol. 1. Oxford University Press on Demand.

18 Hocking, R.K. and Hambley, T.W. (2007). Database analysis of transition metal carbonyl bond lengths: insight into the periodicity of π back-bonding, σ donation, and the factors affecting the electronic structure of the TM−C⋮O moiety. *Organometallics* 26 (11): 2815–2823.

6

Metal–Carbon Multiple Bonded Compounds

Rakesh Kumar Sahu[1,2], Mukesh Kumar Tyagi[3], and Rajesh Haldhar[4]

[1] *BCS GOVT. PG College Dhamtari, Department of Chemistry, Chhattisgarh 493773, India*
[2] *Pt. Ravishankar Shukla University, Raipur, Chhattisgarh 492010, India*
[3] *Atal Bihari Vajpai, Govt. College, Department of Chemistry, Pandatarai, Kabirdham, 491559, India*
[4] *Yeungnam University, School of Chemical Engineering, Gyeongsan 712749, South Korea*

6.1 Introduction

Cacodyl oxide and tetramethyl diarsine and then the first π complex, namely Zeise's salt, $K[PtCl_3(\eta^2\text{-}C_2H_4)]$ discovered in 1827, after which during the last five decades, metal–carbon bonds have grown significantly as a separate field of study [1–4]. Compounds with metal–carbon bonds are called organometallic compounds that have one or more metal–carbon bonds, including compounds with both σ and π bonds between metal atoms and carbon. It contains a variety of chemical compounds and their reactions [1–5].

Transition metal complexes consist of metal–carbon single bonds, double bonds, or triple bonds, involving a significant fraction of organotransition metal chemistry. Their importance stems from the fact that they are intermediates in a wide variety of transition metal-catalyzed processes [4–6]. Over the past 30–40 years, complexes formally consist of a metal–carbon double bond or a metal–carbon triple bond, especially many metal–carbon double bond; organometallic compounds form useful catalysts and have become established as stoichiometric reagents in organic chemistry. Many of these advances have been made over the past decade and active investigations are still being carried out today in many areas and, consequently, are of significant industrial interest. In this chapter, we describe the current details of a variety of organometallic compounds and organic ligands and how they bond to metals [7, 8].

A characteristic feature of metal atoms bonded to organic ligands, especially CO, is that they often exhibit the ability to form covalent bonds with other metal atoms to form cluster compounds. These groups can contain only as many as two or three metal atoms or more. There is no limit to their size or variety. They may have single, double, triple, or quadruple bonds between carbon and metal atoms and, in some cases, ligands that bridge two or more metals [8–12]. Examples of metal cluster compounds containing organic ligands are shown in Figure 6.1.

Organometallic Compounds: Synthesis, Reactions, and Applications, First Edition.
Edited by Dakeshwar Kumar Verma and Jeenat Aslam.

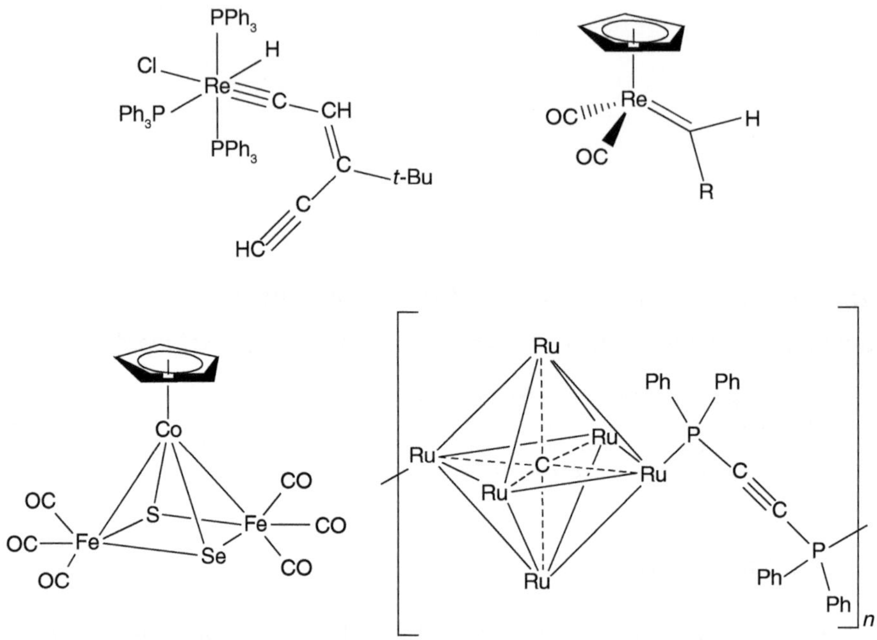

Figure 6.1 Example of organometallic cluster of carbon–metal multiple bond.

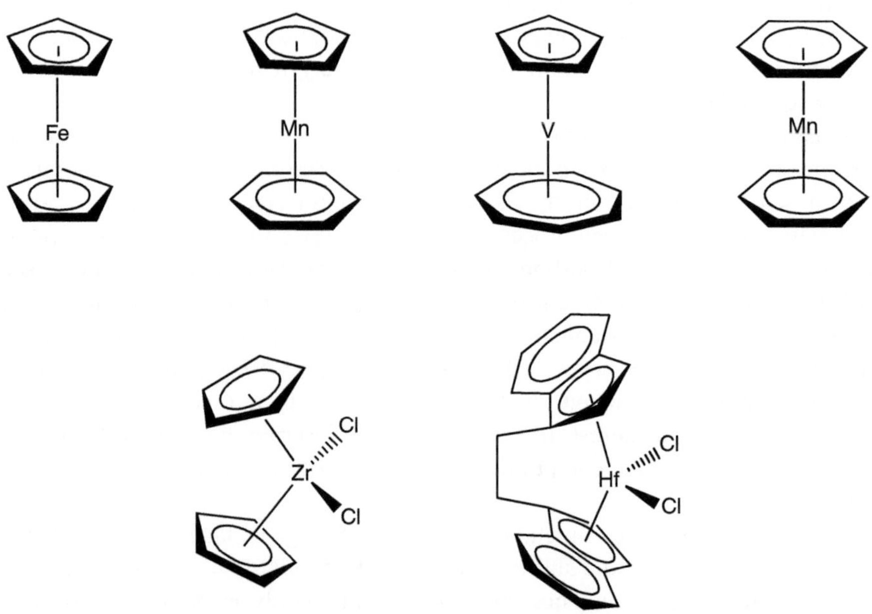

Figure 6.2 Example of sandwich organometallic compounds.

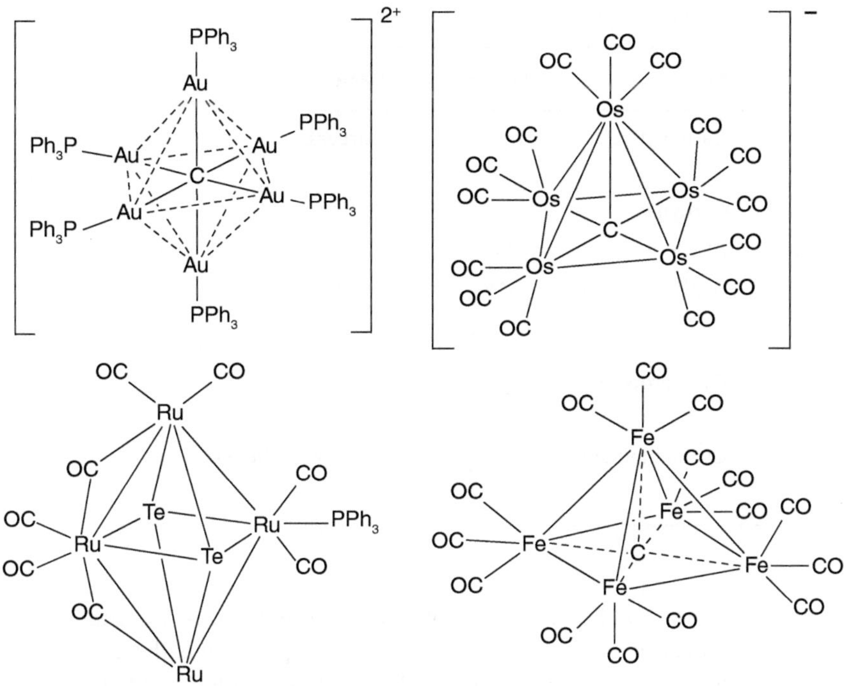

Figure 6.3 Example of carbide cluster.

Some organometallic compounds have similarities to the types of coordination compounds, e.g. $Cr(CO)_6$ and $[Ni(H_2O)_6]^{2+}$, both of which are octagonal. Both CO and H_2O are σ-donor ligands; additionally, CO is a strong π-acceptor. A similar way that can exhibit both behaviors includes CN^-, PPh_3, SCN^-, and several organic ligands. Such cyclic organic ligands that have delocalized π systems can combine with metal atoms to form sandwich compounds, as shown in Figure 6.2.

Carbon itself may play a significantly different role in organic chemistry. Some metal clusters contain carbon atoms, forming carbon-centered clusters, often called carbide clusters; in some instances, carbon is bound to five, six, or more metals. Examples of carbide clusters are shown in Figure 6.3.

This chapter shows many additional examples of the diversity of molecular structures in organometallic compounds, with interesting structures and chemical properties found in [8–14].

6.2 Nomenclature

It is named just as the nomenclature of the coordination compound but many of its ligands can bond to the metal through different numbers of atoms depending on the molecule, as shown in Table 6.1, for which special nomenclature has been devised to designate the way ligands are bound to metal atoms; the atoms of a ligand that bond to the metal are indicated by the Greek letter η(eta) followed by

Table 6.1 General organic ligands.

Ligand	Name	Ligand	Name
CO	Carbonyl	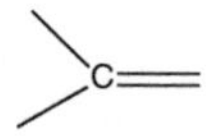	Carbene
$-C\equiv$	Carbyne	$-CR_3$	Alkyl
$H_2C=CH_2$	Ethylene	$HC\equiv CH$	Acetylene
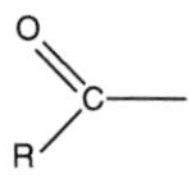	Acyl		Benzene
	Cyclopropenyl		π-allyl
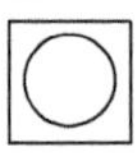	Cyclobutadiene	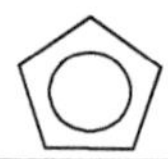	Cyclopentadienyl

the number of atoms bond to the metal, which is represented in the superscript. For example, ferrocenes, bounded to the cyclopentadienyl ligands through all five atoms, are designated as (η^5-C_5H_5). Hence, ferrocene is written as (η^5-C_5H_5)$_2$Fe. Similarly speaking, at the time of η^5-C_5H_5 ligands, it has been designated as pentahaptocyclopentadienyl ligand. Hapto comes from the Greek word for tightness; therefore, pentahapto means "bonding in five places." Hence, the ferrocene will be called bis-pentahaptocyclopentadienyl iron(II) or bis-(η^5-cyclopentadienyl) iron(II) [1, 12, 13].

According to this system, the corresponding formula and name are shown in Table 6.2.

In the case of other coordinating compounds of organometallic chemistry, bridging ligands that are very common are represented by the prefix μ followed

Table 6.2 General organic ligands name and hapticity.

Number of bonding atoms	Name	Formula	Example
1	Monohaptocyclopentadienyl	η^1-C_5H_5	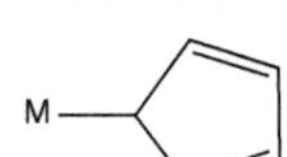
3	Trihaptocyclopentadienyl	η^3-C_5H_5	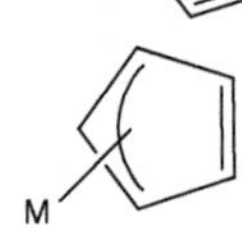
5	Pentahaptocyclopentadienyl	η^5-C_5H_5	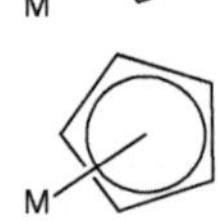

Table 6.3 Bridging carbonyl ligands.

Number of atoms bridged	Formula
Terminal	CO
2	μ_2-CO
3	μ_3-CO

by the number in the subscript to denote the number of metal atoms [1, 12, 13]. For example, bridging carbonyl ligands is represented in Table 6.3.

6.3 Classifications

6.3.1 Metal–alkylidene Complexes

There is a double bond between metals and carbon, and such types of complexes are called carbene or alkylidene. Presently, alkylidene complexes of wide range of ligands have been known for the majority of transition metals.

$$M{=}C{-}R$$

where R = aryl, alkyl, H, SiMe:, NEt_2, PMe_3, SPh, Cl, etc.

IUPAC has recommended that the term "alkylidene" be used to define all complexes containing metal–carbon double bonds [1, 8, 12–14].

6.3.2 Metal–alkylidyne Complexes

In organometallic chemistry, there is a triple bond between metal and carbon, it is called carbyne or alkylidyne complex. It corresponds to alkynes. Until now, numbers of alkylidyne complexes have been well-known. The following are examples of carbyne ligands:

$$M \equiv C{-}R$$

where R = aryl, alkyl, H, SiMe:, NEt_2, PMe_3, SPh, Cl, etc.

IUPAC has recommended that "alkylidyne" can be used to assign complexes containing metal–carbon triple bonds [1, 8, 12–14].

6.4 Structure

6.4.1 Alkylidene (Carbene)

The polarization of the M=C bond can be shown in two ways, indicating the flexibility of the electronic properties of the transition metal complex.

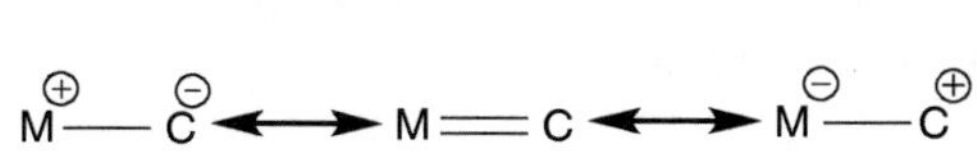

In cationic or neutral complexes, the Pettit type and the Schrock type of carbene can be distinguished. In Pettit type, electrophilic carbenes are coordinated with transition metals to some extent as carbocations, and in Schrock type, nucleophilic carbenes react as ylids in neutral metal complexes (Figure 6.4).

It has been suggested that the electrophilic carbene that forms a complex with metal is a singlet carbene. However, the metal complex obtained by the nucleophilic carbene is a triplet carbene. In the ground state, the free methylene is found in the triplet state and, in the excited state, there is a low-energy single state. The carbene that is associated with the electron-withdrawing substituent, such as Fischer and Arduengo carbene, is singlet, although Pettit's type of iron methylene complex contains electrophilic methylene that can be explained as positive charge and electron-withdrawing carbonyl ligands (Figure 6.5).

In the alkylidene ^{13}C NMR spectroscopy, the carbon resonance peak always appears in a low field (200–300 ppm, reference to $SiMe_4$). Such a large-scale

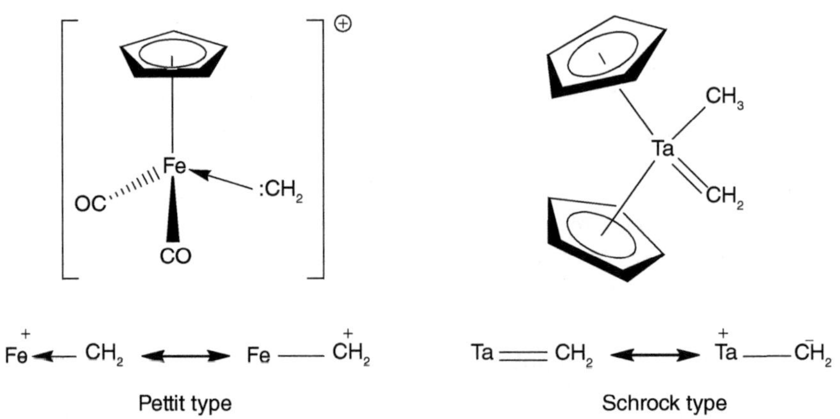

Figure 6.4 Example of Pettit's- and Schrock-type metal complexes.

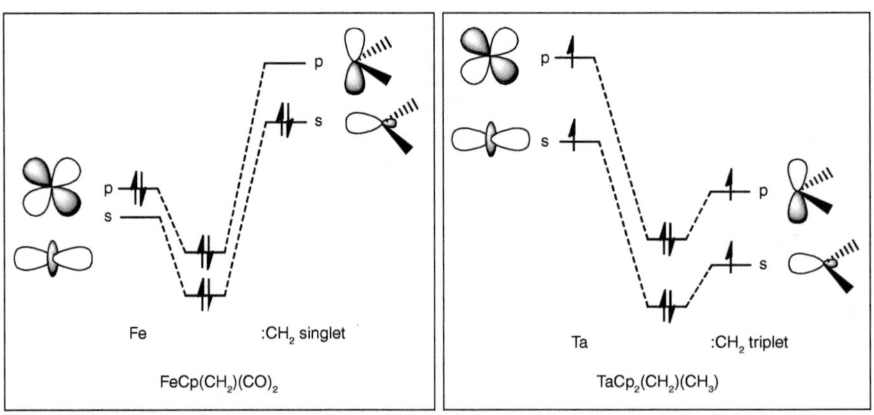

Figure 6.5 MOT diagram of singlet and triplet carbene.

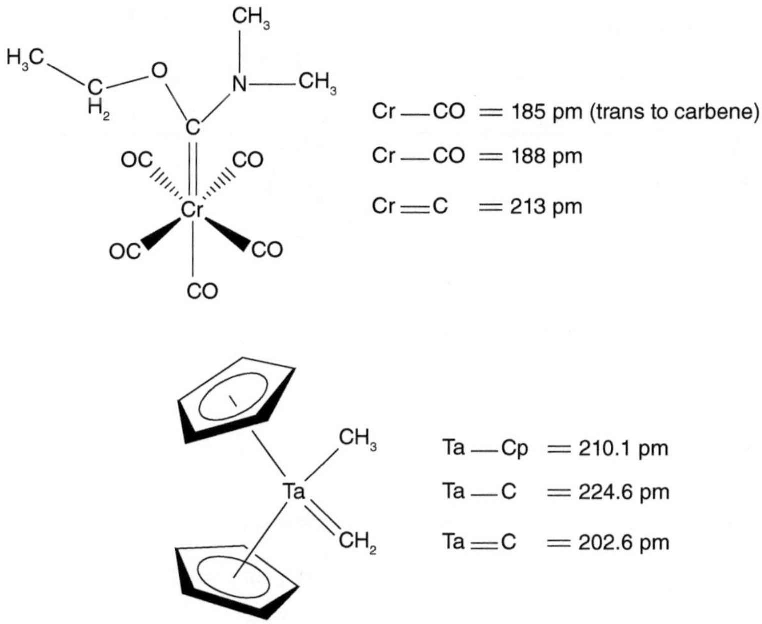

Figure 6.6 Typical alkylidene complexes.

chemical shift is the characteristic of a carbon compound in which carbon forms multiple bonds. It shows that the carbon is bonded to the metals with multiple bonds.

It is observed that the M—C bond order in the alkylidene complexes is generally less than two [1, 8, 12–14]. Typical alkylidene complexes are shown in Figure 6.6.

6.4.2 Carbyne (Alkylidyne)

The carbyne ligand CR can be classified into two types: the first LX, and the second X_3. LX is known as Fischer's carbyne, while X_3 is known as Schrock's carbyne. Both types follow the same logic, which means that both are derived from the electronic structure of the free carbyne.

The alkylidyne or carbyne ligand, a monovalent carbon species $[:C\text{–}R]^+$ containing two empty p-orbitals, is able to accept electron density from a metal in a typical π-acceptor manner as well as a σ-donor, because an alkylidyne or carbyne has two empty p-orbitals in which the metal can accept two sets of π-electrons, resulting a kind of regular bond, such as $[(OC)_5M{\equiv}CR]^+$ (Figure 6.7).

The carbon of the alkylidyne complex is "sp" hybridized; accordingly, it can be assumed that the metal–carbon bond, M≡CR, will have a significantly shorter bond length and M—C—C bond angle of ~180°. X-ray structural studies generally confirm this, although M—C—C bond angle variations have been found in some cases [1, 8, 12–14]. Structures of few typical complexes are shown in Figure 6.8.

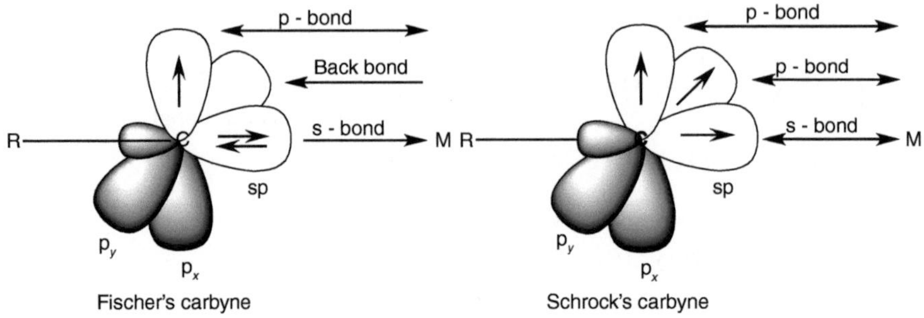

Figure 6.7 Orbital diagram of Fischer's and Schrock's carbyne.

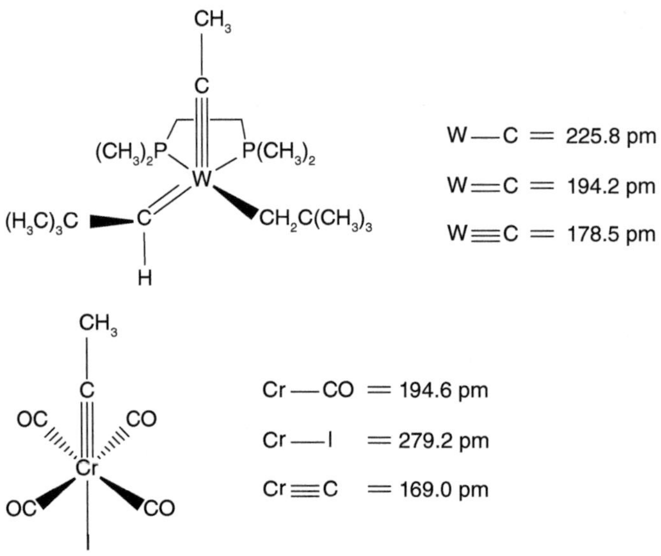

Figure 6.8 Typical alkylidyne complexes.

6.5 Preparation Methods

6.5.1 Metal–alkylidene Complexes

Alkylidene complexes are formed from double bonds between metal and carbon. The carbene complex was first synthesized by Fisher in 1964; now the wide ranges of alkylidene complexes of transition metals are known. Most of such complexes synthesized by Fisher have one or two highly electronegative heteroatoms, such as O, N, or S, directly attached to carbene carbon. Such complexes are commonly referred to as Fischer-type carbene complexes and have been studied extensively. Many years after the synthesis of early Fischer carbene complexes, these have been widely studied by Schrock's research group and many others. These are therefore also sometimes referred to as Schrock-type carbene complexes, commonly designated as alkylidenes.

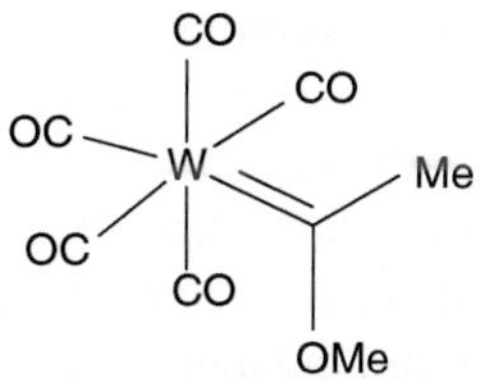

Figure 6.9 First synthesized metal–carbene complex.

In 1964, Fischer and Maasbol formed the first synthesized "metal–carbene" complex (shown in Figure 6.9) by the reaction of phenyllithium with $W(CO)_6$ to form an anionic complex containing the acyl ligand, followed by protonation and treated with diazomethane and proposed its structure. Then, soon hundreds of metal–carbene compounds were formed that contained carbene ligands stabilized by heteroatoms (generally O or N) [1, 8, 12, 15, 16].

6.5.1.1 By Nucleophilic Carbene

The complexes have been synthesized in a significant manner by Schrock and their group in which the metal is in a high oxidation state and is processed by α-hydrogen elimination in an alkyl ligand, which is done as per the following.

By α-elimination of the neopentyl ligand, Schrock obtained the first alkylidene complex $[Ta(CH_2CMe_3)_5]$. The neopentyl ligand is the lack of β-hydrogen as a consequence of the elimination of α-hydrogen, which is due to steric bulk around the metal that results the formation of alkylidene. The α-hydrogen elimination can be explained on the basis of the d^0 electronic configuration of Ta^V atom and sigma bond metathesis that step in whichever after the first or second substitution of Cl by $-CH_2CMe_3$ [1, 8, 12, 17–20].

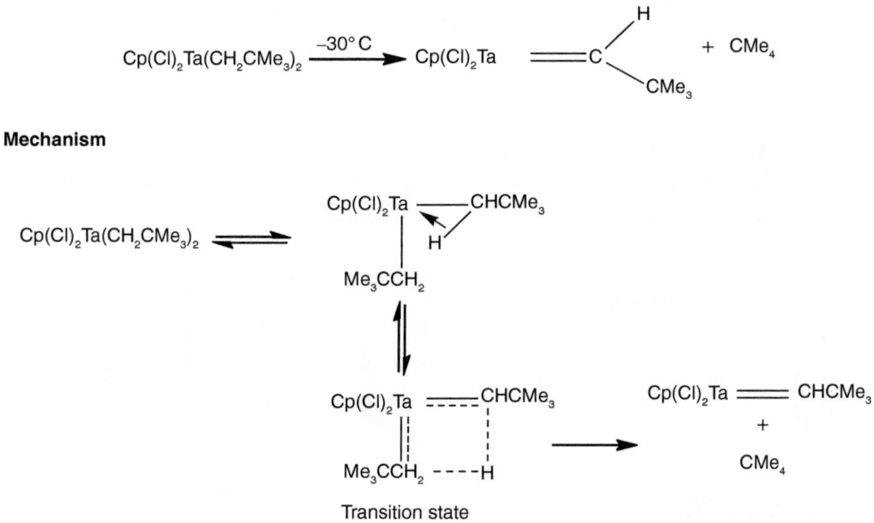

Other examples of complexes of this type were discovered by Schrock's group, wherein α-hydrogen elimination is effectively carried out by the occurrence of the bulky groups around the metal center and, on that occasion, becomes a major foundation of early transition metal chemistry. As follows, it served as a course exhibition for the early transition metal, especially the alkylidene complexes.

Phosphine also induces α-hydrogen elimination by making a bulky group around the Ta atom. The reaction occurs when phosphine is added, which brings the steric bulk around the metal center [19, 20].

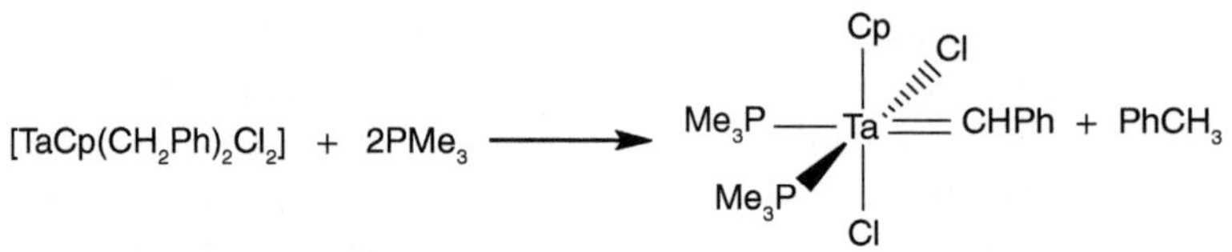

F. N. Tebbe in 1973 discovered the reaction between Cp_2TiCl_2 and excess of $AlMe_3$ to produce methane and **I**. In the structure of **I**, methylene (CH_2) is bridging between Ti and Al, further which give Cp_2TiCH_2 and adduct of AlClMe3–pyridine in the presence of base-like pyridine [17–19].

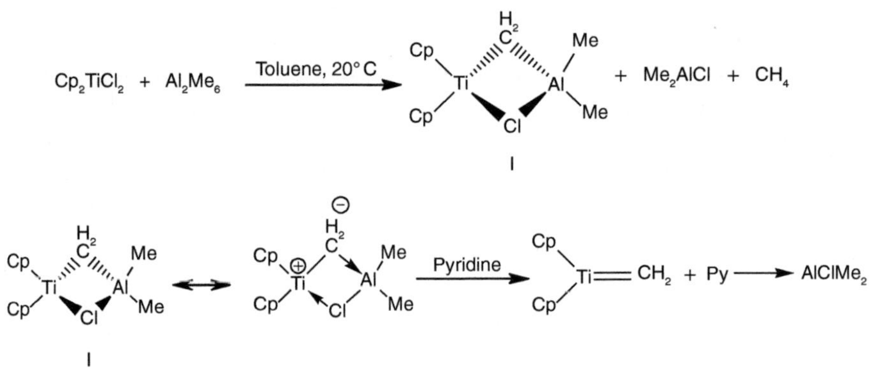

6.5.1.2 By Electrophilic Alkylidenes

Metal–alkyl complexes bear a good leaving group (such as X^-) in the α-position that react with an acid or a Lewis acid to form alkylidenes.

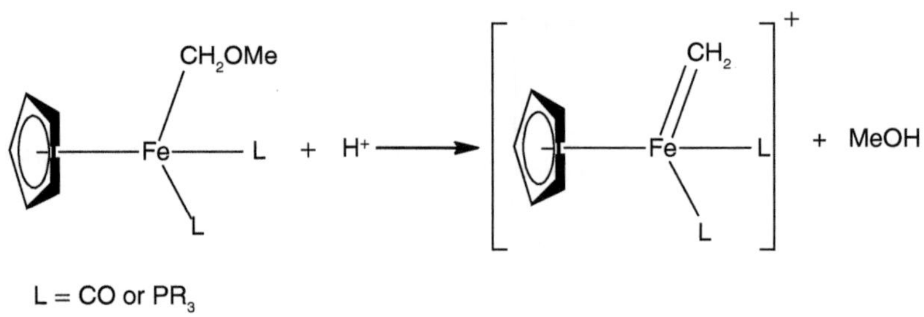

This type of precursor complex also forms the reaction of Fischer carbene with a nucleophile.

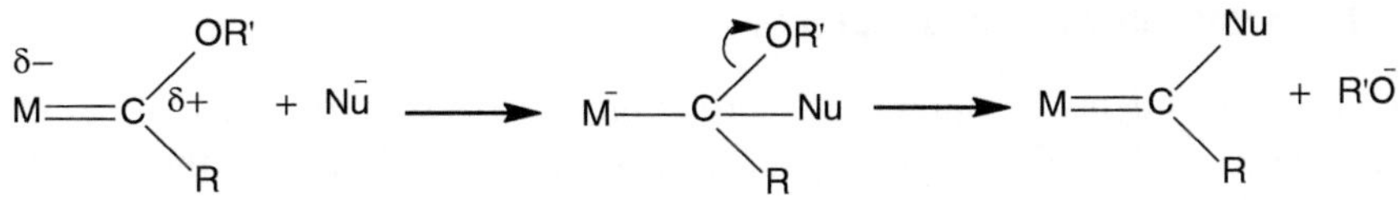

$N\bar{u} = \bar{H}, \bar{R}, R_2\bar{N}, R\bar{S}$

The alkylidene complex obtained from the protonation of the vinyl complex is less stable and thermally rearranged to give an olefin complex.

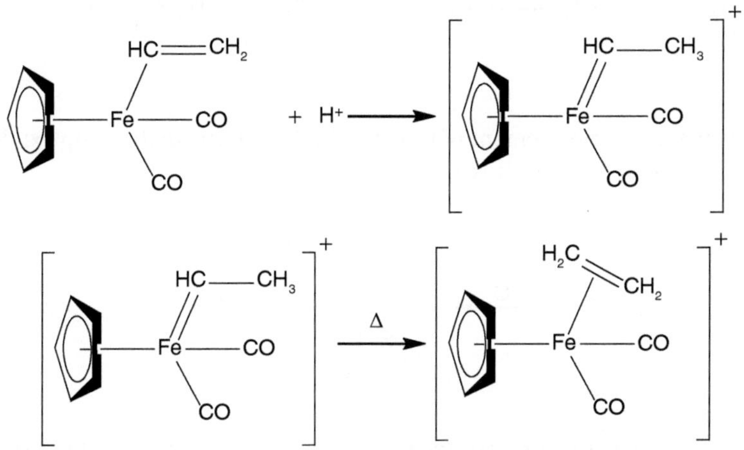

Alkylidene is formed when a complex coordinating a weak ligand as a potential site is reacted with a diazoalkane or another carbene source, as shown in the reaction. Osmium complex reacts with diazoalkane to form Roper's osmium methylene complex, which is at the boundary between electrophilic and nucleophilic [1, 8, 12, 17–20].

$$[Os(PPh_3)_3(Cl)(NO)] + CH_2N_2 \longrightarrow [Os(PPh_3)_2(Cl)(NO)(CH_2)] + PPh_3 + N_2$$

CCl_4 is a source for the dichlorocarbene ligand; in this order, Mansuy illustrated how CCl_4 can be metabolized by cytochrome P-450 and, as a result, revealed the metallocarbenes in the biochemistry aspect for the first time.

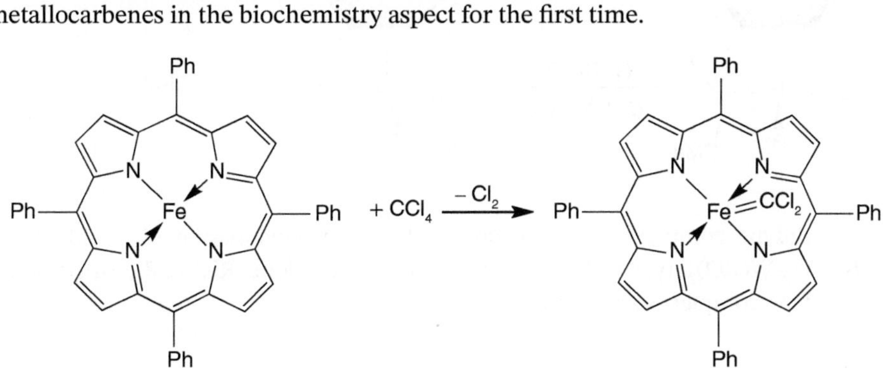

6.5.2 Metal–alkylidyne Complexes

The first alkylidyne complex was serendipity discovered in 1973 by Fischer when they reacted alkylidene complex with Lewis acid, where they tried to replace the methoxy group of the alkylidene complex and attempt to synthesize α halo-substituted alkylidene complexes, although the alkylidyne complex was obtained as a product [1, 8, 12, 17–22].

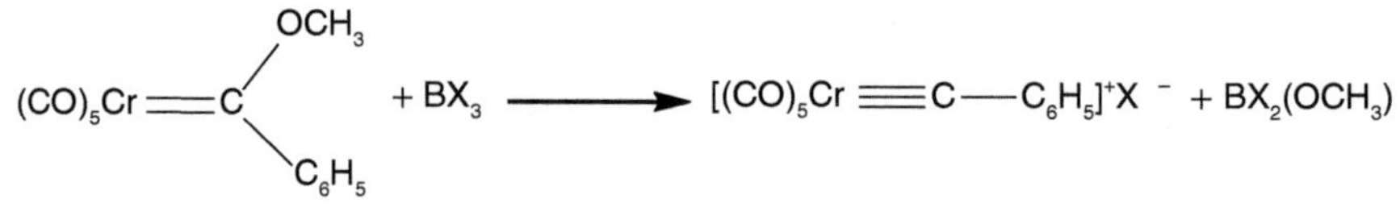

Accordingly, there is a replacement of CO by X in the intermediate complex; the coordination position of X is trans to carbyne.

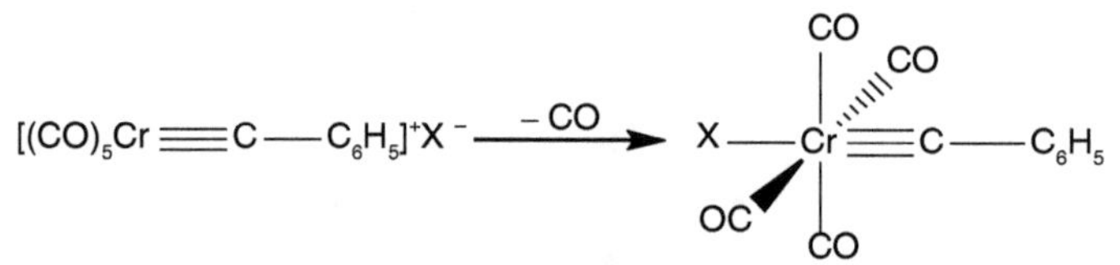

By the dehydrohalogenation process, the primary alkylidene complexes Ln(X)Ta=C(H)R react with a base or reducing agent in the presence of phosphine to form Schrock-type alkylidyne complexes [23–36].

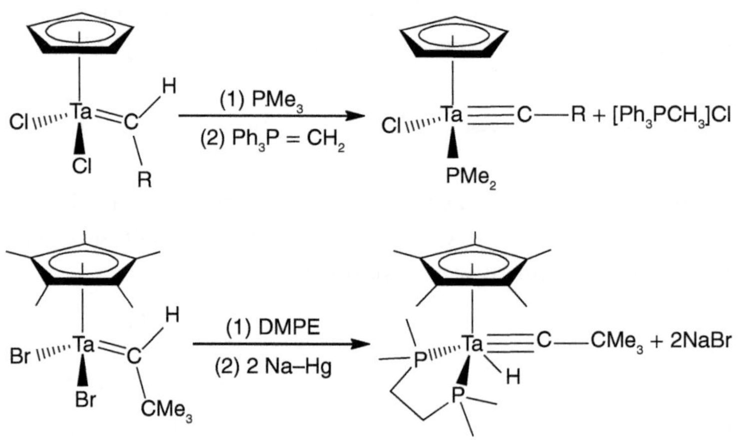

Alkylidyne complexes are also obtained by the bond fission of M≡M. When $(RO)_3W{\equiv}W(RO)_3$ types of complexes react to acetylene R′C=CR′, surprisingly, alkylidyne complexes are formed.

$$(RO)_3W{\equiv}W(RO)_3 + R'C{\equiv}CR' \longrightarrow 2\ (RO)_3W{\equiv}CR'$$

$R = CMe_3$

$R' = -CH_3, -C_2H_5, -C_3H_7$

6.6 Important Reactions

6.6.1 Reaction of Alkylidene Metathesis

Metathesis has previously been studied using homogeneous catalysts, which are made up of Ru and group 6 and 7 transition metals [27–30]. For d^0 Schrock-type catalysts of widespread formulation (X)(Y)M(E)(=CHR_1) (M = Mo or W with E = NR and M = Re with E = CR), whether supported on silica or no longer (Y = siloxy or alkoxy, X = alkoxy, amido, or alkyl), we have formerly proven that the fundamental steps for the catalytic system are alkenes coordination, [2+2]-cycloaddition, and the corresponding reverse steps (cycloreversion–decoordination), and this is regardless of the nature of the metallic (Mo, W, and Re) and the E group (E = imido or alkylidyne). The dissymmetry of the metal center (X ≠ Y) also plays a significant role in increasing the reactivity of these systems since it lowers the transition-state energy for alkenes coordination and raises the energy of metallacyclobutane intermediates (Scheme 6.1) [31–33].

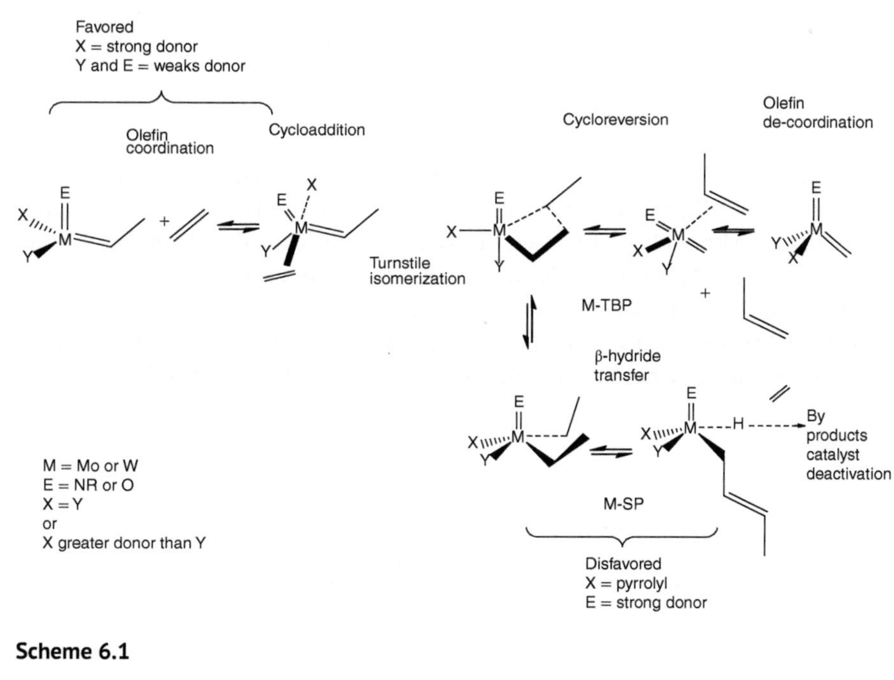

Scheme 6.1

Piers developed an additional type of alkylidene ruthenium compound by protonating ruthenium carbides (complex-1) with a superacid. The obtained 14e–species (complex-2), in contrast to the methylidene active species, were highly stable and could readily be isolated and characterized (Scheme 6.2). Complex 2, which has an empty coordination site trans to the NHC ligand, is very active in simple RCM reactions due to its high initiation rate. For the first time, the ruthenacyclobutane species was observed in the solution when it was reacting with ethylene [34–37].

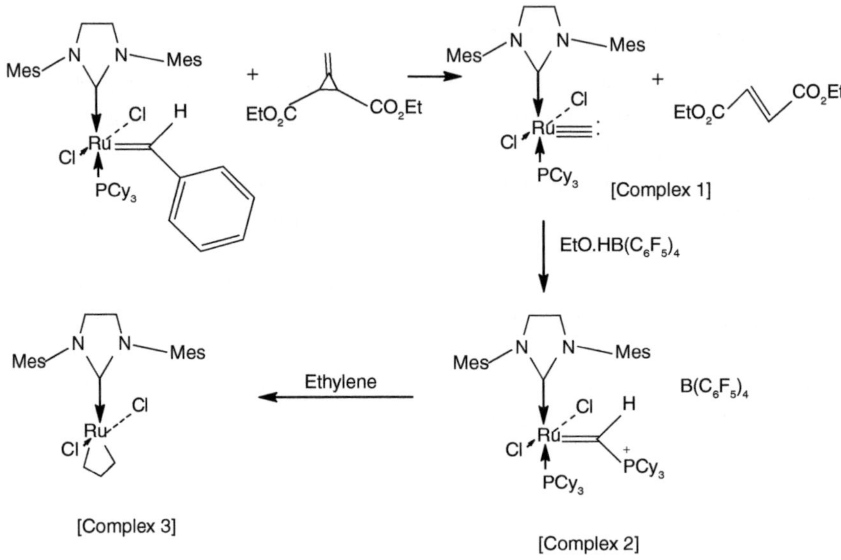

Scheme 6.2

The cross-metathesis (CM) reaction of 1-alkenes with benzylidene compounds can yield simple alkylidene complexes besides the ubiquitous methylidene complex (vide supra). Although olefin metathesis is usually considered an equilibrium process, it may be possible to isolate kinetic products under certain conditions. For example, the complex may be quantitatively converted into the alkylidenes $(PCy_3)_2$-$Cl_2Ru{=}C(H)R$, [R = Me (1), Et (2), *n*Bu (3)] when treated with a tenfold excess of 1-propene, 1-butene, or 1-hexene, respectively (Scheme 6.3) [38–40].

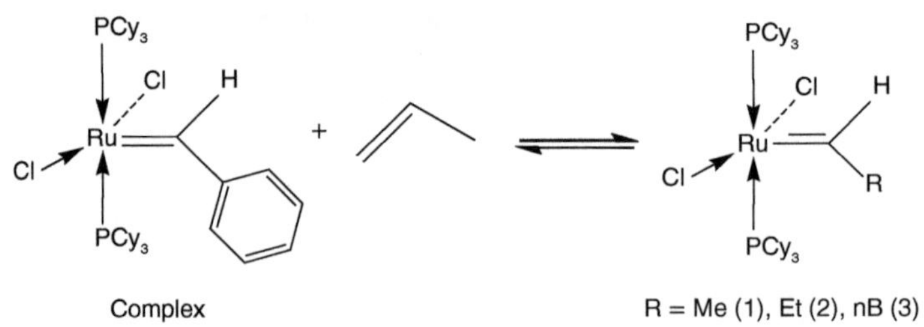

Scheme 6.3

Fischer-type carbenes are another class of alkylidene ligands attached to ruthenium olefin metathesis catalysts. Despite their different reactivity and electronic configurations, Fischer carbenes have been included here because of their role in ruthenium metathesis catalysis. The Grubbs group synthesized a series of electron-rich (ER) carbene complexes at the turn of the millennium [41, 42]. As a result, $(PCy_3)_2Cl_2Ru{=}C(H)R$ was treated with excess $H_2C{=}C(H)ER$ in good yields, yielding Fischer-type carbene complexes (Scheme 6.4).

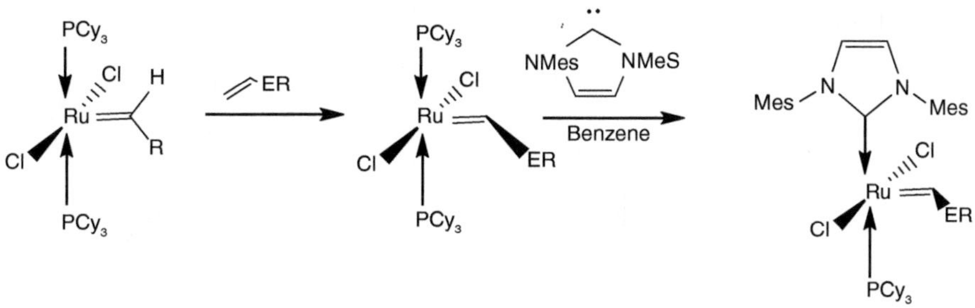

Scheme 6.4

Grubbs' group also developed chelated alkylidenes, such as 1, first prepared by Fürstner, to promote ring expansion metathesis polymerization in this section. With the help of this clever strategy, exclusive cyclic polymers could be prepared (Scheme 6.5) [43].

The addition of excess ethyl vinyl ether to 4 (Scheme 6.6) resulted in the preparation of an important Fischer carbene complex, $(H_2IMes)(PCy_3)Cl_2Ru{=}C(H)OEt$ in 93% yield. A Fischer-type complex predominated over an alkylidene analog at equilibrium. The thermal stability of $(PCy_3)_2Cl_2Ru{=}C(H)ER$ complexes generally followed the trend ER = NR_2 CR_3 SR OR. Furthermore, complexes coordinated with an NHC ligand were thermally more robust than their bis(phosphane) analogous. ROMP was initiated at room temperature with all Ru Fischer-type carbene complexes and quantitative yields of polymer were obtained.

Nevertheless, further studies revealed that the electronic effects may not be as easily predicted due to the intrinsic relationship between steric and electronics and the influence of EWGs and EDGs on both the chelating heteroatom and the alkylidene carbon. Despite the expectations of reduced activity due to an increased electron density on chelating oxygen, electron-donating methoxy groups in the para position improved catalyst stability and activity. A cheap starting material, α-asarone, was used in Scheme 6.7 as a ruthenium metathesis initiator (compound) [44].

(A) formation of a π-complex between the reacting alkene and a coordinatively unsaturated metal ion mechanism, (B) formation of a π-complex followed the transformation of the π-allyl hydride intermediate into a metallacyclobutane (allyl mechanism), (C) formation of a metal-oxo intermediate mediated by a surface Bronsted acid site and subsequent hydrogen shift (H-assisted mechanism), and (D) formation of an oxametallacyclobutane and subsequent elimination of a carbonyl compound (a pseudo-Wittig mechanism).

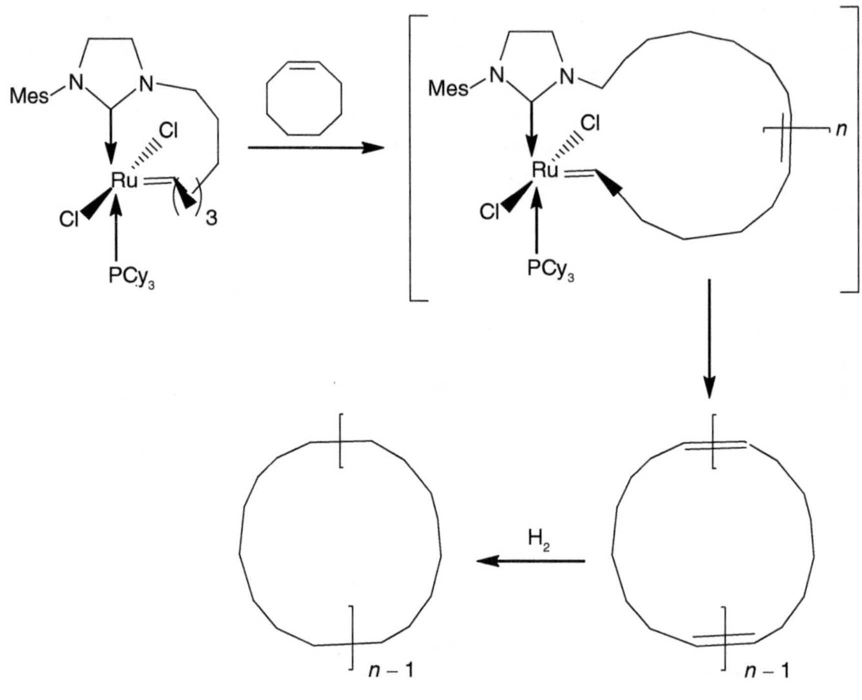

Ring-expansion metathesis polymerization

Scheme 6.5

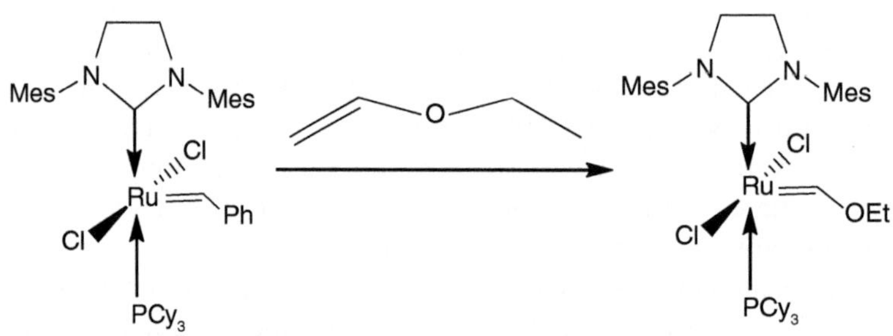

Synthesis of electron-rich carbene complex

Scheme 6.6

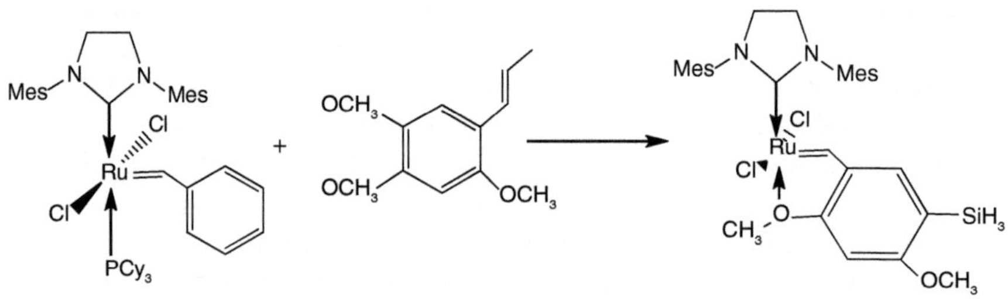

In expensive preparation of a chelated catalyst from natural α-asarone

Scheme 6.7

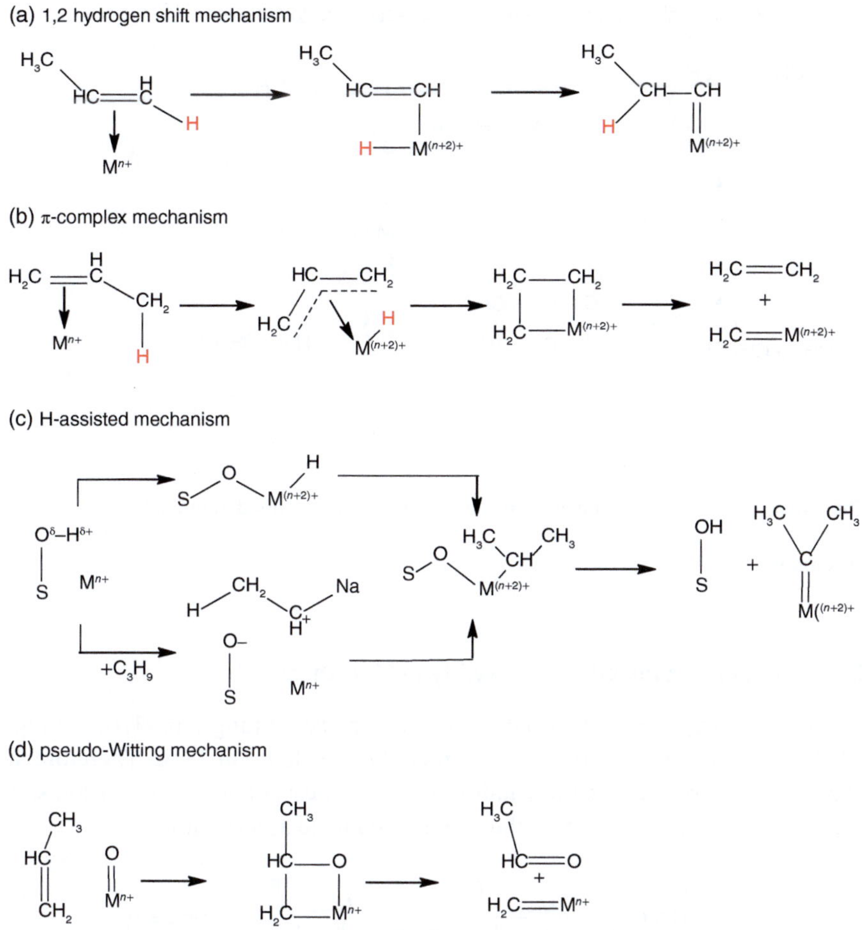

[a]M = active metal center for metathesis reaction, S = support element

Post reaction ethene-d_4 metathesis for active site counting: A modified version of the dynamic active site counting technique originally developed by Handzlik was employed, wherein ethene-d_4 ($CD_2{=}CD_2$) was used instead of nonlabeled ethene as the probe olefin to titrate metal carbene centers after the catalysis. According to the Chauvin mechanism [43] (Scheme 6.8), equal amounts of molybdenum–methylidene ($Mo{=}CH_2$) and molybdenum–ethylidene ($Mo{=}CH{-}CH_3$) species are present under steady-state conditions of propene metathesis on the catalyst surface. The active site counting consists of the quantification of molybdenum–ethylidene ($Mo{=}CH{-}CH_3$) species by titration with ethene-d_4 ($CD_2{=}CD_2$), which results in the liberation of propene-1,1-d_2 ($CD_2{=}CH{-}CH_3$) via metathesis reaction according to reaction and scheme [45].

$$CD_2{=}CD_2 + Mo{=}\underset{H}{C}{-}CH_3 \longrightarrow CD_2{=}\underset{H}{C}{-}CH_3 + Mo{=}CD_2$$

Reaction mechanism of propene metathesis according to Chauvin

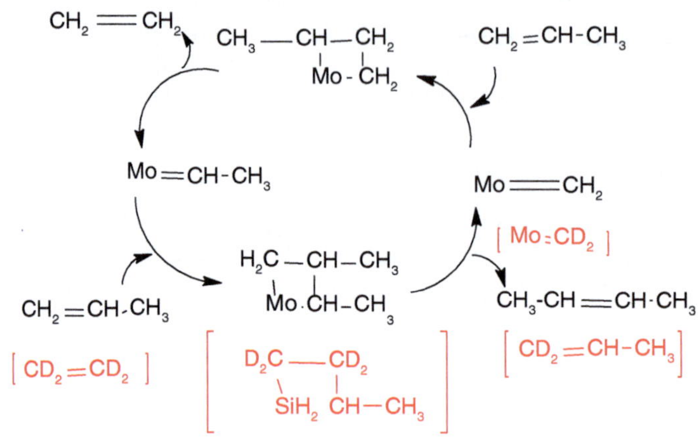

The titration reaction of Mo–ethylidene with ethene-d_4 is described in red color.

Scheme 6.8

6.6.2 Important Reaction of Alkylidyne Metathesis

Recently, we prepared some oxoalkylidene complexes of tungsten(VI) (counting the alkylidene ligand as a dianion), which metathesize olefins at 25 °C. Therefore, if an alkylidyne complex is used to catalyze the metathesis of acetylenes, it too should contain tungsten(VI) (counting the alkylidyne ligand as a trianion).

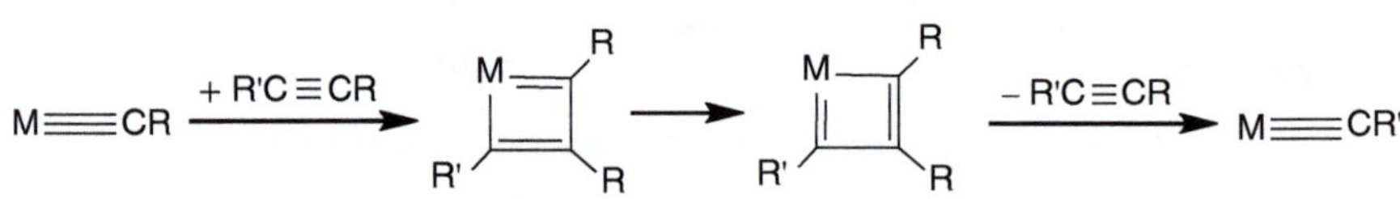

General Procedure: A sealed NMR tube containing 0.5 mL C_6D_6 solution of a complex 1 (10 mM) was injected an excess of 4-(trifluoromethyl)pyridine (7.4 mg, 10 equiv). A total of 5 μL of $PhSiMe_3$ is used as an internal standard. The ratio of the complex and the internal standard was determined by using ^{1}H NMR spectroscopy. The NMR tube was heated at 60 °C and ^{1}H and ^{31}P{^{1}H} NMR spectra were recorded periodically [46].

[1]

Given our ability to generate highly nucleophilic and terminal titanium alkylidynes, it was speculated that nitriles should undergo alkylidyne for nitride exchange. We show that nitriles, such as NtCtBu and NtCAd (Ad) 1-adamantyl), [2+2] cycloadd across the neopentylidyne linkage of intermediate (PNP) TitCtBu (A) (PNP) N[2-P($CHMe_2$)$_2$-4-methylphenyl]$^{2-}$), to afford the first examples of azametallacyclobutadienes [47].

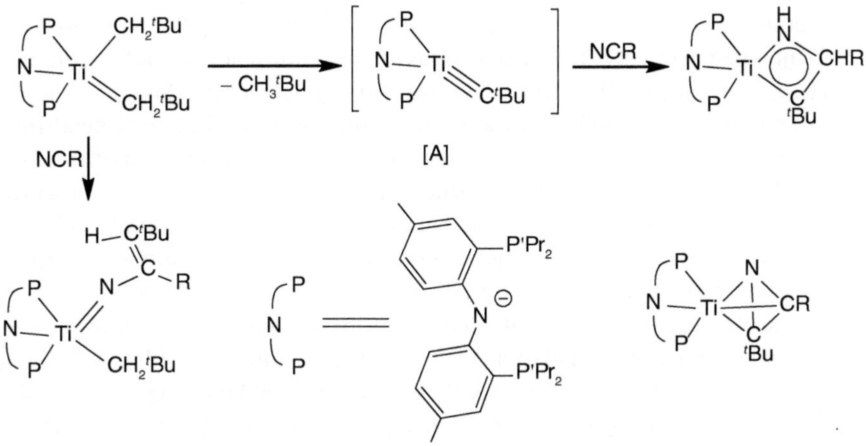

The reversal of nitride vs. alkylidyne ligand preference for these new Mo complexes compared with their W analogs in systems containing both nitrile and alkyne is depicted in Scheme 6.9.

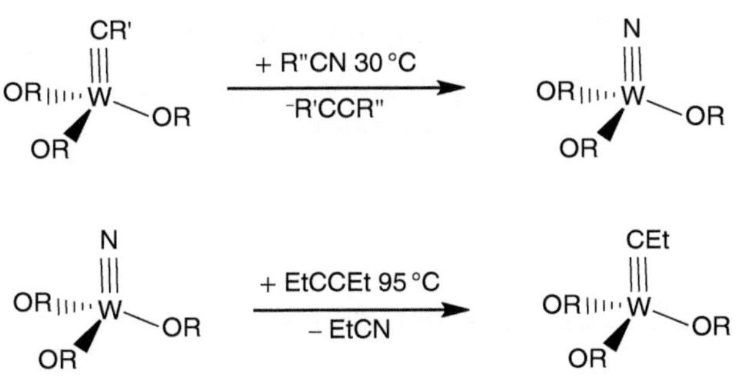

Scheme 6.9

6.7 Applications

(1) Metal–carbon double bonds and bulky supporting ligands are required for alkene and alkyne metathesis reactions, which are highly reactive catalysts for electron-deficient high oxidation state early transition metal complexes.
(2) In spite of the fact that it appears to have multiple aspects, metathesis has emerged as one of the most significant chemical reactions and is now extremely

useful. Having developed inorganic, medicinal, polymer, and materials chemistry beyond the research stage into a familiar tool for these specialists, this area has gone beyond inorganic and organometallic chemistry.

(3) Unsaturated organic fragment protection and stabilization: Protection of arenes, dienes, and alkynes; stabilization of cyclobutadiene, *o*-xylylene, etc.

(4) On the hydrocarbon ligands, there are two types of reactions: nucleophilic and electrophilic. (i) Addition of carbanions to $[Cr^0(\eta^6\text{-arene})(CO)_3]$, $[Fe^{II}(\eta^5\text{-methoxycyclohexadienyl})(CO)_3]^+$, and $[Pd^{II}(\text{allyl})]^+$ (allylic substitution: Tsuji–Trost reaction). (ii) Use of various nucleophilic reactions on the complexes $[Fe(\eta^6\text{-arene})Cp]^+$ (benzylic deprotonation, nucleophilic substitution, heterolytic exocyclic C—O cleavage) for the synthesis of dendrons and dendrimers.

(5) Methods of formation of C—C bonds using the oxidative addition of an organic halide or related electrophile: Due to their non-β-elimination properties, cuprates R_2CuLi readily alkylate halides and related electrophiles RX as in Normant and Alexakis' synthetic process for insect pheromones. The C—C coupling reactions catalyzed by Pd^0 begin with an oxidative addition leading to the $Pd^{II}(R)(X)$ intermediate. The relevant subsequent mechanisms are of two types: (i) transmetallation of an organometallic R′M ($M{=}BR_2''$, SnR_3'', CuR'', ZnR'') on the intermediate $Pd^{II}(R)(X)$ followed by reductive elimination: Suzuki, Stille, Sonogashira, Corriu–Kumada, and Hiyama reactions, etc.; (ii) insertion of an olefin followed by β-elimination: the Heck reaction that, such as the Sonogashira reaction, does not require the preparation of an organometallic compound.

- *Heck reaction*:

$$R^1X + CH_2{=}CHR^2 + NHEt_3 \xrightarrow{\text{cat.}: Pd^0} R^1CH{=}CHR^2 + NEt_3H^+X^-$$

[R¹ = Aryl, Vinyl, Benzyl but not Alkyl]

- *Sonogashira reaction*:

$$ArX + HC{=}CR' + NEt_3 \xrightarrow{\text{cat.}: Pd^0} ArC{=}CR' + NEt_3H^+X^-$$

(6) Extension of Pd catalysis to the formation of C—O and C—N bonds

$$ArX + NuH + B \xrightarrow{\text{cat.}: Pd^0} ArNu + BH^+X^-$$

with NuH = RR′NH or ArOH; B = base

cat: specific bulky phosphines required (or bulky NHC carbene)

(7) Coupling reactions of alkynes with other unsaturated fragments: Vollhardt reaction (cyclization 2+2+2: diyne + alkyne): example of the short synthesis of estrone.

Pauson-Khand reaction (alkyne + alkene + CO ⟶ Cyclopentenone).

Regioselective click reaction of terminal alkynes with azido derivatives catalyzed by CuI in water.

(8) Metal–carbene complexes in organic synthesis:
Dötz reaction, methylenation of ketones, and alkene metathesis (dimerization, cyclization).

(9) Some examples of applications in asymmetric catalysis:

Schrock's asymmetric metathesis catalyst, Sharples's epoxidation of allylic alcohols, Noyori's isomerization of allylic amines; hydrocyanation and carbene transfer: Lewis acid catalysis.

References

1 Didier, A. (2000). *Organometallic Chemistry and Catalysis*. French: Springer.

2 VenkatRamani, S., Ghiviriga, I., Abboud, K.A., and Veige, A.S. (2015). A new ONO^{3-} trianionic pincer ligand with intermediate flexibility and its tungsten alkylidene and alkylidyne complexes. *Dalton Trans.* 44: 18475–18486.

3 Flores, J.A., Cavaliere, V.N., Buck, D. et al. (2011). Methane activation and exchange by titanium-carbon multiple bonds. *Chem. Sci.* 2: 1457–1462.

4 Dötz, K.H., Fischer, H., Hofmann, P. et al. (1983). *Transition Metal Carbene Complexes*. Verlag Chemie.

5 Barnett, H.J. and Hill, A.F. (2021). Benzyne addition to a metal–carbon multiple bond. *Dalton Trans.* 50 (27): 9383–9387.

6 Guerchais, V. and Astruc, D. (1985). New entry to $(C_5Me_5)Fe$ chemistry and the methylene complexes $[Fe(\eta^5\text{-}C_5Me_5)(CO)(L)(=CH_2)]^+$, L = CO or PPh_3. *J. Chem. Soc., Chem. Commun.* 835: 835–837.

7 Delaney, A.R., Frogley, B.J., and Hill, A.F. (2019). Metal coordination to a dimetallaoctatetrayne. *Dalton Trans.* 48: 13674–13684.

8 Schrock, R.R. (2001). Transition metal–carbon multiple bonds. *J. Chem. Soc., Dalton Trans.* 18: 2541–2550.

9 Cavaliere, V.N. and Mindiola, D.J. (2012). Methane: a new frontier in organometallic chemistry. *Chem. Sci.* 3: 3356–3365.

10 O'Reilly, M.E., Ghiviriga, I., Abboud, K.A., and Veige, A.S. (2012). A new ONO^{3-} trianionic pincer-type ligand for generating highly nucleophilic metal–carbon multiple bonds. *J. Am. Chem. Soc.* 134 (27): 11185–11195.

11 Schrock, R.R. (2005). High oxidation state alkylidene and alkylidyne complexes. *Chem. Commun.* 22: 2773–2777.

12 Miessler, G.L., Fischer, P.J., and Tarr, D.A. (2014). *Inorganic Chemistry*. Boston: Pearson.

13 Ebsworth, E.A.V., Rankin, D.W.H., and Cradock, S. (1987). *Structural Methods in Inorganic Chemistry*. Oxford: Blackwell Scientific.

14 Nugent, W.A. and Mayer, J.M. (1988). *Metal-Ligand Multiple Bonds*, 11–16. New York: Wiley-Interscience.

15 Olah, G.A. and Rasul, G. (1997). From Kekulé's tetravalent methane to five-, six-, and seven-coordinate protonated methanes. *Acc. Chem. Res.* 30 (6): 245–250.

16 Schrock, R.R. (1974). Alkylcarbene complex of tantalum by intramolecular. alpha.-hydrogen abstraction. *J. Am. Chem. Soc.* 96 (21): 6796–6797.

17 Schrock, R.R. and Meakin, P. (1974). Pentamethyl complexes of niobium and tantalum. *J. Am. Chem. Soc.* 96: 5288–5290.

18 Tebbe, F.N., Parshall, G.W., and Reddy, G.D. (1978). Olefin homologation with titanium methylene compounds. *J. Am. Chem. Soc.* 100 (11): 3611–3613.

19 Schrock, R.R. (1979). Alkylidene complexes of niobium and tantalum. *Acc. Chem. Res.* 12 (3): 98–104.

20 Hill, A.F. (2002). *Organotransition Metal Chemistry*. Cambridge: *Royal Society of Chemistry*.

21 Filippou, A.C., Woossner, D., Kociok-Koohn, G. et al. (1997). Metal-carbon multiple bonds: synthesis, structure, and electrochemistry of chromium aminocarbyne and phenylcarbyne complexes bearing phosphine or alkyl isonitrile ligands. *J. Organomet. Chem.* 532: 207–218.

22 Bourissou, D., Guerret, O., Gaggai, F.P., and Bertrand, G. (2000). Stable carbenes. *Chem. Rev.* 100: 39–92.

23 Arduengo, A.J. (1999). Looking for stable carbenes: the difficulty in starting anew. *Acc. Chem. Res.* 32 (11): 913–921.

24 Dörwald, F.Z. (1999). *Metal Carbenes in Organic Synthesis*. Weinheim: Wiley-VCH.

25 Sierra, M.A. (2000). Di- and polymetallic heteroatom stabilized (Fischer) metal carbene complexes. *Chem. Rev.* 100 (10): 3591–3638.

26 Schrock, R.R. (2002). High oxidation state multiple metal-carbon bonds. *Chem. Rev.* 102 (1): 145–180.

27 Schrock, R.R. (1986). High-oxidation-state molybdenum and tungsten alkylidene complexes. *Acc. Chem. Res.* 19 (11): 342–348.

28 Schrock, R.R. (1994). Recent advances in the chemistry and applications of high oxidation-state alkylidene complexes. *Pure Appl. Chem.* 66 (7): 1447–1454.

29 Filippou, A.C., Hofmann, P., Kiprof, P. et al. (1993). Mononuclear bis-aminocarbyne complexes of tungsten with Tp′ ligands; theory of metal-centred carbyne-carbyne coupling. *J. Organomet. Chem.* 459: 233–247.

30 Filippou, A.C., Wössner, D., Kociok-Köhn, G., and Hinz, I. (1997). Metal-carbon multiple bonds: half-sandwich phenylcarbyne complexes of chromium-synthesis, structure, electrochemistry and reactions with PMe_3. *J. Organomet. Chem.* 541 (1-2): 333–343.

31 Solans-Monfort, X., Clot, E., Copéret, C., and Eisenstein, O. (2005). d^0 Re-based olefin metathesis catalysts, Re (⋮ CR)(CHR)(X)(Y): the key role of X and Y ligands for efficient active sites. *J. Am. Chem. Soc.* 127 (40): 14015–14025.

32 Schrock, R.R. and Fellmann, J.D. (1978). Multiple metal-carbon bonds. 8. Preparation, characterization, and mechanism of formation of the tantalum and niobium neopentylidene complexes, $M(CH_2CMe_3)_3(CHCMe_3)$. *J. Am. Chem. Soc.* 100 (11): 3359–3370.

33 Ushio, J., Nakatsuji, H., and Yonezawa, T. (1984). Electronic structures and reactivities of metal-carbon multiple bonds; Schrock-type metal-carbene and metal-carbyne complexes. *J. Am. Chem. Soc.* 106 (20): 5892–5901.

34 Carlson, R.G., Gile, M.A., Heppert, J.A. et al. (2002). The metathesis-facilitated synthesis of terminal ruthenium carbide complexes: a unique carbon atom transfer reaction. *J. Am. Chem. Soc.* 124 (8): 1580–1581.

35 Romero, P.E. and Piers, W.E. (2005). Direct observation of a 14-electron ruthenacyclobutane relevant to olefin metathesis. *J. Am. Chem. Soc.* 127 (14): 5032–5033.

36 Wenzel, A.G. and Grubbs, R.H. (2006). Ruthenium metallacycles derived from 14-electron complexes. new insights into olefin metathesis intermediates. *J. Am. Chem. Soc.* 128 (50): 16048–16049.

37 Wadepohl, H., Arnold, U., Kohl, U. et al. (2000). Hydroboration of metal–carbon triple bonds. *J. Chem. Soc., Dalton Trans.* 20: 3554–3565.

38 Schwab, P., Grubbs, R.H., and Ziller, J.W. (1996). Synthesis and applications of $RuCl_2$ $(CHR')(PR_3)_2$: the influence of the alkylidene moiety on metathesis activity. *J. Am. Chem. Soc.* 118 (1): 100–110.

39 Gandelman, M., Rybtchinski, B., Ashkenazi, N. et al. (2001). A new general method for the preparation of metal carbene complexes. *J. Am. Chem. Soc.* 123 (22): 5372–5373.

40 Gandelman, M., Naing, K.M., Rybtchinski, B. et al. (2005). A general method for preparation of metal carbenes via solution-and polymer-based approaches. *J. Am. Chem. Soc.* 127 (43): 15265–15272.

41 Louie, J. and Grubbs, R.H. (2002). Metathesis of electron-rich olefins: structure and reactivity of electron-rich carbene complexes. *Organometallics* 21 (11): 2153–2164.

42 GreenM, M.P.C., Page, E.M., Whittlesey, M.K., and Lynam, J.M. (2001). Complexes containing metal-carbon σ-bonds of the groups titanium to manganese, including carbenes and carbynes. *Organomet. Chem.* 29: 227–288.

43 Boydston, A.J., Holcombe, T.W., Unruh, D.A. et al. (2009). A direct route to cyclic organic nanostructures via ring-expansion metathesis polymerization of a dendronized macromonomer. *J. Am. Chem. Soc.* 131 (15): 5388–5389.

44 Grela, K. and Kim, M. (2003). A good bargain: an inexpensive, air-stable ruthenium metathesis catalyst derived from α-asarone. *Eur. J. Org. Chem.* 2003 (6): 963–966.

45 Chauvin, Y. (2006). Olefin metathesis: the early days (Nobel lecture). *Angew. Chem. Int. Ed.* 45 (23): 3740–3747.

46 Cui, M., Bai, W., Sung, H.H. et al. (2020). Robust alkyne metathesis catalyzed by air stable d^2 Re(V) alkylidyne complexes. *J. Am. Chem. Soc.* 142 (31): 13339–13344.

47 Bailey, B.C., Fan, H., Baum, E.W. et al. (2005). Intermolecular C–H bond activation promoted by a titanium alkylidyne. *J. Am. Chem. Soc.* 127 (46): 16016–16017.

7

Metallocene: Synthesis, Properties, and Structure

Amarpreet K. Bhatia[1], Priyanka Singh[2], and Shippi Dewangan[3]

[1] *Bhilai Mahila Mahavidyalaya, Department of Chemistry, Hospital Sector, Sector-9, Bhilai Nagar, Chhattisgarh 490006, India*
[2] *Govt. Digvijay PG Autonomous College, Department of Chemistry, Rajnandgaon, Chhattisgarh 491441, India*
[3] *Hemchand Yadav University, SW Pukeshwar Singh Bhardiya Govt. College, Department of Chemistry, Nikum, Durg, Chhattisgarh 491221, India*

7.1 Introduction

The metallocene icon of organometallic chemistry originally refers to bis(cyclopentadienyl)metal complexes ("sandwich"), but wider usage is now accepted to include cyclopentadienyl complexes ("half sandwich") and multicyclopentadienyl complexes ("multidecker sandwich") as well as complexes with additional substitution at the metal center [1–3]. Nevertheless, the metallo prefix is restrictive when one considers the more recent development of complexes involving nonmetallic elements such as boron, silicon, or arsenic. In fact, metallocene-like complexes are now known for many elements in the periodic table. Metal–sandwich complexes occupy a major place in organometallic chemistry for various reasons. First, from the historical point of view, the discovery of the ferrocene structure has been a starting point for the boom of the chemistry of π-complexes. Then, ferrocene has provided a very rich chemistry as a superaromatic compound. Its redox properties have led to multiple applications in the field of materials and molecular engineering: molecular ferromagnets, modified electrodes for redox catalysis (titration of glucose in blood), polymers and dendritic electrochemical sensors for molecular recognition, and antitumor drugs and paintings. The family of metallocenes (MCp^*_2) and decamethylmetallocenes (MCp^*_2) now includes all the transition and main-group metals. Metal–arene complexes also have such properties, especially the permethylated ones that are reservoirs of electrons, protons, hydrides, and hydrogen atoms. They are efficient for transition-metal-mediated aromatic synthesis and for the synthesis of dendrimers. Early transition-metal metallocenes, in particular, do not have the sandwich structure, and the Cp rings make an angle of about 130°. The metal bears up to three other ligands. They also have remarkable properties. For instance, $[TiCp_2Cl_2]$ is an antitumor agent and the zirconium and lanthanide derivatives are known as a new generation

Organometallic Compounds: Synthesis, Reactions, and Applications, First Edition.
Edited by Dakeshwar Kumar Verma and Jeenat Aslam.

of initiators of olefin polymerization [4–8]. This chapter provides an overview of the extensive series of compounds in which a d block elements is π-bound to one or more cyclopentadienyl ligands. Synthetic procedures, characterization features, structural types, and bonding are classified in general terms and then the classic examples of complexes involving elements are examined in some detail [2].

7.2 Structure of Metallocene

Metallocenes, in a broad sense, have the composition [MCp_2] and are known for all the transition metals and many main-group and rare-earth metals. The sandwich structure is reserved to the first row of transition metals from Ti to Ni and for the iron column for which the 18-electron rule brings a special robustness. It is also known for a few other metals including main-group ones. Let us first examine these sandwich metallocenes. Ferrocene has two isoenergetic conformations: eclipsed (D_{5h} symmetry) and staggered (D_{5d} symmetry), as shown in Figure 7.1 [1].

X-ray crystallographic and electronic diffraction studies in the gas phase indicate an eclipsed conformation as well as for ruthenocene and osmocene. The rotation barrier about the iron–Cp ring axis is very weak (2–5 kcal mol^{-1}; 8–21 kJ mol^{-1}). Thus, this rotation is free and fast in fluid solution, i.e. there is only one 1,1′-disubstituted isomer (by 1 and 1′, it is meant that the substituents are located on different rings) because this fast rotation makes all the positions on the first ring equivalent for the second ring. In decamethylferrocene, the solid-state conformation is staggered. Ferrocene is an air-stable orange crystalline powder that sublimes at 184 °C under 1 atm and is stable up to 400 °C; it is soluble in hydrocarbons, has a nil dipole moment and thus covalent metal–ring bonds. The molecular orbital (MO) diagram shows various metal–ring interactions (Figure 7.2):σ(Cp→Fe), π(Cp→Fe), and δ(Cp→Fe) [1].

The other metallocenes are represented by the same diagram, although the energy levels of the orbitals may vary from one metal to the other. The occupation of the orbitals does not correspond to a low-spin structure except for ferrocene and cobaltocene. Decamethylmetallocenes [MCp^*_2] and the cations [MCp_2]$^+$ and [MCp^*_2]$^+$ all are low spin [9]. Indeed, the Cp* ligand field is much stronger than

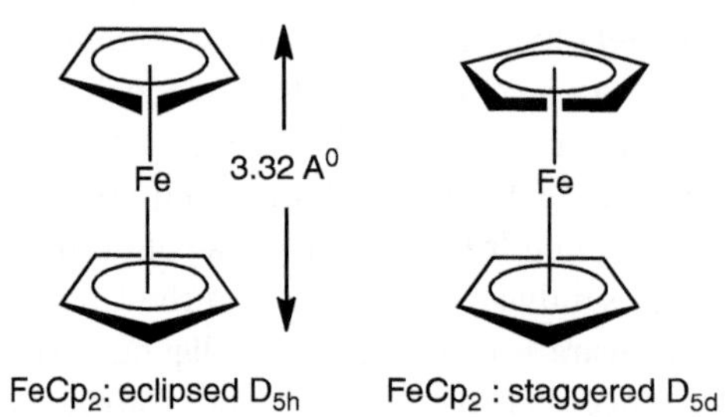

Figure 7.1 Structure of ferrocene [1].

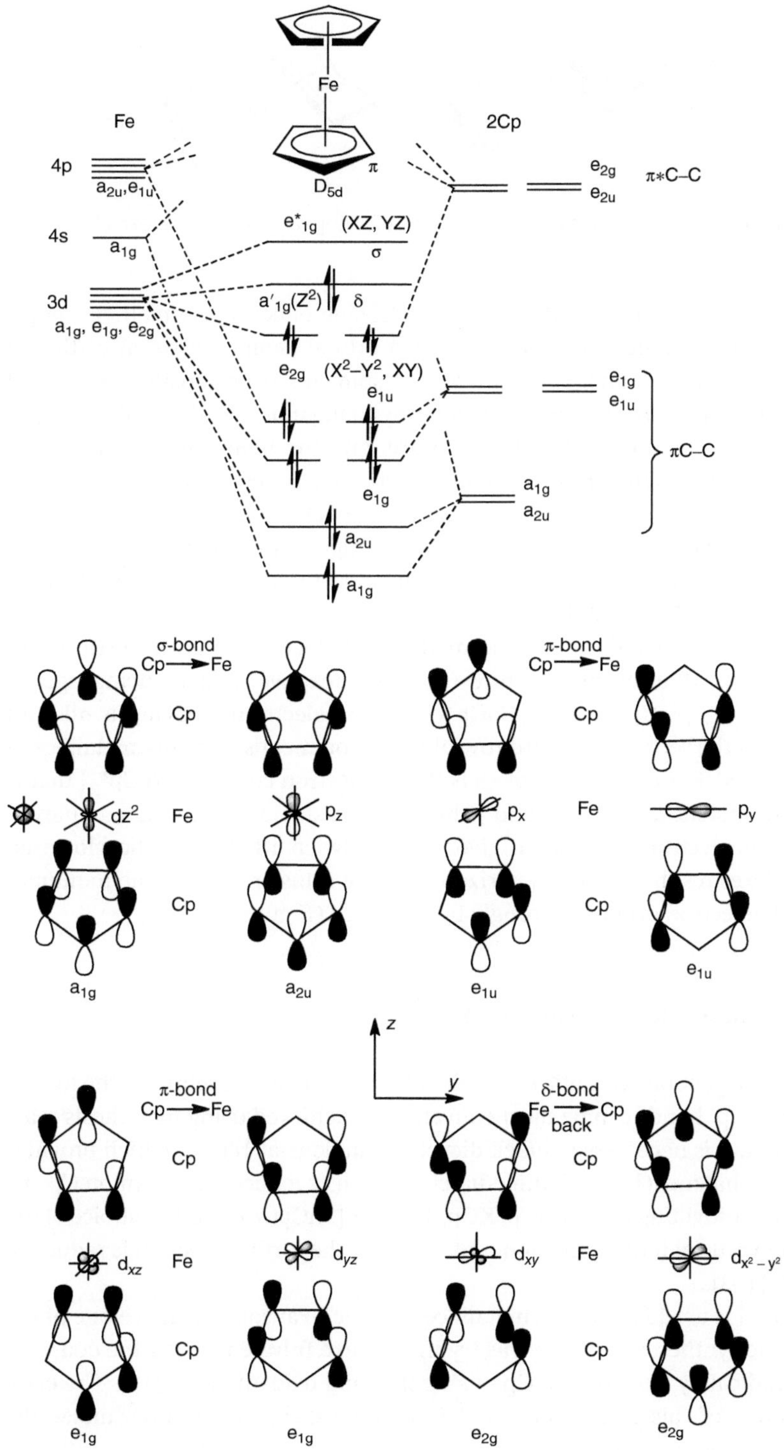

Figure 7.2 Molecular orbital diagram and interactions in ferrocene (staggered conformation) [1].

Figure 7.3 Formation of nonsandwich 18-electron dicyclopentadiene complexes from reduction of $[RhCp_2]^+$ [1, 10, 11].

the Cp ligand field because of the electron-releasing effect of the methyl groups. Manganocene has 5d electrons, i.e. one per d orbital (semioccupation of the 5d orbitals), which means a high-spin structure. A spin crossover transition $5/2 \leftrightarrow 1/2$ is observed for manganocene and 1,1-dimethylmanganocene, the pink low-spin form being the stable one at low temperature, and the high-spin state observed above the transition temperature. In manganocene, cooperative intermolecular interactions are responsible for the antiferromagnetic coupling provoking the spin lowering. The energy difference between high spin and low spin is only 0.5 kcal mol^{-1} (2.1 kJ mol^{-1}). Decamethylmanganocene always stays low spin as $[FeCp_2]^+$ and $[FeCp^*{}_2]^+$ [1–3, 10, 11].

Only one anionic metallocene or permethyl metallocene is known, $[MnCp^*{}_2]^-$, stable due to its 18-electron structure (but very reactive as a reducing agent). Thus, $[MnCp^*{}_2]^-$, $[FeCp^*{}_2]$, and $[CoCp^*{}_2]^+$ are isoelectronic, having a nil spin (NVE = 18). In the second and third row of transition metals, the only metallocene known whose NVE is different from 18 is the 17-electron complex $[ReCp^*{}_2]$ that is low spin. For instance, $[RhCp_2]^+$ and $[RhCp^*{}_2]^+$ are stable 18-electron complexes, but their monoelectronic reduction gives nonsandwich 18-electron dicyclopentadiene complexes resulting from dimerization of the transient rhodocene monomer whose single-electron orbital is strongly ligand based (Figure 7.3).

7.3 Synthesis of Metallocene

Consistent with general synthetic principles of organometallic chemistry, the formation of metal–Cp′ bonds can be most conveniently achieved using metathesis reactions [3]. First, it is necessary to crack dicyclopentadiene at 180 °C, which provides the monomer that must be used immediately as such or reduced to the anion Cp– by deprotonation providing LiCp, NaCp, KCp, TlCp, or $[SnCp(n\text{-}Bu_3]$. The choice of the metal is often crucial if the transition metal used in the further reaction is reducible (Figure 7.4) [1–3].

This method is used for various metallocenes. It led Pauson to achieve one of the first ferrocene syntheses while he was trying to make fulvalene using the coupling of the Cp· radical by oxidation of Cp– using the mild oxidant $FeCl_3$ [12]. The erroneous structure of "bis-cyclopentadienyl–iron" proposed by Pauson resembled the hoped-for one (Figure 7.5) [1, 12].

Figure 7.4 Formation of $[MCp_2]$ [1–3].

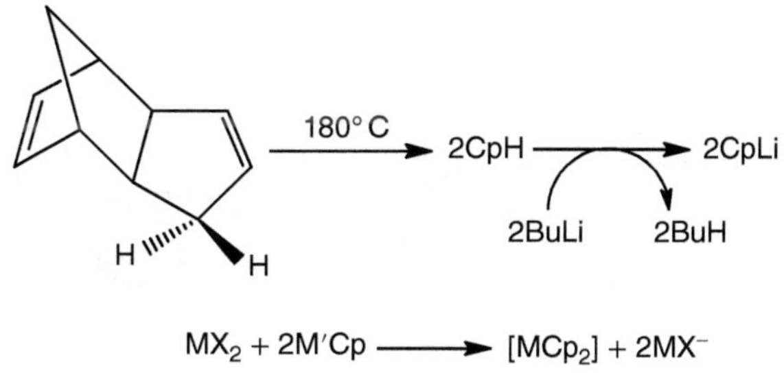

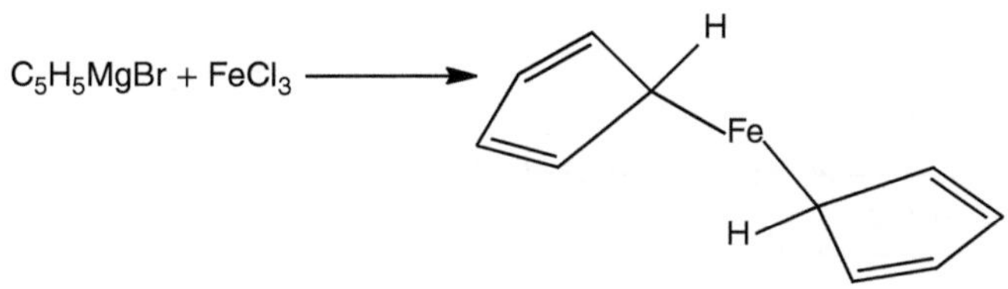

Figure 7.5 Erroneous structure of bis-cyclopentadienyl–iron proposed by Pauson [1, 12].

Figure 7.6 Formation of decaisopropyl ferrocene [1, 13].

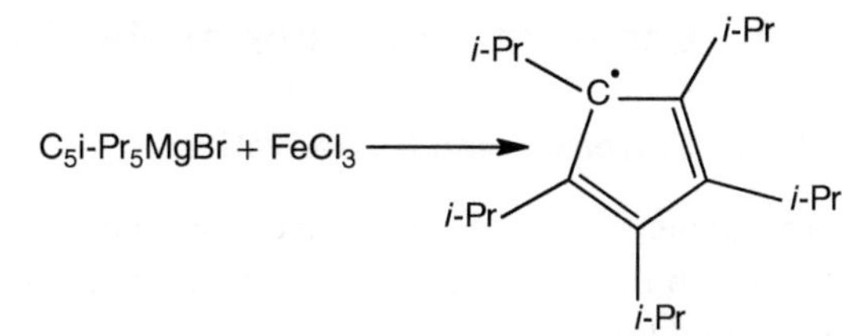

Figure 7.7 Formation of $[RuCp_2]$ [1, 13].

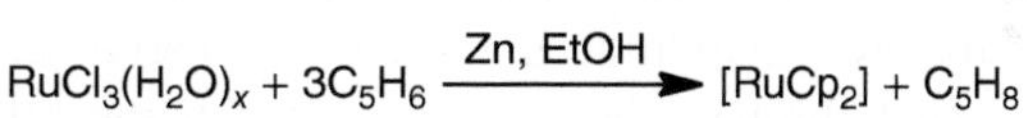

On the other hand, this oxidation of the anion to the radical worked in the case of $C_5(i\text{-}Pr_5)$ for which Sitzmann was trying to make the elusive decaisopropyl ferrocene (Figure 7.6) [1, 13]. Sometimes, the reduction of the transition-metal halide is necessary *in situ*, either by the salt MCp itself or by Zn (Figure 7.7) [1, 13].

This first classic method can also be used sequentially in order to introduce two different cyclopentadienyl ligands in a transition-metal–sandwich complex, especially if one of the Cp derivatives is Cp* (Figure 7.8) [1, 14].

The second method, published by Miller, Tebboth, and Tremaine also in 1951, starts from cyclopentadiene and freshly reduced iron. It works well for instance with the metal-vapor technique that consists in vaporizing metal atoms at low temperature and pressure. In fact, the most practical method for the syntheses of ferrocene and its derivatives consists the *in situ* deprotonation of cyclopentadiene ($pK_a = 15$) or its derivative by diethylamine, the solvent, in the presence of anhydrous $FeCl_2$ or $FeCl_3$. The iron chloride can be generated *in situ* from iron powder and the amine chlorohydrate (Figure 7.9) [1, 15].

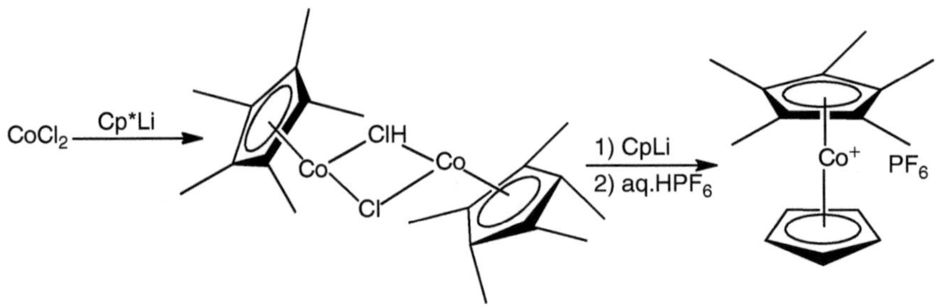

Figure 7.8 Formation of $[CoCp_2]$ [1, 14] .

$$2CpH + Fe \xrightarrow{300°C} [FeCp_2] + H_2$$

$$2CpH + 2Et_2NH + FeCl_2 \longrightarrow [FeCp_2] + 2Et_2NH_2^+Cl^-$$

$$2CpH + Fe \xrightarrow{Et_2NH_2^+Cl^-} [FeCp_2] + H_2$$

Figure 7.9 Formation of $[FeCp_2]$ [1, 15].

7.4 Chemical Properties of Metallocene

7.4.1 Ferrocene and Its Derivatives

Ferrocene undergoes many electrophilic reactions, more rapidly than benzene, although they are limited by oxidation reactions with electrophiles that are strong oxidants (H_2SO_4 or HNO_3). Formylation and carboxylation reactions give only monofunctionalization because the functional group strongly deactivates the ferrocenyl group. Indeed, there is, to a certain extent, a transmission of the electronic effect through the metal center. On the other hand, metallation and acylation reactions can be followed by an identical reaction on the other ring leading to 1,1′-disubstituted derivatives because the deactivation of the second Cp ring by the substituent is only modest (Figure 7.10) [1, 16].

Lewis acids such as $AlCl_3$ form complexes leading to the cleavage of the Fe–Cp ring upon heating. The cleaved ring is protonated *in situ* and becomes an electrophile toward ferrocene. The reaction continues another time to give the cyclopentylene bridge. The syn and anti-isomers of the dibridged derivatives also form, and they have been separated, yielding cage structure in which the iron is encapsulated by four five-carbon rings. In the presence of an aromatic or CO, these reactions are minimized because the main reaction is substitution of a ferrocene ring by an arene or three CO ligands. This reaction is the classic mode of synthesis of the sandwich complexes $[FeCp(arene)]^+$ (Figure 7.11) [1, 17].

The most useful ferrocene derivatives for the development of functional ferrocene chemistry are acetylferrocene, ferrocenyl carbaldehyde, ferrocenyl lithium, and chloromercuriferrocene. Ferrocenylcarboxylic acid reacts with PCl_5 to give chlorocarbonylferrocene, which easily couples with primary amines in the presence of NEt_3 to yield ferrocenyl amides. An essential characteristic of ferrocene chemistry

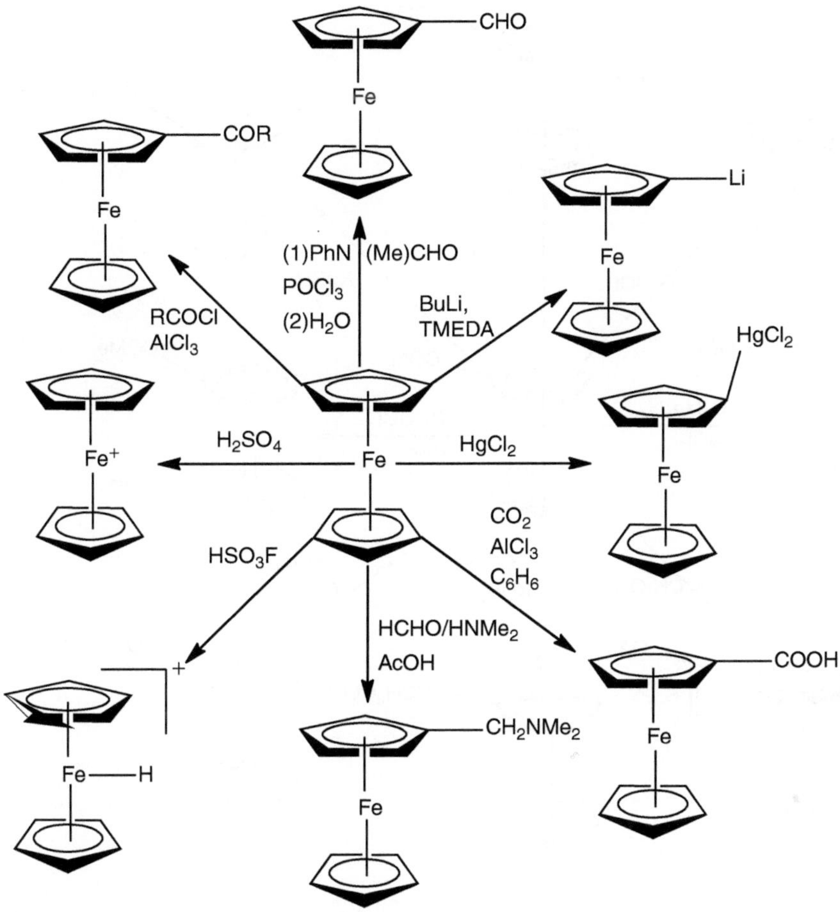

(a)

Figure 7.10 (a) Different chemical reactions of ferrocene [1, 16]. (b) Different chemical reactions of ferrocene derivatives [1, 16]. (c) Different chemical reactions of ferrocene derivatives [1, 16]. (d) Different chemical reactions of ferrocene derivatives [1, 16].

is the stabilization of ferrocenyl carbonium ions. These carbocations are mesomers of the corresponding hexahapto fulvene complexes $[FeCp(\eta^6\text{-fulvene})]^+$. They are even more stable than the trityl cation Ph_3C^+. The stabilization of the α-ferrocenyl carbonium ions explains the acetolysis of vinylferrocene, the hydrolysis of the acetate formed, the ease of nucleophilic substitution in α-position, and the OH– abstraction from the α-ferrocenyl alcohols. This stabilization is still enhanced by going down in the iron column of the periodic table because the size of the d orbital increases, which facilitates their insertion with the "carbocation" and accelerates the solvolysis of acetates (Os > Ru > Fe) [17].

Finally, note that the presence of two different substituents on the same ferrocene ring introduces a planar chirality. Many 1,2- or 1,3-heterodisubstituted ferrocene derivatives (amines, carboxylic acids, ketones, etc.) have indeed been split into

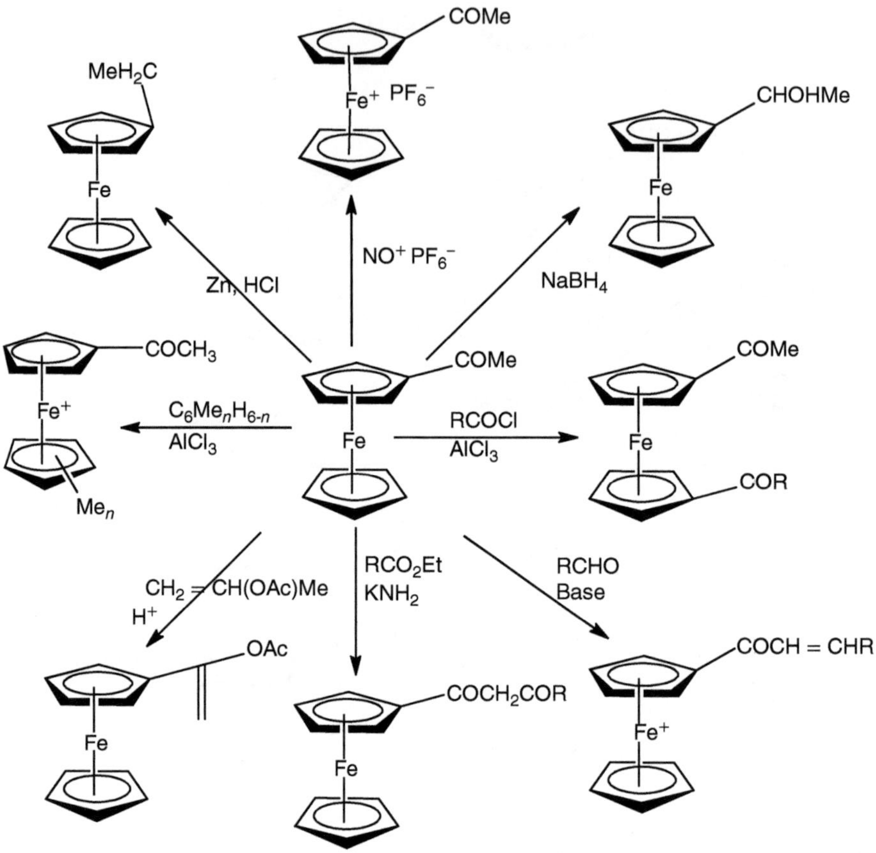

(b)

Figure 7.10 *(Continued)*

enantiomers (recall that the 1,1′-heterodisubstituted derivatives are not chiral because of the fast rotation about the Fe rings axis) (Figure 7.12) [1, 17].

7.4.2 Other Metallocene Sandwiches

Chromocene reacts with two-electron donors, such as CO and $F_3CC{\equiv}CCF_3$, to give 18-electron adducts. Manganocene reacts with CO to give $[MnCp(CO)_3]$. Cobaltocene reacts with oxygen and oxidants, tetrafluoroethylene, and halides. The reaction of cobaltocene with dihalogenomethanes gives ring extension. For instance, borabenzene complexes are accessible in this way by reaction with alkyldihalogenoboranes. The reactions of nickelocene are particular because they usually do not preserve the sandwich geometry due to the presence of two electrons in antibonding orbitals weakening the Ni–Cp bonds [18]. There are essentially two kinds of reactions:

- hydrogenation of one of the Cp rings to allyl bringing the electronic structure to 18 valence electrons;

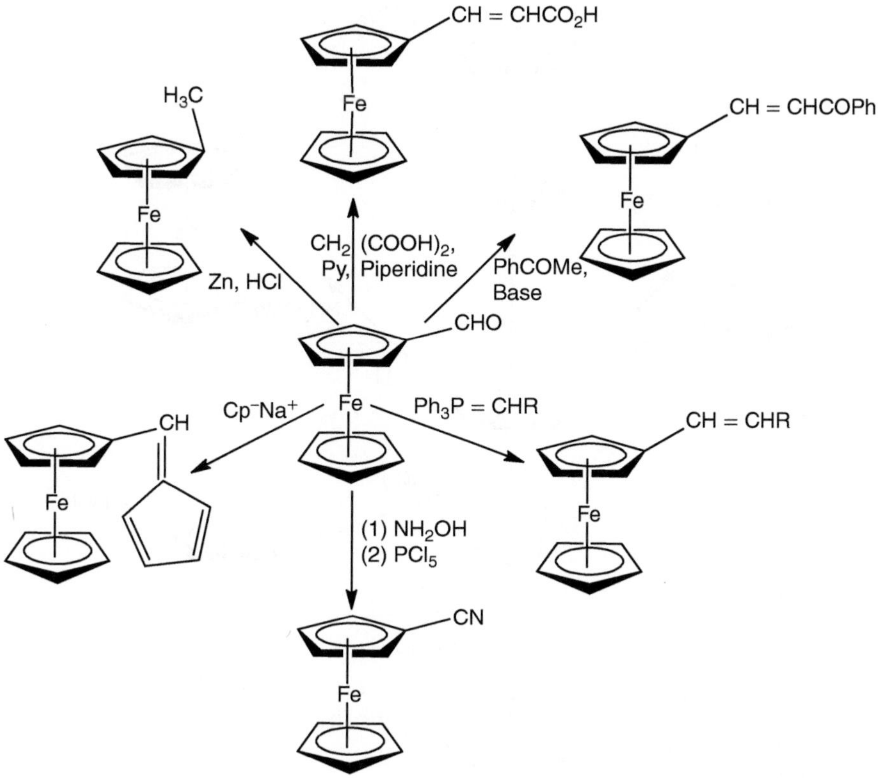

(c)

Figure 7.10 (*Continued*)

- facile thermal loss of a Cp ring, more frequent and interesting, because it yields a source of the very reactive 15-electron fragment that reacts with a large variety of unsaturated substrates to complete the 18-electron structure.

7.4.3 Main-group Metallocene

Finally, the family of metallocenes includes all the main-group metals, and robust examples involve the permethylmetallocenes, a chemistry developed by Jutzi. One Cp ligand is often bonded in a monohapto mode (η^1-C_5H_5) for main-group metallocenes, but many compounds also have the bis-pentahapto-Cp-metal–sandwich structure. For instance, in beryllocene, the energies of the η^5/η^5 and η^5/η^1 structures have been calculated to be about the same, only favoring the mixed structure by a few kcal/mol. Beryllocene is highly fluxional, only one set of resonances being found for the two C_5H_5 groups in the ^{1}H- and ^{13}C-NMR spectra at −135 °C. There are 1,5 sigmatropic shifts of the Be(η^5-Cp) unit around the periphery of the η^1-C_5H_5 ring and a molecular inversion that interchanges the η^1-C_5H_5 and η^5-C_5H_5 rings. The Carmona group reported remarkable examples representative of the monohapto/pentahapto dichotomy with the synthesis of tetra- and pentamethyl Cp

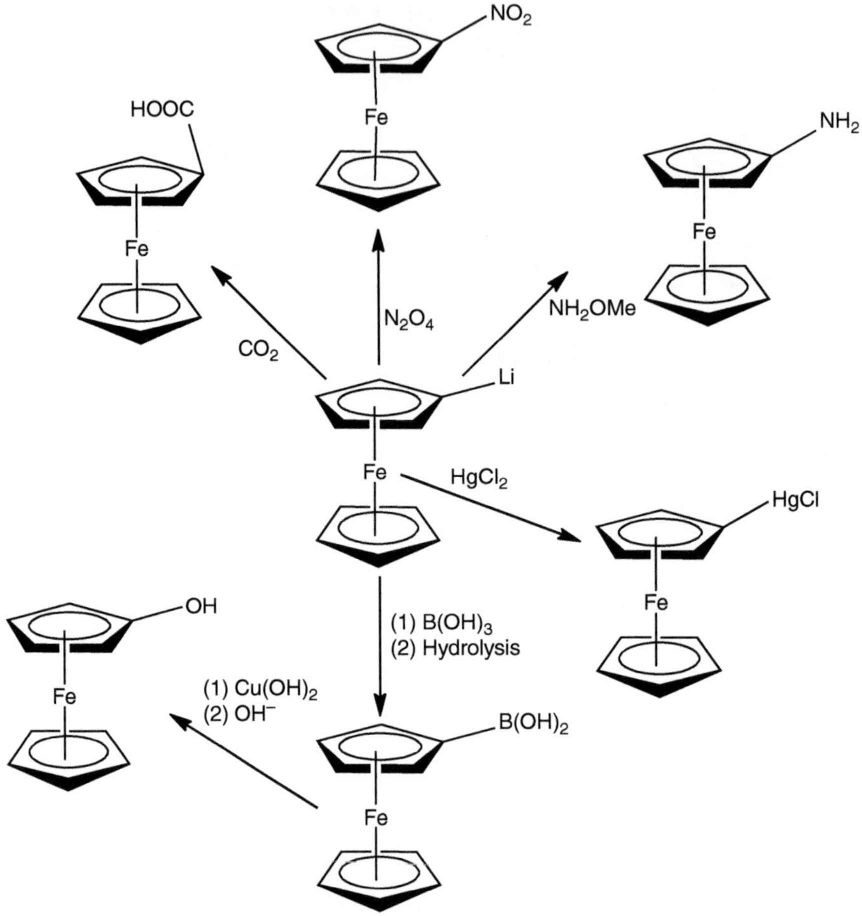

(d)

Figure 7.10 (*Continued*)

sandwich Be complexes below. For these complexes, the X-ray crystal structures have been determined, illustrating the subtile difference between the η^5/η^5 and η^5/η^1 structures (Figure 7.13) [1, 19].

7.4.4 Metal–bis-arene Sandwich Complexes

7.4.4.1 General View

Bis-arene–metal–sandwich complexes are known for all the transition metals, and their structure and chemistry resemble that of metallocenes. Fischer and Hafner have developed the first general synthetic method from metal halides, $AlCl_3$, and Al. This method also leads to the isomerization of the methyl- and alkylaromatics by retro-Friedel–Crafts reaction, especially for the second- and third-row transition metals. It has been applied with success for V, Cr, Fe, Co, and Ni with benzene and its polymethyl derivatives, in particular C_6Me_6 with Co and Ni. The monocations

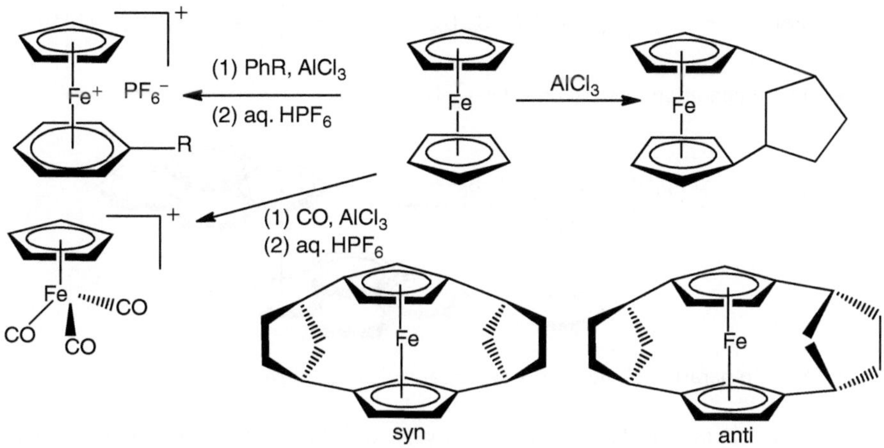

Figure 7.11 Syn and anti-isomers of the dibridged derivatives of $[FeCp_2]$ [1, 17].

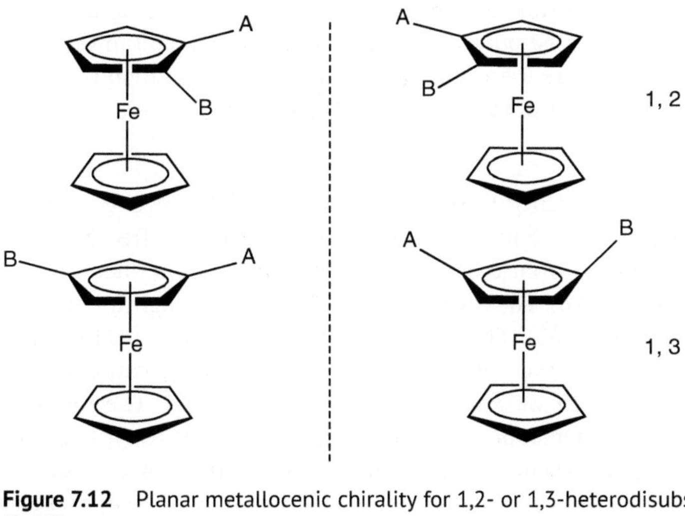

Figure 7.12 Planar metallocenic chirality for 1,2- or 1,3-heterodisubstituted ferrocenes [1, 17].

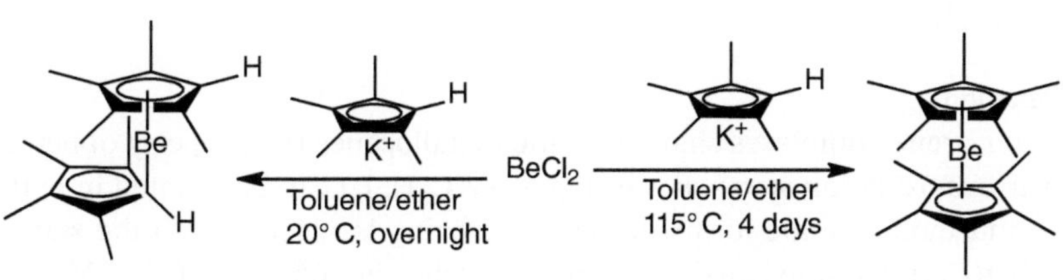

Figure 7.13 Some chemical reactions of metallocene of main-group element (Be) [1, 19].

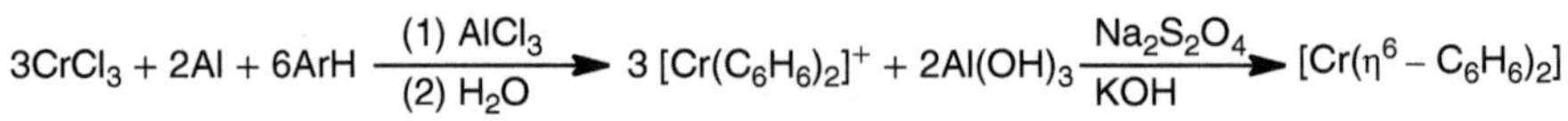

Figure 7.14 Preparation method of [Cr(η^6-C_6H_6)$_2$] [1, 18].

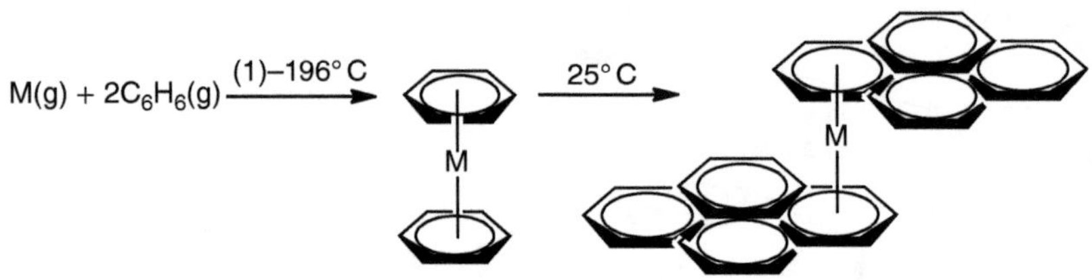

Figure 7.15 Preparation of metal–bis-arene sandwich complexes [1, 18].

[Cr(η^6-arene)$_2$]$^+$ are produced after hydrolysis, and the dications are obtained with the first-row transition metals (Figure 7.14) [1, 18].

The second general method consists of condensing metal vapors with the aromatic at low temperature and pressure. It has been pioneered by the groups of Timms, Green, and Cloke and applied to all the transition metals. Eschenbroich and Kundig have also applied it to condensed polyaromatics, and the metals bind to the rings located at the periphery because these rings are more electron rich than the other ones due to the reduced number of common bonds (Figure 7.15) [1, 18].

7.4.4.2 Structure

The sandwich structure of [Cr(η^6-C_6H_6)$_2$] has two parallel benzene rings in which the C—C bonds are all equal and slightly longer than in free benzene, reflecting the δ backbonding from the metal d orbitals into the antibonding e_{2g} orbitals of benzene (Figure 7.16). The covalent Cr–C_6H_6 bonds are slightly polarized (δ^+ Cr = +0.7; δ^- C_6H_6 = −0.35). The bond energy (40 kcal mol^{-1}; 167 kJ mol^{-1}) is smaller than in ferrocene (52 kcal mol^{-1}; 217 kJ mol^{-1}), and the rotation barrier about the Cr-ring axis is very weak (<1 kcal mol^{-1}; 4.2 kJ mol^{-1}). The MO diagram includes the orbital interactions and electronic structure of [Cr(η^6–C_6H_6)$_2$] (Figure 7.16). This diagram resembles that of ferrocene but the benzene orbitals have lower energies than the Cp orbitals because the arene ring is larger, more delocalized than Cp. Thus, the energies of the e_{2g} benzene orbitals are closer to those of Cr, which strengthens the δ backbonding from Cr to benzene by comparison with ferrocene [1, 18].

7.4.4.3 Reactions

The metal–bis-arene complexes share with the metallocenes the property of being found in various oxidation states. For instance, [V(C_6H_6)$_2$] can be isolated in both its neutral and monoanionic forms; these forms being isoelectronic to the stable complexes [Cr(C_6H_6)$_2$]$^+$ and [Cr(C_6H_6)$_2$], respectively. The complexes [M(C_6Me_6)$_2$] (M = Fe, Co, Ni) exist as neutral, mono-, and dicationic complexes (although the structures of complexes that would have more than 20 valence electrons in the fully

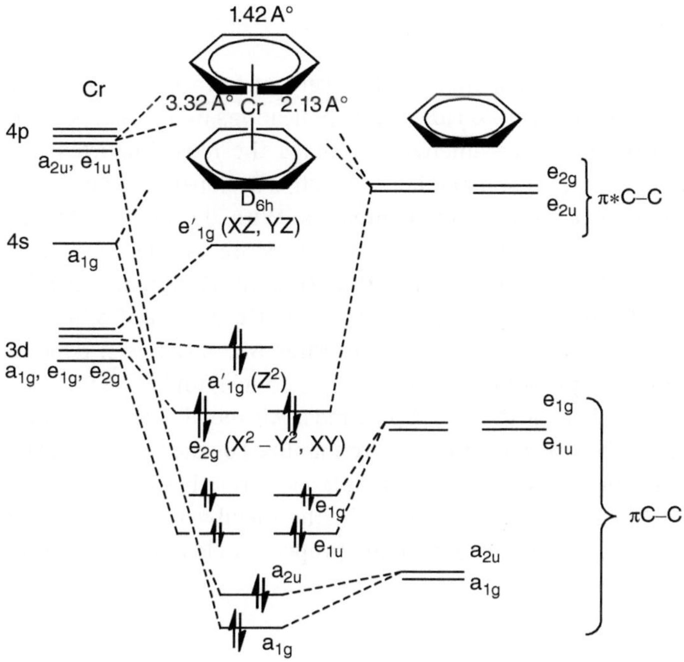

Figure 7.16 Molecular orbital (MO) diagram of $[Cr(\eta^6\text{-}C_6H_6)_2]$ and interactions of the benzene orbital with the Cr orbitals of appropriate symmetry to form $[Cr(\eta^6\text{-}C_6H_6)_2]$ [1, 18].

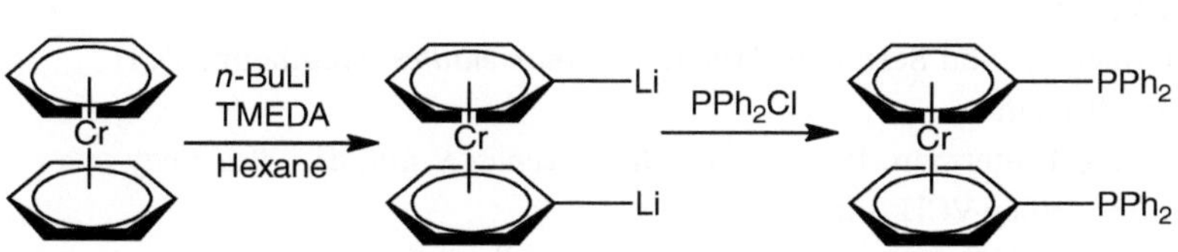

Figure 7.17 Some electrophilic substitution reaction of $[Cr(\eta^6\text{-}C_6H_6)_2]$ [1, 18].

sandwich form do not exist as such but partial decoordination reduces their valence electron count). In the iron–bis-arene series, the neutral, mono-, and dicationic complexes are also isolable, and these three complexes are thermally stable in the case of C_6Me_6.

Moreover, it is also possible to transfer one or two hydrides, one or two protons, and one or two hydrogen atoms. Thus, these systems are reservoirs of electrons, protons, hydrides, and hydrogen atoms. The complex $[Cr(\eta^6\text{–}C_6H_6)_2]$ cannot undergo electrophilic additions because it is too easily oxidized ($E^0 = -0.70$ V vs. SCE in DME, which is almost as negative as the oxidation potential of cobaltocene, $E^0 = -0.89$ V vs. SCE and much more negative than the oxidation potential of ferrocene, $E^0 = +0.40$ V vs. ECS). On the other hand, metallation is easier than for free benzene and allows the functionalization (Figure 7.17) [1, 18].

7.5 Conclusion

Metallocenes are known with most metals (discovery of ferrocene in 1951 and report of its sandwich structure in 1952) and have applications in organic, polymer, and medicinal chemistry. Cyclopentadienyl is perhaps the most important and widespread ancillary ligand in organometallic chemistry. Cyclopentadienyl ligands stabilize a wide variety of organometallic complexes. With the MCp_2 structure, they are sandwich complexes in which the two Cp rings are parallel (ferrocene). The sandwich MCp_2 structure is very robust with NVE = 18 (Fe, Ru, and Os). The neutral metallocenes of the first row with 15–20 e^- (V, Cr, Mn, Fe, Co, and Ni) correspond to the filling of the 5d orbitals split in three MO levels (e_{1g}: bonding, double; a_{1g}: nonbonding, simple, and e^*_{1g}; antibonding, double) under the influence of the pseudo-octahedral field of the two Cp ligands. All metallocenes are more fragile as their NVE is further from 18, but they are stabilized by the Cp* (η^5–C_5Me_5) ligand. However, metal–bis-arene sandwich complexes, accessible by reactions of the Fischer type (MCl_3 + arene + $AlCl_3$ + Al) or by the metal-vapor condensation technique, have electronic structures and redox properties close to those of the metallocenes.

References

1 Astruc, D. (2007). *Organometallic Chemistry and Catalysis*. Berlin Heidelberg: Springer-Verlag.

2 Elschenbroich, C. and Salzer, A. (1989). *Organometallics*. Weinheim: VCH Verlagsgesellschaft.

3 Togni, A. and Halterman, R.L. (1998). *Metallocenes: Synthesis – Reactivity – Applications*. Wiley-VCH – libgen.lc.

4 Wilkinson, G. (1975). The iron sandwich. A recollection of the first four months. *J. Organomet. Chem.* 100: 273.

5 Togni, A. and Hayashi, T. (ed.) (1995). *Ferrocenes*. Weinheim: VCH.

6 Astruc, D. (1995). *Electron Transfer and Radical Processes in Transition Metal Chemistry*, 142. New York: VCH.

7 Kahn, O. (1994). *Molecular Magnetism*, 294. New York: VCH.

8 Astruc, D. (1992). The use of organoiron compounds in aromatic chemistry. *Top. Curr. Chem.* 160: 47.

9 Ammeter, J.H., Bucher, R., and Oswald, N. (1974). On the low-spin-high-spin equilibrium of manganocene and dimethylmanganocene. *J. Am. Chem. Soc.* 96: 7833.

10 Mugnier, Y., Moïse, C., Tirouflet, J., and Laviron, E. (1980). Réduction électrochimique du ferrocène. *J. Organomet. Chem.* 186: C49.

11 Bard, A.J., Garcia, E., Kukharenko, S., and Strelets, V.V. (1993). Electrochemistry of metallocenes at very negative and very positive potentials. Electrogeneration of Cp_2Co^{2+}, Cp_2Co^{2-} and Cp_2Ni^{2+} species. *Inorg. Chem.* 32: 3528.

12 Keally, T.J. and Pauson, P.L. (1951). A new type of organoiron compound. *Nature* 168: 1039.

13 Sitzmann, H., Bock, H., Boese, R. et al. (1993). Pentaisopropylcyclopentadienyl: singlet anion, doublet radical and triplet cation of a carbocyclic System. *J. Am. Chem. Soc.* 115: 12003.

14 Herberich, G.E., Gaffke, A., and Eckenrath, H.J. (1998). [FeCp*CpR] derivatives. A new and highly selective synthesis. *Organometallics* 17: 5931.

15 Miller, S.A., Tebboth, J.A., and Tremaine, J.F. (1952). Dicyclopentadienyliron. *J. Chem. Soc.* 632.

16 Gmelin, L. and Fluck, E. *Fe Organic Compounds*, Gmelin Handbook of Inorganic and Organometallic Chemistry, 8e, Vols. A1-A8 (1974, 1977, 1978, 1980, 1981, 1985, 1989, 1991).

17 Rockett, B.W. and Marr, G. (1991). Ferrocene chemistry. *J. Organomet. Chem.* 416: 327. and cited references.

18 Muetterties, E.L., Bleeke, J.R., Wucherer, E.J. et al. (1982). Structural, stereochemical and electronic features of arene-metal complexes. *Chem. Rev.* 82: 499.

19 Hanusa, T.P. (1993). Ligand influences on structure and reactivity in organoalkaline earth chemistry. *Chem. Rev.* 93: 1023; Jutzi, P. and Burford, N. (1999). Chem. Rev. 99, 969; Jutzi, P. and Burford, N. (1998). *Metallocenes* (A., Togni and R.L., Haterman Eds), Weinheim: Wiley-VCH, Chapter 2.

8

σ-Complexes, π-Complexes, and η^n-C_nR_n Carbocyclic Polyenes-Based Organometallic Compounds

Shippi Dewangan[1], Jeenat Aslam[2], and Amarpreet K. Bhatia[3]

[1] *Hemchand Yadav University, SW Pukeshwar Singh Bhardiya Govt. College, Department of Chemistry, Nikum, Durg, Chhattisgarh 491221, India*
[2] *Taibah University, College of Science, Department of Chemistry, Yanbu, Al-Madina 30799, Saudi Arabia*
[3] *Bhilai Mahila Mahavidyalaya, Department of Chemistry, Bhilai Nagar, Chhattisgarh 490006, India*

8.1 Introduction

Organometallic chemistry is defined as the chemistry of metal–carbon bonds. More generally, it concerns the transformations of organic compounds using metals from the main groups, transition series, lanthanides, and actinides. Organometallic chemistry is a part of coordination chemistry and inorganic chemistry. The boom in organometallic chemistry occurred essentially in the United States, England, and Germany during the third quarter of the twentieth century. Some organometallic compounds have been known for well over a hundred years (e.g. the alkyls of zinc, mercury, and arsenic and Zeise's salt. The first transition-metal compound, $K[Pt(C_2H_4)Cl_3)$, the development of the subject, especially of the organometallic chemistry of the transition elements, is much more recent. The study of organometallic compounds has often contributed significantly to both chemical theory and practice. The chance synthesis of ferrocene and the determination of its structure in the following year opened up a field of research of hitherto unforeseen variety, which has contributed greatly to our understanding of chemical bonding. The rapid subsequent development of this area owes much to new methods of structure determination, especially nuclear magnetic resonance spectroscopy and X-ray crystallography [1, 2].

Organotransition-metal complexes are now finding wide application as reagents for specific organic syntheses. They are also involved as intermediates in many catalytic processes, some of which are used for the large-scale conversion of carbon monoxide, hydrogen, alkenes, and other small molecules into useful organic chemicals. Most organometallic compounds resemble organic rather than inorganic compounds in their physical properties. Many possess discrete molecular structures and, hence, exist at ordinary temperatures as low-melting crystals, liquids, or gases. Commonly, they are soluble in weakly polar organic solvents, such as toluene, ethers, or dichloromethane. Their chemical properties vary widely and their

Organometallic Compounds: Synthesis, Reactions, and Applications, First Edition.
Edited by Dakeshwar Kumar Verma and Jeenat Aslam.

thermal stability, for example, depends markedly on their chemical composition. Thus, tetramethylsilane (Me_4Si) is unchanged after many days at 500 °C, whereas tetramethyltitanium decomposes rapidly at room temperature. Similarly, there are wide differences in their kinetic stability to oxidation; some (e.g. Me_4Si, Me_2Hg, and *Cp_2Fe) are not attacked at room temperature by the oxygen in air, whereas others (e.g. Me_3B, Me_2Zn, and Cp_2Co) are spontaneously inflammable. Organometallic compounds may conveniently be classified by the type of metal–carbon bonding that they contain. It may be ionic, volatile covalent, hydrogen bridged, and metal like. The organic compounds of the d-block transition elements often involve not only σ- but also π- or δ-bonding, which is not commonly found among compounds of the main-group elements [1, 2].

8.2 σ-Bond Containing Organometallic Compounds

8.2.1 Metal Carbonyl

8.2.1.1 General Overview

The family of metal carbonyls is one of the very most important ones in inorganic and organometallic chemistry. This is due to the interest in their structures, applications in organic and organometallic synthesis, and essential role in catalysis. The remarkable ability of CO to give π-backbonding is responsible for the stability of metal carbonyls with oxidation states that are nil (neutral metal carbonyls) or negative (anions). A large variety of mixed complexes containing carbonyls and other ligands is also available, the best known being those that also contain phosphines, cyclopentadienyls, arenes, or other unsaturated hydrocarbon ligands. Because the carbonyl ligands are strongly bonded to transition metals, the variations of the frequency (around 2000 cm^{-1} for terminal CO) of the intense infrared carbonyl absorption allow to measure the electronic effects of the other ligands. The low-nuclearity binary metal–carbonyls always have 18 valence electrons on the metal except the 17-electron radical [$V(CO)_6$] whose structure is stabilized by the octahedral geometry that largely contributes to prevent radical-type reactions. The high-nuclearity metal–carbonyl clusters [$M_6(CO)_{16}$] and beyond often escape the 18-electron rule to obey Wade's rules. Thus, mononuclear neutral metal carbonyls are known for the metals that have an even number of d electrons, whereas neutral dimers with a single metal–metal bond are known for metals that have an odd number of d electrons. The most current metal carbonyls are compiled in Table 8.1 [3, 4].

8.2.1.2 Syntheses of Metal Carbonyls

The synthesis of metal–carbonyl complexes proceeds industrially. It is sometimes possible to directly use the metal, but the most frequent synthetic method consists in reducing a metal salt in the presence of CO. In the laboratory, some syntheses classically employ photochemistry or redox reagents. The use of alkali metals leads to metal–carbonyl anions starting from neutral metal carbonyls [1, 5].

Table 8.1 Some examples of stable metal carbonyls.

Types	Metal carbonyls
Monomeric	$V(CO)_6$, $Cr(CO)_6$, $Mo(CO)_6$, $W(CO)_6$, $Fe(CO)_5$, $Ni(CO)_4$, $Ru(CO)_5$, $Os(CO)_5$
Dimeric	$Mn_2(CO)_{10}$, $Tc_2(CO)_{10}$, $Re_2(CO)_{10}$, $Fe_2(CO)_9$, $Co_2(CO)_8$
Polymeric	$Fe_3(CO)_{12}$, $Co_6(CO)_{16}$, $Co_4(CO)_{12}$, $Ru_3(CO)_{12}$, $Rh_4(CO)_{12}$, $Os_3(CO)_{12}$, $Ir_4(CO)_{12}$, $Ir_6(CO)_{16}$

Direct method	$Ni + 4CO \longrightarrow [Ni(CO)_4]$ $Fe + 5CO \longrightarrow [Fe(CO)_5]$
Reduction method	$CrCl_3 + Al + 6CO \longrightarrow [Cr(CO)_6] + 1/2\ Al_2Cl_6$ $WCl_6 + 2Et_3Cl + 6CO \longrightarrow [W(CO)_6] + 2\ Al_2Cl_6 + 3C_4H_{10}$ $Re_2O_7 + 17CO \longrightarrow [Re_2(CO)_{10}] + 7CO_2$ $2CoCO_3 + 2H_2 + 8CO \longrightarrow [Re_2(CO)_{10}] + 7CO_2$
Photochemical method	$2[Fe(CO)_5] \xrightarrow{CH_3COOH,hv} [Fe_2(CO)_9] + CO$
By the use of redox reagents	$2[Fe(CO)_5] + 2OH^- \longrightarrow [HFe(CO)_4]^- + HCO_3^-$ $3[HFe(CO)_4]^- + 3MnO_2 \longrightarrow [Fe_3(CO)_{12}] + 3OH^- + 3MnO$ $[Fe(CO)_5] + 2Na/Hg \longrightarrow [Fe(CO)_4]^{2-}(Na^+)_2 + CO + 2Hg$

8.2.1.3 Structure of Metal Carbonyls

CO is an σ-donor (two-electron L ligand) and a π-acceptor. The corresponding orbital interactions are the following.

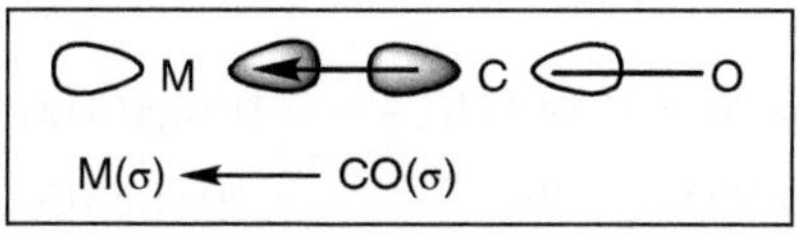

The two mesomer forms below (a neutral- and an ionic one) can be used to describe the formalism of the M–C–O chain.

The σ-donation from the carbonyl to the metal weakens the electron density at the carbon atom, and the backbonding enriches both the carbon and the oxygen atoms of the carbonyl. The overall result is that the carbon atom is more electron poor upon coordination, and the oxygen atom is more electron rich. The ligand is thus polarized upon coordination and undergoes nucleophilic attacks on the positively polarized carbon atom. The more important the π-backbonding from the metal, the more populated is the π* antibonding carbonyl orbital, the more the C—O bond is weakened and lengthened, the more the metal–carbon bond is strengthened and shortened. It is possible to correlate the bond length determined by X-ray diffraction. For instance, for $[Mo(CO)_3(NR_3)_3]$, the length of a single Mo—C bond would be 233 pm, and the experimental length, 193 pm, indicates a strong double bond character, i.e. important π-backbonding. The most practical method for these correlations is infrared spectroscopy, however, because the weakening of the C—O bond, provoked by the π-backbonding, results in the lowering of the absorption frequency. The frequency domain is between 1820 and 2150 cm^{-1} for a terminal carbonyl. The correlation between the vibration frequency and the bond order is possible because the elongation frequency of the C—O bond is, in the first approximation, independent from the other vibrations of the molecule [1, 6, 7].

8.2.1.4 Reactions of Metal Carbonyls [6–11]

Ligand substitution reactions	$[Fe(CO)_5] + 2\,L/L_2 \xrightarrow{C_6H_6(\text{reflux})} [Fe(CO)_3L_2] + 2CO$ L = phosphane, L_2 = diene $[Cr(CO)_6] + \text{arene} \xrightarrow{\text{n-}Bu_2O)} [Cr(\text{arene})(CO)_3] + 3CO$
Nucleophilic attacks	$[Fe(CO)_5] + Na^+OH^- \longrightarrow (CO)_4Fe^- — COOH$ $\longrightarrow Fe((H)CO)_4{}^-Na^+ + CO_2$
Electrophilic addition on the oxygen atom	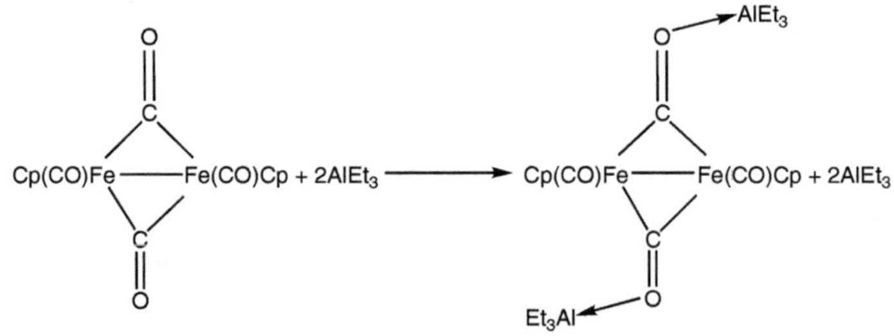
Insertion	$[FeCp(CO)_2CH_3] + PPh_3 \longrightarrow [FeCp(CO)(PPh_3)COCH_3$
Oxidation	$[FeCp^*(CO)_2]_2 + Br_2 \xrightarrow{CH_2Cl_2, 20^\circ C} 2[FeCp^*(CO)_2Br]$
Reduction	$[VCp(CO)_4] + Na/Hg \longrightarrow [VCp(CO)_3]^-Na^+ + Hg$
Dismutation	$[L_nM\text{-}ML_n] \xrightarrow{PR_3} [M(PR)_3]^+[L_nM]^-$

8.2.2 Metal–Alkyl, –Vinyl, and –Hydride Complexes

8.2.2.1 Metal Alkyls

8.2.2.1.1 General Overview

The interest of chemists for transition-metal–alkyl complexes in the 1950s was partly due to the fact that the attempts to synthesize them had failed during the first half of the twentieth century. These failures had been attributed for a long time to the supposed intrinsic weakness of the metal–carbon σ-bond. It is now known that the strengths of these bonds are in the range between 30 and 65 kcal mol^{-1}, and an increase of strength being observed upon going down in the periodic table as well

Figure 8.1 Representation of agnostic bond.

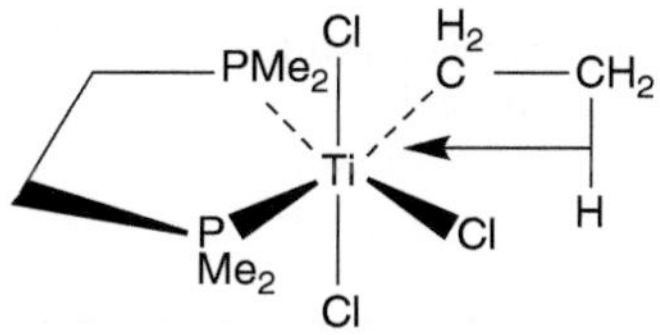

as with perfluorinated ligands, such as C_6F_5. The bond between tetrahedral carbon and transition metals mainly exists in the form of terminal metal–alkyl complexes, such as $[W(CH_3)_6]$. Metal–alkyl complexes are stable if the β-elimination path is blocked either because the complexes have 18 valence electrons or have less valence electrons, but no β-hydrogen. $[Ta(CH_2Ph)_5]$, $[Ti(CH_2CMe_3)_4]$, $[Cr(CH_2SiMe_3)_6]$, $[TaMe_5]$, and $[FeCp(CO)_2(Et)]$ are the examples of some stable metal–alkyl complexes. Metal–alkyl complexes are stabilized by the agostic C—H—M bond, as shown in Figure 8.1. The agostic bond was discovered by M. Brookhardt and M.L.H. Green, and formally consists in using the electron pair of a C—H bond of one of the pre-coordinated ligands as an additional L ligand if a d metal orbital is empty, i.e. if the metal has otherwise less than 18 valence electrons. Agostic bonds are structurally characterized by X-ray diffraction, the β-carbon atom being located closer to the metal than expected if the agostic bond was absent. The β-hydrogen interaction with the metal, often difficult to see among the X-ray data, is sometimes characterized by neutron diffraction of a deuterium-labeled complexes [8, 12, 13].

8.2.2.1.2 Synthesis and Properties of Metal–alkyl Complexes [12–15]

(A) Synthesis of metal–alkyl complexes

Transmetallation	$[TiCl_4] + 4PhCH_2MgCl \longrightarrow [Ti(CH_2Ph)_4] + 4MgCl_2$
Oxidative addition	$[FeCp^*(CO)_2]_2 \xrightarrow{Na/Hg} 2[FeCp^*(CO)_2]^-Na^+ \xrightarrow{CH_3OCH_2Cl} 2\left[Fe(Cp^*CO)_2(CH_2OCH_3)\right]$
Olefin and carbene insertion into metal hydrides	$[FeCp(CO)_2(H)] + CH_2 = CH - CH = CH_2 \longrightarrow [FeCp(CO)_2(CH_2 = CH - CH = CH_3)]$
Nucleophilic addition	$[FeCp(CO)_2(CH_2 = CMe_2)]^+ + NaBH_4 \longrightarrow [FeCp(CO)_2(t\text{-}Bu)]$

(B) Properties of metal–alkyl complexes

The insertion of CO, NO, SO_2, olefins, and alkynes in metal–alkyl bonds

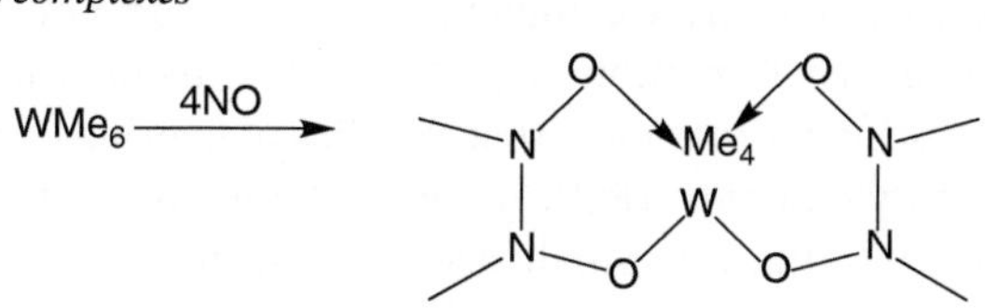

$$WMe_6 \xrightarrow{O_2} [W(O)Me_6]$$

The reaction with electrophiles	$[Fe(CO)_4(CH_3)^-Na^+ + PhCOCl$ $\longrightarrow$ "$Fe(CO)_4(CH_3)(COPh)$" $+ Na^+, Cl^-$ $\longrightarrow PhCOCH_3 +$ "$Fe(CO)_4$" $[FeCp(CO)_2(C_2H_5)] + Br_2 \longrightarrow [FeCp(CO)_2Br + C_2H_5Br$

8.2.2.2 Metal Vinyls

The vinyl complexes are accessible by transmetallation, oxidative addition of a vinyl halide, addition of an acid on a neutral alkyne complex, insertion of an alkyne in a metal hydride, reduction of a vinylidene complex, or nucleophilic attack of a cationic alkyne complex. The vinyl ligand can also coordinate to a second metal through its double bond. The syntheses of various types of metal vinyls are as follows [1].

$$MCl_n + nLi(RC + CR'_2) \longrightarrow M(RC{+}CR'_2)_n \quad M{=}Zr, R{=}Ph, R'{=}Me, n{=}4$$

$$L_nM + RCH{=}CHBr \longrightarrow L_nM(Br)(CH{=}CHR) \quad ML_n = Pt(PPh_3)_3, R = Ph$$

$$L_nM(RC{\equiv}CR) + HX \longrightarrow L_nM(X)(CR{=}CHR) \quad ML_n = Pt(PPh_3)_2, R = Me, X = Cl$$

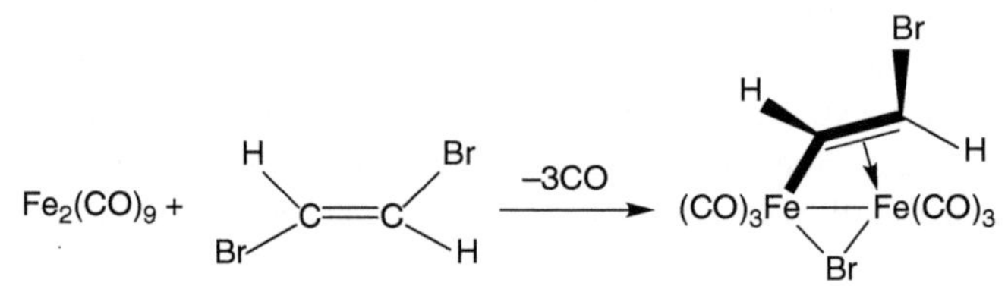

8.2.2.3 Metal Hydrides

8.2.2.3.1 General Overview

Transition-metal hydrides occupy an essential place in inorganic chemistry not only because of the originality of the H ligand due to its very small size but also the applications of the insertion reactions in the M—H bonds in synthesis, catalysis, and polymers science. The H ligand is most of the time terminal, but many complexes are known in which the H ligand bridges two or three metals or even more. The H atoms are sometimes interstitial in inorganic clusters, such as $[Rh_{13}(H)_3(CO)_{24}]_2^-$. It is possible to count electrons brought to the metal or to two metals. In bimetallic complexes, the bridging hydride is an X (one-electron) ligand for a metal, and this M—H bond is a two-electron L ligand for the second metal. Beyond two metals, the count can be carried out only for the whole cluster. There are many metal–carbonyl–hydride complexes with one or several hydride ligands due to inter alia to their easy formation by protonation of metal–carbonyl anions. According to the same principle, various other families of hydrides can be synthesized. The only stable binary complex, $[R_eH_9]_2^-$, is well known for its exceptionally high valence ($C = 9$) due to the very small size of the hydrogen atom. Different types of metal hydrides are shown in Figure 8.2 [16, 17].

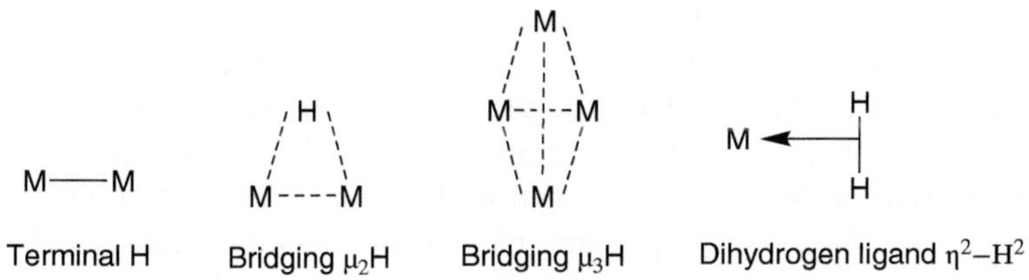

Figure 8.2 Different types of metal hydrides.

8.2.2.3.2 Synthesis of the Hydrides [1, 16, 17]

Protonation	$[CoCp(PMe_3)_2] + NH_4^+PF_6^- \xrightarrow{Et_2O} [CoCp(H)(PMe_3)_2]^+PF_6^- + NH_3$
Hydriding	$[FeCp^*(CO)_2Br] + Na^+BH_4^- \xrightarrow{THF} [FeCp^*(CO)_2H] + Na^+Br^- + BH_3$
H atom transfer from the medium	$[Fe^{I}Cp(arene)] + 2PR_3 \xrightarrow{THF\ or\ Toluene} [Fe^{II}Cp(PR_3)_2(H)] + arene$ R = Ph, Me
Oxidative addition of H_2 on 16-electron complexes	$[IrCl(CO)(PPh_3)_2] + H_2 \longrightarrow [IrCl(CO)(PPh_3)_2(H)_2]$
Hydrogenolysis of d^0 metal–alkyl complexes by σ-bond metathesis	$WMe_6 + 3PMe_2Ph + 6H_2 \longrightarrow [WH_6(PMe_2Ph)_3] + 6CH_4$
Decomposition of a ligand	$Fe(CO)_5 + OH^- \rightarrow Fe(CO)_4(COOH)^- \longrightarrow [Fe(CO)_4(H)^- + CO_2$
Deprotonation of H_2 complexes	$[FeCp(dppe)(H_2)]^+PF_6^- + NEt_3 \longrightarrow [FeCp(dppe)H] + [NHEt_3]PF_6^-$

8.2.2.3.3 Properties of the Hydrides

The hydride ligands are easily detected by ^{1}H NMR because they are usually found between 0 and −20 ppm. The hydride can be coupled with the metal when the latter has a spin 1/2 and with the phosphorus atom of phosphanes coordinated to the same metal in cis ($J = 15$–30 Hz) or trans ($J = 90$–150 Hz) position. The infrared stretching M—H bond near 1500–2200 cm^{-1} is not intense enough to be reliable. The hydrogen atoms are also detected with difficulty by X-ray because of their very small size. Neutron diffraction is a much better technique, although it is a heavier one and requires relatively large crystals (at least 1 mm^3). The hydride complexes can be either good hydride donors (hydridic property) or acids. This hydridic vs. acidic property depends on the charge, nature of the ligands, and location of the metal in the periodic table. $[VH(CO)_6]$ and $[CoH(CO)_4]$ are strong acids. The acidity decreases as one goes down in the periodic table, or if one or several carbonyls are replaced by more donor ligands. There is also, of course, a continuum of intermediate situations where some hydrides, such as for instance $[WCp(CO)_3(H)]$, are able to behave as H^+ or H^- donors depending on the substrate. The charge plays a key role for the acidity and hydride-transfer property of metal–hydride complexes. Tilset [18] has shown that the monoelectronic oxidation of a metal hydride lowers its pK_a by 20.6 units for a variety of structures. The H–donor property of transition-metal hydrides can be paralleled with that of n-Bu_3SnH, which is very much used in organic chemistry in spite of the toxicity of molecular tin compounds [1, 16–18].

8.2.2.4 Metal–Carbene Complexes

8.2.2.4.1 General Overview

The complexes containing a double carbon-element bond, i.e. metal–carbenes M=C, are key species that are important from a theoretical point of view in biochemistry, organic synthesis and catalysis of metathesis, dimerization, polymerization, and oxidation of olefins. The M=C bond can be polarized in one way or the other, given the large flexibility of electronic properties of transition-metal fragments [19, 20].

$$M^+ - C^- \longleftrightarrow M = C \longleftrightarrow M^- - C^+$$

The cationic or neutral complexes of electrophilic carbene of Pettit type tend to behave somewhat as carbocations coordinated to transition metals, and the neutral metal complexes of nucleophilic carbene complexes of Schrock type tend to react as ylids. It has been suggested that complexes of electrophilic carbenes be considered as the combination of a metal fragment with a singlet carbene, an L ligand, whereas the complexes of nucleophilic carbenes are the combination of a metal fragment with a triplet carbene, an X_2 ligand (Figure 8.3) [19, 20].

Pettit's methylene complex could be characterized only by its reactivity or spectroscopically upon attenuating the positive charge by introducing phosphines (instead of the two COs) or C_5Me_5 (instead of Cp). These ligands strengthen the $Fe^+{=}CH_2$ bond, which is characterized by a lower value of the rotation barrier around this bond (for instance, $\Delta G^\ddagger = 10.6\,kcal\,mol^{-1}$ for $[Fe(C5Me_5)(CO)(PPh_3)(=CH_2)]^+$. It is Schrock's methylene complex that was isolated and characterized first, this complex being more stable partly because of a weaker polarity, which allowed the determination of its X-ray crystal structure that confirmed the $Ta{=}CH_2$ double bond. This double bond is also apparent in ^1H NMR, and the two methylene protons remain nonequivalent up to at least 100 °C, the decomposition temperature ($\Delta G\ddagger > 22\,kcal\,mol^{-1}$) [19, 20].

8.2.2.4.2 Syntheses of Metal Carbenes [21–24]

Complexes of nucleophilic carbenes	1. Dehydrohalogenation of a bulky complex 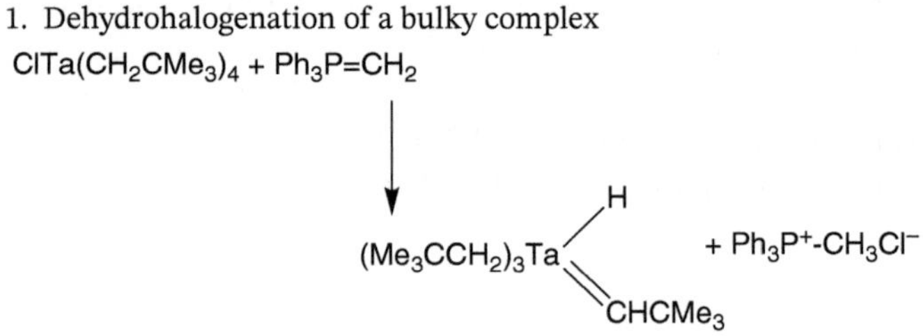

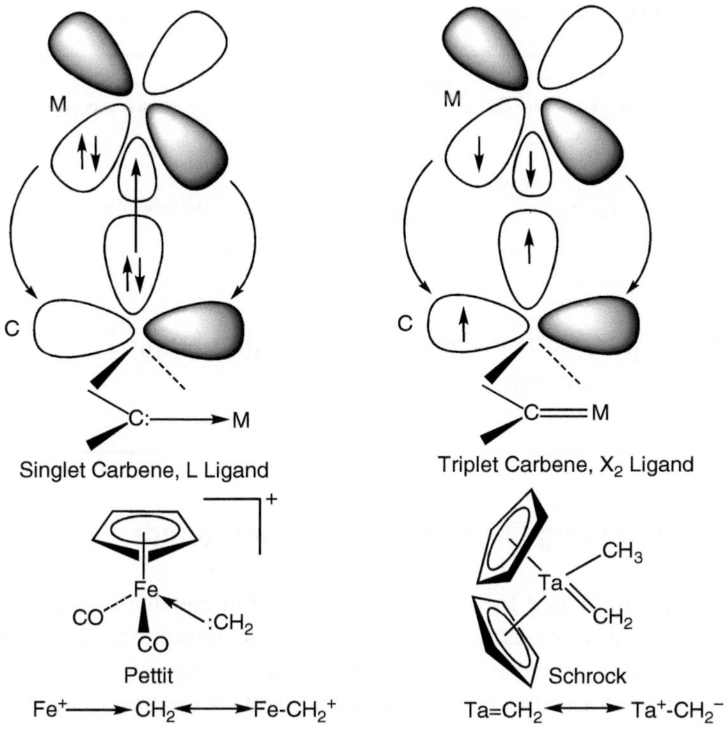

Figure 8.3 Singlet and triplet carbene.

---

2. The transmetallation of an alkylidene ligand

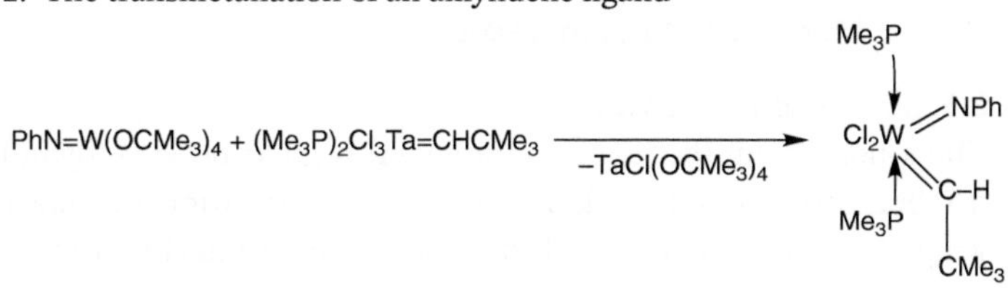

Complexes of electrophilic carbenes

1. The reaction of an acid (a Lewis acid) with a metal-alkyl complex bearing a good leaving group as X− in α−position

$$[RuCl(CO)L_2(CF_3)] + Me_3SiCl \longrightarrow [RuCl(CO)(L_2)(=CF_2)] + Me_3SiF$$

2. The protonation of a vinyl complex

$$[M - CH = CH_2] + H^+ \longrightarrow [M(= CHMe)]^+ \xrightarrow{\text{Heat}} [M\left(\eta^2 - CH_2 = CH_2\right]^+$$

$$M = FeCp(CO)_2$$

8.2.2.4.3 Reactions of Metal Carbenes

Reactions of complexes of nucleophilic carbenes with electrophiles and acids	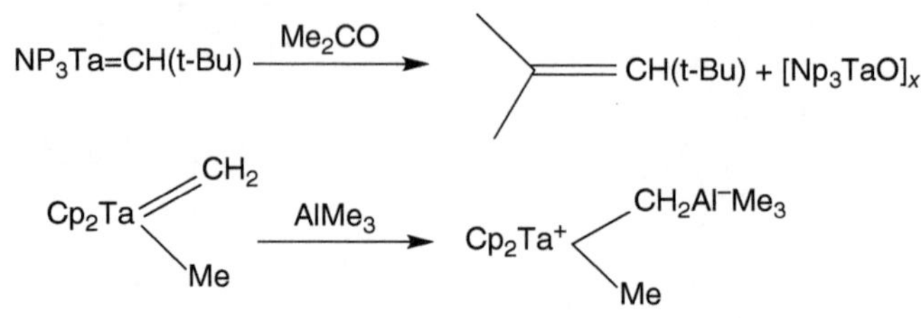
Reactions of complexes of electrophilic carbenes with nucleophiles and bases	$[FeCpL_2(CH_2)]^+ + PPh_3 \longrightarrow [FeCpL_2(CH_2PPh_3)]^+$ $L = CO, PPh_3$

8.3 π-Bond Containing Organometallic Compounds

The complexation of unsaturated hydrocarbons by transition metals is a powerful activation method that plays a fundamental role in their stoichiometric and catalytic transformation. The bonds formed are governed by π-backbonding. Olefins, dienes, and arenes are usually two-electron L, four-electron L_2, and six-electron L_3 ligands, respectively, and alkynes are two-electron L or four-electron L_2 depending on the needs of the metal. The odd-electron ligands are the three-electron LX allyl radical and five-electron L_2X dienyl radicals [25].

8.3.1 Metal–Olefin Complexes

8.3.1.1 General Overview

The π-bond from the olefin to the metal is the σ-bond (L ligand), whereas the filled d metal orbitals give back an electron density from the side into the antibonding π*ligand orbitals, which makes the π-backbonding (Figure 8.4). This bond requires

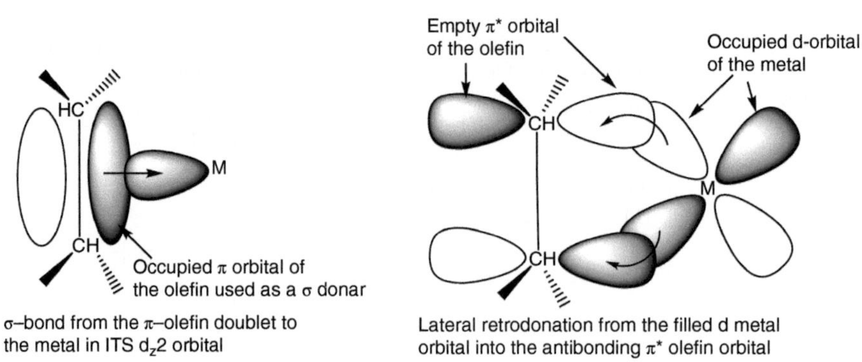

Figure 8.4 σ- and π-bonding in metal–olefin complex.

nonbonding d electrons in the metal valence shell. Thus, d^0 complexes, such as those of Sc^{III}, Ti^{IV}, or V^{V}, cannot give stable olefin complexes. With electron-rich late transition metals in low oxidation states (Ni^0, Pd^0, and Pt^0 with phosphine ligands) and/or electron-poor olefins, the backbonding is so important that it is formalized as an oxidative addition of the olefin, i.e. giving a metallacyclopropane. Therefore, X-ray crystallography that allows to accurately determine the exact C–C distance is useful in order to give a measure of the π-backbonding [25, 26].

8.3.1.2 Syntheses of Metal–Olefin Complexes

The olefin complexes are usually obtained by addition or substitution, eventually using a substrate able to abstract X^-, such as Ag^+ or a reductant. It is also possible to protonate an allyl complex or to abstract H^- from an alkyl complex [21–27].

$$\begin{aligned} AgOSO_2CF_3 + C_2H_4 &\longrightarrow [(C_2H_4)AgOSO_2CF_3] \\ [PtCl_4]^{2-} + C_2H_4 &\longrightarrow [PtCl_3(C_2H_4)]^- + Cl^- \\ [FeCp(CO)_2I] + C_2H_4 + AgBF_4 &\longrightarrow [FeCp(CO)_2I(C_2H_4)][BF_4] + AgI \end{aligned}$$

8.3.1.3 Reactions of Metal–Olefin Complexes

The main reactions of olefin complexes are the facile exchange of the olefin ligand with another ligand, the insertion of the olefin ligand into an M—H or M—alkyl bond, the addition of nucleophiles, and the activation of allylic C—H bonds [21–27].

$$[Pt(PPh_3)_2(C_2H_4)] + C_{60} \longrightarrow [Pt(PPh_3)_2(\eta^2\text{–}C_{60})] + C_2H_4$$

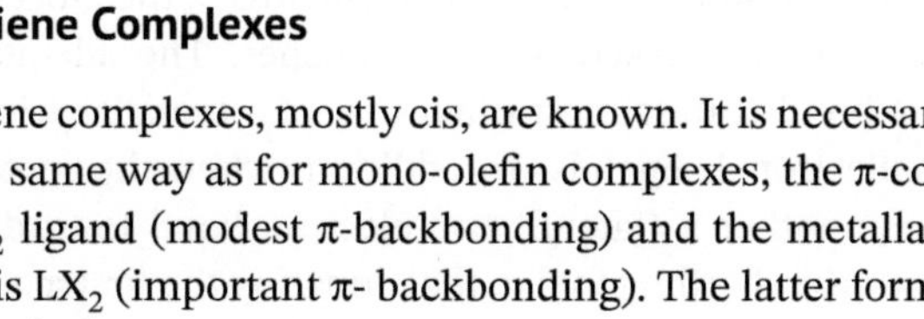

8.3.2 Metal–Diene Complexes

Many 1,3-butadiene complexes, mostly cis, are known. It is necessary to involve two limit forms in the same way as for mono-olefin complexes, the π-complex in which the diene is an L_2 ligand (modest π-backbonding) and the metallacyclopentene in which the ligand is LX_2 (important π- backbonding). The latter form corresponds to oxidative coupling [28].

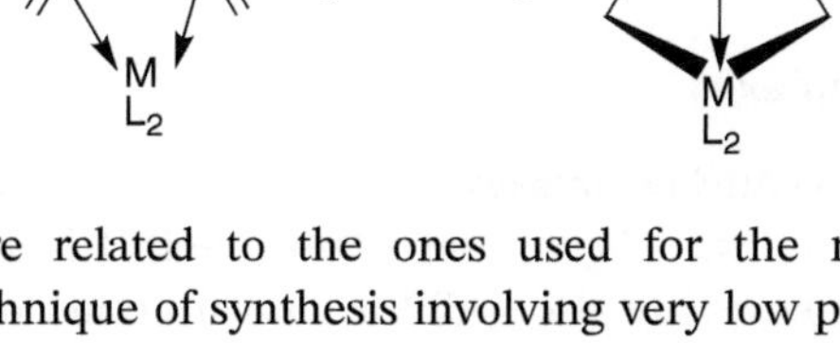

The syntheses are related to the ones used for the monoene complexes. The metal-vapor technique of synthesis involving very low pressure of metals has been used to make binary complexes, such as molybdenum (0) tris-1,3-butadiene. The diene ligand can be protonated or acylated. Subsequently, a free coordination site is liberated and can then be filled by a ligand present in the reaction medium or by the oxygen atom of the acyl group. In the absence of such a ligand, this site is occupied by a C—H bond of the dienyl ligand, which provides an agostic bond. The complex [$Fe(CO)_3(\eta^4$-1,3-cyclohexadiene)], useful in organic synthesis, is obtained by complexation–isomerization of the 1,4-isomer by reaction with [$Fe(CO)_5$] [28].

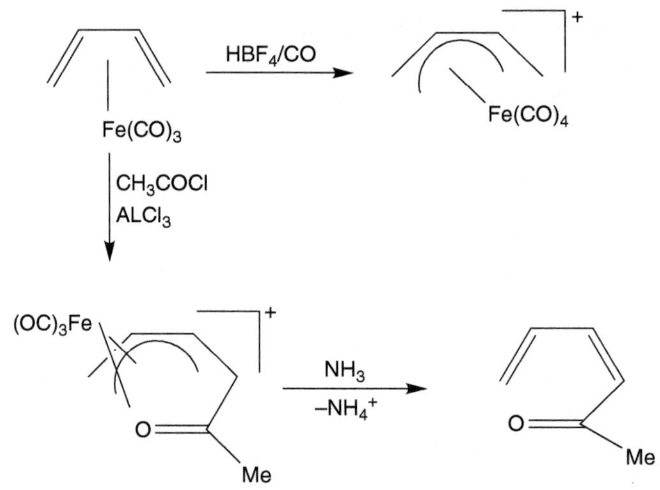

8.3.3 Metal–Alkyne Complexes

The organometallic chemistry of alkynes is very rich, including oxidative addition and coupling, insertion, and metathesis, and leads to a variety of applications in organic synthesis, such as functionalization, C—H and C—C bond cleavage and formation, dimerization, oligomerization, and polymerization of alkynes. Alkynes being much easier to reduce than alkenes (much lower energy of the π* low unoccupied molecular orbital molecular [LUMO]), the π-backbonding of metal d-orbitals is much more marked than for alkenes. The adequate formalism for the metal–alkyne bond is thus most often that of a metallacyclopropene. Alkynes usually are two-electron L ligands. In addition, when the metal is still electron deficient, there is another π (lateral) two-electron bond toward a vacant d metal orbital from the π-orbital orthogonal to that donating the first π doublet forming the σ-bond. Alkynes are also excellent ligands bridging two metals, with one π-orbital for each metal (two orthogonal σ-bonds). Under these conditions, the alkyne is an L ligand for each of the two metals (Figure 8.5) [28].

8.3.4 π-Allyl Complexes

8.3.4.1 Structure of π-Allyl Complexes

π–allyl palladium complexes play an important role in Pd-catalyzed organic reactions but π–allyl complexes are also well known with other transition metals. The

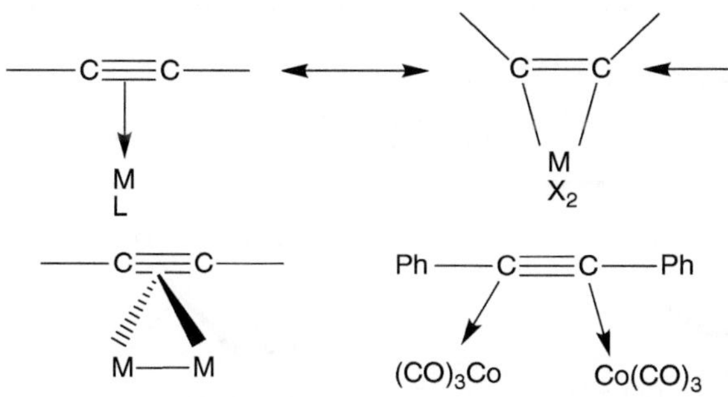

Figure 8.5 Various types of metal–alkyne complexes.

allyl ligand can bind to a metal in a monohapto fashion like an alkyl, in which case it is a one-electron X ligand. Most of the time, however, it is bonded to the metal as a π three-electron LX trihapto ligand [29, 30].

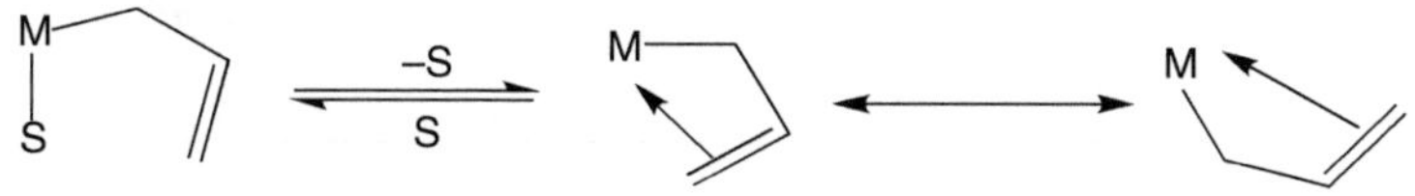

The two forms rapidly interconvert, which is facilitated by a weakly coordinated solvent. This interconversion probably explains the equilibrium between the syn and antiprotons that are nonequivalent in ^{1}H NMR and the reactivity that is due to the temporary monohapto coordination. For static η^3-allyl complexes, the antiprotons are closer to the metal than the syn protons; thus, these antiprotons are shielded and appear at high field (typically between 1 and 3 ppm) in 1 H NMR. The syn protons appear between 2 and 5 ppm and the central protons between 4 and 6.5 ppm [29, 30].

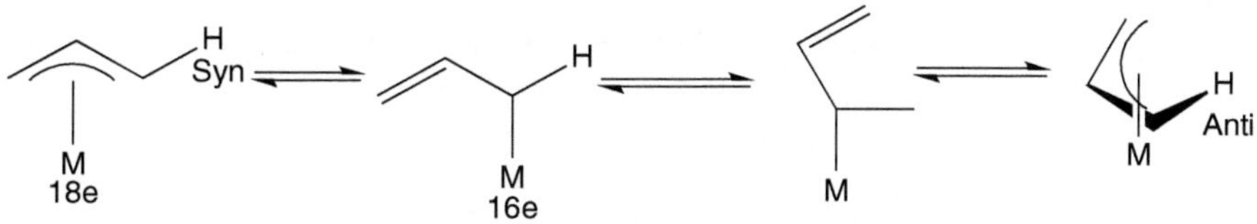

8.3.4.2 Syntheses of π-Allyl Complexes [29, 30]

Allyl transfer using a source of anionic or cationic allyl complexes	$CH_2{=}CHCH_2Cl + Mn(CO)_5^- \xrightarrow{-Cl^-} [Mn(\eta^1\text{-}CH_2{=}CHCH_2)(CO)_5]$
Chelation of a monohapto allyl ligand	$[Mn(\eta^1\text{-}CH_2{=}CHCH_2)(CO)_5] \xrightarrow[-CO]{Heat} [Mn(\eta^3\text{-}CH_2CHCH_2)(CO)_4]$

Oxidative addition

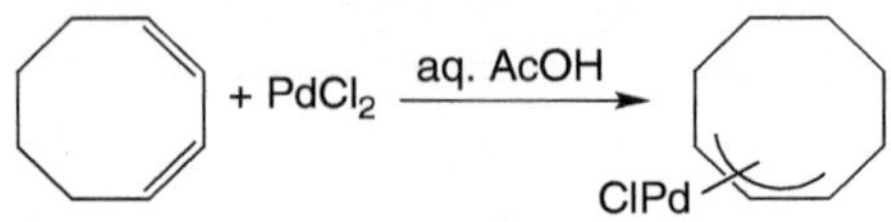

Nucleophilic addition

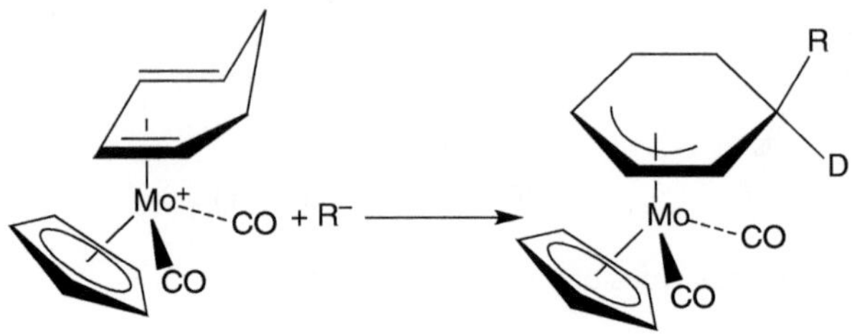

Insertion of a mono- or diene ligand

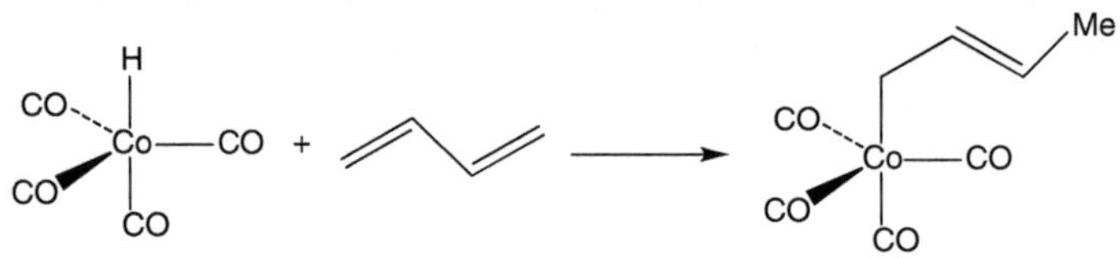

8.3.4.3 Reactions of π-Allyl Complexes [29, 30]

Nucleophilic additions

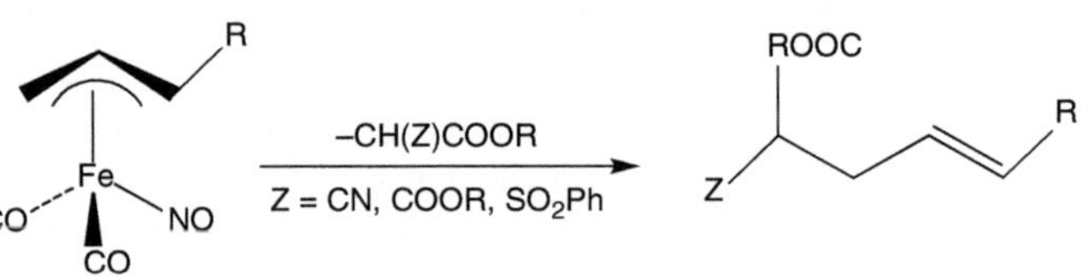

Oxidative addition

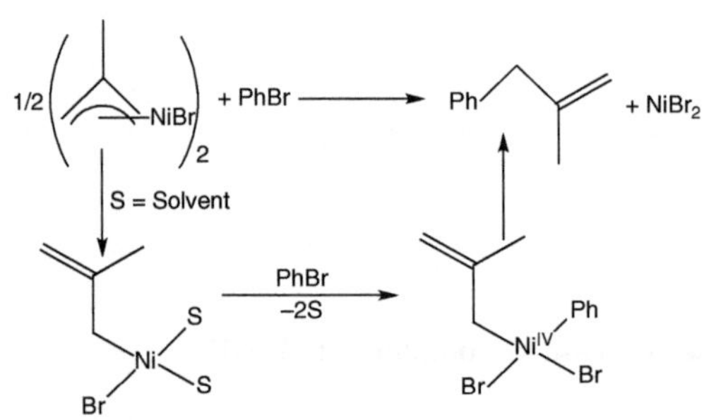

Insertion reactions

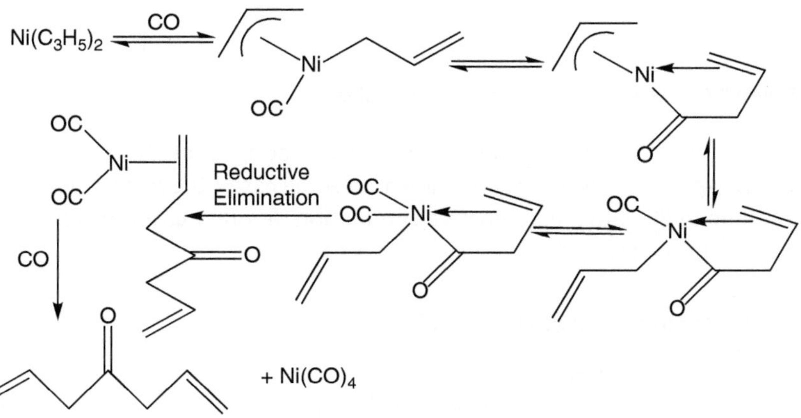

8.4 η^n-C_nR_n Carbocyclic Polyenes Containing Organometallic Compounds

8.4.1 Cyclopropenyls, η^3-C_3R_3

The cation $[C_3H_3]^+$ is the simplest aromatic (2π) carbocyclic system. However, it has only three p-orbitals with which to construct (π) frontier orbitals. Cyclopropenyl halides readily ionize to provide aromatic (2π) cyclopropenium salts (isolated with noncoordinating anions). Alternatively, a hydride may be abstracted from the sp^3 carbon of cyclopropenes (e.g. with $[Ph_3C]BF_4$) to provide isolable cyclopropenium salts, among which the triphenyl derivative has been the most widely studied in an organometallic context. These and the halides serve as the most common means for introducing the cyclopropenyl ligand via oxidative addition or nucleophilic attack by carbonyl metallates. η^1 cyclopropenyl complexes may be isolated as intermediates, while in others, migration to a carbonyl ligand may be followed by the CO group into the (strained) carbocycle, resulting in the formation of η^3 cyclobutenyl ligands [1, 2].

8.4.2 Cyclobutadienes, η^4-C_4R_4

As early as 1956, it was suggested that the ($4\pi \neq 4n+2$) molecule cyclobutadiene might be stabilized by coordination to a transition metal. This, coupled with the known ability of the $Fe(CO)_3$ unit to stabilize butadiene coordination, and the previous isolation of complexed substituted cyclobutadienes led Pettit to design and isolate $Fe(CO)_3(\eta$-$C_4H_4)$ from the reaction of dichlorocyclobutene with $Fe_2(CO)_9$. In this reaction, the iron carbonyl acts as both a reductant and a trap. More generally, the majority of cyclobutadiene complexes arise from, prior to or as a result

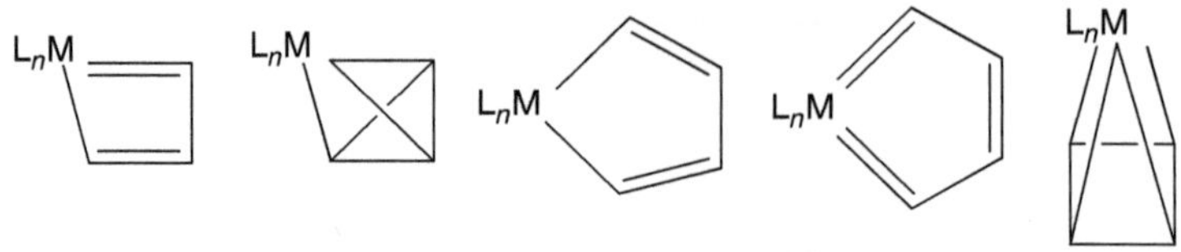

Figure 8.6 Different types of η^4-C_4R_4 complexes.

of coordination. Many examples of both metallacyclopentadienes and cyclobutadiene complexes (Figure 8.6) are known. In some cases, the former lies synthetically en route to the latter via reductive elimination. In other cases, this final step is not easily induced [1, 2].

8.4.3 Cyclopentadienyls, η^5-C_5R_5

8.4.3.1 General Overview

Complexes of the cyclopentadienyl ligand fall loosely into three classes, namely metallocene, bent metallocene, and half-sandwich complexes. For metallocenes, interest focuses on the chemistry of the cyclopentadienyl rings, while for bent metallocenes and half sandwich complexes, the cyclopentadienyl ligands usually serve as innocent spectator ligands. A special case of bent metallocenes involves those wherein the two rings are joined by a bridge, namely ansa metallocene and metallocenophanes, which may impart steric (and resulting electronic) constraints upon the complex geometry [31].

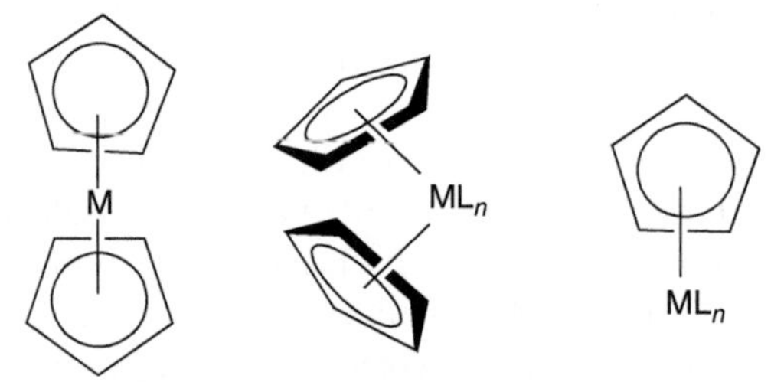

Metallocene Bent metallocene Half sandwich

Metal–sandwich complexes occupy a major place in organometallic chemistry for various reasons. First, from the historical point of view, the discovery of the ferrocene structure by Wilkinson [32] has been a starting point for the boom of the chemistry of π-complexes. Then, ferrocene has provided a very rich chemistry as a superaromatic compound. The family of metallocenes (MCp_2) and decamethylmetallocenes (MCp^*_2) now includes all the transition and main-group metals. Then, ferrocene has provided a very rich chemistry as a superaromatic compound. Its redox properties have led to multiple applications in the field of materials and molecular engineering: molecular ferromagnets, modified electrodes for redox catalysis (titration of glucose in blood), polymers and dendritic electrochemical sensors for molecular recognition, and antitumor drugs and paintings [32].

8.4.3.2 Structure of Metallocene

Metallocenes, in a broad sense, have the composition [MCp_2] and are known for all the transition metals and many main-group and rare-earth metals. The sandwich

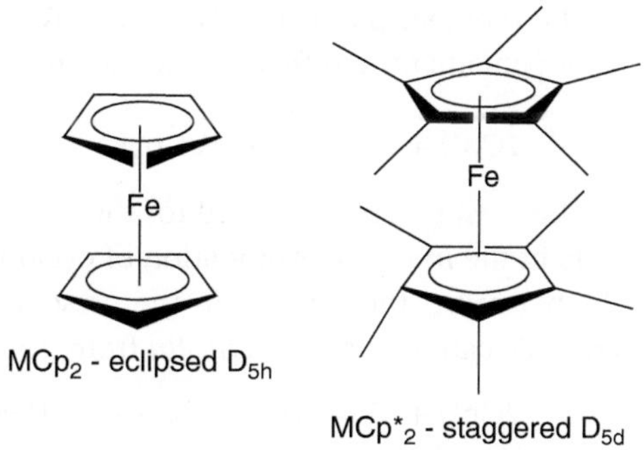

Figure 8.7 Representation of eclipsed and staggered ferrocene.

structure is reserved to the first row of transition metals from Ti to Ni and for the iron column for which the 18-electron rule brings a special robustness. It is also known for a few other metals, including main-group ones. Early transition-metal metallocenes, in particular, do not have the sandwich structure, and the Cp rings make an angle of about 130°. The metal bears up to three other ligands. Ferrocene is a prime example of metallocene. Ferrocene has two isoenergetic conformations: eclipsed (D_{5h} symmetry) and staggered (D_{5d} symmetry) (Figure 8.7). X-ray crystallographic and electronic diffraction studies in the gas phase indicate an eclipsed conformation as well as for ruthenocene and osmocene. The rotation barrier about the iron–Cp ring axis is very weak (2–5 kcal mol^{-1}1). Thus, this rotation is free and fast in fluid solution, i.e. there is only one 1,1′-disubstituted isomer (by 1 and 1′, it is meant that the substituents are located on different rings) because this fast rotation makes all the positions on the first ring equivalent for the second ring. In decamethyl ferrocene, the solid-state conformation is staggered [33].

Ferrocene is an air-stable orange crystalline powder that sublimes at 184 °C under 1 atm and is stable up to 400 °C; it is soluble in hydrocarbons, has a nil dipole moment and thus covalent metal–ring bonds. The molecular orbital (MO) diagram below shows various metal–ring interactions: σ (Cpto Fe), π (Cpto Fe), and δ (Fe to Cp). The other metallocenes are represented by the same diagram, although the energy levels of the orbitals may vary from one metal to the other [33].

8.4.3.3 Syntheses of Metallocene

First, it is necessary to crack dicyclopentadiene at 180 °C, which provides the monomer that must be used immediately as such or reduced to the anion Cp^- by deprotonation providing LiCp, NaCp, KCp, TlCp, or [SnCp(n-Bu_3)] [34–36].

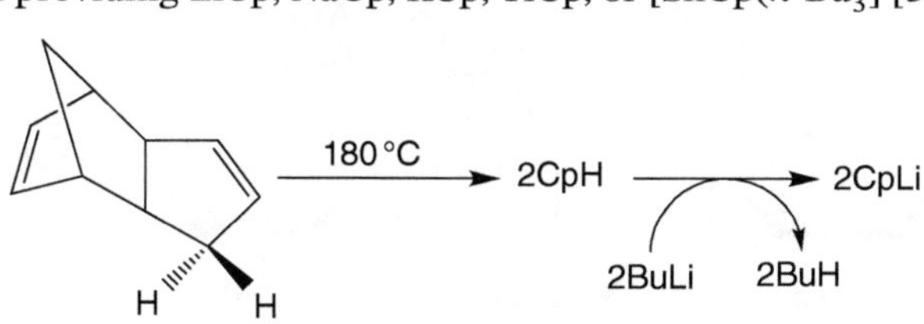

$$MX_2 + 2M'Cp \longrightarrow [MCp_2] + 2M'X^-$$

The reduction of the transition-metal halide is necessary *in situ*, either by the salt MCp itself or by Zn.

$$RuCl_3(H_2O)_x + 3\,C_5H_6 \xrightarrow{\text{Zn,EtOH}} [RuCp_2] + C_5H_8$$

The method, published by Miller, Tebboth, and Tremaine in 1951, starts from cyclopentadiene and freshly reduced iron.

$$2CpH + Fe \xrightarrow{300°C} [FeCp_2] + H_2$$

The most practical method for the syntheses of ferrocene and its derivatives consists in the *in situ* deprotonation of cyclopentadiene ($pK_a = 15$) or its derivative by diethylamine, the solvent, in the presence of anhydrous $FeCl_2$ or $FeCl_3$. The iron chloride can be generated *in situ* from iron powder and the amine chlorohydrate.

$$2CpH + 2Et_2NH + FeCl_2 \longrightarrow [FeCp_2] + 2Et_2NH_2^+Cl^-$$

$$2CpH + Fe \xrightarrow{Et_2NH_2^+Cl^-} [FeCp_2] + H_2$$

8.4.3.4 Chemical Properties of Metallocene

Ferrocene undergoes many electrophilic reactions, more rapidly than benzene, although they are limited by oxidation reactions with electrophiles that are strong oxidants (H_2SO_4 or HNO_3). Formylation and carboxylation reactions give only monofunctionalization because the functional group strongly deactivates the

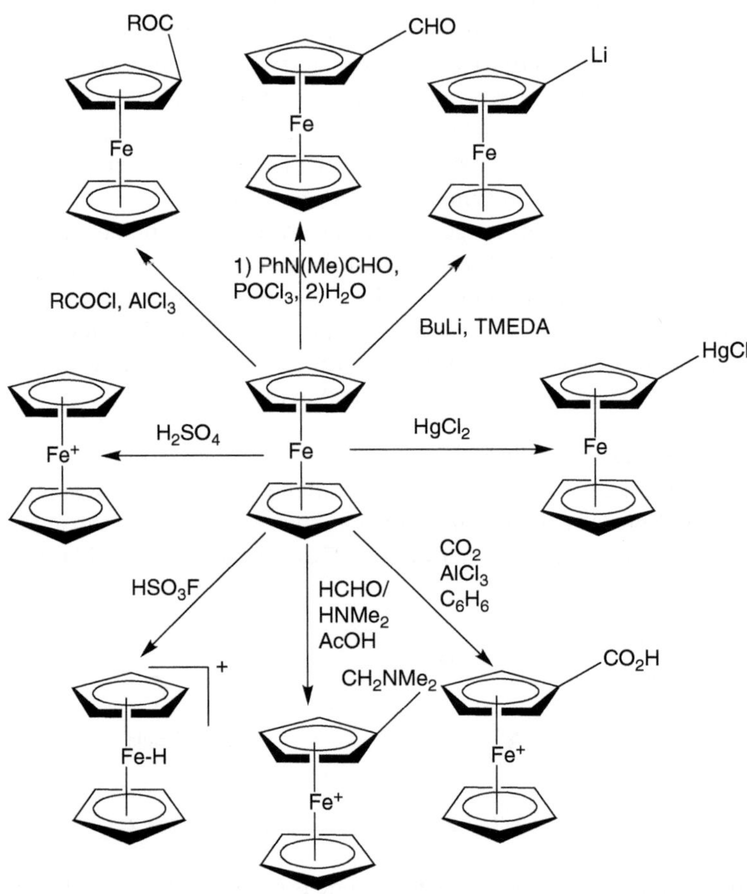

ferrocenyl group. Indeed, there is, to a certain extent, transmission of the electronic effect through the metal center. On the other hand, metallation and acylation reactions can be followed by an identical reaction on the other ring leading to 1,1′-disubstituted derivatives because the deactivation of the second Cp ring by the substituent is only modest [37–40].

Lewis acids such as $AlCl_3$ form complexes leading to the cleavage of the Fe–Cp ring upon heating. The cleaved ring is protonated *in situ* and becomes an electrophile toward ferrocene. The reaction continues another time to give the cyclopentylene bridge. The syn and anti-isomers of the dibridged derivatives also form, and they have been separated, yielding cage structure in which the iron is encapsulated by four five-carbon rings. In the presence of an aromatic or CO, these reactions are minimized because the main reaction is substitution of a ferrocene ring by an arene or three CO ligands. This reaction is the classic mode of synthesis of the sandwich complexes $[FeCp(arene)]^+$ [37–40].

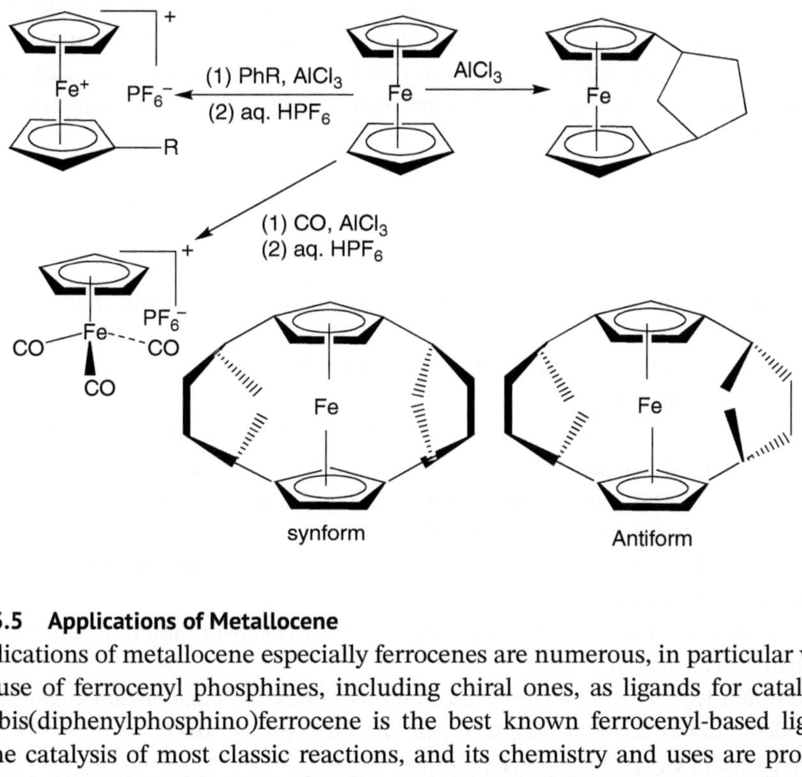

8.4.3.5 Applications of Metallocene

Applications of metallocene especially ferrocenes are numerous, in particular with the use of ferrocenyl phosphines, including chiral ones, as ligands for catalysis. 1,1′-bis(diphenylphosphino)ferrocene is the best known ferrocenyl-based ligand in the catalysis of most classic reactions, and its chemistry and uses are prolific. The redox property of ferrocene has been used to attach it to macrocycles (below), cryptands, calixarenes, and other endoreceptors for sensing, a chemistry beautifully developed by Paul Beer at Oxford. Ferrocenes are used in materials science, for instance to assemble charge-transfer complexes (decamethylferrocene) and thermotropic liquid crystals. Ferrocenes are key components of polymers. For instance, Manners disclosed remarkable ring-opening polymerization of ferrocenophanes. Ferrocene-containing polymers have also been synthesized using classic routes.

cationic, radical, or ROMP polymerization. They are used inter alia to derivatize electrodes, which can provide excellent redox sensors. Finally, ferrocenes, as many other cyclopentadienyl–metal complexes, have been widely studied as antitumor drugs or other therapeutic agents [37–40].

8.5 Conclusion

Compounds with at least one bond between carbon and metal are known as organometallic compounds. This chapter has explored the different syntheses routes, properties, and applications of carbonyls, alkyls, carbenes, hydrides, and carbocyclic ligands that contain organometallic compounds. Carbonyls, alkyls, carbenes, and ferrocenes ligands containing complexes are such important players of organometallic chemistry and offer a very facile route to the synthesis of many other organometallic compounds. There are numerous applications of these complexes such as constituent parts of elastomers, adhesives, electron transfer resins, semiconductors, UV absorbers, solid-propellant additives, combustion activators, heat transfer fluids, lubricants, and antioxidants. These compounds also offer a versatile platform for the design of therapeutics from antimicrobials to anticancer agents [1, 2].

References

1 Astruc, D. (2007). *Organometallic Chemistry and Catalysis*. Berlin Heidelberg: Springer-Verlag.

2 Powell, P. (1988). *Principles of Organometallic Chemistry*, 2e. Springer-Science + Business Media, B. V.

3 Cotton, F.A. (1976). Metal carbonyls: some new observations in an old field. *Prog. Inorg. Chem.* 21: 1.

4 Ellis, J.E. (1990). Highly reduced metal carbonyl anions: synthesis, characterization and chemical properties. *Adv. Organomet. Chem.* 31: 1.

5 Dobson, G.R. (1976). Trends in reactivity for ligand exchange reactions of octahedral metal carbonyls. *Acc. Chem. Res.* 9: 300.

6 Basolo, F. (1990). Kinetics and mechanisms of CO substitution of metal carbonyls. *Polyhedron* 9: 1503.

7 Werner, H. (1990). Complexes of carbon monoxide: an organometallic family celebrating its birthday. *Angew. Chem. Int. Ed. Engl.* 29: 1077.

8 Wilkinson, G. (1959). *Pure Appl. Chem.* 71: 627.

9 Marks, T.J. (1990). *Bonding Energetics in Organometallic Compounds*, ACS Symp. Series. United States, Washington DC: American Chemical Society.

10 Usón, R. and Forniés, J. (1992). Synthesis and structures of novel types of Pt-M (M = Ag, Sn or Pb) neutral or anionic organometallic complexes. *Inorg. Chim. Acta* 198: 165.

11 Brookhart, M. and Green, M.L.H. (1983). Carbon-hydrogen-transition metal bonds. *J. Organomet. Chem.* 250: 395.

12 Schrock, R.R. and Parshall, G.W. (1976). σ-Alkyl and -aryl complexes of the group 4-7-transition metals. *Chem. Rev.* 76: 243.

13 Davison, P.J., Lappert, M.F., and Pearce, R. (1976). Metal σ-hydrocarbyls, MRn. Stoichiometry, structures, stabilities, and thermal decomposition pathways. *Chem. Rev.* 76: 219.

14 Caulton, K.G. and Hubert-Pfalzgraf, L.G. (1990). Synthesis, structural principles and reactivity of heterometallic alkoxides. *Chem. Rev.* 90: 969.

15 Sharp, P.R., Astruc, D., and Schrock, R.R. (1979). Niobium and tantalum mesityl complexes and the role of the mesityl ligand in hydrogen abstraction reactions. *J. Organomet. Chem.* 182: 477.

16 Dedieu, A. (ed.) (1992). *Transition Metal Hydrides*. Weinheim: VCH.

17 Schollhorn, R. (1984). *Intercalation Compounds, in Inclusion Compounds*, 249–349. London: Academic Press.

18 Parker, V.D., Handoo, K.L., Roness, F., and Tilset, M. (1991). Electrode potentials and the thermodynamics of isodesmic reactions. *J. Am. Chem. Soc.* 113: 7493. and 115, 7493, 1993.

19 Von, K.H., Dötz, H., Fischer, P. et al. (1983). *Transition Metal Carbene Complexes*. Weinheim: VerlagChemie.

20 Schrock, R.R. (1979). Alkylidene complexes of niobium and tantalum. *Acc. Chem. Res.* 12: 98; Science 13, 219, 1983.

21 Fischer, E.O., Schubert, U., and Fisher, H. (1978). Selectivity and specificity in chemical reactions of carbene and carbyne metal complexes. *Pure Appl. Chem.* 50: 857.

22 Brothers, P.J. and W.R. (1988). Roper – transition-metal dihalocarbene complexes. *Chem. Rev.* 88: 1293.

23 Guerchais, V. and Astruc, D. (1985). New Entry to (C_5Me_5)Fe chemistry and the methylene complexes [Fe(η^5-C_5Me_5)(CO)(L)(=CH_2)]$^+$, L = CO or PPh_3. *J. Chem. Soc., Chem. Commun.* 835.

24 Taylor, T.E. and Hall, M.B. (1984). Theoretical comparison between nucleophilic and electrophilic transition-metal carbenes using generalized MO and CI methods. *J. Am. Chem. Soc.* 106: 1576.

25 Dewar, M.J.S. (1951). A review of the π-complex theory. *Bull. Soc. Chim. Fr.* 18: C79.

26 Albright, T.A., Hoffmann, R., Thibeault, J.C., and Thorn, D.L. (1979). Ethylene complexes. bonding, rotational barriers, and conformational preferences. *J. Am. Chem. Soc.* 101: 3801.

27 Balch, A.L., Catalano, V.J., Lee, J.W. et al. (1991). (.eta.2-C70) Ir (η^2-C_{70})(CO)(Cl)(PPh_3)$_2$: the synthesis and structure of an iridium organometallic derivative of a higher fullerene. *J. Am. Chem. Soc.* 113: 8953.

28 Trost, M.B. (1980). New rules of selectivity: allylic alkylations catalyzed by palladium. *Acc. Chem. Res.* 13: 385.

29 Baker, R. (1973). Allyl metal derivatives in organic synthesis. *Chem. Rev.* 73: 487.

30 Green, M.L.H. (1964). Allyl metal complexes. *Adv. Organomet. Chem.* 2: 325.

31 Hill, A.F. (2002). *Organotransition Metal Chemistry*. The Royal Society of Chemistry.

32 Wilkinson, G. (1975). The iron sandwich. A recollection of the first four months. *J. Organomet. Chem.* 100: 273.

33 Togni, A. and Hayashi, T. (ed.) (1995). *Ferrocenes*. Weinheim: VCH.

34 Astruc, D. (1995). *Electron Transfer and Radical Processes in Transition Metal Chemistry*, 142. New York: VCH.

35 Kahn, O. (1994). *Molecular Magnetism*, 294. New York: VCH.

36 Gmelin, L. and Fluck, E. *Gmelin Handbook of Inorganic and Organometallic Chemistry*, Fe Organic Compounds, 8e, Vols. A1-A8 (1974, 1977, 1978, 1980, 1981, 1985, 1989, 1991).

37 Rockett, B.W. and Marr, G. (1991). Ferrocene Chemistry. *J. Organomet. Chem.* 416: 327. and cited references.

38 Herberich, G.E. (1983). A novel and general route to metal (η^5-borole) complexes: compounds of manganese, ruthenium and rhodium. *Angew. Chem. Int. Ed. Engl.* 22: 996.

39 Buchholz, D., Fillaut, J.-L., Cotrait, M., and Astruc, D. (1995). Mono- and bis-pentaisopropylcyclopentadienyl cobalt and rhodium sandwich complexes and other decabranched cyclopentadienyl complexes. *Chem. Eur. J.* 1: 374.

40 Carmona, E. (2000). *Angew. Chem. Int. Ed. Engl.* 39, 1949; Chem. Commun., 2916, 2000; Synthesis, solid-state structure, and bonding analysis of the octa, nona, and decamethylberyllocenes, Chem. Eur. J. 9, 4452, 2003.

9

Organometallic Complexes of the Lanthanides and Actinides

Shippi Dewangan[1], Amarpreet K. Bhatia[2], and Priyanka Singh[3]*

[1] *Hemchand Yadav University, Durg, Sw. Pukeshwar Singh Bhardiya Govt. College, Department of Chemistry, Nikum, Durg, Chhattisgarh 491221, India*
[2] *Hemchand Yadav University, Durg, Bhilai Mahila Mahavidyalaya, Department of Chemistry, Sector-1, Bhilai, Chhattisgarh 490006, India*
[3] *Hemchand Yadav University, Durg, Govt. Digvijay PG Autonomous College, Department of Chemistry, Killapara, Rajnandgaon, Chhattisgarh 491441, India*

9.1 Introduction

Lanthanides and actinides are the 28 elements following lanthanum and actinium, respectively, in the periodic table. They are also known as the inner transition or f-block metals. However, rare-earth metals include scandium, yttrium, and lanthanum along with the lanthanides. All the lanthanides from lanthanum to lutetium are represented by the symbol Ln. Similarly, all the actinides from actinium to lawrencium are represented by the symbol An. Actinoids are generally radioactive and hence unstable. The f-block elements show steady decline in size throughout the series due to lanthanoid and actinide contractions. They have characteristic +3 oxidation state. Organometallic compounds of the lanthanide and actinide metals are among the very newest organometallic compounds. Despite the extreme air and moisture sensitivity of organometallic compounds of rare earths, this is a rapidly expanding research area. An exciting aspect of organolanthanoid and organoactinide chemistry is the discovery of many efficient catalysts for organic transformations. In this chapter, we will be discussing the organometallic compounds of lanthanides and actinides (Tables 9.1 and 9.2).

9.2 Methods of Preparation

Several methods are available for the preparation of organometallic complexes lanthanides and actinides. It would be beyond the scope of this book to discuss all those methods. Here, we will be discussing some of the important methods that include the following.

*Corresponding author.

Organometallic Compounds: Synthesis, Reactions, and Applications, First Edition.
Edited by Dakeshwar Kumar Verma and Jeenat Aslam.

Table 9.1 List of lanthanides.

Element name	Symbol	Atomic number	Atomic weight	Electronic configuration (Ln)	Electronic configuration (Ln^{3+})
Lanthanum	La	57	138.905	$[Xe]6s^2 5d^1$	$[Xe]4f^0$
Cerium	Ce	58	140.116	$[Xe]4f^1 6s^2 5d^1$	$[Xe]4f^1$
Praseodymium	Pr	59	140.908	$[Xe]4f^3 6s^2$	$[Xe]4f^2$
Neodymium	Nd	60	144.243	$[Xe]4f^4 6s^2$	$[Xe]4f^3$
Promethium	Pm	61	144.913	$[Xe]4f^5 6s^2$	$[Xe]4f^4$
Samarium	Sm	62	150.36	$[Xe]4f^6 6s^2$	$[Xe]4f^5$
Europium	Eu	63	151.964	$[Xe]4f^7 6s^2$	$[Xe]4f^6$
Gadolinium	Gd	64	157.25	$[Xe]4f^7 6s^2 5d^1$	$[Xe]4f^7$
Terbium	Tb	65	158.925	$[Xe]4f^9 6s^2$	$[Xe]4f^8$
Dysprosium	Dy	66	162.500	$[Xe]4f^{10} 6s^2$	$[Xe]4f^9$
Holmium	Ho	67	164.930	$[Xe]4f^{11} 6s^2$	$[Xe]4f^{10}$
Erbium	Er	68	167.259	$[Xe]4f^{12} 6s^2$	$[Xe]4f^{11}$
Thulium	Tm	69	168.934	$[Xe]4f^{13} 6s^2$	$[Xe]4f^{12}$
Ytterbium	Yb	70	173.055	$[Xe]4f^{14} 6s^2$	$[Xe]4f^{13}$
Lutetium	Lu	71	174.967	$[Xe]4f^{14} 6s^2 5d^1$	$[Xe]4f^{14}$

Table 9.2 List of actinides.

Element name	Symbol	Atomic Number	Atomic Weight	Electronic configuration (Ln)	Electronic configuration (Ln^{3+})
Actinium	Ac	89	227	$[Rn]6d^1 7s^2$	$[Rn]5f^0$
Thorium	Th	90	232.04	$[Rn]6d^2 7s^2$	$[Rn]5f^1$
Protactinium	Pa	91	231.04	$[Rn]5f^2 7s^2 6d^1$	$[Rn]5f^2$
Uranium	U	92	238.03	$[Rn]5f^3 7s^2 6d^1$	$[Rn]5f^3$
Neptunium	Np	93	237	$[Rn]5f^4 7s^2 6d^1$	$[Rn]5f^4$
Plutonium	Pu	94	244	$[Rn]5f^6 7s^2$	$[Rn]5f^5$
Americium	Am	95	243	$[Rn]5f^7 7s^2$	$[Rn]5f^6$
Curium	Cm	96	247	$[Rn]5f^7 7s^2 6d^1$	$[Rn]5f^7$
Berkelium	Bk	97	247	$[Rn]5f^9 7s^2$	$[Rn]5f^8$
Californium	Cf	98	251	$[Rn]5f^{10} 7s^2$	$[Rn]5f^9$
Einsteinium	Es	99	252	$[Rn]5f^{11} 7s^2$	$[Rn]5f^{10}$
Fermium	Fm	100	257	$[Rn]5f^{12} 7s^2$	$[Rn]5f^{11}$
Mendelevium	Md	101	258	$[Rn]5f^{13} 7s^2$	$[Rn]5f^{12}$
Nobelium	No	102	259	$[Rn]5f^{14} 7s^2$	$[Rn]5f^{13}$
Lawrencium	Lr	103	262	$[Rn]5f^{14} 7s^2 6d^1$	$[Rn]5f^{14}$

9.2.1 Salt Elimination

This method can be used to introduce organic ligands to the lanthanide and actinide coordination spheres through elimination of alkali halide salt. Organometallic reagents of alkali metals are generally used for this purpose. This method is convenient due to the large lattice enthalpies of alkali halides and their insolubility. But it has a disadvantage of salt occlusion which will be discussed in details under the properties section. Following reaction shows a general method of forming Ln–C σ-bonds.

Reaction of three moles of methyl lithium with $LnCl_3$ (Ln = Y, La) in THF (Scheme 9.1) yielded probably $Ln(CH_3)_3$ contaminated with LiCl and whose IR spectra showed the presence of THF (Scheme 9.2). If an excess of CH_3Li is added to MCl_3 in THF, in the presence of a chelating ligand (L–L), either tetramethylethylenediamine (tmed or tmeda) or 1,2-dimethoxyethane (dme) complexes $[Li(tmed)]_3[M(CH_3)_6]$ or $[Li(dme)]_3[M(CH_3)_6]$ are obtained for Sc, Y, and all lanthanides except Eu (Scheme 9.3).

$$LnCl_3 + 3LiR \longrightarrow LnR_3 + 3LiCl$$

Scheme 9.1 Reaction of alkyl lithium with $LnCl_3$.

$$MCl_3 + 6MeLi + 3L\text{-}L \longrightarrow [Li(L\text{-}L)_3\,[M(CH_3)_6] + 3LiCl$$

Scheme 9.2 Reaction of CH_3Li to MCl_3 in THF.

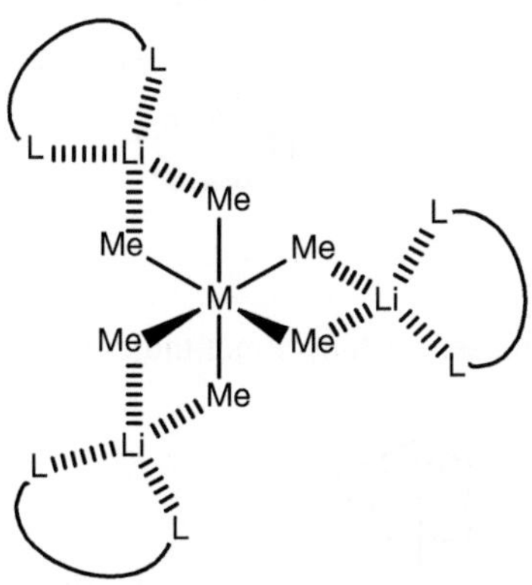

Scheme 9.3 The structure of $[Li(L–L)]_3\ [M(CH_3)_6]$.

Structures I and II represent the mono- and dimetallic complexes of lanthanum, respectively (Scheme 9.4). The preparation of organometallic complex of samarium and uranocene (Schemes 9.5 and 9.6) is done through the following schemes:

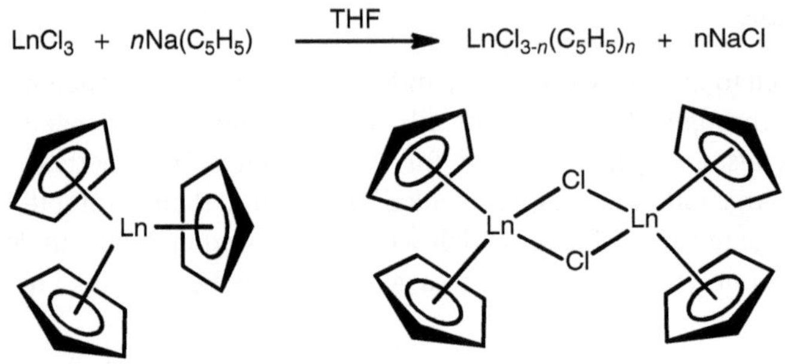

I II

Scheme 9.4 Reaction of $LnCl_3$ and $nNa(C_5H_5)$ and structure of mono- and dimetallic complexes of lanthanum.

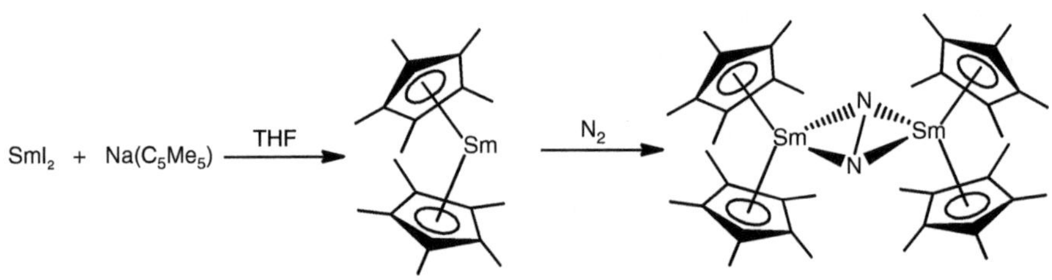

Scheme 9.5 Preparation of organometallic complex of samarium.

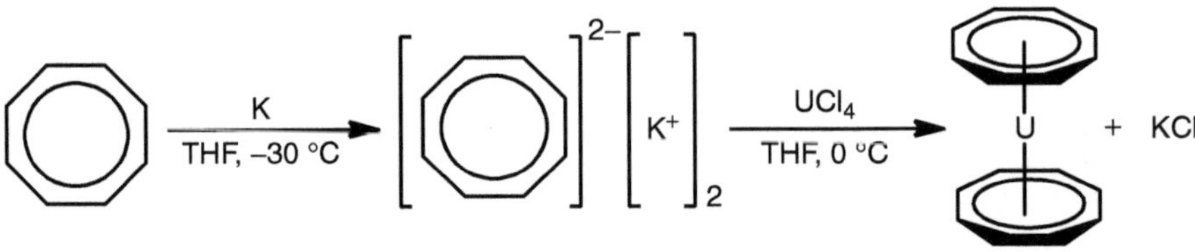

Scheme 9.6 Preparation of uranocene.

Trivalent rare-earth elements form $M(Cp)_3$ complexes with cyclopentadiene.

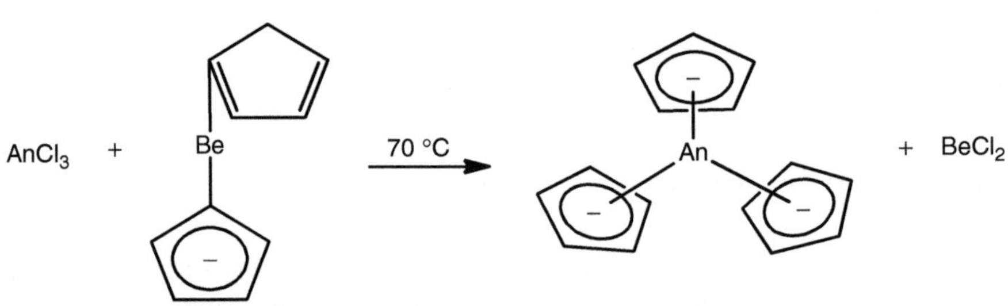

An = Th, U, Np, Pu, Am, Cm, Bk, Cf

Sandwich complexes of actinides are also obtained using cyclooctatetraene reagents (Scheme 9.7). They have atomic radii larger than transition metals;

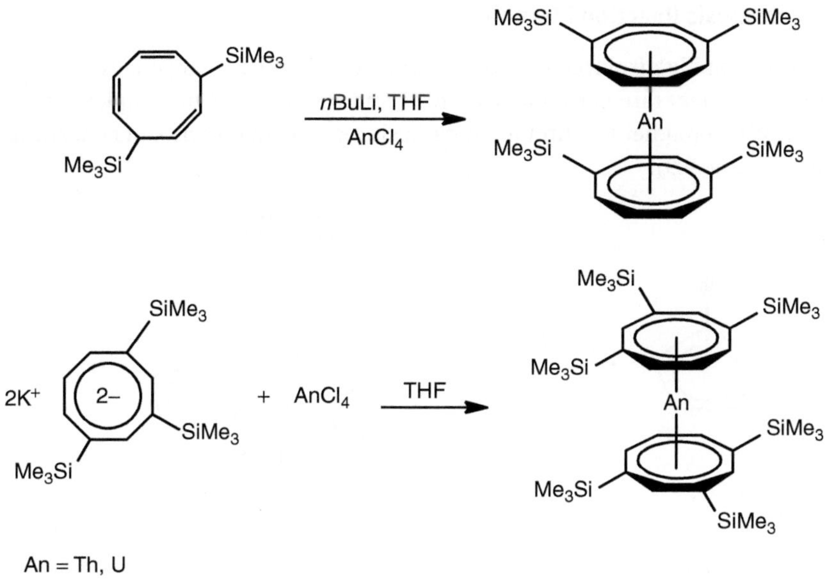

Scheme 9.7 Preparation of sandwich complexes of actinides.

hence, they prefer to form complexes with $C_8H_8^{2-}$ ions instead of cyclopentadienyl ligands.

9.2.2 Metal Vapor Synthesis Method

In this method, metal atoms are vaporized under high vacuum, followed by their condensation along with prospective ligands on cold surface. It is generally used for the preparation of *bis*(arene) organometallic complexes of lanthanides and actinides (Scheme 9.8).

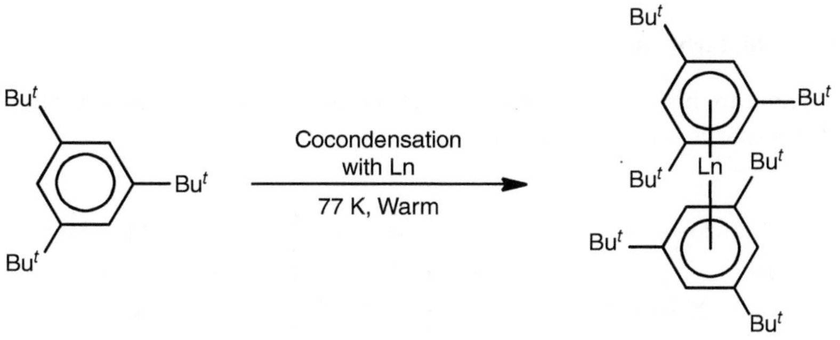

Scheme 9.8 Preparation of *bis*(arene) organometallic complexes of lanthanides through metal vapor synthesis method.

9.2.3 Metathesis Reaction Method

Metathesis or double displacement reaction involves bond exchange between two reactants. Tetravalent rare-earth metals (Th, U, and Np) form MCp_4 complexes from potassium cyclopentadienide through metathesis reaction in presence of benzene as a solvent (Scheme 9.9).

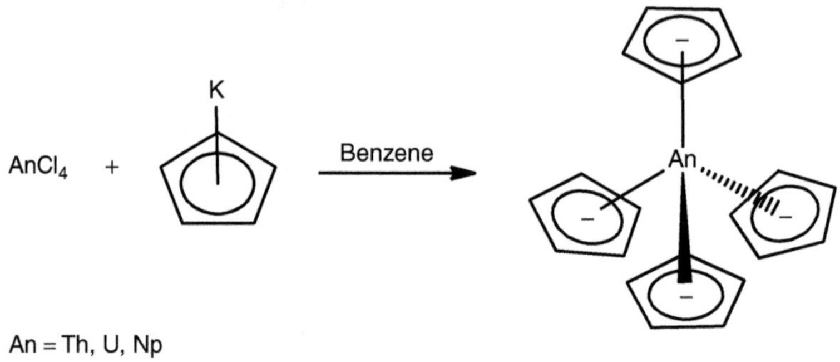

Scheme 9.9 Preparation of organometallic complexes of actinides through metathesis reaction method.

Tris(cyclopentadienyl) lanthanide compounds can be synthesized by a metathesis reaction between NaCp/KCp and anhydrous $LnCl_3$ or $LnI_3(THF)_x$ at reflux in THF for all of the lanthanides with the exception of Eu (Scheme 9.10).

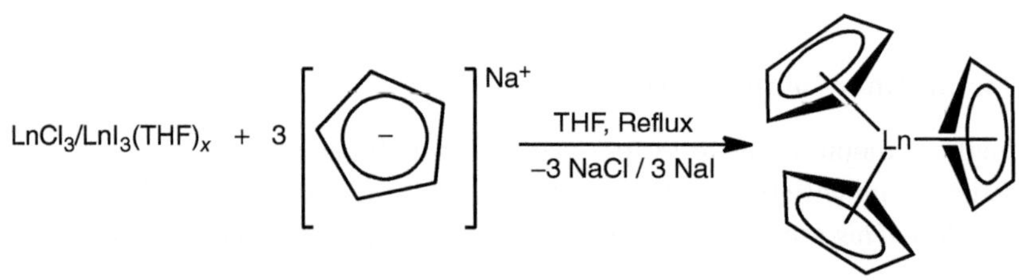

Scheme 9.10 Preparation of tris(cyclopentadienyl) lanthanide compounds.

9.2.4 σ-Bond Metathesis

Rare-earth metals are highly polarizing; hence, their derivatives have high reactivities. σ-bond metathesis is illustrated by the following Scheme 9.11:

Scheme 9.11 Illustration of σ-bond metathesis.

9.2.5 Acid–Base Method

Rare-earth ligand bonds are polar and labile, hence are prone toward acid–base reactions. The pK_a values play important role. Amides and hydrocarbyls are used for preparation of organolanthanides and organoactinides (Scheme 9.12).

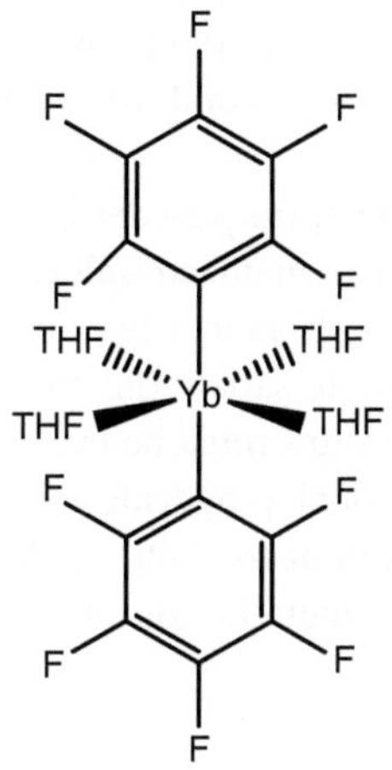

Scheme 9.12 Structure of rare earth metal complex prepared through acid–base method.

9.3 Organometallic Compounds of Lanthanides

In contrast to the extensive carbonyl chemistry of the d-block metals, lanthanoid metals do not form complexes with CO under normal conditions. Unstable carbonyls, such as $Nd(CO)_6$, have been prepared by matrix isolation. Since organolanthanoids and organoactinides are usually air and moisture sensitive and may be pyrophoric, handling the compounds under inert atmospheres is essential. With the exception of cerium, which forms dicyclooctatetraenyl cerium(IV) and europium, samarium, and ytterbium, which form cyclopentadienyl and pentamethylcyclopentadienyl lanthanide(II) derivatives, all the other rare-earth metals form only organometallic compounds with the metal in the oxidation state Ln^{3+}. Most of them have unsubstituted or substituted cyclopentadienyls bonded to the metal along with hydrogen, halogen, alkyl groups, or alkoxides as additional ligands.

9.3.1 Types

Compounds of these metals involving either σ- or π-bonds to carbon are generally much more reactive to both air and water than those of the d-block metals. Thus, there is no lanthanide equivalent of ferrocene, an unreactive air, and heat stable compound. They are often thermally stable to 100 °C or more but are usually decomposed immediately by air (and are not infrequently pyrophoric). Within these limitations, lanthanide organometallic compounds have their own special features, often linked with the large size of these metals.

9.3.1.1 σ-Bonded Complexes

These are the compounds in which carbon atom of the organic ligand is bonded to the metal by a two-electron two-center (2e–2c) covalent bond. Most of the elements with the values of electronegativity are higher than 1. Generally, lanthanoids are formed by LnR_3-type complexes with alkyl group that have Ln–C σ-bond. In order to stabilize LnR_3, bulky alkyl substituents must be used.

Alkyl complexes of lanthanide have been greatly researched over the past decades. Trimethylsilyl alkyl is a major ligand class used in these rare-earth metal complexes. Trialkyl lanthanide complexes with sterically small groups, such as methyl, have been found to be thermally unstable because of their inability to satisfy the coordination number of the lanthanide cation. In addition, alkyl groups must be free of β-hydrogen atoms in order to prevent ligand decomposition through β-hydride elimination. Therefore, bulky (trimethylsilyl)alkyl groups were sought to stabilize the metal center in early efforts to create trialkyl lanthanide organometallic complexes. There are three types of (trimethylsilyl)alkyl groups:

1. (trimethylsilyl)methyl [$-CH_2SiMe_3$];
2. bis(trimethylsilyl)methyl [$-CH(SiMe_3)_2$]; and
3. tris(trimethylsilyl)methyl [$-C(SiMe_3)_3$].

The first yttrium alkyl(trimethylsilyl) complex was synthesized by Lappert and coworkers, while lutetium, ytterbium, thulium, erbium, and terbium complexes were synthesized soon thereafter. Several synthetic procedures exist, but all make use of stoichiometric salt metathesis reactions to form the lanthanide trialkyl complexes as shown in Scheme 9.13.

$$LnCl_3 + 3LiCH_2SiMe_3 \xrightarrow[0\ °C/RT]{THF,\ Pentane,\ Et_2O} Ln(CH_2SiMe_3)_3(THF)_x + 3LiCl$$

Scheme 9.13 Stoichiometric salt metathesis reactions to form the lanthanide trialkyl complexes.

Since the (trimethylsilyl)methyl ligands do not provide sufficient steric bulk, solvent molecules, such as THF, are coordinated to the metal center. The number of THF molecules increases with the size of the metal's ionic radius. A significant deterrent to using these complexes is their thermal instability. At ambient temperature, $Ln(CH_2SiMe_3)_3$ complexes have been shown to decompose in hours to form insoluble products and $SiMe_4$. Furthermore, the (trimethylsilyl)methyl ligand can only be used for middle to small rare-earth metal elements because of ligand degradation for larger size lanthanides. A second drawback to using (trimethylsilyl)methyl ligands is their ability to form lanthanates or $[Li]^+ [Ln(R)_4]^-$. The lanthanates exhibit no solubility in nonpolar solvents but reasonable solubility in coordinating solvents, such as diethyl ether and THF. Salt formation of this type can be avoided through

$$\mathrm{Ln[(OC_6H_3{}^tBu_2\text{-}2,6)]_3 + 3LiCH_2SiMe_3 \xrightarrow{\text{Pentane, 25 °C}} Ln(CH_2SiMe_3)_3 + 3LiOC_6H_3{}^tBu_2\text{-}2\text{-}6}$$

$$\mathrm{Ln = La\ and\ Sm}$$

Scheme 9.14 Stoichiometric reaction of lanthanide aryloxides with $LiCH_2SiMe_3$.

the stoichiometric reaction of lanthanide aryloxides with $LiCH_2SiMe_3$ as shown in Scheme 9.14.

9.3.1.2 π-Bonded Organometallic Compounds

These are alkene, alkyne, or some other carbon group that have a system of electron in π-orbitals, overlapping of these π-orbitals with vacant orbitals of the metal atom gives rise to arrangement in which the metal atom is bonded to several carbon atoms instead of one.

9.3.1.2.1 Aryl Organolanthanoid Compounds

In an effort to find suitable complexes to use as a universal entry point for the generation of early lanthanide species, benzyl ligands [–CH_2Ph] were investigated. The first definitive report of tribenzyl lanthanide complexes came about from the salt metathesis reaction between an anhydrous metal halide and benzyl potassium (Scheme 9.15).

$$\mathrm{LnBr_3(THF)_3 + 3KCH_2Ph \xrightarrow{\text{THF, 0 °C}} [Ln(CH_2Ph)_3(THF)_3 + 3KBr}$$

$$\mathrm{Ln = La,\ Nd}$$

Scheme 9.15 Stoichiometric reaction of lanthanide complexes with benzyl ligands.

The $Ln(CH_2Ph)_3(THF)_3$ series was later expanded to include Ce, Pr, Sm, Gd, Dy, and Er. Crystal structures of these complexes revealed that the benzyl and THF ligands each adopt a FAC arrangement. Although the geometry of each lanthanide complex is the same, there are subtle differences in the coordination environment of the benzyl ligand. For the early lanthanides (La–Nd), the ipsocarbon coordinates to the metal resulting in an η^2–interaction. Due to the lanthanide contraction, the ipsocarbon no longer coordinates to the metal center for middle to late lanthanides (Sm–Lu) giving rise to an η^1-interaction. The tribenzyl complexes have been subjected to a number of reactions to investigate their reactivity and properties. Protonolysis reactions with various amidines, ArN=CRNHAr (Ar = C_6H_3-2,6-^{i}Pr_2; R = Ph, ^{t}Bu), led to the formation of the respective monoamidate–bisbenzyl complexes. Benzyl groups with amines, such as [$CH_2C_6H_4$-*o*-NMe_2] and [$CH(NMe_2)C_6H_5$], have recently gathered attention as suitable donor ligands for the stabilization of rare-earth metal complexes.

One notable feature apparent in the solid-state structures of these complexes is that as the ionic radius of the metal decreases, one of the benzyl ligands will reposition itself “upside down” to alleviate steric repulsions resulting from the decrease in the Ln–C σ-bond length, ultimately destroying the threefold symmetry observed with the larger metals. Two constitutional isomers exhibit this behavior:

(o-Me_2N-C_6H_4-$CH_2)_3La$ and $La[CH(NMe_2)C_6H_5]_3$. Both La complexes maintain a threefold symmetry at low temperature, but when the metal center is changed to yttrium {(o-$Me_2NC_6H_4$-$CH_2)_3Y$ and $Y[CH(NMe_2)C_6H_5]_3$}, both lose their threefold symmetry, even at low temperature. The reactivity of these amino-stabilized benzyl ligands has been probed through various protonolysis and acid–base reactions with weak Brønsted acids. In addition, such complexes have been used as precursors for polymerization reactions and cross coupling of terminal alkynes with isocyanides. The first trivalent organometallic phenyl lanthanide complex was produced from the salt metathesis reaction between the metal chloride and phenyl lithium (Scheme 9.16).

$$\mathrm{LnCl_3 + 3LiPh} \xrightarrow{\text{THF, 0 °C/RT}} \mathrm{[Ln(Ph)_3] + 3LiCl}$$
$$\mathrm{Ln = Sc, Y}$$

Scheme 9.16 Salt metathesis reaction between the metal chloride and phenyl lithium.

Its crystal structure was determined 30 years later with a slight modification to the original procedure (THF/Et_2O solvent mixture). Salt metathesis reactions with Ln^{3+} chlorides and aryllithium reagents (Scheme 9.17) still remain the dominant method for the production of 17 triphenyl-substituted products. The product produced in this type of reaction is dependent on the size of the metal and the stoichiometric amount of the lithiated species used. In some cases, the solvent coordinates to the lanthanoid metal, e.g. $TmPh_3(THF)_3$ (Scheme 9.18).

$$\mathrm{LnCl_3 + 3LiPh^R} \xrightarrow{\text{THF/Et}_2\text{O, 0 °C/RT}} \mathrm{Ln(Ph^R)_3(THF)_2 + 3LiCl}$$
$$\mathrm{Ln = Sc, Lu}$$
$$\mathrm{Ph^R = C_6H_5,\ C_6H_4\text{-}p\text{-}Me,\ C_6H_4\text{-}p\text{-}Et}$$

$$\mathrm{LnCl_3 + 4\ LiPh} \xrightarrow{\text{THF/Et}_2\text{O, 0 °C/RT}} \mathrm{LiLn(Ph)_4}$$
$$\mathrm{Ln = La, Pr}$$

$$\mathrm{Tm + HgPh_2} \xrightarrow[\text{in presence of TmCl}_3]{\text{THF}} \mathrm{TmPh_3(THF)_3}$$

Scheme 9.17 Salt metathesis reactions with Ln^{3+} chlorides and aryllithium reagents.

9.3.1.2.2 *Tris(cyclopentadienyl) Systems*

Cyclopentadienyl (and substituted variants thereof) has been the most versatile ligand used in organolanthanide chemistry. Unlike the situation with the d-block metals, where a maximum of two pentahapto-(η^5-)cyclopentadienyls can coordinate, up to three η^5-cyclopentadienyls can be found for the lanthanides, in keeping with the higher coordination numbers found for the f-block elements. In terms

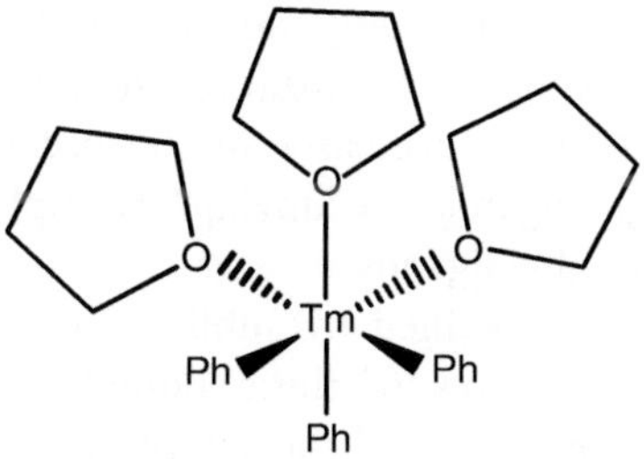

Scheme 9.18 Structure of $TmPh_3(THF)_3$.

of the space occupied, a η^5-cyclopentadienyl takes up three sites in the coordination sphere. Three types of compound can be obtained, depending upon the stoichiometry of the reaction mixture (Scheme 9.19). The mono(cyclopentadienyl) complexes all have pseudo-octahedral structures. Scheme 9.20 shows the structure of $CpLnCl_2(THF)_3$.

$$LnCl_3 + NaCp \xrightarrow{THF} LnCpCl_2(THF)_3 + NaCl\ (Ln = Sm\ to\ Lu)$$

$$LnCl_3 + 2NaCp \xrightarrow{THF} LnCp_2Cl + 2\ NaCl\ (Ln = Gd\ to\ Lu)$$

$$LnCl_3 + 3NaCp \xrightarrow{THF} LnCp_3 + 3NaCl\ (all\ Ln)$$

Scheme 9.19 Preparation of tris(cyclopentadienyl) systems based on stoichiometry of the reaction mixture.

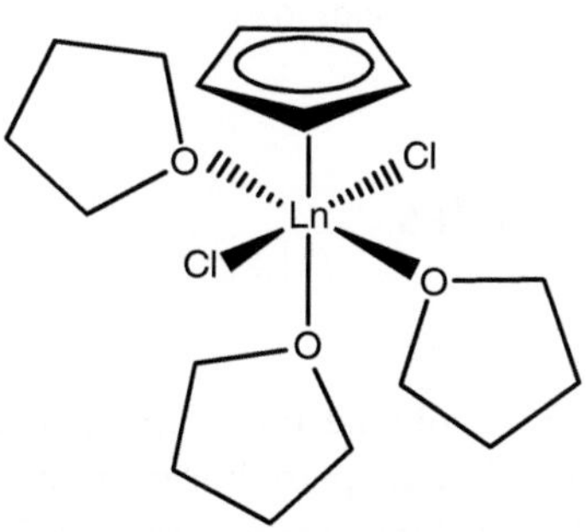

Scheme 9.20 Structure of $CpLnCl_2(THF)_3$.

The development of lanthanide organometallic chemistry involving the coordination of carbanions to the lanthanide metal center began with the synthesis of lanthanide tris(cyclopentadienyl), $Ln(Cp)_3$, complexes by Birmingham and Wilkinson in 1954. These organolanthanide compounds with three π-bonded, η^5-coordinated cyclopentadienyl anions preceded the synthesis of lanthanide σ-bonded alkyl carbanion, $Ln(CH_2SiMe_3)_3$ and $Ln[CH(SiMe_3)_2]_3$ solvates, and ionic "ate" complexes by 20+ years. In the majority of cases, the challenge was to find or design and synthesize ligands that were sufficiently sterically bulky to saturate the lanthanide metal coordination sphere. Stabilization requires sterically bulky anionic ligands but carbanion stability decreases in the order: primary > secondary > tertiary, so "bulky" alkyl ligands occupying multiple coordination sites were often delocalized, e.g. Cp or allyl ligands.

Solvent-free crystals of these compounds can be obtained by sublimation at c. 200–250 °C under vacuum. The solid-state structures of these homoleptic tris(cyclopentadienyl)lanthanide complexes (Scheme 9.21) of the simplest and smallest cyclopentadienyl ring are infinitely polymeric, $[Cp_3Ln]_\infty$, because the lanthanide coordination sphere is unsaturated and lanthanide cations form ionic interactions over distances that exceed the length of chemical bonds. The intermolecular interactions connecting the Cp_3Ln molecules in the polymer are η^1-interactions. One cyclopentadienyl ring of each Cp_3Ln unit has an η^5 interaction with one Ln atom and an η^1 interaction with the Ln atom in the neighboring Cp_3Ln unit. These compounds are only soluble in THF, a good donor solvent, in which they are stable.

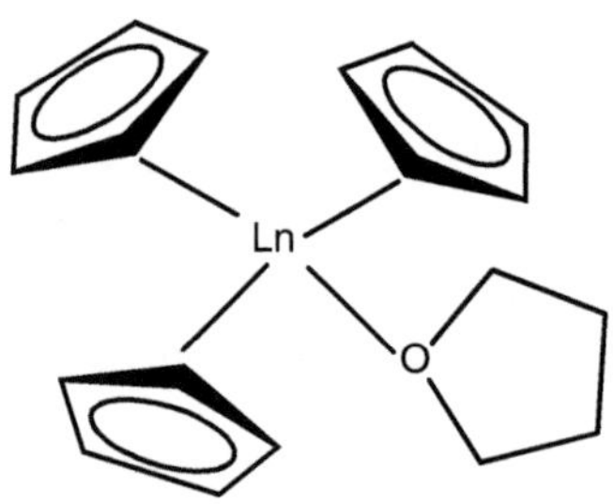

Scheme 9.21 Structure of tris(cyclopentadienyl)lanthanide compounds in THF.

9.4 Organometallic Compounds of Actinides

9.4.1 Types

They are often soluble in aromatic hydrocarbons, such as toluene and in ethers (e.g. THF) but are generally destroyed by water. Sometimes, they are pyrophoric on exposure to air. Although organometallic complexes are known for all the early actinoids, compounds of Th and U far exceed those of the other metals. Most of the synthetic works have been carried out with Th and U; this is partly due to the ready availability of MCl_4 (M = Th, U) and also because of the precautions that have to be taken in handling compounds of other metals, especially Pu and Np.

9.4.2 σ-Bonded Complexes

The leading organometallic chemist of the day, Henry Gilman, investigated the synthesis of simple alkyls and aryls, concluding that such compounds did not exist or at least were unstable. But some difficulty was originally encountered in preparing homoleptic σ-bonded alkyl or aryl complexes of the actinoids. But the use of chelating ligand led to the formation of stable σ-bonded organometallic compounds of actinides. Even today, just one neutral homoleptic alkyl of known structure exists, $[U\{CH(SiMe_3)_2\}_3]$. This $[U\{CH(SiMe_3)_2\}_3]$ (Scheme 9.22) is prepared by the reaction with aryloxide by a route ensuring no contamination with lithium halide.

$$3\ LiCH(SiMe_3)_2 + U(O\text{-}2,6\text{-}Bu^t_2C_6H_3)_3 \longrightarrow [U\{CH(SiMe_3)_2\}_3] + 3Li(O\text{-}2,6\text{-}Bu^t_2C_6H_3)$$

Scheme 9.22 Preparation of $[U\{CH(SiMe_3)_2\}_3]$.

If $LiCH(SiMe_3)_2$ is treated directly with $UCl_3(THF)_3$ in a salt-elimination reaction, the alkylate salt $[(THF)_3Li\text{-}Cl\text{-}U\{CH(SiMe_3)_2\}_3]$ is obtained (Scheme 9.23).

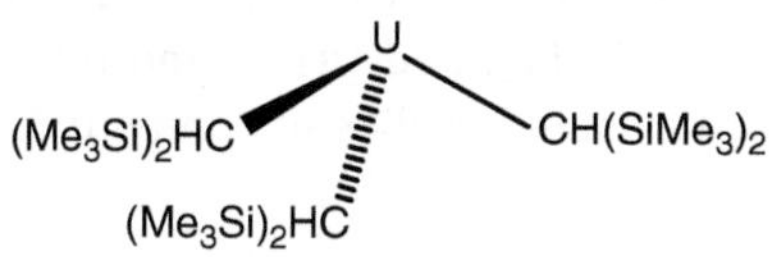

Scheme 9.23 Structure of $U\{CH(SiMe_3)_2\}_3$

$[U\{CH(SiMe_3)_2\}_3]$ has a trigonal pyramidal structure as shown above. There are three short U-C_{methyl} contacts (309 pm), and these may contribute to the deviation from planarity as in $[Ln\{CH(SiMe_3)_2\}_3]$ and also like the isoelectronic amides $[M\{N(SiMe_3)_2\}_3]$ (M = U, lanthanides). In a similar way, compounds of Np and Pu$[M\{CH(SiMe_3)_2\}_3]$ can also be prepared. MO calculations for hypothetical $M(CH_3)_3$ (M = U, Np, Pu) molecules also predict pyramidal structures, as a result of better 6d involvement (not 5f) in the M—C bonding orbital's in pyramidal geometries.

The benzyl $[Th(CH_2Ph)_4]$ has been synthesized and the pale yellow dimethylbenzyl compound $Th(1,3,5\text{-}CH_2C_6H_3Me_2)_4$ has also been reported, but their structures are not known (and may have multihapto coordination of benzyl to thorium). The heptamethylthorate anion $[ThMe_7]^{3-}$ is found in the salt $[Li(TMEDA)]_3[ThMe_7]$ where use of the chelating ligand TMEDA ($Me_2NCH_2CH_2NMe_2$) was the key to stabilizing the Li^+ salt of $[ThMe_7]^{3-}$ (Scheme 9.24). The salt $[Li(TMEDA)]_3[ThMe_7]$ contains thorium in monocapped trigonal prismatic coordination; one methyl is terminal (Th–C 2.571 Å), while the other six methyl groups pairwise bridge the lithium and thorium (Th–C 2.655–2.765 Å). The methyl groups are, however, equivalent in solution (NMR). In a similar way, hexa alkyls of type $Li_2UR_6.7TMEDA$ can also be isolated.

$$ThCl_4 + \text{Excess MeLi} \xrightarrow{Et_2O.TMEDA} [Li(TMEDA)]_3[ThMe_7]$$

Scheme 9.24 Preparation of salt $[Li(TMEDA)]_3[ThMe_7]$.

9.4.3 π-Bonded Complexes

9.4.3.1 Cyclopentadienyl Derivatives

These are regarded as derivatives of the $C_5H_5^-$ ligand; substituted rings also occur. Some of these were the first organoactinide compounds to be reported, shortly after the discovery of ferrocene. The most important compounds are in the +4 state but there is a significant chemistry of U(+3) as well as examples in oxidation states (+5) and (+6). One factor that should be noted is that replacing one or more hydrogens in the cyclopentadienyl ring can have a significant effect upon the stability and isolability of complexes made from them, through both steric and electronic effects. Thus, three or four Cp ligands can be accommodated around uranium (IV), whereas, with Cp*, two is the norm. In another example, $[UCp_3]$ gives no sign of reacting with carbon monoxide; $[U\{C_5H_3(SiMe_3)_2\}_3]$ forms a compound that can be detected in solution; and $[U(C_5Me_4H)_3]$ forms $[U(C_5Me_4H)_3(CO)]$, which can be isolated in the solid state. Cyclopentadienyl derivatives are plentiful among organometallic complexes of Th(IV), Th(III), U(IV), and U(III), and reactions 27.55–27.58 give methods of synthesis for the main families of compounds (M = Th, U).

9.4.3.1.1 Oxidation State (VI)

In organometallic chemistry, high oxidation states are not possible, but a remarkable uranium (VI) imide (Scheme 9.25) has been prepared.

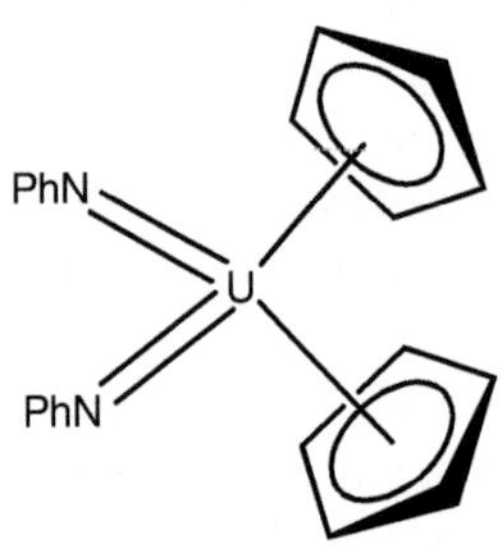

Scheme 9.25 Structure of uranium (VI) imide.

The reaction takes place via an intermediate $[UCp^*_2(\eta^2\text{-PhN–NPh})]$ (Scheme 9.26). The visible spectrum of the above compound does not show f–f transitions, confirming its U^{VI}–f^0 character. The above compound shows the expected pseudo-tetrahedral geometry; the short U–N bond length (1.952 Å) confirms its multiple-bond character and the U=N—Ph linkage is virtually linear.

$$U(\eta^5\,Cp)_2\,(CH_3)Cl + Li[PhNN(H)Ph] \longrightarrow U(\eta^5\,Cp)_2\,(=NPh)_2 + CH_4 + LiCl$$

Scheme 9.26 Formation of intermediate $[UCp^*_2(\eta^2\text{-PhN–NPh})]$.

9.4.3.1.2 Oxidation State (V)

Rare organoimides U(Me-η^{5}Cp)$_3$(=NR) have been made by oxidation of U(Me-η^{5}Cp)$_3$(THF) with RN_3 (R = Me_3Si and Ph) (Scheme 9.27). The U—N bond length of 2.109 Å is regarded as having multiple-bond character (cf. the U=O bond in uranyl compounds).

$$\mathrm{U(Me\text{-}\eta^5\,Cp)_3(THF)} + \mathrm{RN_3} \longrightarrow \mathrm{U(Me\text{-}\eta^5\,Cp)_3({=}NR)} + \mathrm{THF} + \mathrm{N_2}$$

Scheme 9.27 Preparation of rare organoimides U(Me-η^{5}Cp)$_3$(=NR).

9.4.3.1.3 Oxidation State (IV)

Cyclopentadienyl derivatives of actinoids with oxidation state (IV) are divided into three families of compounds whose structures can be regarded as derived from a tetrahedral MX_4 molecule with one or more halogens replaced by one or more cyclopentadienyl rings; [Cp_4M], [Cp_3MX]; [$CpMX_3$] (X = halogen, most usually Cl) (Scheme 9.28). Since halogens have less steric influence than a cyclopentadienyl group, a $CpMX_3$ molecule would be coordinatively unsaturated, so $CpMX_3$ actually exist as solvates, such as $CpMX_3$ (THF)$_2$, where THF is known as tetrahydrofuran. Cp_2MX_2 is unknown for X = halogen, attempts to prepare them resulting in mixtures of [Cp_3MX] and [$CpMX_3$], but they are known for X = NEt_2 and BH_4.

$$\mathrm{MCl_4} + 4\,\mathrm{KCp} \xrightarrow{\mathrm{C_6H_6}} (\eta^5\,\mathrm{Cp})_4\mathrm{M} + 4\mathrm{KCl} \quad (\mathrm{M = U, Th, Pa, Np})$$

$$\mathrm{MCl_4} + 3\,\mathrm{TlCp} \xrightarrow{\text{1,2-dimethoxyethane}} (\eta^5\,\mathrm{Cp})_3\mathrm{MCl} + 3\mathrm{TlCl} \quad (\mathrm{M = U, Th, Np})$$

$$\mathrm{MCl_4} + \mathrm{TlCp} \xrightarrow{\mathrm{THF}} (\eta^5\mathrm{Cp})\mathrm{MCl_3(THF)_2} + \mathrm{TlCl}$$

Scheme 9.28 Preparation of [Cp_4M], [Cp_3MX], and [$CpMX_3$].

On studying the structure of the above synthesized compound, [(η^{5}Cp)$_4$M] has regular tetrahedral geometry (Th–C = 287 pm and U–C = 281 pm). However, tetrahedral structures are also observed for [(η^{5}Cp)$_3$MCl] with the halogen occupying a position corresponding to the centroid of a fourth ring and [(η^{5}Cp)MCl$_3$(THF)$_2$] has a pseudo-octahedral structure (Scheme 9.29).

There is evidence for a significant covalent contribution to the bonding in some of these actinide (IV) compounds. How to describe the metal–ligand bonding in these and other Cp derivatives of the actinides is the subject of much theoretical debate. The current picture suggests involvement of the metal 6d atomic orbitals with the 5f orbitals being fairly unperturbed. [(η^{5}Cp)$_3$MCl] is the most useful of these compounds as the halogen can be replaced by a number of groups (e.g. NCS, BH_4, acac, Me, Ph), which shows bonding role for the 6d rather than 5f atomic orbitals (Scheme 9.30). UCp_3Cl does not react with $FeCl_2$ forming ferrocene, whereas MCp_3 do (M, e.g. Ln, U). It appears to undergo ionization in aqueous solution as

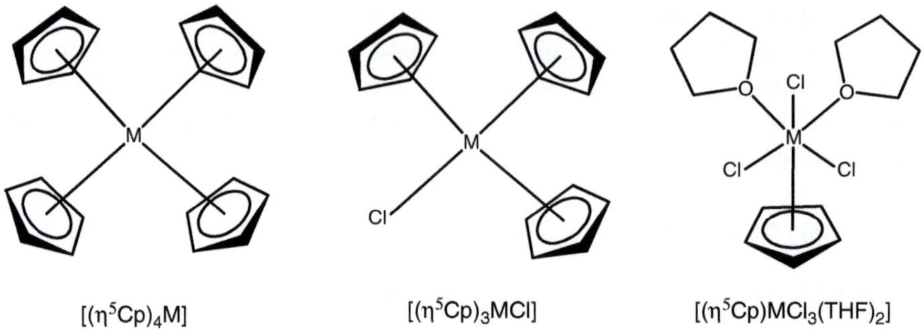

Scheme 9.29 Structure of $[(\eta^5Cp)_4M]$, $[(\eta^5Cp)_3MCl]$, and $[(\eta^5Cp)MCl_3(THF)_2]$.

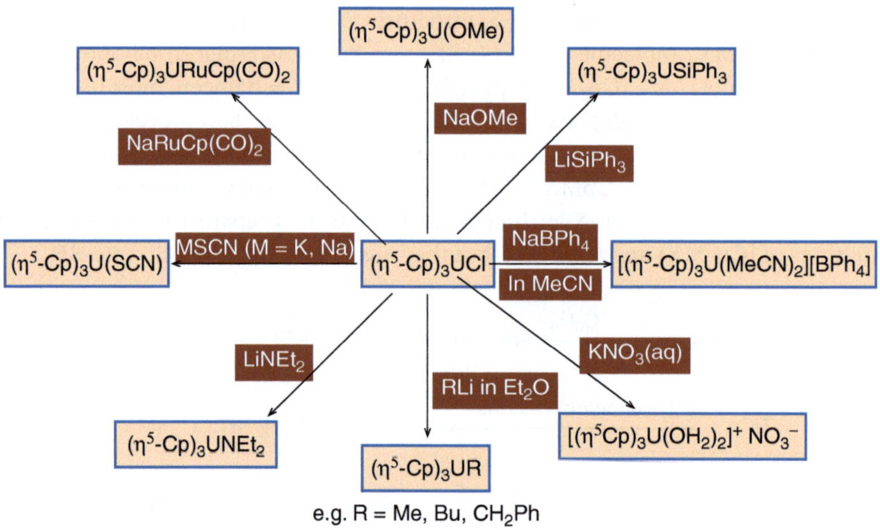

Scheme 9.30 Selected reactions of $[(\eta^5Cp)_3MCl]$.

a stable green $[(\eta^5Cp)_3U(OH_2)_n]^+$ ($n \approx 2$) cation. There is also some evidence from Mossbauer spectra of neptunium (III) and (IV) cyclopentadienyl compounds that, in a Np(IV) compound, such as $NpCp_4$ (but not Np(III) compounds), there is a greater shielding of the 6s shell than in an ionic compound, such as $NpCl_4$, leading to enhanced covalence in the bond. Bonds are reasonably strong; the mean bond dissociation energy in UCp_4 has been determined to be 247 kJ mol^{-1}, compared with a value of 297 kJ mol^{-1} for $FeCp_2$.

9.4.3.1.4 Oxidation State (III)

A variety of syntheses are possible to these compounds, including the expected salt-elimination route (Scheme 9.31).

Reduction with sodium naphthalenide is often possible (Scheme 9.32).

The cyclopentadienyl products were sublimed out of the reaction mixture and characterized by powder diffraction data, showing they were isostructural with the

$$UCl_3 + 3\,NaCp \xrightarrow{C_6H_6} (\eta^5\,Cp)_3U + 3NaCl$$

Scheme 9.31 Preparation of $(\eta^5Cp)_3U$ through salt elimination route.

$$M(\eta^5Cp)_3Cl + Na \xrightarrow{THF} (\eta^5Cp)_3M + NaCl \quad (M = U, Th, Np)$$

Scheme 9.32 Preparation of $(\eta^5Cp)_3M$ through reduction with sodium naphthalenide.

corresponding $LnCp_3$ (Ln = Pr, Sm, Gd). Similar compounds have been made with substituted cyclopentadienyl groups; their properties strongly resemble those of the corresponding $LnCp_3$. The unsolvated $U(C_5H_4SiMe_3)_3$ is known to have trigonal planar geometry, as is $U(C_5Me_4H)_3$ (X-ray). Lewis bases, e.g. THF, $C_6H_{11}NC$, and py, form pseudo-tetrahedral adducts UCp3.L {X-ray, e.g. for $[UCp_3(THF)]$} in the same way that lanthanides do. Some adducts are known with rather soft bases, such as tertiary phosphines, e.g. $[U(MeCp)_3(PMe_3)]$, that might not have been expected for a fairly "hard" metal, such as U(III).

Evidence from UV–visible spectra (Cm and Am compounds) and Mossbauer spectra (Np compounds) indicates that these compounds have probably rather more covalent contributions to their bonding than do the corresponding lanthanide compounds but are largely ionic compared with cyclopentadienyls in the +4 state. Analogous compounds $AnCp_3$ exist for other actinides, e.g. Th, Np, Pu, Am, Bk, Cf, and Cm. The thorium(III) compound is especially noteworthy, on account of the rarity of this oxidation state; the structure of a similar $Th[C_5H_3(SiMe_3)_2]_3$ is known to have trigonal planar geometry around thorium (Scheme 9.33).

$$3\,Th[C_5H_3(SiMe_3)_2]Cl_2 + Na/K\ (alloy) \longrightarrow Th + 2\,Th[C_5H_3(SiMe_3)_2]_3 + 6Na/KCl$$

Scheme 9.33 Preparation of $Th[C_5H_3(SiMe_3)_2]_3$.

These Th(III) compounds are EPR active; spectra indicate a $6d^1$ ground state for $Th[C_5H_3(Si\,Me_3)_2]_3$, despite the fact that the free Th^{3+} ion has a $5f^1$ configuration.

9.5 Stability

The stability of organometallic lanthanide and actinide complexes can be improved by the following.

9.5.1 Maintaining Anhydrous and Anaerobic Conditions

Organometallic complexes of rare-earth metals are sensitive toward air and moisture. Therefore, it becomes mandatory to maintain dry and inert atmosphere. Nitrogen or argon can be utilized through glove box and/or Schlenk line methods. Lanthanoid and actinoid complexes in +2 oxidation state are generally strongly reducing and hence they might reduce dinitrogen and incorporate it as N_2^{-2}. Therefore, care must be taken while choosing the inert atmosphere. Dried and degassed solvents and reagents should be used. Inert gases must be passed through drying or deoxygenation columns in order to remove traces of O_2 and moisture.

9.5.2 Avoiding Elimination Reactions

Organometallic complexes of main-group metals possess appreciable stability up to 100 °C due to filled 5d orbitals, complete octet, and high energy σ anti-bonding orbitals. In case of transition metals, empty d-orbitals offer low-energy pathways for decomposition. Rare-earth metals with highly electropositive nature can polarize bonds and the kinetically labile ligands aid easy decomposition. Therefore, alkyl ligands lacking β-hydrogens have gained importance due to higher thermal stability of complexes. But M–alkyl bonds have labile nature in case of lanthanoid and actinoid complexes. A variety of alkyl ligands capable of imparting stability to the organometallic complexes or rare earths have been developed.

9.5.3 Blocking the Coordination Sites

The coordination environment can be closed at the metal center in organometallic complexes. This would prevent the otherwise kinetically facile decompositions. Ligands with bulky groups can be used for this purpose. Use of cyclopentadienyl ligand lead to the formation of highly coordinated complexes that are formed. Mono, bis, and tris(cyclopentadienyl) complexes occupy three, six, and nine coordination sites at metal centers. Multidentate ligands, such as DME, TMEDA, and phosphines, may occupy the coordination sites at the metal centers.

9.5.4 Donor-functionalized Chelating Pendant Donor Ligands

The use of pendant donor chelating groups has become popular because of their ability to occupy the vacant coordination sites present in the rare-earth complexes. This prevents various oligo or polymerization reactions, salt occlusion, and solvent molecule coordination. The pendant donor groups enable the formation of entropically favorable and stable chelate rings, thus enhancing the complex stability.

9.6 Properties

The physicochemical properties of organometallic complexes of the rare earths will be discussed under the following heads.

9.6.1 Thermodynamic and Kinetic Considerations

Organometallic linkages of rare earths are not weak but ionic bonding makes them labile. An example of *bis*(pentamethylcyclopentadienyl) samarium has been illustrated below in Scheme 9.34:

$$[(C_5Me_5)_2Sm{-}R] + X_2 \longrightarrow [(C_5Me_5)_2Sm{-}X] + RX$$

$$[(C_5Me_5)_2Sm{-}X] \longrightarrow [(C_5Me_5)_2Sm] + 1/2\ X_2$$

Scheme 9.34 Lability of *bis*(pentamethylcyclopentadienyl) samarium.

Bond disruption enthalpies have been obtained through calorimetric experiments. The results show that organometallic linkages of rare earths are weaker than that of their halide and oxo-bonds. The ionic nature of the M—R bond makes it highly polar and labile that can easily be broken due to kinetically facile pathways. This provides high reactivity during catalytic reactions but can easily be poisoned by the presence of trace amount of O_2, moisture, and impurities, such as halides and sulfur.

9.6.2 Solvation

Many polar solvents have lone pair of electrons and hence they can act as dative ligands to the rare-earth metal centers. If the coordinated ligands do not have high steric requirements in case of organometallic complexes of rare-earth metals, coordination sphere may have some space left for the Lewis bases, such as diethyl ether, tetrahydrofuran, and dimethoxyethane. The coordination of Lewis bases might block the accessibility of the reaction centers, thus reducing the Ln–R bond reactivity through saturation of steric requirements, depolarization, and competitive reactions.

9.6.3 Aggregation

Rare-earth metals are electropositive and electron deficient. Therefore, they form aggregates to increase their electron density to maximum. This takes place when the metal coordination sphere is not blocked sterically; Lewis base is not available and possibilities of oligomerization or polymerization and occlusion have been avoided. It can be prevented by formation of solvated and lower order aggregates or use of ligands with high steric demands.

9.6.4 Donor–Acceptor Interactions

The metal center Ln(III) acts as a hard Lewis acid center. The electron deficiency and steric unsaturation lead to adduct formation for stabilization of complexes. Here are

$$[Dy(OCHtBu)_3]_2 + 4\ CH_3CN \xrightarrow{\text{Hexane}} 2\ Dy(OCHtBu)_3(CH_3CN)_2$$

$$Ln[N(SiMe_3)_2]_3 + O{=}CPh_2 \xrightarrow{\text{Hexane}} Ln[N(SiMe_3)_2]_3(O{=}CPh_2)$$

Scheme 9.35 Addition reaction of Lewis base ($O{=}CPh_2$) and strongly donating ligands (acetonitrile).

some examples of addition of Lewis base ($O{=}CPh_2$) and strongly donating ligands (acetonitrile) (Scheme 9.35).

9.6.5 Salt Occlusion or Ate Complexation

If salt elimination is used for synthesis of organometallic complexes of rare earths, then some alkali halide might stay in the primary coordination sphere of the metal. This happens if the coordination sphere has some space left. The alkali halide fragment might not be removed, instead halide anion remains bonded with rare-earth metal center and the alkali metal as bridge, as illustrated by the following Scheme 9.36:

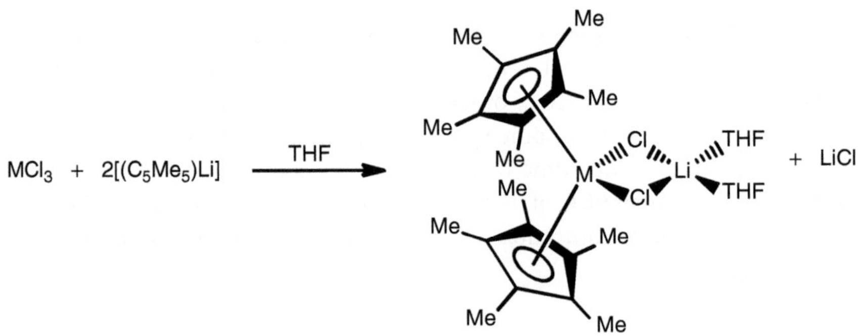

Scheme 9.36 Salt occlusion of alkali halide in organometallic complexes of rare earths.

These complexes are known as the "ate" complexes and are known for all the rare earths. They are soluble in organic solvents and can be used for the formation of hydrocarbyl derivatives that can be used as precatalysts or precursors in certain reactions.

9.6.6 Neutral π-Donor Ligation

Neutral π-donors, such as arenes and olefina, coordinate wih the electron-deficient metal ion. η^2 complexes of rare earths having olefins and alkynes as ligands (Scheme 9.37) are unstable, owing to absence of dπ–pπ backbonding. But increase in the Lewis basicity of olefin or chelation increases the stability of complexes. η^6 arene complexes are comparatively more stable.

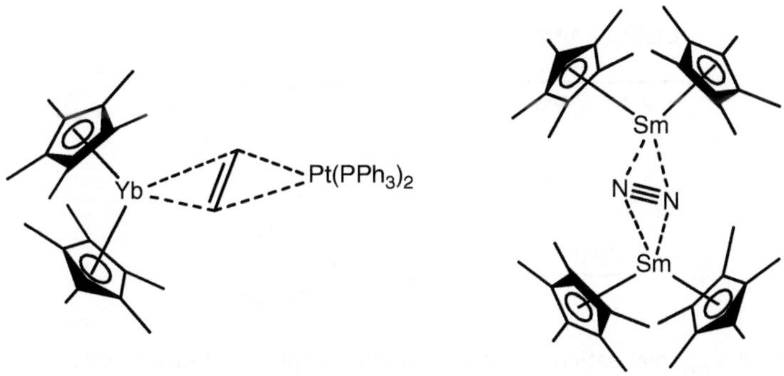

Scheme 9.37 Structure of η^2 complexes of rare earths having olefins and alkynes as ligands.

9.6.7 Agostic Interactions

Agostic interactions stand for two-electron three-center bonds. It is a type of binding found in highly electron-deficient metal centers with steric unsaturation. It is found in complexes having the fragments "CH," "SiMe," and "SiH." These are weak chelate-type intramolecular electrostatic interactions. It is generally absent in solution but has important role in structure and reactivity of molecules. Some of the examples are shown below in Scheme 9.38:

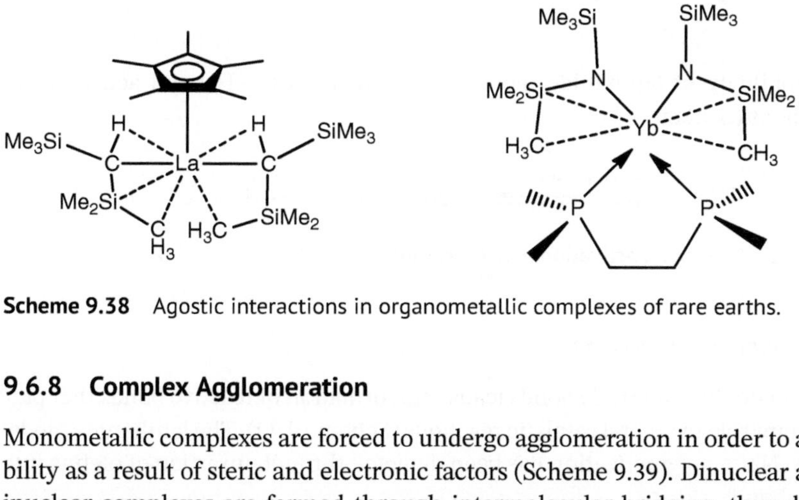

Scheme 9.38 Agostic interactions in organometallic complexes of rare earths.

9.6.8 Complex Agglomeration

Monometallic complexes are forced to undergo agglomeration in order to attain stability as a result of steric and electronic factors (Scheme 9.39). Dinuclear and multinuclear complexes are formed through intermolecular bridging, thus decreasing reactivity. Higher agglomerated systems involving self assembly to form rings and clusters are also observed. The following examples show the formation of dinuclear complexes.

9.6.9 Ligand Exchange and Redistribution Reactions

High ligand exchangeability is seen in organolanthanide complexes playing an important role in catalytic processes. Counter ligand exchange takes place by

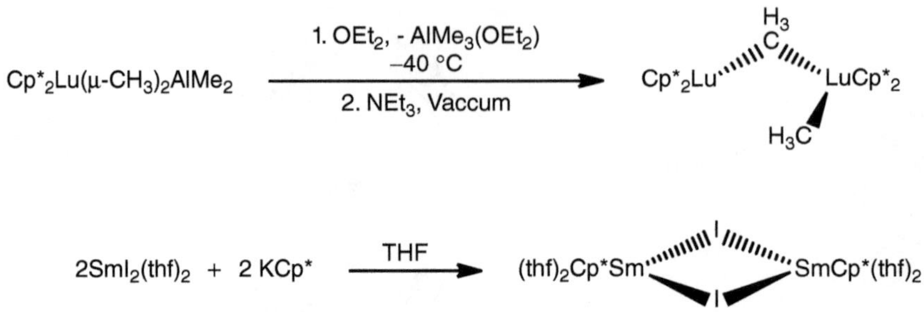

Scheme 9.39 Complex agglomerization in organometallic complexes of rare earths.

σ-bond metathesis through four-center transition state. Hydrogenolysis reaction involving Si–H/Ln–C transposition of hydrosilylation reaction has been shown as an example in Scheme 9.40.

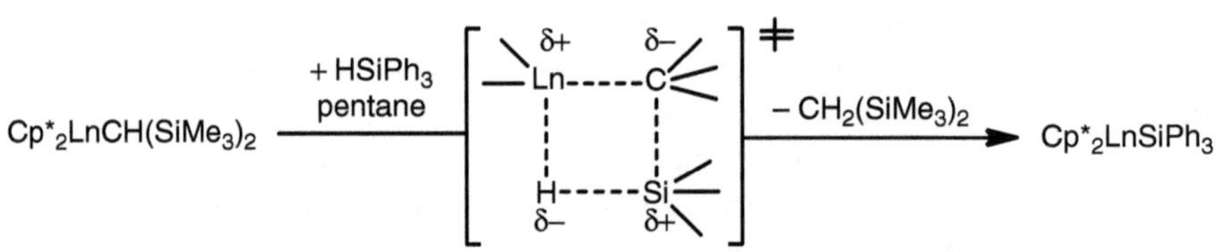

Scheme 9.40 Ligand exchange and redistribution reactions in organometallic complexes of rare earths.

Electrophilicity of lanthanide metal center provokes C—H bond activation in hydrocarbons (Scheme 9.41).

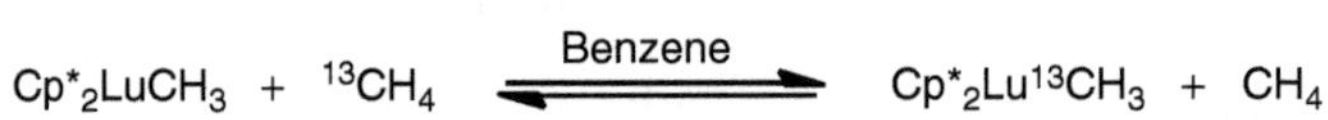

Scheme 9.41 C—H bond activation in hydrocarbons.

9.6.10 Insertion Reactions

Insertion in double and triple bonds leads into formation of reactive bonds that play an important role in several catalytic reactions (Scheme 9.42). The lanthanide model of Ziegler–Natta polymerization has been depicted through insertion of olefins into Lu—CH_3 bond.

9.6.11 Elimination Reactions

Organometallic complexes of rare earths undergo β-hydride and β-alkyl elimination in case of metal alkyl complexes to give metal hydride and alkene, as shown below in Scheme 9.43:

β-hydride and β-alkyl elimination in $Cp^*_2LuCH_2CH(CH_3)_2$ led to the formaion of two-methyl propene and propene, respectively (Scheme 9.44).

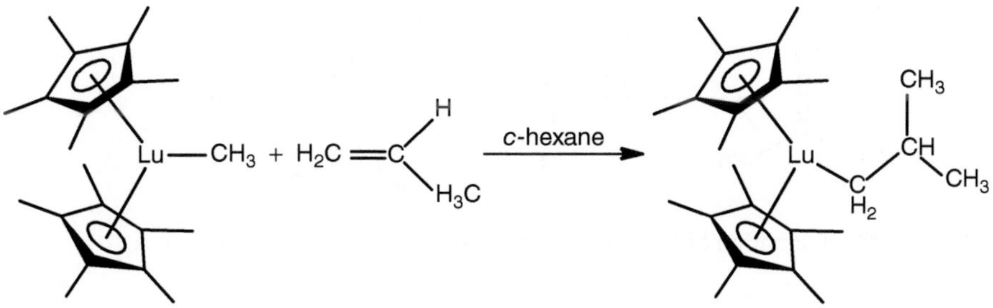

Scheme 9.42 Insertion reactions in organometallic complexes of rare earths.

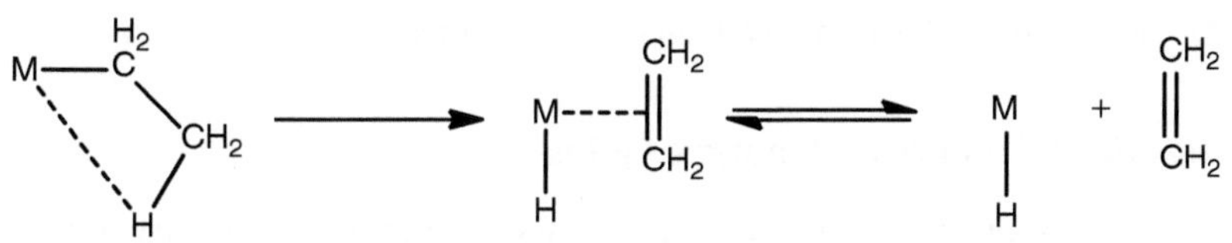

Scheme 9.43 β-hydride elimination in organometallic complexes of rare earths.

$$Cp^*_2LuCH_2CH(CH_3)_2 \xrightleftharpoons{c\text{-hexane}} Cp^*_2LuH + H_2C{=}C(CH_3)_2$$

$$Cp^*_2LuCH_2CH(CH_3)_2 \xrightleftharpoons{c\text{-hexane}} Cp^*_2LuCH_3 + H_2C{=}CH(CH_3)$$

Scheme 9.44 β-hydride and β-alkyl elimination in $Cp^*_2LuCH_2CH(CH_3)_2$.

9.6.12 Redox Chemistry

Organolanthanides show redox reactions, for example, $Cp^*_2Sm(thf)_2$ is oxidized to an oxo-bridged dimer in presence of oxygen. However, $Sm(TpMe_2\text{-}3,5)_2$ gives a super oxo-complex in presence of oxygen (Scheme 9.45).

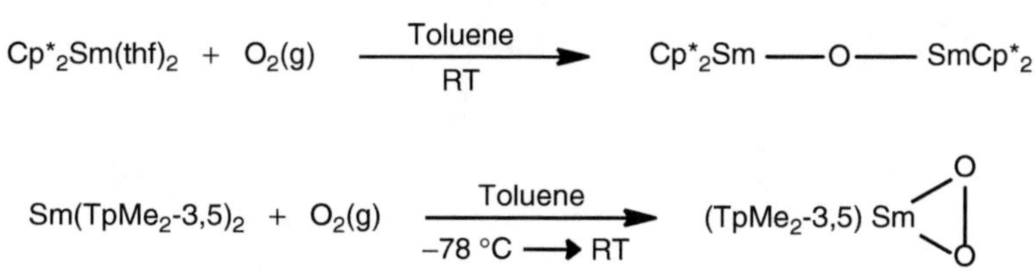

Scheme 9.45 oxidation of $Cp^*_2Sm(thf)_2$ to form an oxo-bridged dimer in presence of oxygen and formation of superoxo complex from $Sm(TpMe_2\text{-}3,5)_2$.

The formation of lanthanoid–imine azametallacyclopropane complex has been shown below in Scheme 9.46:

Homoleptic Ce(III) alkoxide upon oxidation with benzoquinone produces dinuclear heteroleptic Ce(IV) complex (Scheme 9.47).

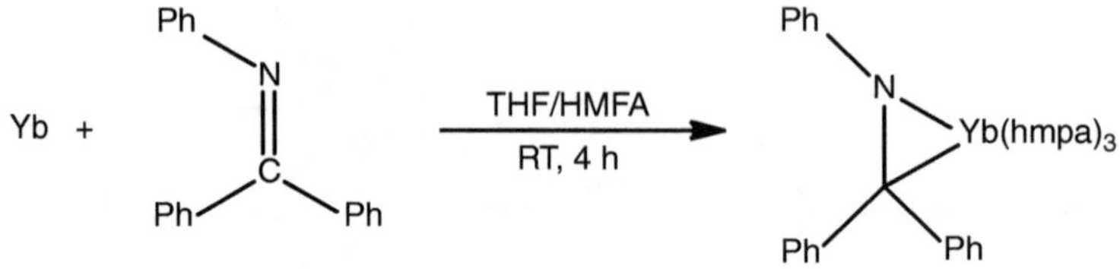

Scheme 9.46 Formation of lantanoid-imine azametallacyclopropane complex.

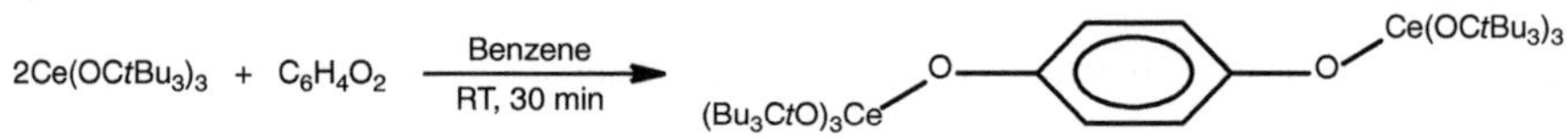

Scheme 9.47 Formation of dinuclear heteroleptic Ce(IV) complex.

9.6.13 Reaction Sequences – Catalytic Cycles

Organometallic compounds of rare earths play important roles in organic catalysis. Hydroamination has been discussed here as an example in Scheme 9.48.

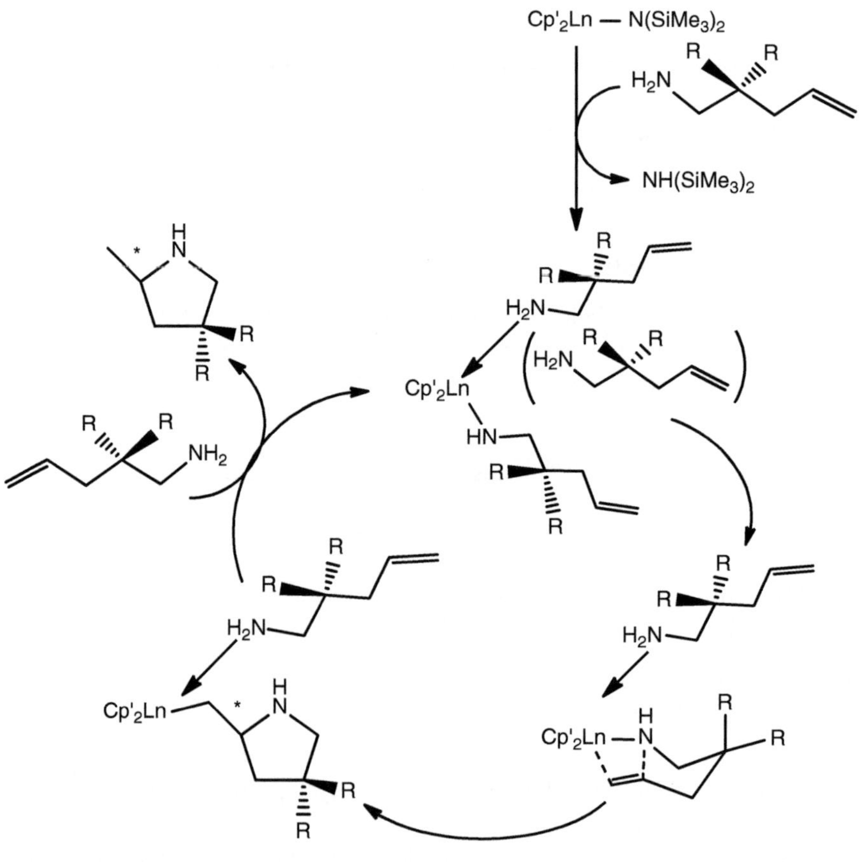

Scheme 9.48 Steps involved in hydroamination.

9.7 Applications of Organolanthanoids and Organoactinoids

Organoderivatives of lanthanoids as such do not have broad practical applications. These compounds can be widely used in chemical industry, metallurgy, mechanical engineering, medicine, electronics, and other fields of engineering. Moreover, it should be noted that, practically in all cases, where salts, alloys, or free rare eath metals are applied as the catalysts for organic synthesis, their organic derivatives are formed at intermediate stages.

9.7.1 Catalytic Applications

Considerable study is being made of the ability of organometallic compounds of the lanthanides and actinides to catalyze reactions of organic substrates, much of this work associated with the name of Tobin J. Marks. The hydride $[Cp^*_2Lu(\mu\text{-}H)_2LuCp^*_2]$ ($Cp^* = C_5Me_5$) is an exceptionally active catalyst for the homogeneous hydrogenation of alkenes and alkynes. A suggested mechanism, involving dissociation to the coordinatively unsaturated monomer $[Cp^*_2LuH]$, is shown below in Scheme 9.49.

A whole range of reactions are catalyzed by such complexes. The mechanism proposed in a recent study of hydroamination, catalyzed by $[Cp^*_2La\{CH(SiMe_3)_2\}_2]$, has been discussed below in Scheme 9.50. It indicates a two-stage mechanism, cyclization to form La—C and C—N bonds, followed by La—C protonolysis. The alkene inserts into the La—C bond via a four-center transition state, followed by protonolysis by a second substrate molecule and dissociation of the cyclized amine, which regenerates the catalyst. The rate-determining step involves a highly organized seven-membered chair-like cyclic transition state.

9.7.1.1 Catalysis of Polymerization

The main area of application of the rare earth metal compounds is catalytic processes, among which the most important and most studied process appears to be the stereospecific conjugated diene polymerization. The first reports on the high catalytic activity and high stereospecificity in these processes of the systems consisting of lanthanoid(III) chlorides and trialkylaluminum appeared in 1964 and have been developed in many other papers by Chinese chemists. An essential contribution to studying the problem has been made by a group of investigators from Ufa (Russia) and a group of Dolgoplosk. All rare earths, except Sm(II), Eu(II), and Yb(II), indicate a noticeable catalytic activity in the diene polymerization. Pathways for the polymer formation have to be assumed by taking $(Me_5C_5)_2Ln\ (Et_2O)$ and $Ln/MO_2O_3/Al_2O_3$ (Scheme 9.51).

9.7.1.2 Catalysts and Reagents for Reduction Processes

The complexes formed on the cocondensation of vapors of rare earth metal and alkynes RC=CR′ catalyze the hydrogenation of alkenes and alkynes by dihydrogen under ambient conditions (Scheme 9.52). La, Sm, Nd, Er, and Lu indicate

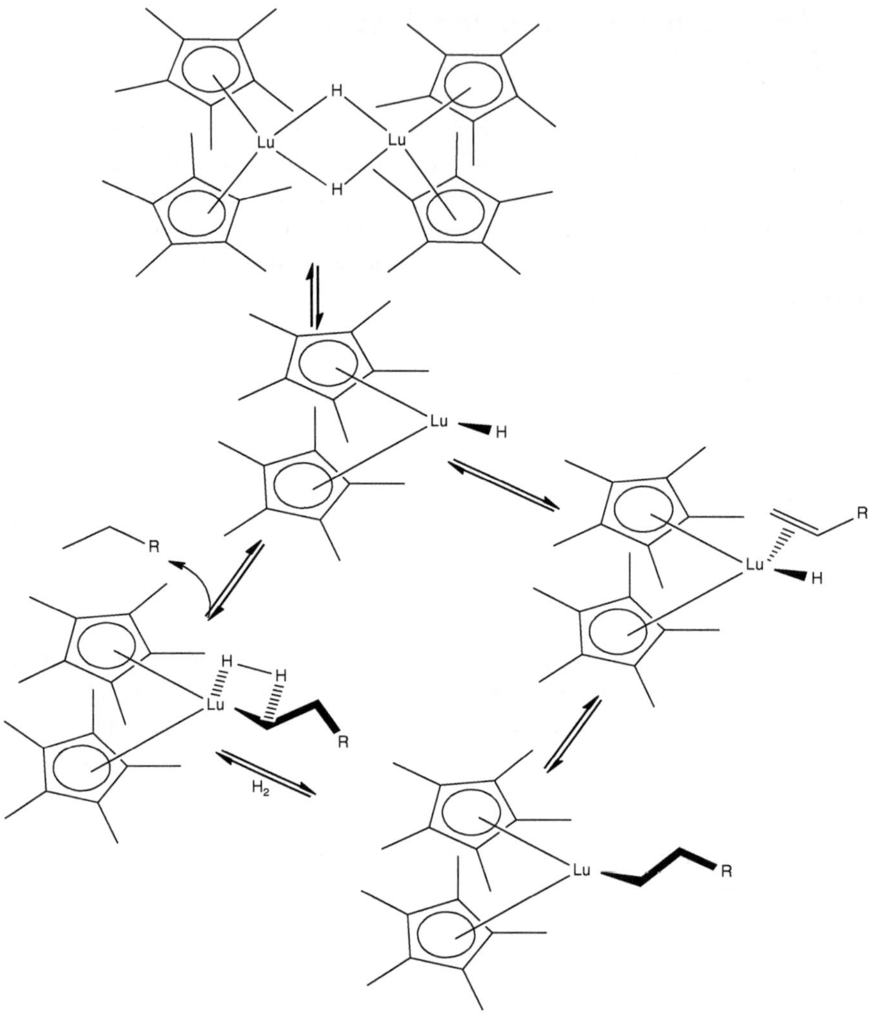

Scheme 9.49 Catalytic hydrogenation using [Cp*$_2$LuH] catalyst.

practically the same catalytic activity in the hydrogenation of three alkynes. It is suggested that the REM hydride is formed at the initial stage of the reaction. Cyclopentadienyl complexes of lanthanoids can be used as catalysts for the hydroamination/cyclization of arninoolefines, the hydroformylation of olefines, the cyclization of alkynes, and other processes. SmI_2 and YbI_2 appear to be effective catalysts for the Meerwein–Ponndorf–Verley reactions. High catalytic activity of ytterbium di-iodide and especially samarium di-iodide in many other organic syntheses – the reduction of ß-heterosubstituted ketones, the Oppenauer oxidations, the reduction of alkyl halides in alkanes, the transformation of epoxides into the corresponding olefines, and the pinacol formation from aromatic aldehydes – has also been reported. Cerium(Ill) and Sm(llI) chlorides also promote the pinacol formation. The treatment of aldehydes, ketones, and carboxylic acids by alkyl

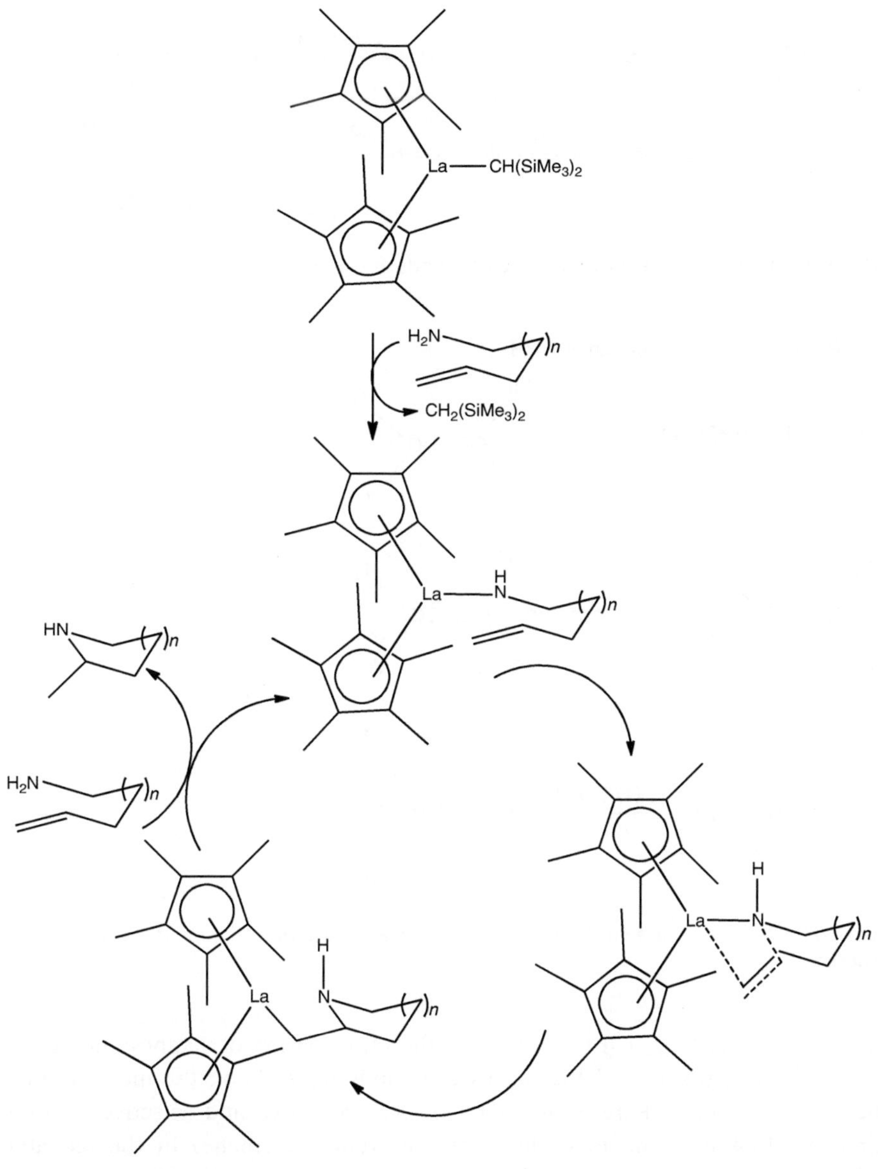

Scheme 9.50 Hydroamination catalyzed by $[Cp^*_2La(CH(SiMe_3)_2)]$.

halides in the presence of $SrnI_2$ in a THF solution affords the products of the carbonyl groups alkylation.

9.7.1.3 Catalysts and Reagents for Oxidation Processes

Lanthanoid compounds are active in oxidation reactions. The t-$BuOSrnI_2$, promotes the alkylation of carbonyl compounds, catalyzes the reverse reaction as well as the oxidation of alcohols to aldehydes and ketones (the Oppenauer reaction). Heterometallic oxides of the general composition $La_{1-x}M_xM'O_3$ (M = Ce, Sr;

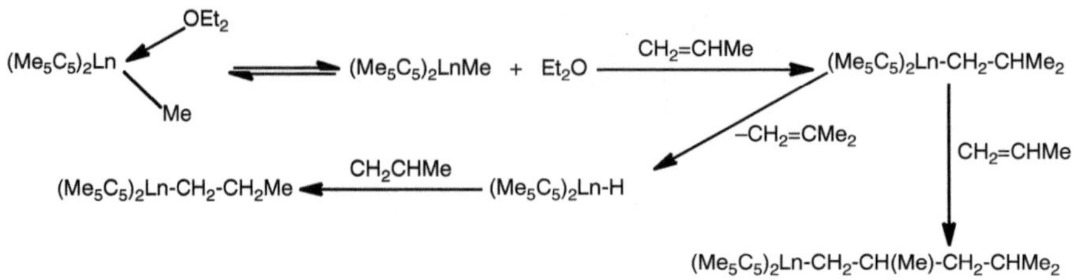

Scheme 9.51 Use of rare earth metal compounds in catalysis of polymerization.

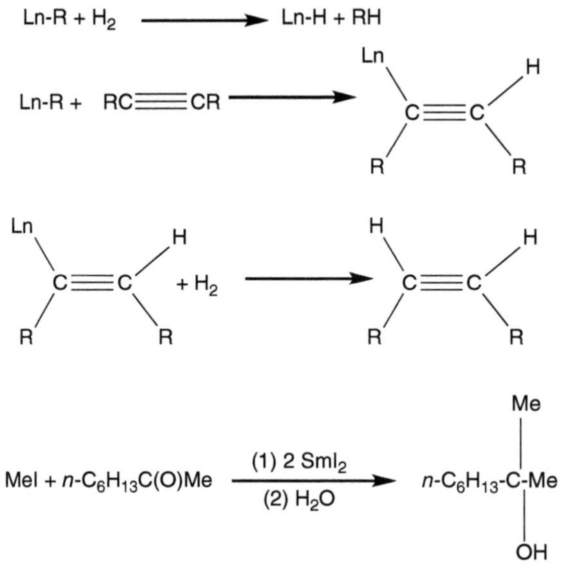

Scheme 9.52 Use of rare earth metal compounds as catalysts and reagents for reduction processes.

M′ = Fe, Co), $LaMn_{1-x}Cu_xO_3$ accelerate the O_2 oxidation of alkanes and carbon oxide. The effectiveness of the oxides essentially depends on the molar ratio of the metals in them. Rare earth metal oxides are active and selective catalysts for converting methane to C_2-hydrocarbons (ethane, ethene) in the oxidative dehydrogenation and coupling of methane. Among the tested oxides, Sm_2O_3 and especially the lanthanum/aluminum oxide $LaAlO_3$ indicated the highest activity and selectivity. La_2O_3 and Al_2O_3 are less active than heterometallic $LaAlO_3$. This suggests that $LaAlO_3$ is not a simple mixture of the component oxides. The same compound doped by platinum group metals catalyzes the dehydrohalogenation of 3,4-dichlorobut-2-ene at 500 °C to chloroprene. The exploitation of La, Ce, Pr, Nd, and Sm nitrosonaphthol complexes in this reaction has allowed to decrease the reaction temperature down to 150 °C.

9.7.2 Miscellaneous Applications

Among the other application fields of REM organoderivatives, the exploitation of substituted $(RC_5H_4)_3Ln$ and simple CP_3Ln cyclopentadienyl complexes for doping a rare earth into semiconductors by metal–organic chemical vapor deposition. The deposition is carried out at 460–700 °C in a hydrogen atmosphere. The rare-earth-doped 13–15 semiconductors are used for the preparation of laser materials. Erbium-doped GaAs has been used in light-emitting diodes. One of the most important application fields of REM organoderivatives is the preparation of a thin film of warm superconductors. The metal alkoxides $M(OR)_x$. are widely used as the precursors for metal oxides in thermodecomposition or hydrolysis processes. In reference to lanthanoids, this method offers opportunities for obtaining homogeneously mixed bimetallic and trimetallic REM oxides of high purity. Isopropyl–cyclopentadienyl complexes of lanthanum, neodymium, and holmium are used for the synthesis of heterogenic catalysts of the Fischer–Tropsch reaction. For this purpose, the silicon oxide containing OH groups on the surface is treated with a THF solution of $(i\text{-}C_3H_7C_5H_4)_3Ln$. This leads to the fixation of REM ions on the surface of the carrier. Organic groups are removed by heating in vacuum and then the catalyst is doped with palladium. Simple reactions of hydrogen sulfide with CP_2Ln (Ln = Sm,Eu,Yb) in THF can be used for the preparation of sulfides LnS, which are interesting as semiconductors. This route is evidently suitable for Ln(III) sulfides.

9.8 Conclusion

Lanthanides and actinides have several unique properties in contrast to other elements. The development of organometallic chemistry of lanthanides and actinides still remains rather new. Inspite of their exciting chemistry, organolanthanides and organoactinides still remain unexplored. Further efforts need to be made in order to explore the scope of this class of chemicals for application in various fields.

Further Reading

Anwander, R. (1999). *Principles in Organolanthanide Chemistry*. Berlin Heidelberg: Springer-Verlag.

Atwood, J.L., Hunter, W.E., Rogers, R.D. et al. (1978). Neutral and anionic silylmethyl complexes of group 3A and lanthanoid metals. X-ray crystal and molecular structure of $Li(THF)_4Yb[CH(SiMe_3)_2]3Cl$ (THF = tetrahydrofuran). *J. Chem. Soc., Chem. Commun.* 140–142.

Bambirra, S., Meetsma, A., and Hessen, B. (2006). Lanthanum tribenzyl complexes as convenient starting materials for organolanthanum chemistry. *Organometallics* 25: 3454–3462.

Bambirra, S., Perazzolo, F., Boot, S.J. et al. (2008). Strategies for the synthesis of lanthanum dialkyl complexes with monoanionic ancillary ligands. *Organometallics* 27: 704–712.

Behrle, A. C. (2012). The study of lanthanides for organometallic and separations chemistry. Ph.D. Thesis, The University of Toled.

Behrle, A.C. and Schmidt, J.A.R. (2011). Synthesis and reactivity of homoleptic αmetalated N,N-dimethylbenzylamine rare-earth-metal complexes. *Organometallics* 30: 3915–3918.

Cotton, S. (2006). *Lanthanide and Actinide Chemistry*. Wiley.

Evans, W.J. (1987). *Polyhedron* 6: 803.

Harder, S. (2005). Syntheses and structures of homoleptic lanthanide complexes with chelating o-dimethylaminobenzyl ligands: key precursors in lanthanide chemistry. *Organometallics* 24: 373–379.

Harder, S., Ruspic, C., Bhriain, N.N. et al. (2008). Benzyl complexes of lanthanide(II) and lanthanide(III) metals: trends and comparisons. *Z. Naturforsch., B: Chem. Sci.* 63: 267–274.

Hart, F.A. and Saran, M.S. (1968). Triphenylscandium and related compounds. *Chem. Commun.* 1614.

Hitchcock, P.B., Lappert, M.F., Smith, R.G. et al. (1988). Synthesis and structural characterization of 1st neutral homoleptic lanthanide metal(III) alkyls – LnR_3 (Ln = La or Sm, R = $CH(SiMe_3)_2$). *J. Chem. Soc., Chem. Comm.* 1007–1009.

Housecroft, C.E. and Sharpe, A.G. (2018 (print and electronic)). *Inorganic Chemistry*, 5e. Pearson Education Limited.

Jaroschik, F., Shima, T., Li, X. et al. (2007). Synthesis, characterization, and reactivity of mono(phospholyl)lanthanoid(III) bis(dimethylaminobenzyl) complexes. *Organometallics* 26: 5654–5660.

Jones, C. (2017). Lanthanide complexes of bulky hybrid ligands, Ph.D. Thesis, Newcastle University.

Lappert, M.F. and Pearce, R. (1973). Stable silylmethyl and neopentyl complexes of scandium(III) and yttrium(III). *J. Chem. Soc., Chem. Commun.* 126.

Lappert, M.F. and Pearce, R. (1973). *J. Chem. Soc. Chem. Commun.* 126.

Li, X., Nishiura, M., Mori, K. et al. (2007). Cationic scandium aminobenzyl complexes. Synthesis, structure and unprecedented catalysis of copolymerization of 1-hexene and dicyclopentadiene. *Chem. Commun.* 4137–4139.

Manzer, L.E. (1977). New chelated C,N-benzyllithium reagent and some organometallic derivatives of selected early transition metals. *J. Organomet. Chem.* 135: C6–C9.

Marks, T.J. (1982). Actinide Organometallic Chemistry. *Science* 217: 989–997.

Miessler, G.L., Fischer, P.J., and Tarr, D.A. (2014). *Inorganic Chemistry*, 5e. Pearson Education Limited.

Petrov, A.R., Rufanov, K.A., Harms, K., and Sundermeyer, J. (2009). Reinvestigation of ortho-metalated N,N-dialkylbenzylamine complexes of rare-earth metals. First structurally characterized arylates of neodymium and gadolinium Li(LnAr)4. *J. Organomet. Chem.* 694: 1212–1218.

Putzer, M.A., Rogers, J.S., and Bazan, G.C. (1999). Intramolecular nucleophilic substitution on coordinated borabenzenes: a new entry into borabenzene complexes. *J. Am. Chem. Soc.* 121: 8112–8113.

Schumann, H. (1984). *Angew. Chem.* 475; Angew. Chem. Int. Ed. Engl. 1984, 23, 474.

Schumann, H., Albrecht, I., Gallagher, M. et al. (1988). Organometallic compounds of the lanthanides-xl.* Recent developments in organolanthanide chemistry. *Polyhedron* 7 (22/d): 2307–2315.

Schumann, H., Freckmann, D.M.M., and Dechert, S. (2002). Organometallic compounds of the lanthanides. 157 The molecular structure of tris(trimethylsilylmethyl)samarium, -erbium, -ytterbium, and -lutetium. *Z. Anorg. Allg. Chem.* 628: 2422–2426.

Streitwieser, A., Kinsley, S.A., Rigsbee, J.T. et al. (1985). *J. Am. Chem. Soc.* 107: 7786.

Wilkinson, G. and Birmingham, J.M. (1954). *J. Am. Chem. Soc.* 76: 6210.

Wooles, A.J., Mills, D.P., Lewis, W. et al. (2010). Lanthanide tri-benzyl complexes: structural variations and useful precursors to phosphorus-stabilised lanthanide carbenes. *Dalton Trans.* 39: 500–510.

Xing, Y., Shi, Z., Li, G.H., and Pang, W.Q. (2003). Hydrothermal synthesis and structure of $[C_2N_2H_{10}][La_2(H_2O)_4(SO_4)_4]\bullet 2H_2O$, a new organically templated rare earth sulfate with a layer structure. *Dalton Trans.* 940–943.

Zimmermann, M. and Anwander, R. (2010). *Chem. Rev.* 110: 6194.

10

Bioorganometallic Chemistry

Priyanka Tiwari

Atal Bihari Vajpayee University, Government J.P.Verma P.G. Arts and Comm. College, Department of Chemistry, Ratanpur road, Koni, Bilaspur, Chhattisgarh 495009, India

10.1 Introduction

Inorganic elements play an important role in life processes. Bioinorganic chemistry includes the discipline at the interface of biology and inorganic chemistry. Bioorganometallic chemistry in principle only concerns systems containing metal–carbon bonds [1]. Bioorganometallic chemistry is a relatively new area of research that extends organometallic chemistry into biology, medicine, and bioanalytics [2]. It covers the enzymes and cofactors that are organometallic, which possesses at least one direct metal–carbon bond. In particular, the naturally occurring vitamin B_{12} family (cobalamins), hydrogenases, the CO dehydrogenases (CODHs), acetyl-CoA synthetases, and methyl coenzyme M reductase are covered. Second, medicinal applications of organometallic compounds (anticancer, antimicrobial, NO/CO drugs, and rapido pharmaceuticals) and toxicological aspects are treated. The use of organometallic complexes, which are covalently linked to biomolecules (amino acids, peptides, proteins, DNA and derivatives, sugars, lipids, and receptor ligands), for biological studies is described. Thus, bioorganometallic chemistry has emerged as a significant subtopic in bioinorganic chemistry. Wolfgang Beck, the first recipient of the Lavoisier medal for seminal studies in bioorganometallic, and his coworkers in Munich were among the first to study the reactions of organometallic complexes with bioligands [3, 4]. Initially, the only examples of naturally occurring bioorganometallic compounds were cobalamin cofactors (vitamin B_{12}) in its various forms [5]. In the recent years, there is a discovery of many naturally occurring systems containing carbon–metal bonds. It is being increasingly relevant to pharmacology and medicine [6].

Living organisms store and transport transition metals both to provide appropriate concentration of them for use in metalloproteins or cofactors and to protect themselves against the toxic effects of metal excess [7]. Metalloproteins and metal cofactors are found in plants, animals, and microorganisms. The metals are generally found either by bounding directly to proteins or in cofactors, such as porphyrins

Organometallic Compounds: Synthesis, Reactions, and Applications, First Edition.
Edited by Dakeshwar Kumar Verma and Jeenat Aslam.

or cobalamins, or in clusters that are, in turn, bound by protein [7]; the ligands are usually O, N, S, and C. Proteins with which transition metals and zinc are most commonly associated catalyze intramolecular or intermolecular rearrangement of electrons. Although redox properties of metals are important in many reactions, in others, metal appears to contribute to the structure of active stage [7].

10.2 Cobalamin: Vitamin B_{12}-Coenzyme

10.2.1 Occurrence and Function

B_{12}-coenzymes are conceivably the nature's most complex and most physiologically broadly relevant organometallic cofactors. B_{12}-derivatives, therefore, catalyze unique enzymatic reactions that directly depend upon the reactivity of cobalt coordinated organic ligand and they hold an exceptional position in the area of bioorganometallic chemistry [8]. B_{12}-dependent enzymes catalyze the metabolism of most of the living organisms. In humans, deficiency of vitamin B_{12} causes pernicious anemia.

10.2.2 Structure

In coenzyme-B_{12}, cobalt is bound to a tetraazamacrocyclic ligand [9]. The cobalt atom lies approximately in the plane of corrin ligand. Complexes that possess the α-D-ribofuranose-3-phosphate and the terminal 5,6-dimethylbenzimidazole as an axial ligand are called cobalamins [7]. The upper or β-position is occupied by another ligand, which may be water, OH^-, CN^-, or an alkyl group. The cyano derivative is vitamin B_{12}. 5′-deoxyadenosylcobalamin is called coenzyme B_{12} [7] (Figure 10.1).

The Co—C bond in adenosyl cobalamin is weak (29 ± 5 kcal mol^{-1} or 123 ± 21 kJ mol^{-1}) and easily undergoes reversible hemolytic cleavage at an ambient temperature to give adenosyl radical RCH_2 and the 17-electron Co^{II} radical [10]. This cleavage is basis of enzymatic mechanisms of vitamin B_{12} coenzyme [7].

$$[(L_4X)Co^{III} - CH_2R]^+ \longrightarrow [(L_4X)Co^{II}]^+ + RCH_2$$

The weakly bonded adenosyl ligand is easily substituted *in vivo* by other ligands in axial position, such as H_2O (aquocobalamin or B_{12}a) or CH_3 (methylcobalamin).

It has been proposed that the flexibility of the corrin ligand is the reason why nature does not use the porphyrin ligand in vitamin B_{12} [7]. In an alternative explanation, the weakening of the Co—C bond would be an electronic effect associated with the labilization of the Co—N bond [7].

The electronic spectrum of the metal-free corrin resembles that of metal derivatives; it seems therefore the bands are essentially π–π^* transitions modified by the central atom and by the axial ligands [7]. The cob(III)alamins are red, whereas the cob(II)alamins, which are brown, show an additional band at 600 nm. The latter have an EPR spectrum typical of an unpaired electron in the $d_z{}^2$ orbital with some

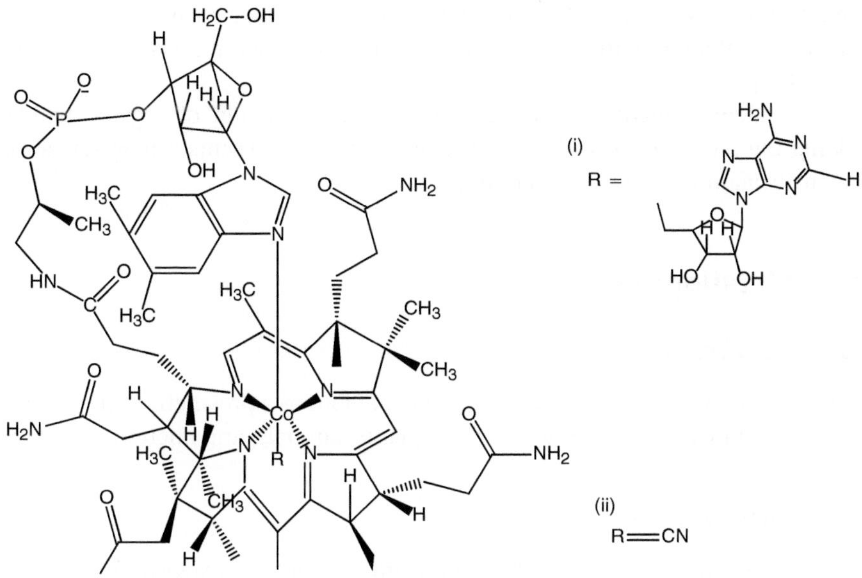

Figure 10.1 Structure of (i) coenzyme B_{12} 5′-deoxyadenosylcobalamin and (ii) vitamin B_{12} cyanocobalamin.

4s mixing. Both cobalt(I)- and cobalt(II)-containing cobalamins readily react with alkyl derivatives to give alkylcob(III)alamins [7].

$$\text{Cob(II)alamin} + CH_3I \longrightarrow CH_3 - \text{Cob(III)alamin} + \frac{1}{2}I_2$$
$$\text{Cob(I)alamin} + CH_3I \longrightarrow CH_3 - \text{Cob(III)alamin} + I^-.$$

These can formally be regarded as complexes of cobalt(III) with a carbanion. These are rare examples of naturally occurring organometallic compounds.

10.2.3 B_{12}-dependent Enzymes

It was first isolated by Folkers and Smith in 1948 that vitamin B_{12} (B_{12}, Cobalamin, 1,) has been recognized as a cofactor for enzymes that catalyze a range of biological process, including isomerization, methyltransferases, and dehalogenation [11].

10.2.4 Occurrence

B_{12}-dependent enzymes are, therefore, divided into three groups: (a) 5′deoxyadenosylcobalamin-dependent isomerases; (b) MeCbl-dependent transferases, present in mammalian cells; and (c) reductive dehalogenases, found only in organohalide respiring bacteria [11].

10.2.5 Function

AdoCb1-dependent enzymes catalyze isomerization (e.g. glutamate mutase, methylmalonyl-CoA mutase, and dioldehydrase) and reduction reactions (e.g.

ribonucleotide reductase) [11]. MeCbl-dependent enzyme, such as methionine synthetase, methane synthetase, or DNA methylase, operates via Co—C bond cleavage [11].

B_{12}-dependent reductive dehalogenases catalyze reduction of organic halides involving the halogen–cobalt bond formation, which represents a new activation mode in cobalamin-catalyzed reactions [11].

10.3 Metalloproteins

10.3.1 Occurrence

Metalloproteins are the proteins containing metal ions as a prosthetic group. Metalloproteins and metal cofactors are found in plants, animals, and microorganisms.

10.3.2 Functions

The prosthetic group usually plays a significant part in the biological functions of proteins. It is mainly the redox activity that makes metal ions important as biological cofactors [12]. Thus, redox reactions play an important role in many biological processes, such as respiration and photosynthesis. Three types of oxidation–reduction (redox) centers are found in biology: protein-side chains, small molecules, and redox cofactors [7]. A variety of small molecules, both organic and inorganic, can function as redox reagents in biological systems.

Metalloproteins containing a single type of redox cofactor can be divided into two general classes: electron carriers and proteins involved in the transport or activation of small molecule [7].

Proteins that function as electron transferases typically place their prosthetic groups in a hydrophobic environment and may provide hydrogen bonds (in addition to ligands) to assist in stabilizing both the oxidized and the reduced forms of the cofactor [7]. Metal–ligand bonds remain intact upon electron transfer to minimize inner-sphere reorganization. Many of the complex multisite metalloenzymes contain redox centers that function as intramolecular electron transferases, shuttling electrons to/from other metal centers that bind exogenous ligands during enzymatic turnover [7].

10.3.3 Electron Transferases

There are four classes [13, 14] of electron transferases, each of which contains many members that exhibit important structural differences: flavodoxins, blue copper proteins, iron–sulfur proteins, and cytochromes.

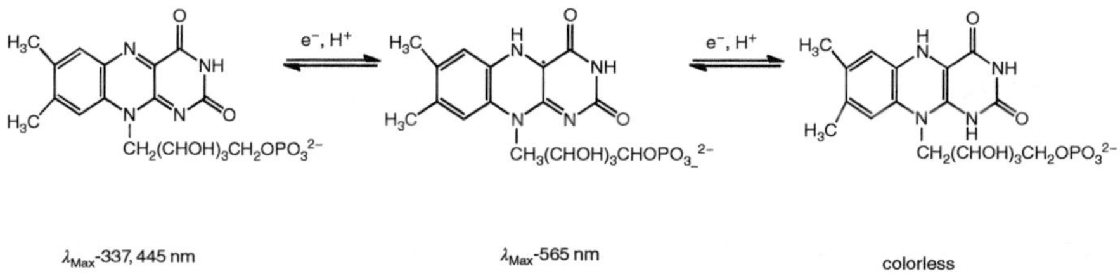

Figure 10.2 Reduction of FMN.

10.3.3.1 Flavodoxins

10.3.3.1.1 Occurrence

The flavodoxins are typical electron-transfer proteins in which they contain an organic redox cofactor, flavin mononucleotide (FMN) [7]. Thus, flavodoxins are proteins that have molecular weights in the 8–13 kDa range and are found in many species of bacteria and algae [7].

10.3.3.1.2 Function

FMN can act as either a one- or two-electron redox center. In solution, semiquinone form of free FMN is unstable and disproportionate to quinine (oxidized) and hydroquinone (reduced) forms [7]. Hence, free FMN can function in effect as a two-electron reagent. FMN in flavodoxins, on the other hand, can function as a single-electron carrier [7] (Figure 10.2).

10.3.3.2 Blue Copper Proteins

10.3.3.2.1 Structure

The blue copper proteins are characterized by intense S(cys)⟶Cu charge-transfer absorption near 600 nm; an axial EPR spectrum displaying an unusually small hyperfine coupling constant and a relatively high reduction potential [7]. X-ray structures of several blue copper proteins indicate that the geometry of copper site is approximately trigonal planar, as illustrated by Alcaligenes denitrificans azurin structure [7]. In these proteins, three ligands (one cys, two His) bind tightly to the copper in a trigonal arrangement [7].

10.3.3.3 Iron–Sulfur Proteins

Occurrence and Function

The iron–sulfur proteins play important roles as electron carriers in virtually all living organisms and participate in plant photosynthesis, nitrogen fixation, steroid metabolism, oxidative phosphorylation, as well as many other processes [7]. Iron sulfide proteins involved in electron transfer are called ferredoxins and rubredoxins.

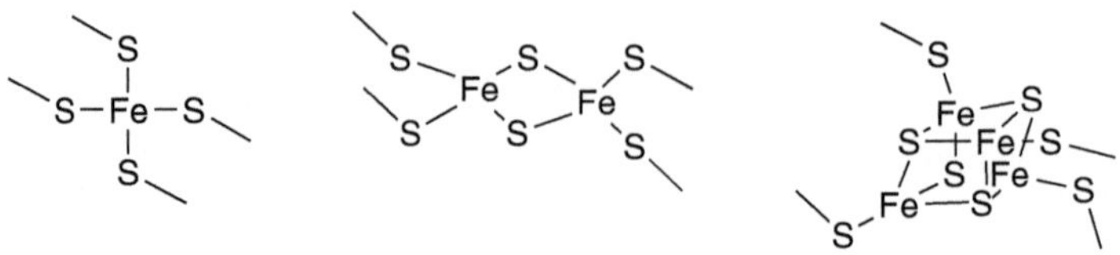

Figure 10.3 Active site of nonheme iron–sulfur proteins serving as biological mediators.

Ferredoxins are distinguished from rubredoxins by their possession of acid-labile sulfide. Rubredoxins have no acid-labile sulfide and generally have iron in a more or less isolated site [7] (Figure 10.3).

Rubredoxins

Occurrence

Rubredoxins are primarily found in anaerobic bacteria [7].

Structure

Rubredoxins are small proteins (6 kDa) and contain iron ligated to four cys sulfur in a distorted tetrahedral arrangement [7]. The E^0 value for Fe(III/II) couple in water is 770 mV, that of *Clastridium pasteurianum* rubredoxin is −57 mV [7]. The reduction potential of iron–sulfur proteins is quite negative, indicating a stabilization of oxidized form of redox couple as a result of negatively charged sulfur ligands [7].

Ferredoxins

[2Fe–2S] Ferredoxins: Occurrence and Structure

The [2Fe–2S] ferredoxins (10–20 kDa) are found in plant chloroplasts and mammalian tissue [7]. The structure of spirulina platensis ferredoxin is based on EPR and Mossbauer studies in which iron atoms are present in a spin-coupled [2Fe–2S] cluster structure [7].

Four-iron Clusters [4Fe–4S]: Occurrence and Structure

Four-iron clusters[4Fe–4S] are found in many strains of bacteria [7]. In many of these ferredoxins, two such clusters are present in protein [7]. These proteins have reduction potentials in −400 mV range and are rather small (6–10 kDa). Each of the clusters contain four-iron centers and form sulfides at alternate corners of a distorted cube. Each iron is coordinated to three sulfides and one cysteine thiolate. The iron is paramagnetic when reduced by one electron [7]. $[4Fe\text{–}4S(SR)4]_n$-clusters exist in three physiological oxidation states [7]. The bacterial ferredoxins and High Potential Iron–Sulfur Proteins (HiPIPs) all possess tetracubane clusters containing thiolate ligands [7].

10.3.3.4 Cytochromes

Occurrence and Function

Cytochromes are found in plants and animals. They function as electron carriers.

The cytochromes [7] are the most thoroughly characterized proteins of the electron transferases. A cytochrome contains one or more heme cofactors. These proteins have distinctive optical properties, an intense absorption in 410–430 nm region (called the soret band) [7].

Classification

Cytochromes are typically classified on the basis of heme type. Heme a possesses a long phytyl "tail" and is found in cytochrome c oxidase; heme b is found in b-type cytochromes and globins; and heme c is covalently bound to c-type cytochromes via two thioether linkages [7].

Cytochrome nomenclature presents a real challenge. Some cytochromes are designated according to the historical order of discovery, e.g. cytochrome c_2 in bacterial photosynthesis. Others are designated according to the λ_{max} of the α band in the absorption spectrum of the reduced protein (e.g. cytochrome c_{551}) [7].

Ambler [15] divided cytochromes into three classes on the basis of structure.

Class I Cytochrome c contains axial His and Met ligands, with the heme located near the N-terminus of the protein. These proteins are globular and X-ray studies of class I cytochrome c from a variety of eukaryotes and prokaryotes clearly demonstrated an evolutionarily preserved "cytochrome fold" with the edge of heme solvent exposed. Cytochromes possess positive reduction potentials (200–320 mV) [7].

Class II Cytochromes c are found in photosynthetic bacteria. These are high spin, and the iron is five coordinated, with an axial His ligand. These proteins are four-α-helix bundles [7].

Class III Cytochrome c contains four hemes, each ligated by two axial histidines. These proteins are found in a restricted class of sulfate-reducing bacteria and may be associated with cytoplasmic membrane. The low molecular weights of cytochromes c_3 (~14.7 kDa) require that the four hemes be much more exposed to solvents that may be in part responsible for their negative (−200 to −350 mV) reduction potentials. These proteins possess many aromatic residues and short heme–heme distances, two properties that could be responsible for their anonymously large solid-state electrical conductivity [7] (Figure 10.4).

10.4 Oxidoreductase

The oxidoreductase includes two categories: the metalloporphyrins and the other nonhemic enzymes (monooxygenases).

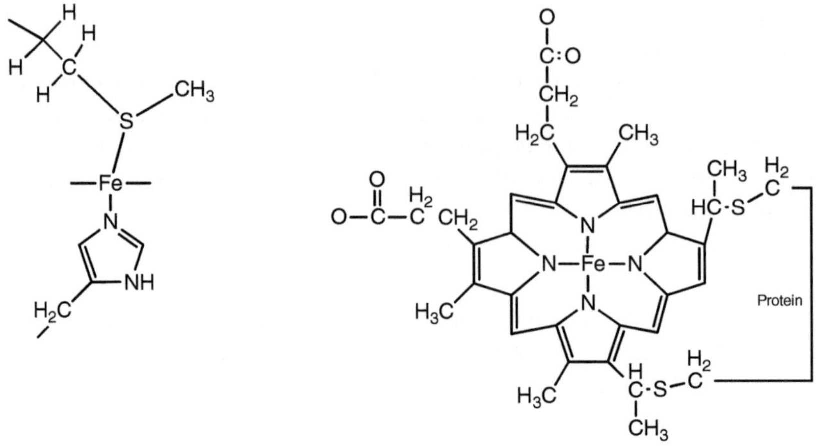

Figure 10.4 Cytochrome c with, on the left, the axial iron ligands: imidazole and thioether methionine.

10.4.1 Iron–Porphyrin Complexes

Occurrence

Metal porphyrins, most notably the iron porphyrins, are one of the most important groups of organometallic compounds [16, 17]. They occur in many proteins and enzymes that are responsible for three major activities: electron transfer, oxygen transfer, and photosynthesis.

Structure

Iron porphyrins serve as prosthetic group in haemoglobin (Hb) and myoglobin (Mb), as well as other heme proteins, such as peroxidases, catalases, and cytochromes. One of the most important properties of iron porphyrins is strong attraction of central Fe to molecules in addition to porphyrin. This extra coordination is responsible for attachment of active iron porphyrin to carrier proteins. The active center of Hb and Mb, and deoxyheme give rise to a five-coordinate high-spin ($s = 2$) Fe^{II} [18] upon oxygenation at the opposite distal face, a diamagnetic $Fe^{III}O_2^-$ complex forms reversibly [19, 20].

10.4.2 Monooxygenases

Function

Monooxygenases are the oxidoreductase enzymes. They incorporate one atom of oxygen molecule into the substrate, while second oxygen is reduced to water.

Structure

Metal-containing monooxygenase enzymes are known that contain heme iron, non-heme iron, or copper at their active sites [7]. The monooxygenase enzymes that strongly resemble well-characterized reversible dioxygen-carrying proteins are the following.

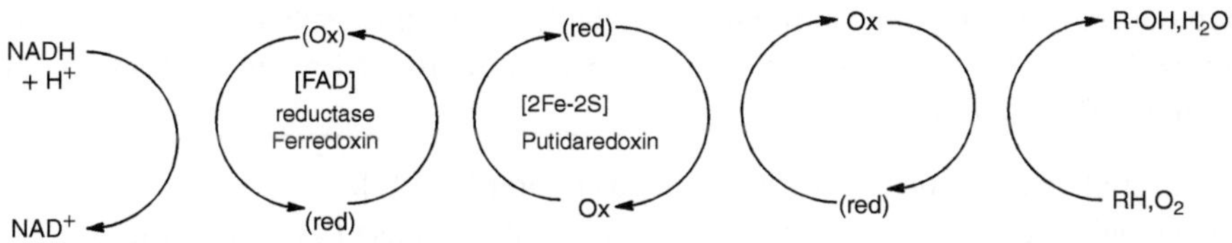

Figure 10.5 Electron transport chain.

(1) Cytochrome P-450, a heme containing protein whose active site resembles dioxygen binding sites of myoglobin or haemoglobin in many respects, except that the axial ligand to iron is a thiolate-side chain from cysteine rather than an imidazole-side chain from histidine [7].
(2) Tyrosine, which contains two copper ions in close proximity in its active site and which has deoxy, oxy and met states that closely resemble comparable states of hemocyanin in their spectroscopic properties [7].
(3) Methane monooxygenase, which contains two nonhemerythrin in many of its spectroscopic properties [7].

Most of the presently known metal-containing mono- and dioxygenases are multicomponent, requiring the involvement of additional proteins (electron transferases) to shuttle electrons from common biological reductant (usually NADH or NADHP) to the metallo-oxygenase. Cytochrome P-450 serves as an excellent example [7] (Figure 10.5).

Cytochrome P-450

Occurrence

Cytochrome P-450 is found in mammals, insects, fishes, yeasts, and plants. In the human body, it is located in various tissues and organs, particularly in the liver [21].

Structure

It is named so because its complexes with common ligands, such as CO and pyridine, show an absorption maximum around 450 nm [21]. Haemoglobin and myoglobin have the same iron–porphyrin structure of the active site as cytochrome P-450 but with a histidine bound to iron by a nitrogen atom of an imidazole [21]. The extraordinary difference of reactivity toward oxygen between cytochrome P-450 and these two hemoproteins is thus solely due to the nature of the axial ligand, i.e. due to its trans effect on the reactivity of bound oxygen [21].

10.5 Nitrogenases-catalyzing Nitrogen Fixation

Nitrogen is an essential element present in many living organisms and is necessary to sustain life. Due to inert nature of nitrogen, most organisms are unable to metabolize, although it is abundantly available in the atmosphere. Nitrogen fixation is the process of conversion of nitrogen to ammonia. Nitrogen fixation occurs in three different ways:

(1) through geochemical processes, such as lightening [22];
(2) biologically through the action of enzyme, nitrogenase [23, 24], found only in selected group of microorganisms [25, 26];
(3) industrially through the Haber–Bosch process [27–29].

Nitrogen fixation has a profound agronomic, economic, and ecological impact owning to the fact that the availability of fixed nitrogen represents the factor that most frequently limits agricultural production throughout the world [27].

Occurrence

Two kinds of nitrogen fixing bacteria are known: free-living or nonsymbiotic bacteria, including cyanobacteria (BGA), anabaena and nostoc, and genera, such as Azotobacter, Clostridium, or symbiotic bacteria, such as Rhizobium associated with leguminous plants and certain Azospirillum species, associated with cereal grasses.

In 1934, Burk coined the term "nitrogenase" [23, 24] for the enzyme that catalyzes the conversion of nitrogen to a bio accessible form of nitrogen and initiated the first meaningful studies of nitrogenase in living cells. Intensive efforts by various investigators have been done to reveal the structure and catalytic functions of nitrogenase.

There are different types of nitrogenases found in various nitrogen fixing bacteria: molybdenum (Mo) nitrogenase, vanadium (V) nitrogenase, and iron-only (Fe) nitrogenase [30]. Molybdenum nitrogenase, which can be found in diazotrophs, such as legume-associated rhizobia [31, 32], is the nitrogenase that has been studied the most extensively and this is the most well characterized [30].

Structure

Nitrogenase is a two component system [33, 34] composed of the MoFe protein (also called dinitrogenase or component I) and the electron-transfer Fe protein (also called dinitrogenase reductase or component II) [35–39]. A reducing source and MgATP are required for catalysis [40–43]. Fe protein and MoFe protein associate and dissociate in a catalytic cycle involving single-electron transfer and MgATP hydrolysis [36]. MoFe protein contains two metal clusters: the iron–molybdenum cofactor (FeMo-co) [44, 45], which provides the active site for substrate binding and reduction, and p-cluster, which is involved in electron transfer from the Fe protein to FeMo-co [37, 46–49].

$$N_2 + 8H^+ + e^- \overset{\text{Nitrogenase}}{\rightarrow} 2NH_3 + H_2$$

Molybdenum can change its oxidation states from Mo^{III} to Mo^{IV} in order to accommodate the various intermediates in the nitrogen fixation cycle resulting from successive electron-transfer/proton-transfer sequences [21] (Figure 10.6).

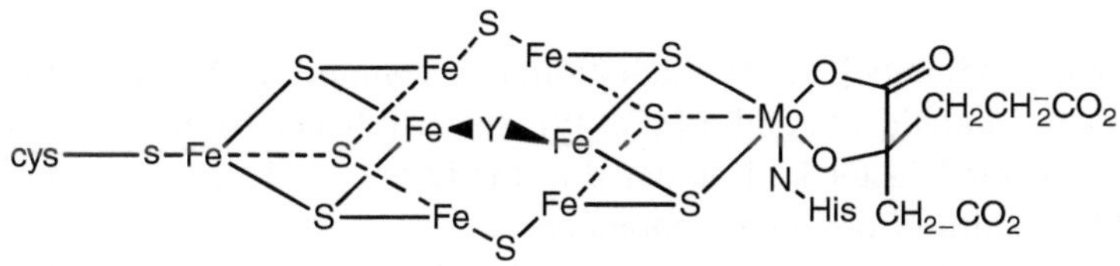

Figure 10.6 Cofactor (Mo–Fe-co) of a nitrogenase enzyme containing the active site of nitrogen fixation.

Despite various model systems, chemists have not yet been able to prepare a catalytic system that can complete with nineteenth century Haber–Bosch process (1895) that is still used in industry to synthesize ammonia [21].

Alternative Nitrogenases [7]

Vanadium Nitrogenase

The unique essentiality of molybdenum for nitrogen fixation went unchallenged until 1980. In 1986, two groups [7] isolated alternative nitrogenase component proteins from different species of Azotobacter and demonstrated that one component contained vanadium and that none of the components contained molybdenum.

Structure

One of the two main components of vanadium nitrogenase system is extremely similar to Fe protein of nitrogenase. This similarity is evident in the isolated proteins from *Azotobacter vinelandii* [7]. Both Fe proteins have an α_2 subunit structure and contain a single Fe_4S_4 cluster that is EPR active in its reduced state. A major difference between V and Mo enzymes lies in substrate specificity and product formation [7].

All-iron Nitrogenase

This nitrogenase has Fe–Fe protein and its cofactor Fe–Fe-co [7]. This nitrogenase seems to be poorest of the set in reducing nitrogen and makes ethane from ethylene.

10.6 Nickel Enzymes: CODH

Function

CODH catalyzes oxidation of CO to CO_2 along with dehydrogenation of water and release of protons and electrons. It is important in the oxygen-based respiratory process in hydrogenogenic bacteria [12]. The chemical process catalyzed by carbon monoxide dehydrogenase is similar to the water–gas shift reaction.

$$CO + H_2O \leftrightarrow CO_2 + 2H^+ + 2e^-$$

Occurrence

There are two types of CODHs: one is Mo-based CODHs with a mono-Mo factor found in aerobic organisms and the other is Ni-containing CODHs with a Ni–Fe–S cluster as well as multiple Fe–S clusters found in anaerobic organisms [50–52].

Structure

Ni CODHs are β2 homodimers [53, 54]. Each monomer contains an Ni–Fe–S cluster (cluster c) as the catalytic site and a [4Fe–4S] cluster (cluster B). In addition, another [4Fe–4S] cluster (cluster D) is situated at the interface of two monomers and coordinated by residues from both monomers.

The structure of Rhodospirillum rubrum NiCODHs [54]and *Moorella thermoacetica* ACS/CODH complex [55] shows cluster C as [Ni–4Fe–4S], coordinated by five cysteines and one histidine from the protein. The nickel is also coordinated by an external nonprotein ligand.

Hydrogenase Enzymes

Function

Hydrogenase enzymes [21]allow some bacteria to feed with hydrogen and some others to eliminate hydrogen. They catalyze the following reaction in the forward or backward direction:

$$H_2 \xleftrightarrow{\text{Hydrogenase}} 2H^+ + 2e^-$$

Structure

The majority of hydrogenase enzymes contain nickel and iron that are seemingly bridged by a carbonyl and a dithiolate ligand. The active site is a nickel atom, probably Ni^{III} [21].

A minority of hydrogenase enzymes only contains iron in a 6Fe–6S cluster [21]. An electron-poor transition metal center binds hydrogen without oxidative addition, and that the complexes M-(H_2) are consequently acidic [21]. The activation process is completed by transfer of electrons from nickel, which considerably increases the acidity of nickel hydride species [21].

$$Ni^{II} \xrightarrow{H_2} Ni^{II}(H_2) \xrightarrow{-H^+-e^-} Ni^{III}(H) \xrightarrow{-H^+-e^-} Ni^{II}$$

The nickel bacteria and the archaebacteria are primary bioorganisms in the evolution, being able to survive with gases as elementary as hydrogen, nitrogen, carbon monoxide, and carbon dioxide [21].

10.7 Conclusion

The other important branches of bioorganometallic chemistry includes anticancer therapy, toxicology, medical imaging, and supramolecular interactions of organometallic compounds with amino-acid, DNA, and RNA. The above discussion shows the significance of bioorganometallic chemistry.

The topics discussed in this chapter are helping to expand bioorganometallic chemistry from a discipline that arouse chiefly from spectroscopic analysis of metal centers in proteins to a discipline where fundamental knowledge about metal functions and the applications of metal as diagnostic and chemotherapeutic agents are making important contribution to medicine. The knowledge of fundamental bioorganometallic chemistry will continue to grow enormously through these efforts. Thus, bioorganometallic chemistry has made a major contribution to understanding of biological systems containing metals. It has a vital role to create new paradigms for basic research for betterment of society. The future looks extremely promising for bioorganometallic chemistry.

References

1 Crabtree, R.H. (1994). *The Organometallic Chemistry of the Transition Metals*, 2ed. New York: Wiley.
2 Metzler-Nolte, N. (2007). *Comprehensive Organometallic Chemistry III*. Elsevier.
3 Kramer, R., Polbon, K., Robl, C., and Beck, W. (1992). *Inorg. Chim. Acta* 414: 198–200.
4 Severin, K., Bergs, R., and Beck, W. (1998). *Angew. Chem. Int. ed.* 37: 1086. and references therein.
5 White, J.G., Prosen, R.J., Kenneth, N.T. et al. (1955). *Nature* 176 (4477): 325–328.
6 Jaouen, G. (2006). *Bioorganometallics: Biomolecules, Labeling, Medicine*. Weinheim: Wiley-VCH.
7 Bertini, I., Gray, H.B., Lippard, S.J., and Valentine, J.S. (1998). *Bioinorganic Chemistry*, 1e st South Asian ed. Suasalito, CA: University Science Books.
8 Butler, P. and Krautler, B. (2006). *Top Organomet. Chem.* 17: 1–55.
9 Brink-Shoemaker, C., Cruickshank, D.W.J., Hodgkin, D.C. et al. (1964). *Proc. Roy. Soc. London Ser. A* 1: 278.
10 Halpern, J. (1983). *Pure Appl.Chem.* 55: 1059.
11 Giedyk, M., Goliszewska, K., and Gryko, D. (2015). *Chem. Soc. Rev.* 44: 3391.
12 LiuJ, C.S., HosseinzadehP, Y.Y., Patrik, I. et al. (2014). *Chem. Rev.* 114: 4366–4469.
13 Adman, E.T. (1979). *Biochim. Biophys. Acta* 549: 107–144.
14 Meyer, T.E. and Cusanovich, M.A. (1989). *Biochim. Biophys. Acta* 975: 1–28.
15 Ambler, R.P. (1980). *From Clyclotrons to Cytochromes* (ed. A.B. Robinson and N.D. Kaplan). Academic Press.
16 Berezin, B.D. (1981). *Coordination Compounds of Porphyrins and Phthalocyanines*. New York: Wiley.
17 Lever, A.B.P. and Gray, H.B. (ed.) (1983). *Iron Porphyrins*. Massachusetts: Addison-Wesley Publishing Company Inc.
18 Scheidt, W.R. and Reed, C.R. (1981). *Chem. Rev.* 81: 543.
19 Collman, J.P., Brauman, J.I., Halbert, T.R., and Suslick, K. (1976). *Proc. Natl. Acad. Sci. USA* 73: 3333.
20 Collman, J.P., Fu, L., Zing, A., and Diederich, F.J. (1997). *Chem. Soc., Chem. Commun.* 193.
21 Astruc, D. (2007). *Organometallic Chemistry and Catalysis*. Berlin, Heidelberg, New York: Springer.
22 Gruber, N. and Galloway, J.N. (2008). *Nature* 451: 293.
23 Burk D Ergeb., Enzyme for Sch. 1934:3:23.

24 Burk, D., Lineweaver, H., and Horner, C.K. (1934). *J. Bacteriol.* 27: 325.
25 SE, M.G., Boyol, E.S., Peters, J.W., and Orphan, V.J. (2013). *Front. Microbiol.* 3: 419.
26 Dos Santos, P.C., Fang, Z., Mason, S.W. et al. (2012). *BMC Genomics* 13: 162.
27 Smil, V. (2004). Enriching the Earth. In: *And the Transformation of World Food Production* (ed. H. Fritz and C. Bosch). Cambridge, MA: MIT Press.
28 Haber, F. (1922). *Nat. Wiss.* 10: 1041.
29 Haber, F. (1923). *Nat. Wiss.* 11: 339.
30 Peter, J.W. and Szilagyi, R.K. (2006). *Current opinion in chemical microbiology Bioinorganic Chemistry/Biocatalysis and biotransformation* 10 (2): 101–108.
31 Rubio, L.M. and Ludden, P.W. (2008). *Annu. Rev. Microbiol.* 62 (1): 93–111.
32 Franche, C., Lindstorm, K., and Elmerich, C. (2008). *Plant and Soil* 321 (1-2): 35–59.
33 Bulen, W.A. and JR, L.C. (1966). *Proc. Natl. Acad. Sci. U.S.A.* 56: 979.
34 Mortenson, L.E. (1966). *Biochim. Biophys. Acta* 127: 18.
35 Winter, H.C. and Burris, R.H. (1976). *Annu. Rev. Biochem.* 45: 409.
36 Hageman, R.V. and Burris, R.H. (1978). *Proc. Natl. Acad. Sci. U.S.A.* 75: 2699.
37 Dean, D.R., Biolin, J.T., and Zheng, L.J. (1993). *Bactriol* 175: 6737.
38 Howard, J.B. and Rees, D.C. (1994). *Annu. Rev. Biochem.* 63: 235.
39 Kim, J. and Rees, D.C. (1994). *Biochemistry* 33: 389.
40 Bulen, W.A., Burns, R.C., and Lecomte, J.R. (1965). *Proc. Natl. Acad. Sci. U.S.A.* 53: 532.
41 Burns, R.C. and Bulen, W.A. (1965). *Biochim. Biophys. Acta* 105: 437.
42 Mortenson, L.E. (1964). *Biochim. Biophys. Acta* 81: 473.
43 Mortenson, L.E. (1964). *Proc. Natl. Acad. Sci. U.S.A.* 52: 272.
44 Burgess, B.K. (1990). *Chem. Rev.* 90: 1377.
45 Shah, V.K. and Brill, W.J. (1977). *Proc. Natl. Acad. Sci. U.S.A.* 74: 3249.
46 Peters, J.W., Fisher, K., Newton, W.E., and Dean, D.R. (1995). *J. Biol. Chem.* 270: 27007.
47 Kim, J. and Rees, D.C. (1992). *Science* 257: 1677.
48 Ma, L., Brosius, M.A., and Burgess, B.K. (1996). *J.Biol.Chem.* 271: 10528.
49 Lowe, D.J., Fisher, K., and Thorneley, R.N.F. (1993). *Biochem. J.* 292: 93.
50 Ferry, J.G. (1995). *Annu. Rev. Microbiol.* 49: 305.
51 Ragsdale, S.W. and Kumar, M. (1996). *Chem. Rev.* 96: 2515.
52 Ermler, U., Grabarse, W., Shima, S. et al. (1998). *Curr. Opin. Struct. Biol.* 8: 749.
53 Doblek, H., Svetlitchnyi, V., Gremer, L. et al. (2001). *Science* 293: 1281.
54 Drennan, C.L., Heo, I., Sintchak, M.M. et al. (2001). *Proc. Natl. Acad. Sci. U.S.A.* 98: 11973.
55 Gong, W., Hoo, B., Wei, Z. et al. (2008). *Proc. Natl. Acad. Sci. U.S.A* 105: 9558.

11

Important Reactions of Organometallic Compounds

Sandeep K. Vaishnav[1] *and Priyanka Singh*[2*]

[1] *State Forensic Science Laboratory, Department of Narcotics and Toxicology, Pachpedi Naka, Raipur, Chhattisgarh, 492013, India*
[2] *Hemchand Yadav University, Govt. Digvijay PG Autonomous College, Department of Chemistry, Killapara, Rajnandgaon, Durg, Chhattisgarh, 491441, India*

11.1 Introduction

Organometallic compounds go through several reactions, such as gain and/or loss of ligands, of metal–metal bonds formation or cleavage, and molecular rearrangement. Organometallic compounds are also used as commercial catalysts. Here, we will discuss about the most common and useful organometallic reactions outlined as follows.

1. Reactions involving gain or loss of ligands
   1.1. Associative and dissociative substitution
   1.2. Oxidative addition (OA) and C–H bond activation
   1.3. Nucleophilic displacement
   1.4. Reductive elimination (RE) and Pd-catalyzed cross coupling
   1.5. Application of pincer ligands
2. Reactions involving modification of ligands
   2.1. Insertion
      2.1.1. Carbonyl insertion (alkyl migration)
      2.1.2. Olefinic insertion
      2.1.3. Other insertion
   2.2. Hydride elimination (β-hydride transfer)
   2.3. Abstraction

11.2 Reactions Involving Gain or Loss of Ligands

Organometallic compounds involve alteration in coordination number of metal center either by gaining or losing ligands. If only coordination number changes with

*Corresponding author.

Organometallic Compounds: Synthesis, Reactions, and Applications, First Edition.
Edited by Dakeshwar Kumar Verma and Jeenat Aslam.

the retention of oxidation state of metal center, then these reactions are referred to as addition or dissociation reactions. On the contrary, if the oxidation state of metal changes along with the coordination number, then these reactions are referred as OA or RE reactions.

S. no.	Type of reaction	Coordination number	Formal oxidation state of metal	Electron count
1	Addition Reaction	Increases	Does not change	Does not change
2	Dissociation reaction	Decreases	Does not change	Does not change
3	Oxidative addition reaction	Increases by 2	Increases by 2	Increases by 2
4	Reductive elimination reaction	Decreases by 2	Decreases by 2	Decreases by 2

11.2.1 Associative and Dissociative Substitution

11.2.1.1 Dissociation of Carbonyls

Carbonyls are unsaturated ligands acting as soft π-acceptors due to their ability to accept dπ electrons of metal through backbonding. Metal binds to carbon because the HOMO of carbonyl ligand is the lone pair of carbon atom. The electron count of carbonyl ligand does not alter upon the transition of terminal to bridging carbonyl. A terminal CO ligand donates $2e^-$ to the metal center. However, bridging ketone-like CO ligand (μ^2-CO) donates $1e^-$ to each metal center and $1e^-$ comes from M—M bond. CO group might be dissociated thermally or photochemically, resulting in rearrangement of the molecule or substitution of CO group by another ligand.

The first reaction shows molecular rearrangement upon the dissociaton of CO group (Scheme 11.1), whereas the second reaction shows the replacement of CO ligand by $-P(CH_3)_3$ (Scheme 11.2). It can be used for introduction of new ligands in the organometallic compounds. Generally, thermal CO substituton reactions are first order with respect to metal complex and the rate of such reactions do not depend on ligand concentration. CO dissociation involves slow loss of CO group, which is then followed by fast reaction with another ligand.

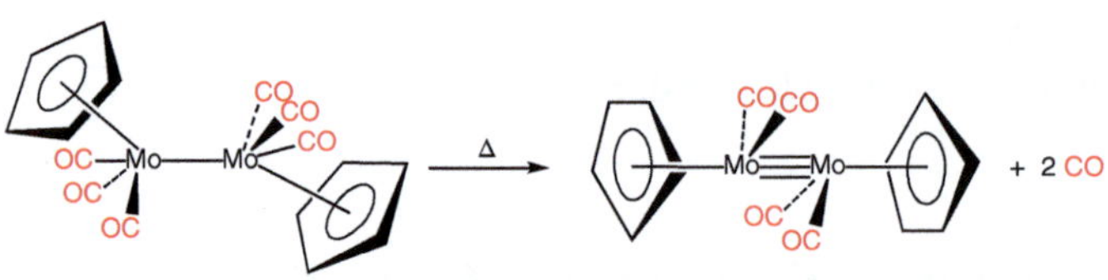

Scheme 11.1 Molecular rearrangement upon the dissociaton of CO group.

Scheme 11.2 Replacement of CO ligand by $-P(CH_3)_3$.

11.2.1.2 Dissociation of Nitrosyls

Nitrosyl (NO^+) is a $2e^-$ donor ligand, which is isoelectronic with CO. It binds linearly and carries a positive charge. It is a better π-acceptor ligand as compared with CO due to its positive charge and electronegative nature of nitrogen. Nitrosyl can also be taken as $3e^-$ donor ligand because it lowers the oxidation state of metal by one unit while forming linear NO (*lin*-NO) complex as it provides an electron to metal center in order to bind as NO^+. The anionic nitrosyl ligand binds in a bent fashion. These are generally prepared from $[NO]BF_4$ salt (Scheme 11.3).

$$CpMo(CO)_2(\eta^3\text{-allyl}) \xrightarrow{NOBF_4} [CpMo(CO)(NO)(\eta^3\text{-allyl})]BF_4$$

Scheme 11.3 Binding of NO group.

Let us understand the redox-active nature of nitrosyl ligand through the following example:

Here, an internal redox reaction takes place and two electrons are transferred from Co(I) to *lin*-NO to give *bent*-NO complex, which increases the oxidation state of metal by two units (Scheme 11.4). $18e^-$ Co(I) changes to $16e^-$ Co(III). The uncoordinated lone pair of nitrogen in *bent*-NO ligand helps to stabilize the $16e^-$ complex. The *lin*-NO complex shows absorption at a frequency at 1750 cm^{-1}, whereas *bent*-NO complex shows absorption at a frequency at 1650 cm^{-1}. These two absorption frequencies do not differ much and overlap. The $18e^-$ configuration can not always be decided based upon the nature of nitrosyl ligand, whether linear or bent.

$$\underset{18\ e^-,\ Co(I)}{CoCl_2L_2(\textit{lin}\text{-}NO)} \rightleftharpoons \underset{16\ e^-,\ Co(III)}{CoCl_2L_2(\textit{bent}\text{-}NO)}$$

Scheme 11.4 Internal redox reaction in $CoCl_2L_2$(*lin*-NO).

11.2.1.3 Dissociation of Phosphine

Phosphines act as monodentate terminal ligand by donating the lone pair of phosphorus to metal, thus forming M—PR_3 bond. They also act as mild π acids, whose strength depends upon the nature of R groups attached to it. The σ^* antibonding orbitals accept the backbonding from metal center. The dissociation of ligands other than CO depends on metal–ligand bond strength that itself depends on electronic effects and steric effects. These steric effects have been investigated for many ligands but especially for neutral donor ligands, such as phosphines. Cone angle is the apex angle of a cone representing the Van der Waals radii of the outermost atoms of ligand and percent buried volume ($\%V_{bur}$), the percentage volume of potential coordination sphere around the metal occupied by the ligand is used to study the steric effects. Tolman found admirable correlation between the cone angles and $\%V_{bur}$ for tertiary phosphines [1–3]. Bulky ligands around the metal lead to increase in metal–ligand bond length that further results into speedy ligand dissociation. Phosphine (PR_3) dissociation rate (initiation step) affects the action of ruthenium olefin metathesis catalyst (Scheme 11.5). The loss of phosphine group increases with an increase in cone angle and bulkiness of the ligand [4, 5].

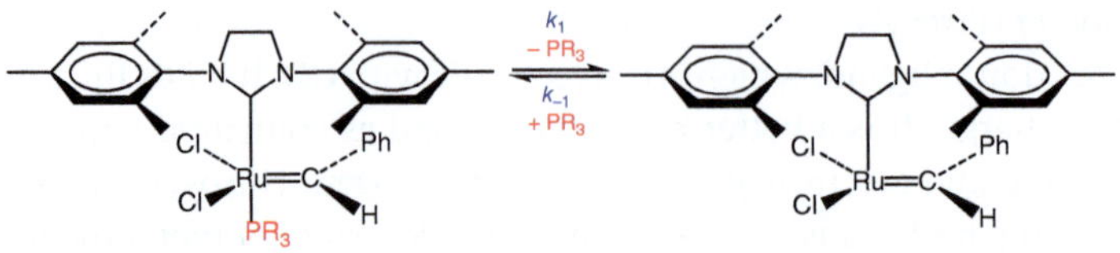

Scheme 11.5 Association and dissociation of phosphine in ruthenium complex.

11.2.2 Oxidative Addition and C—H Bond Activation

The *OA* reactions lead to increase in coordination number, formal oxidation state of metal, as well as electron count of the metal center. However, *RE* reactions lead to decrease in coordination number, formal oxidation state of metal, as well as electron count of the metal center. Both these reaction types are essential steps of several catalytic processes and can be represented as follows in Scheme 11.6:

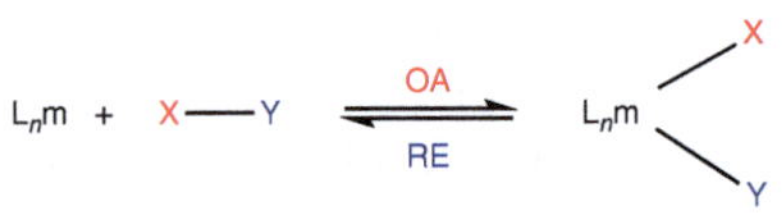

Scheme 11.6 Schematic representation of *Oxidative addition (OA)* and *eductive elimination (RE)*.

The following conditions need to be fulfilled for the successful completion of OA reaction.

(i) The complex L_nm should possess atleast two vacant coordinaton sites for the formation of M–X and M–Y bonds.
(ii) The metal center should have nonbonding electron density.
(iii) The metal center should have energetically and symmetrically suitable orbitals for bond formation.
(iv) The metal center should possess atleast two oxidation states differing by two units having comparable stability.
(v) The complex should either be $16e^-$ or must loose a ligand in case of $18e^-$ complex.

Let us understand these reactions with the example of reaction between hydrogen and Vaska's complex (*trans*-IrCl(CO)[P($C_6H_5)_3]_2$) (Scheme 11.7). The forward reaction represents OA, where two hydride ligands get attached to Vaska's complex and the formal oxidation state of iridium changes from Ir(I) to Ir(III). Trigonal bipyramidal dihydrogen complex is formed as an intermediate. H—H bond break takes place

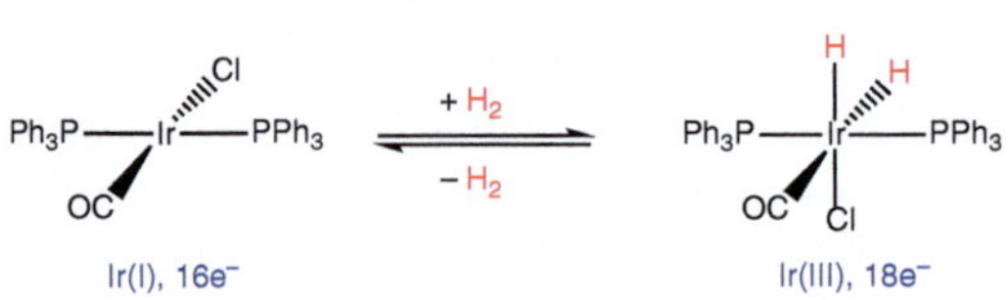

Scheme 11.7 Reaction between hydrogen and Vaska's complex (*trans*-IrCl(CO)[P($C_6H_5)_3]_2$).

owing to back donation of electron into the σ antibonding orbital of H_2 molecule that results into the formation of an octahedral complex [6]. The reactant complex possesses 16 valence electrons with coordination number four and the product complex possesses 18 valence electron with coordination number six. The backward reaction represents RE, where two hydride ligands get detached from Vaska's complex and the formal oxidation state of iridium changes from Ir(III) to Ir(I) [7]. Scheme 11.8 shows another example of oxidative addition of iodine to $Fe(CO)_4$ (Scheme 11.8).

Scheme 11.8 Addition of I_2 to $Fe(CO)_5$.

11.2.2.1 Mechanism

Following types of mechanisms can be followed in case of OA depending upon the nature of metal and ligand complex.

a. *Concerted Mechanism*: Nonpolar molecules, such as H_2, follow this method through three center transition states (Scheme 11.9).

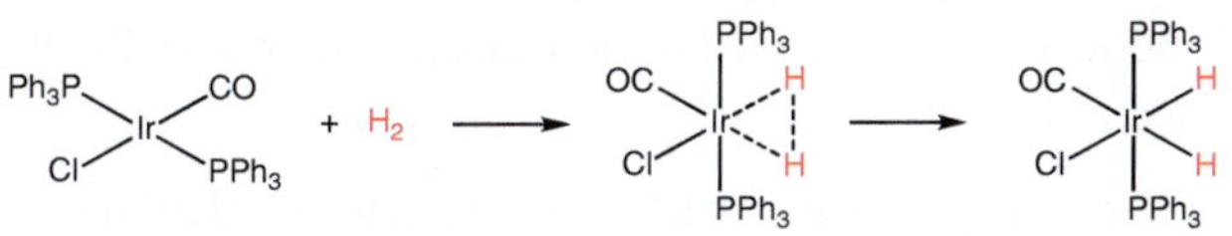

Scheme 11.9 Concerted mechanism of reaction between hydrogen and Vaska's complex (*trans*-$IrCl(CO)[P(C_6H_5)_3]_2$).

b. $S_N{}^2$ *Mechanism*: Polar molecules, such as CH_3I, follow this method through two electrons transfer (Scheme 11.10).

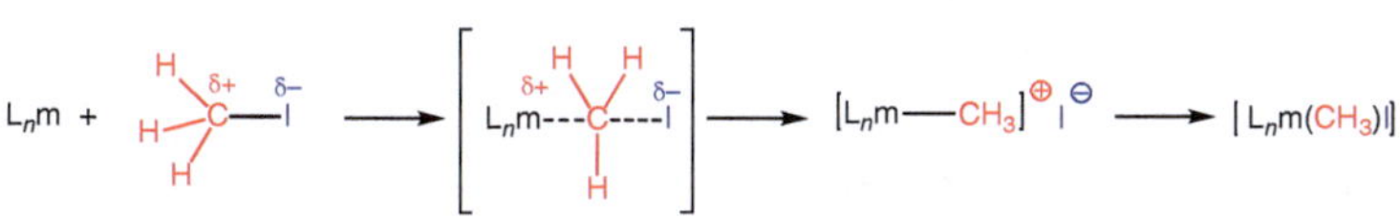

Scheme 11.10 $S_N{}^2$ mechanism of reaction between hydrogen and Vaska's complex (*trans*-$IrCl(CO)[P(C_6H_5)_3]_2$).

c. *Free Radical Mechanism*: Some alkyl halides are added to $Pt(PPh_3)_3$ through this mechanism (Scheme 11.11).

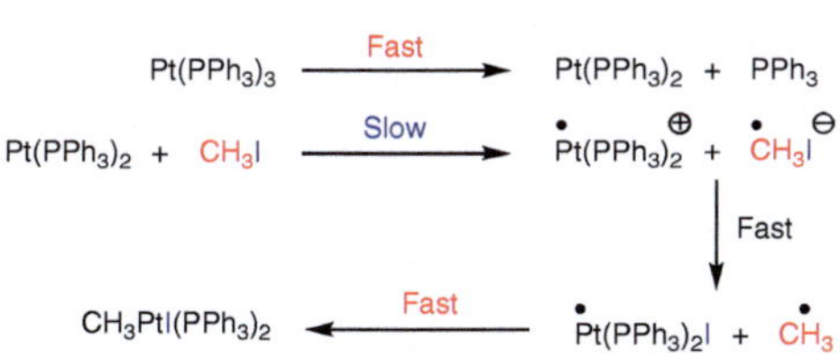

Scheme 11.11 Free radical mechanism of reaction between hydrogen and Vaska's complex (*trans*-$IrCl(CO)[P(C_6H_5)_3]_2$).

d. *Ionic Mechanism*: Hydrogen halides are added to the complexes of transition metals in the presence of polar solvent that follow this mechanism (Scheme 11.12).

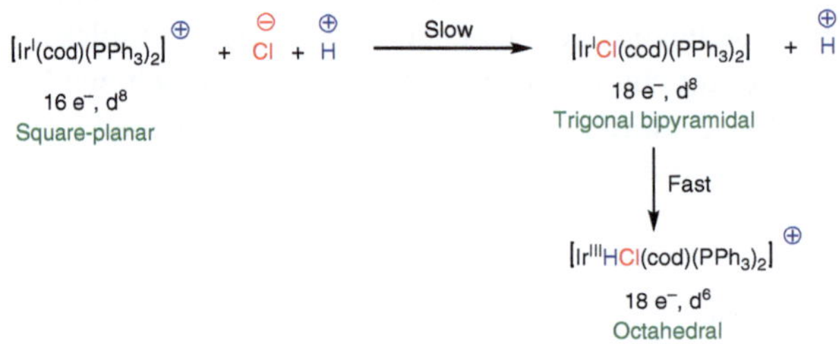

Scheme 11.12 Ionic mechanism of reaction between hydrogen and Vaska's complex (*trans*-$IrCl(CO)[P(C_6H_5)_3]_2$).

Let us discuss another example of *cis*-$I_2Fe(CO)_4$, which is formed from $Fe(CO)_5$ as a result of OA in the presence of I_2 in the following two steps.

(i) Dissociation of CO group gives four coordinated Fe(0) as intermediate.
(ii) Fe(0) is formally oxidized to Fe(II) and coordination number increases by two due to the addition of two I^- ligands.

In the OA of square–planar d^8 complex *trans*-$Ir(CO)Cl(PEt_3)_2$, the oxidation state of iridium changes from +1 to +3 (Scheme 11.13). The new ligands show *cis* or *trans* addition depending on the mechanism. The coordination number of metal

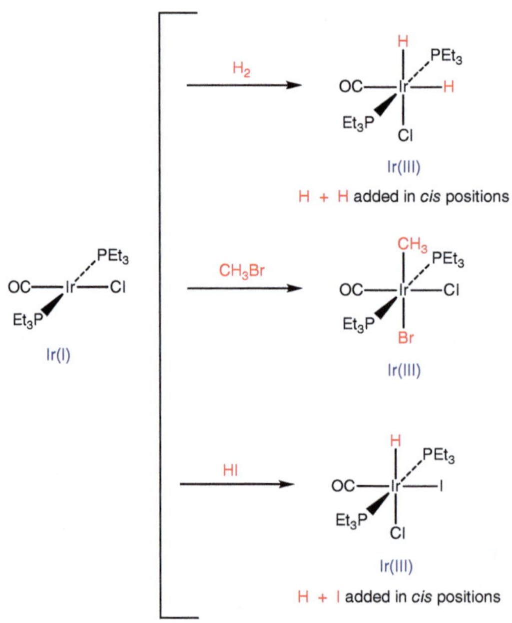

Scheme 11.13 oxidative addition ofsquare-planar d^8 complex *trans*-$Ir(CO)Cl(PEt_3)_2$.

center increases to 6 from 4. This decreases the distance between the new and original ligands, thus enabling reactions between them. These reactions are commonly observed in catalytic mechanisms.

C—H activation reactions involve the cleavage of strong C—H bond by transition-metal complexes. Unfunctionalized hydrocarbons are changed to organometallic complexes. The C–H activation reaction reported by Bergman follows the following mechanism [8] (Scheme 11.14):

(i) Cyclohexane is removed from six coordinated Ir(III) (ML_3X_3) through RE to give a four coordinated Ir(I) (ML_3X) as intermediate. Oxidation state as well as coordination number of metal center decrease by two.
(ii) Benzene is added to four coordinated Ir(I) (ML_3X) to six coordinated Ir(III) (ML_3X_3) through OA resulting in C—H bond activation. Oxidation state as well as coordination number of metal center increase by two.

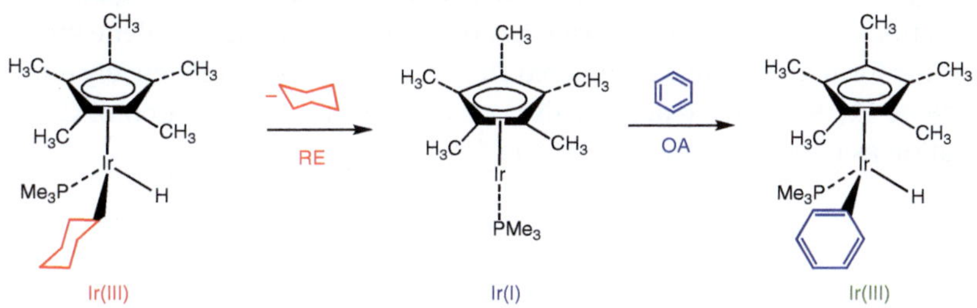

Scheme 11.14 C–H activation reaction reported by Bergman.

11.2.2.2 Binuclear Oxidative Addition

In case of binuclear OA, each metal center changes its oxidation state, coordination number, as well as electron count by increase of one unit. This type of behavior is generally seen in $17e^-$ complexes or binuclear $18e^-$ complexes having M—M bond and metal with a stable oxidation state differing by one positive unit. Net transfer of two electrons takes place into σ antibonding orbital of the molecule X—Y bond. The two σ electrons of X–Y get distributed between both the metal centers leading to the cleavege of X—Y bond and the formation of M—X and M—Y bonds (Schemes 11.15 and 11.16).

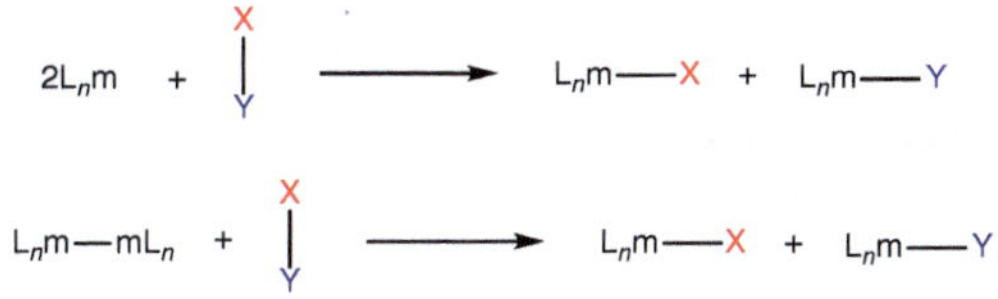

Scheme 11.15 **Schematic representation of** Binuclear oxidative addition.

Schenck and coworkers in 1985 have studied a bimetallic system, incapable of forming metal–metal bond and the metals were firmly held at their positions. They reported that RE might be inducedat one metal through OA at the other metal center [9].

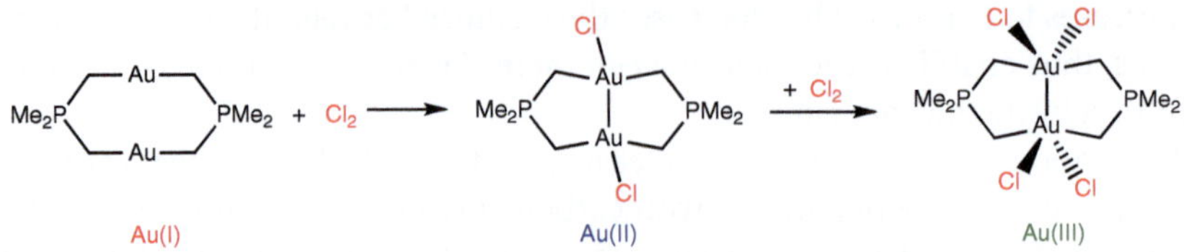

Scheme 11.16 **Example of** Binuclear oxidative addition.

11.2.2.3 Cyclometallations

These reactions involve integration of metals in organic rings. OA at the *ortho* site of aromatic ring also known as *orthometallations* is frequent. In the following reaction, OA of iridium takes place at the carbon atom *ortho* to the phosphorus of PPh_3 ligand and the hydrogen atom of the ortho C–H gets added to iridium (Scheme 11.17). This results into the formation of four membered rings involving iridium center and aromatic ring. In the second reaction, platinum complex also has PPh_3 ligand, but orthometallation is not preferred. Instead, OA of C—H bond of naphthalene group results into the formation of five membered rings and methane is reductively eliminated (Scheme 11.18). These reactions also come under the classification of C—H bond activation as metal facilitates breaking of C—H bonds.

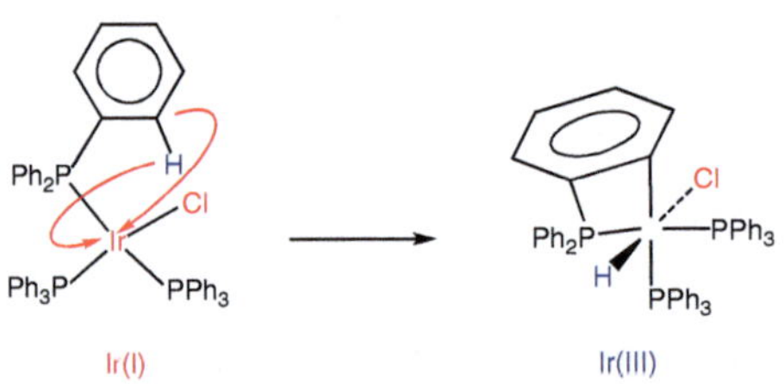

Scheme 11.17 Oxidative addition in iridium complex.

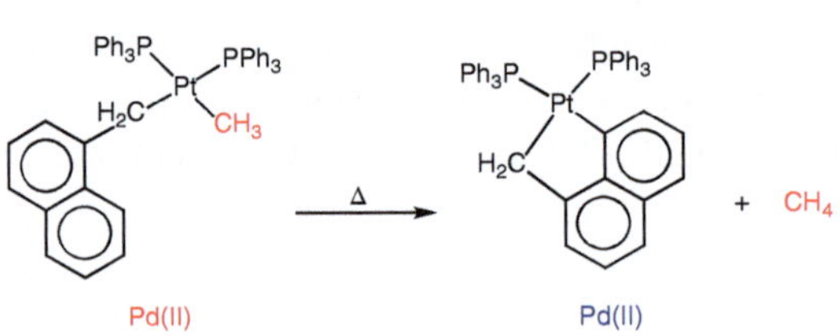

Scheme 11.18 Oxidative addition in platinum complex.

11.2.2.4 Nucleophilic Displacement

Nucleophilic displacement does not strictly fulfills the requirement of OA but is still studied under this class of reactions of organometallic compounds. Organometallic complexes with negative charge act as nucleophiles and ligand displacement in such complexes is termed as nucleophilic displacement or substitution reactions.

Consider a reaction between $[(\eta^5\text{-}C_5H_5)Mo(CO)_3]^-$ and methyl iodide (Scheme 11.19). The organometallic complex acts as a nucleophile and displaces

Scheme 11.19 Concerted mechanism of reaction between hydrogen and Vaska's complex (*trans*-IrCl(CO)[P(C_6H_5)$_3$]$_2$).

iodide methyl iodide. Formal oxidation state of Mo changes from Mo(0) ([ML_5X]$^-$) to Mo(II) (ML_5X_2) and coordination number rises by 1.

11.3 Reductive Elimination and Pd-catalyzed Cross Coupling

As already mentioned, coordination number, formal oxidation state of metal, as well as electron count of the metal center decrease in case of RE. An inverse relationship exists between RE and OA. The following example illustrates this:

Forward and backward reactions represent OA and RE, respectively (Scheme 11.20). Formal oxidation state and coordination number of Ta increase in forward reaction, whereas both formal oxidation state and coordination number of Ta decrease in backward reaction. Reductive elimination depends on electron density, steric hinderance, coordination number, geometry, nature of metal, and participating ligands.

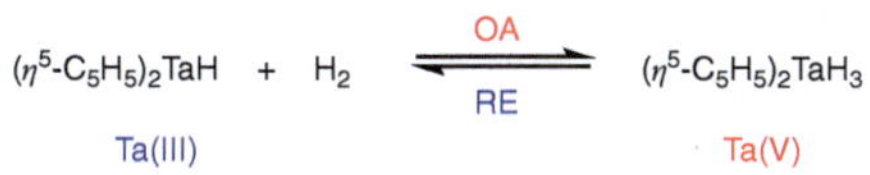

Scheme 11.20 Oxidative addition and reductive elimination in tantalum complex.

Pd-catalyzed cross coupling involves RE as an important step (Scheme 11.21). The following catalytic cycle is an example of Pd-catalyzed cross-coupling reaction (Scheme 11.22).

(1) Alkyl or aryl halide is added oxidatively to Pd(0) complex leading to the formation of square–planar Pd(II) complex.
(2) The square–planar product is a *cis*-isomer that quickly undergoes *cis–trans* isomerization in order to avoid the presence of two good donors R and L in *trans* position to each other.
(3) It involves *trans metallation*, a ligand is substituted by R′ (alkyl or aryl group) at the Pd(II) center.
(4) It involves *cis–trans* isomerization if the product of trans metallation (Step 3) is *trans*.
(5) It involves the RE of an R–R′ present in *cis* position to each other. The pace of this step increases with increase in bulkiness of ligand L because decrease in coordination number leads to decrease in steric hindrance.

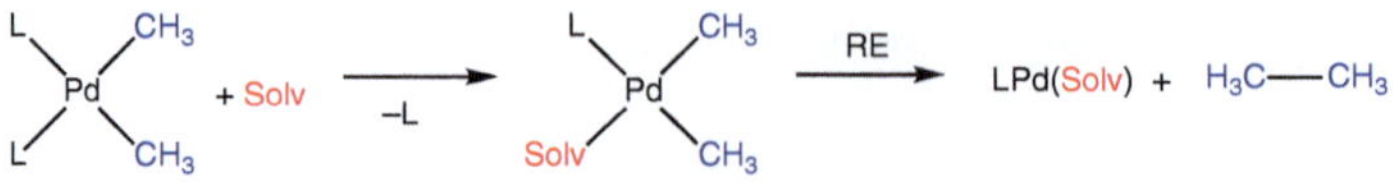

Scheme 11.21 Solvation in palladium complex.

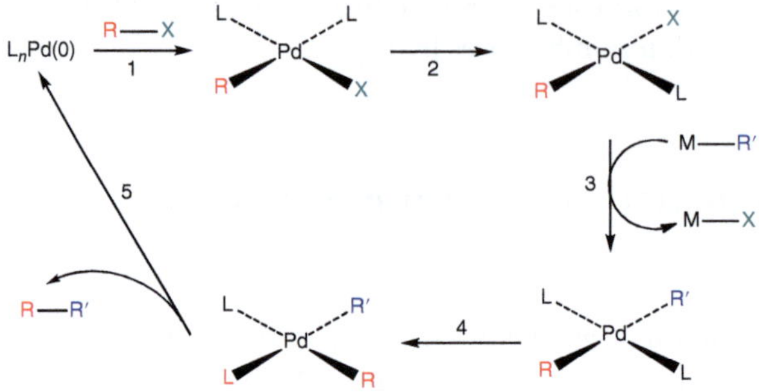

Scheme 11.22 Catalytic mechanism of Pd-catalyzed cross-coupling reaction.

The effect of steric hindrance on rate of reaction can be explained by the following example. The three phosphine ligand-based *cis* dimethyl complexes of Pd (I, II, and III) with varying steric hinderances have been studied (Scheme 11.23).

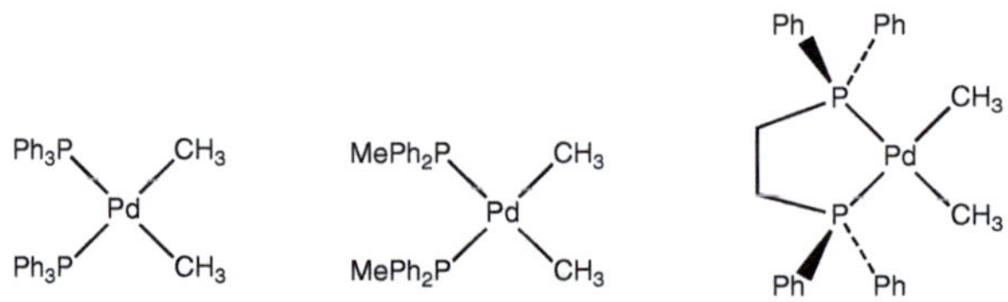

I II III

Scheme 11.23 Three phosphine ligand based *cis* dimethyl complexes of Pd.

The complexes show displacement of posphine group with solvent molecule. This is followed by RE and release of an ethane molecue in each case. Complex I has the highest rate of reaction followed by complexes II and III. This is due to the highest steric hinderance in case of complex I and removal of ethane by RE is the easiest.

11.3.1.1 Binuclear Reductive Elimination

In case of binuclear RE, each metal center changes its oxidation state as well as coordination number by decrease of one unit. However, the d-electron count increases by one unit (Scheme 11.24).

It is observed in transition metals but is common among the metals of first row as they prefer one-unit change in oxidation state [10, 11]. Some of the examples are shown as follows (Scheme 11.25):

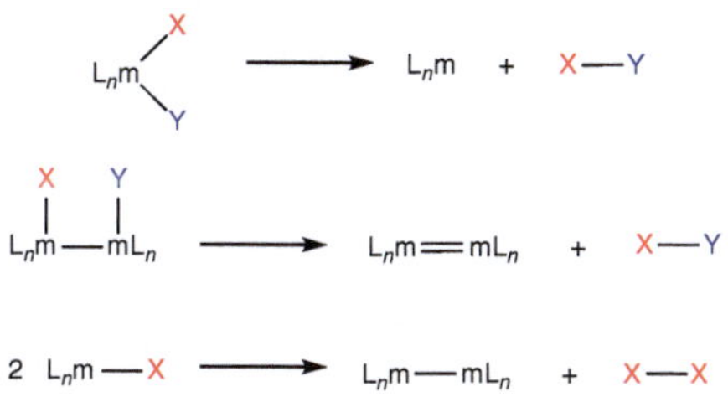

Scheme 11.24 Schematic representation of binuclear reductive elimination.

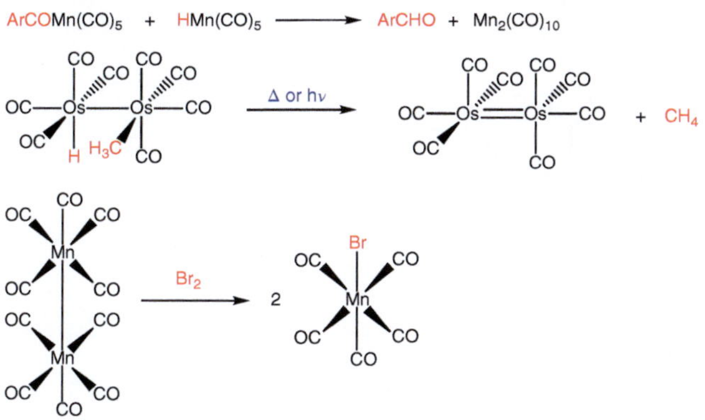

Scheme 11.25 Examples of binuclear reductive elimination.

11.4 Reactions Involving Modification of Ligands

A ligand or molecule fragment seems to insert into a metal–ligand bond in numerous reactions. Many "insertion" reactions do not include a direct insertion phase, despite the fact that some of these reactions are thought to proceed by direct single-step insertion. Carbonyl insertions are the well studied of these reactions.

11.4.1 Insertion

In many organometallic reactions, one ligand (L) is inserted into a metal–hydride or metal–alkyl/aryl bond(Scheme 11.26). The hydride, alkyl, or aryl group usually migrates onto L, which is typically CO, an alkene, or an alkyne. An anionic ligand X and a neutral unsaturated ligand Y are often in cis position in an insertion reaction, and the two ligands couple to create another anionic ligand X–Y that is bound to the same metal [12–16].

This results in a metal vacancy or a coordination site on the metal that is not occupied by a ligand. The anionic ligand is more likely to move in an insertion reaction. The reaction is reversible, and chemical equilibrium is usually achieved. Deinsertion is the opposite reaction. One can either add more neutral ligand or a different neutral ligand to shift the chemical equilibrium to the right side. The neutral ligand then binds to the vacancy, preventing the process from reversing. There are two

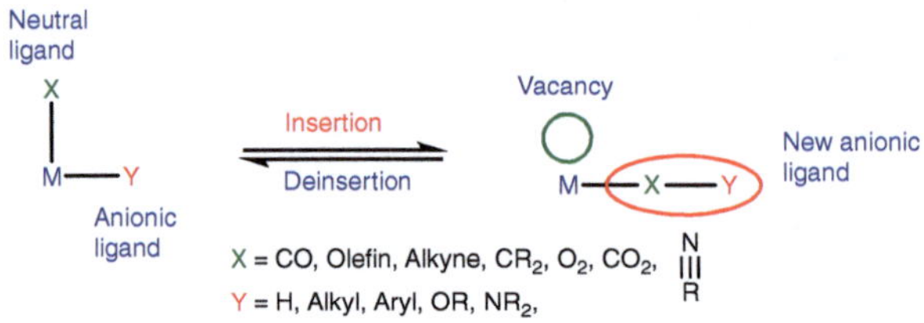

Scheme 11.26 Schematic representation of insertion reactions.

types of insertion reactions: 1,1- or α insertions [17] and 1,2- or β-insertions [18] (Scheme 11.27).

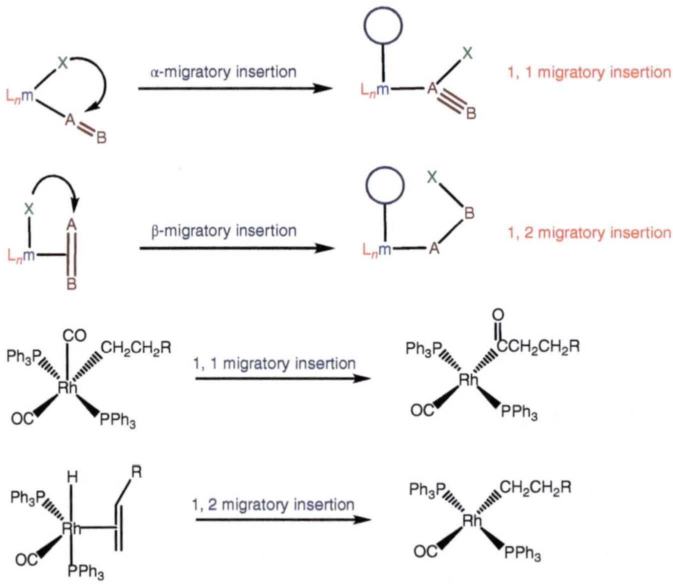

Scheme 11.27 Schematic representation of 1,1- and 1,2-migratory insertion with examples.

CO, olefins, alkynes, carbenes, dioxygen, carbon dioxide, and nitriles are typical neutral unsaturated ligands. Hydrido, alkyl, aryl, alkoxy, and amido ligands are typical anionic ligands. Carbon monoxide and olefins, which are archetypical cases of 1,1- and 1,2-insertions, respectively, are the most important industrial molecules capable of undergoing insertion reactions.

11.4.1.1 Carbonyl Insertion (Alkyl Migration)

Carbonyl insertions into alkyl groups, such as methyl groups, are one of the most prevalent insertions. The reaction of CO with an alkyl complex to produce an acyl [iC("O)R] product has been extensively researched. The reaction can be driven by the presence of a neutral ligand L, such as free CO, which fills the unoccupied site. An excellent illustration is the reaction of $CH_3Mn(CO)_5$ with CO (Scheme 11.28).

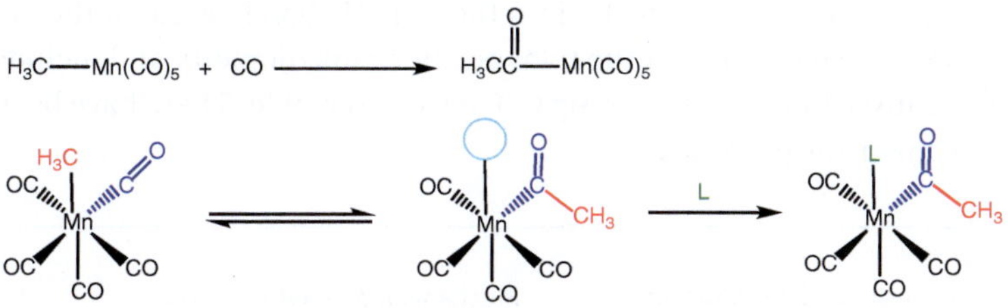

Scheme 11.28 Alkyl migration in manganese complexes.

The reactions illustrated below are known as 1,1-insertions because both bonds to the inserted molecule are to the same atom (Scheme 11.29). For example, both Mn and CH_3 are attached to the sulfur of the added SO_2 and CO in the reaction.

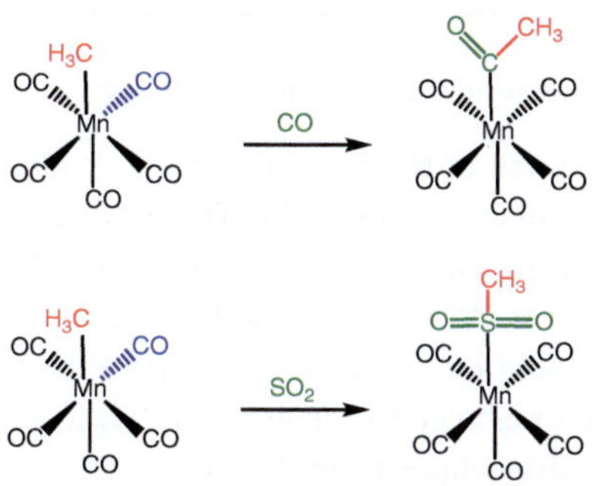

Scheme 11.29 1,1 insertions in manganese complexes.

In 1,2-insertions, bonds are formed between neighboring atoms and the inserted molecule. For example, the $Co(CO)_4$ group is connected to one carbon and the H is attached to the nearby carbon in the reaction of $HCo(CO)_4$ with tetrafluoroethylene (Scheme 11.30).

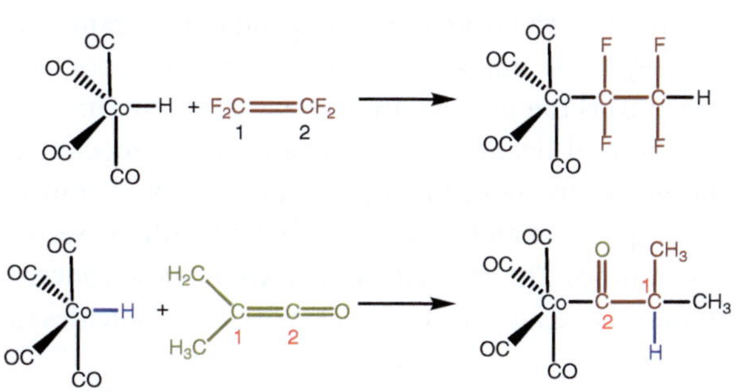

Scheme 11.30 Examples of 1,2 insertions in $HCo(CO)_4$.

The insertion of CO into a metal–carbon bond in alkyl complexes is of particular interest because of its potential applications in organic synthesis and catalysis, and its mechanism should be thoroughly investigated.

We could expect the CO to insert directly into the $MnCH_3$ bond based on the net equation. Other mechanisms that offer the total reaction stoichiometry and include steps other than the insertion of an incoming CO are also possible. There have been three plausible mechanisms proposed.

Mechanism 1: CO insertion	**Mechanism 2: CO migration**	**Mechanism 3: alkyl migration**
Direct insertion of CO into a metal–carbon bond.	Intramolecular CO insertion by migration of a CO ligand. This would result in a five-coordinate intermediate with a vacant site for an incoming CO to be attached to.	In this case, the alkyl group would migrate, rather than the CO, and attach itself to a CO *cis* to the alkyl. This would also give a five-coordinate intermediate with a vacant site available for an incoming CO.

The mechanism of these reactions are depicted in schematic form as follows:

Intramolecular migration is thought to occur to one of the migratory group's closest neighbors in cis orientations in both mechanisms 2 and 3 $CH_3(CO)Mn(CO)_5$ (Scheme 11.31). The product $CH_3Mn(CO)_5$ has the labeled CO entiron when this reaction is carried out using ^{13}C in the acyl position. When ^{13}C is used in a carbonyl ligand that is cis to the acyl group, the reverse reaction produces a 2 : 1 ratio of cis to trans product (cis and trans refer to the position of labeled CO relative to CH_3 in the product). In this reaction, some tagged CO also dissociates from the Mn.

It is now possible to evaluate the mechanisms. The first experiment definitively rules out Mechanism 1 (Scheme 11.32). Direct insertion of ^{13}CO into the acyl ligand should result in ^{13}C; however, none is detected. Mechanisms 2 and 3 are both consistent with the findings of this study. Any reversible reaction must have similar forward and reverse paths, merely proceeding in opposite directions, according to the microscopic reversibility principle.

If carbonyl migration is the forward reaction (mechanism 2), the reverse reaction must first lose a CO ligand, then migrate CO from the acyl ligand to the empty site. Because a trans position is unlikely to be migrated, the entire product should be cis. If alkyl migration (mechanism 3) is the process, the reverse reaction must first lose a CO ligand, then migrate the acyl ligand's methyl group to the unoccupied site. Again, all the product should be cis. Both processes 2 and 3 would transfer tagged CO in the acyl group to a cis position and are thus consistent with the second experiment's experimental observations. The mechanism depicted below concerns the reverse reaction for CO migration and alkyl insertion (1) where C* indicates the location of ^{13}C (Scheme 11.33).

CO is transferred from the acyl ligand to the empty site. As a result, 25% of the product should not have a ^{13}CO label and 75% should have the CO cis to the alkyl identified. Alkyl migration (mechanism 3) on the other hand should give 25% without a label, 50% with a label cis to the alkyl, and 25% with a label trans to the alkyl. The evidence supports mechanism 3, the established reaction pathway, because this is the ratio of cis to trans found in the experiment. The mechanism depicted below

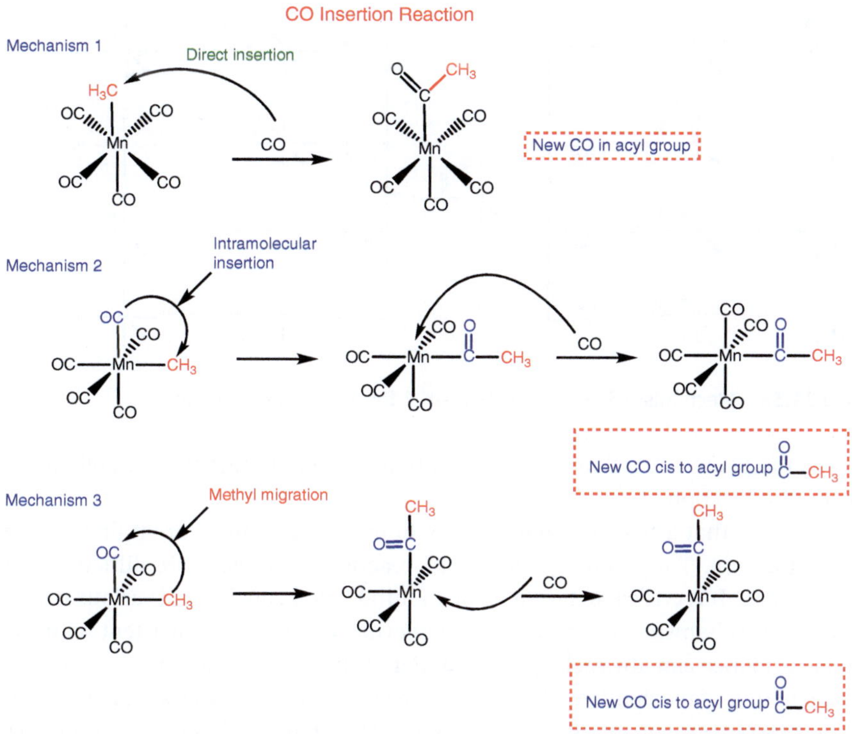

Scheme 11.31 Different mechanisms for CO insertion reactions in $CH_3(CO)Mn(CO)_5$.

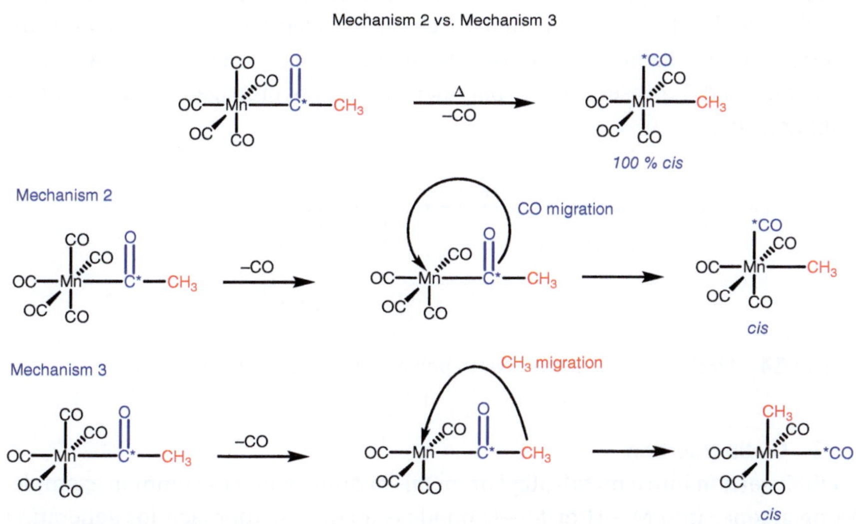

Scheme 11.32 Mechanism 2 versus Mechanism 3 for CO insertion reactions in $CH_3(CO)Mn(CO)_5$.

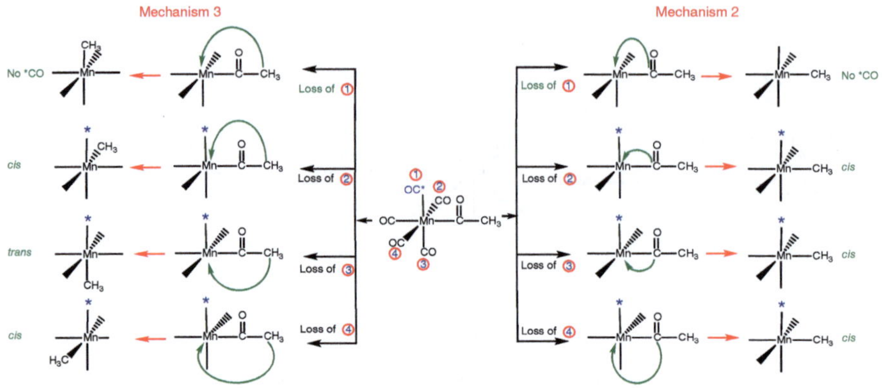

Scheme 11.33 Mechanism 3 versus Mechanism 2 for manganese complex.

concerns the reverse reaction for CO migration and alkyl migration (2), where *C indicates the location of ^{13}C alkenes.

We establish that a reaction that appears to involve CO insertion at first glance actually does not. The "carbonyl insertion" reaction could be more difficult than explained here. It is critical to retain an open mind when considering various mechanisms. No mechanism can be proven but alternatives that are consistent with the available evidence can always be proposed. For example, it was assumed in the earlier explanation of mechanisms 2 and 3 that the intermediate was a square pyramid with no rearrangement to alternative geometries (such as trigonal–bipyramidal). Other labeling experiments with tagged $CH_3Mn(CO)_5$ and phosphines have also suggested a square pyramidal intermediate [19].

The formation of polymers is a key application of 1,2-insertions of alkenes into metal–alkyl bonds. The Cossee–Arlman mechanism suggested that the Ziegler–Natta polymerization of alkenes [19] is one such mechanism (Scheme 11.34). As a result of repeated 1,2-insertions, a polymer chain can propagate into an empty coordination site.

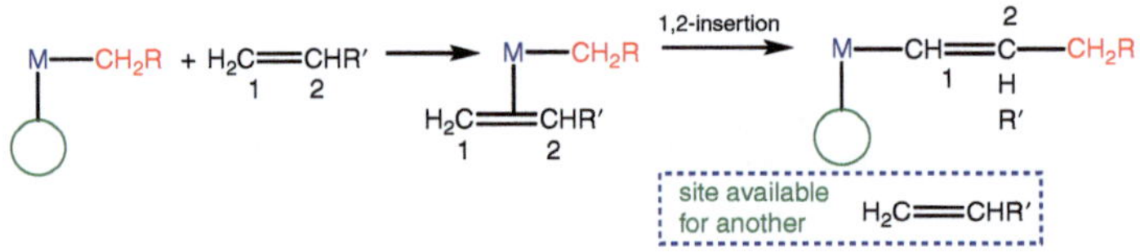

Scheme 11.34 Mechanism of Ziegler–Natta polymerization of alkenes.

11.4.1.2 Olefin Insertion

An olefin insertion into a metal–alkyl or metal–hydride bond is a common example. Inserting alkenes into M—H or M—C bonds is a popular approach for generating metal–alkyl complexes and is a critical step in a number of catalytic processes, including the Ziegler–Natta polymerization of olefins [20–27]. A coordinated alkene undergoes a concerted 1,2-insertion via a planar cyclic transition state in these intramolecular reactions, leaving an open coordination site behind [28, 29]. Polymer chain growth is caused by the binding of a new olefin molecule and subsequent cycles of alternate migratory insertion and alkene coordination (i.e. the Cossee mechanism) (Scheme 11.34). [30] Agostic complexes [31, 32] could also play a vital role as intermediates by decreasing the activation barrier for the insertion step and contributing to the stereochemistry of the final product [33].

Insertion of olefins into metal hydrides is normally an equilibrium process (Scheme 11.35). The reaction may lie totally to the right for extremely electron-withdrawing olefins. Although there are few examples of a single olefin insertion into a metal alkyl, it represents the foundation for a variety of catalytically significant processes, such as olefin polymerization. The reaction is thermodynamically favorable (unlike the formyl example explained above under CO insertions), and this experimental rarity is due to kinetics.

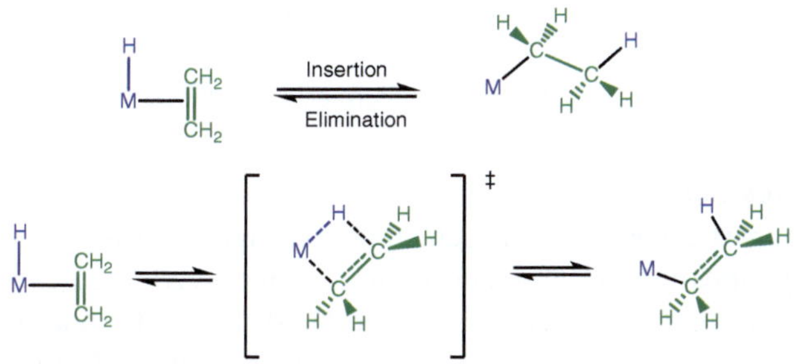

Scheme 11.35 Insertion and elimination of olefins into metal hydrides.

Another major use of olefin insertion chemistry is the hydrozirconation of alkenes with Cp2ZrHCl (Schwartz's reagent) (Scheme 11.36) [34].

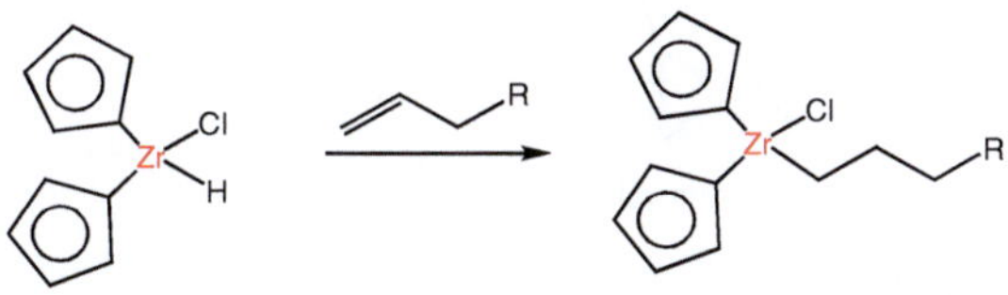

Scheme 11.36 Insertion of alkenes in Cp_2ZrHCl.

The Zr–H bond in the Cp2ZrHCl zirconocene derivative reacts with terminal alkenes at the minimum sterically hindered site (i.e. an anti-Markovnikov addition) and the ensuing zirconium alkyl complex can easily be functionalized to different alkyl halides and alcohols [35–37]. Correspondingly, 1,3-dienes and terminal alkynes also react with Cp_2ZrHCl to provide cis-vinyl and γ, δ-unsaturated alkyl complexes, respectively. Similarly, one of the most versatile and widely investigated procedures for the conversion of olefins to alcohols is the insertion of an alkene into one of the B—H bonds in diborane (B_2H_6) to create an organoborane derivative, which is commonly oxidized to an alcohol using hydrogen peroxide in aqueous base (i.e. hydroboration reactions) [38–42].

Some palladium acyl complexes depict insertion of alkenes in Pd—C bonds [43, 44]. This important reaction is effectively utilized for the synthesis of polyketones with general formula [-$CH_2CH_2C(O)$-]$_n$ by copolymerization of ethylene and CO (Scheme 11.37) [45–49].

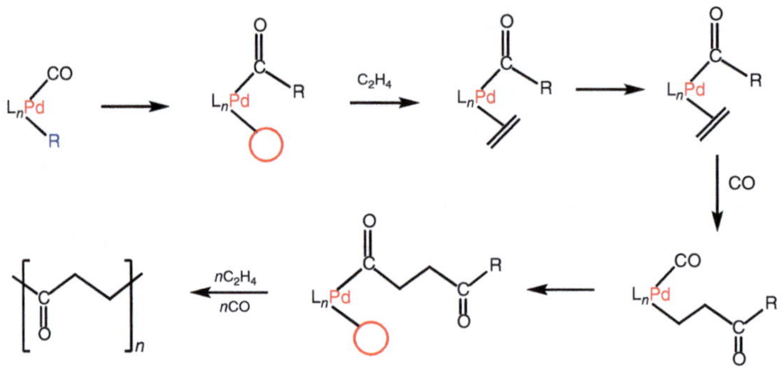

Scheme 11.37 insertion of alkenes in Pd-C bonds of palladium acyl complexes.

11.4.1.3 Other Insertion Reaction

However, alkenes and CO are the most common substrates for migratory insertions, a wide range of other compounds are also accessible to similar processes [50, 51]. For instance, alkyl and aryl isonitriles (RNC) easily insert into M—H or M—C bonds to produce aminoacyl or formimidoyl derivatives [52–55]. Despite the fact that CO and RNC are isoelectronic species, the latter is more reactive that leads to more repeated insertions into M—C bonds than acyl complexes. Similarly, an alkyne inserted into an M—C bond produces a vinyl derivative (Scheme 11.38).

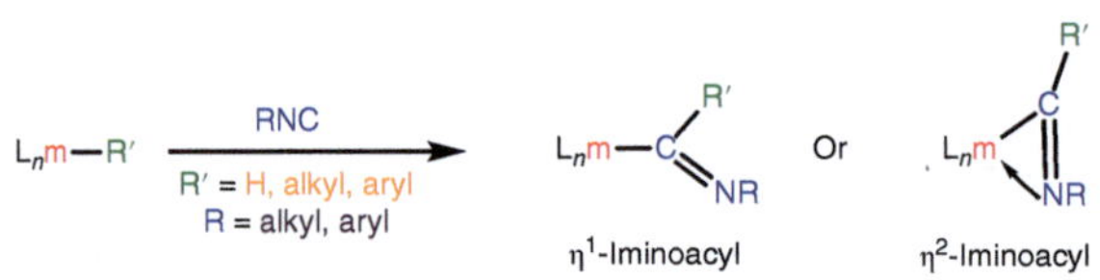

Scheme 11.38 insertion into an M−C bond.

The coordination chemistry and synthesis of carbon dioxide complexes have received keen interest throughout the years [56, 57]. Reactivity of these complexes includes various types of 1,2-migratory insertion reactions [58–61]. Accordingly, CO_2 inserts into a various M—X bonds (X, H, C, O, N, P, S, Si) to give the corresponding M-OC(O)X products, which generally follow the initial formation of complex (Scheme 11.39).

CO_2 reacts with a metal hydride to yield formate complexes M-OC(O)H rather than formyl derivatives M-C(O)OH, and insertion into M—C bonds provides the proper carboxylate molecules M-OC(O)R. Similarly, the reactions with M-OH and M-OR (R1–4alkyl, aryl) produce the bicarbonate M-OC(O)OH and carbonate M-OC(O)OR species, respectively. In biological systems, the interaction of CO_2

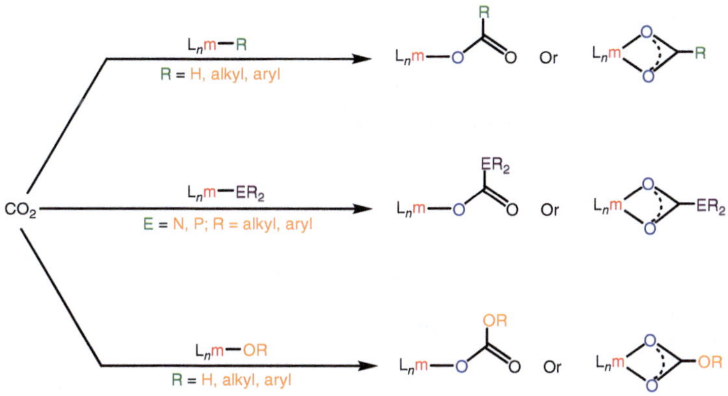

Scheme 11.39 Insertion of CO_2 into various M–X bonds.

with a zinc hydroxide moiety is particularly essential, as it catalyzes the reversible hydration of CO_2 to HCO_3 in carbonic anhydrases [62, 63]. Furthermore, the insertion of CO_2 into M—O bonds is thought to be necessary for the copolymerization of CO_2 and epoxides, as well as the creation of cyclic carbonates and polycarbonates [64, 65]. Similarly, the insertion of CO_2 into the M—N bond of both main-group and transition-metal dialkylamides (M–NR_2) results in the formation of carbamate complexes M–NR_2 (O_2CNR_2) [66].

Insertion reactions occur in other electrophilic heterocumulenes (e.g. CS_2, RNCO, RNCS, and SO_2). As a result, CS_2 insertion into the M—N bond in metal dialkylamides creates dithiocarbamate derivatives [67], and SO_2 insertion into M—C bonds forms sulfinate complexes by either 1,1-insertion [M–OS(O)R] or 1,2-insertion [M–S(R)O_2] [68–70]. Rare cases of dioxygen insertion into Pt—H bonds have also been documented, resulting in the extremely stable hydroperoxide derivatives LnPt–OOH [71, 72].

11.4.2 Hydride Elimination

Hydride elimination reactions are characterized by hydrogen atom transfer from a ligand to a metal. The most famous and ubiquitous type of β-elimination is β-hydride elimination, with a proton in a and b position * on an alkyl ligand transferred to the metal by way of an intermediate in which the metal, the a and b carbons, and the hydride are coplanar. It involves the formation of a π bond and an M—H bond. An example of β-elimination – the reverse of 1,2-insertion – is being discussed here (Scheme 11.40).

The transfer of hydrogen atoms from a ligand to a metal characterizes hydride elimination reactions. β-hydride elimination is the most well-known and widespread type of β-elimination in which a proton in the β position on an alkyl ligand is transferred to the metal via an intermediate in which the metal, the α and β carbons, and the hydride are all coplanar. It entails the creation of a π bond as well as an M—H bond. On the β-carbon, the alkyl must have hydrogens. For this reaction

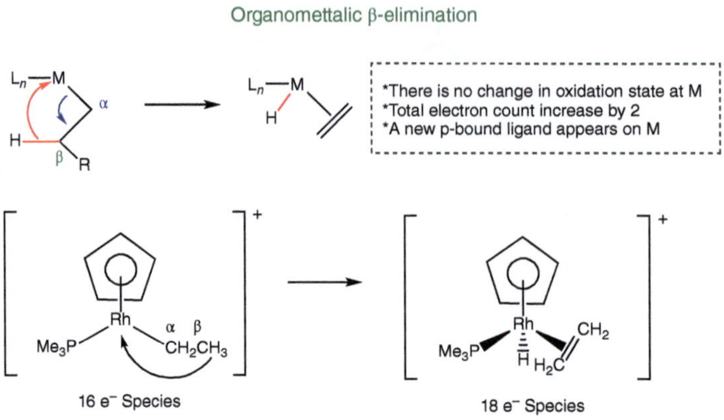

Scheme 11.40 Schematic representation of Organometallic β-elimination with example.

to occur, the metal complex must have an unoccupied (or vacant) site cis to the alkyl group and an approachable vacant orbital on the metal center. Furthermore, a d electron pair is required for donation into the C–H bond's σ* orbital for simple cleavage. As a result, d^0 metal alkyls are more resistant to β-hydride elimination than d^2 and higher metal alkyls, and they can form isolable agostic complexes whenever an empty coordination site is available [73]. It is worth noting that the metal has picked up one more ligand after β-elimination; elimination requires an empty site for that ligand. As a key orbital interaction in this process, we might envisage hydride "attacking" the vacant orbital on the metal center.

The M—Cα and Cβ—H bonds must be able to align in a syn coplanar configuration. The M—Cα and Cβ—H bonds are on the same side of the Cα—Cβ bond (a dihedral angle of 0°) and all four atoms are on a plane. That conformation can be seen in the diagram above. The C—H bond departing from the ligand is best aligned with the vacant orbital on the metal center in the syn coplanar arrangement. β-hydride elimination does not occur in hindered or cyclic complexes that cannot acquire this configuration. The requirement for a syn coplanar conformation has significant consequences for eliminations that could result in diastereomeric olefin formation: the β-elimination is stereospecific. The (E)-olefin is produced by one diastereomer, whereas the (Z)-olefin is produced by the other.

The total number of electrons in the complex must be 16 or less. Notice that how the complex's total electron count rises by two during β-hydride elimination. Because the result would have 20 total electrons, complexes with 18 total electrons do not undergo β-elimination. Of course, while evaluating the likelihood of β-elimination in a complex, keep in mind that the dissociation of a loose ligand can readily form a 16-electron complex. Although ligand dissociation is sometimes reversible, β-hydride elimination is virtually always irreversible.

At least 2d electrons must be carried by the metal to perform backbonding because the σ C–H → M orbital interaction is not enough to encourage β-elimination on its own; an M → σ*C–H interaction is also required.

β eliminations are essential in various catalytic processes (Scheme 11.41). The β-hydride elimination can either be a vital step in a reaction or an unproductive side reaction. β-hydride elimination is utilized in the shell of higher olefin process to create α olefins, which are used to make detergents. In Ziegler–Natta polymerization, β-hydride elimination results in polymers with a lower molecular weight, which is an example of sometimes unwanted β-hydride elimination.

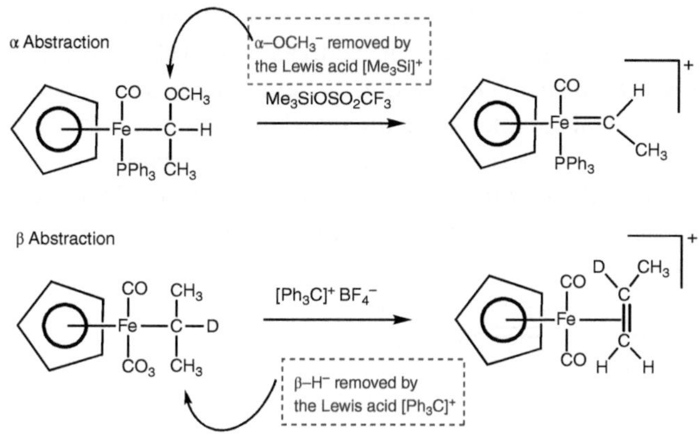

Scheme 11.41 Comparison of α and β abstraction.

The β-hydride elimination can reduce the yield in nickel and palladium-catalyzed aryl halide couplings with alkyl Grignard reagents. Chain walking is required for the formation of branched polymers from ethylene where β-hydride removal is a crucial step.

β-hydride removal is sometimes the first of a series of steps. An intermediate alkoxide complex undergoes a β-hydride elimination to create the hydride ligand and the π-bonded aldehyde, which is subsequently transformed into the carbonyl (carbon monoxide) ligand in the synthesis of $RuHCl(CO)(PPh_3)_3$ from ruthenium trichloride, triphenylphosphine, and 2-methoxyethanol.

11.4.2.1 Avoidance of β-hydrogen Elimination

Because these reactions can only take place in complexes with β hydrogens, alkyl complexes without β hydrogens are more stable than those with β hydrogens (however, the latter may undertake other types of reactions). Butyl groups, for example, can undergo this reaction, whereas methyl groups cannot. Furthermore, complexes containing β hydrogens that are coordinatively saturated are more stable than complexes with empty coordination sites because the elimination mechanism requires the transfer of hydrogen to a vacant coordination site. Finally, elimination reactions of other kinds are also known, for example the elimination of hydrogen from α and γ sites. [74]

11.4.3 Abstraction

Abstraction reactions are elimination reactions where the metal's coordination number remains unchanged. In general, they entail an exogenous reagent, such as a Lewis acid, to remove a substituent from a ligand. Two types of abstractions: α and β, which entail removing substituents from the α and β sites (with respect to the central metal atom) of ligands, respectively, have been depicted below. Carbyne complexes are synthesized through α-abstraction.

The transfer of a hydrogen atom from a ligand's α position to the metal is considered as α-hydride elimination. The metal center is oxidized by two electrons; thus, the process can be regarded as a form of OA reaction (Scheme 11.42). α-elimination cannot happen in a d^0 or d^1 metal complex since the reaction requires a formal oxidation of the metal. α-abstraction is a variation that can occur in certain situations. The α-hydrogen is transported directly to neighboring ligand instead of the metal center, resulting in no change in oxidation state (Scheme 11.42).

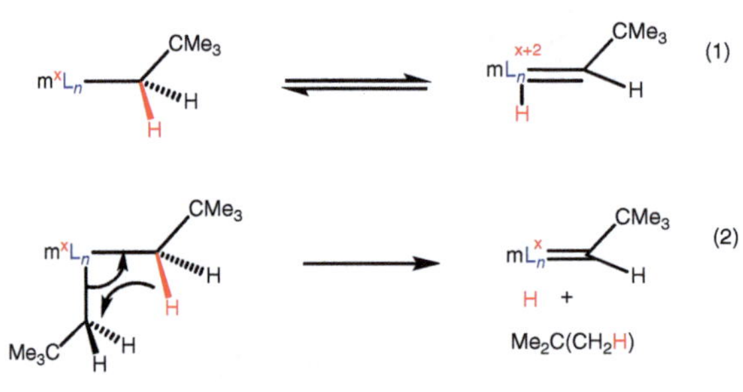

Scheme 11.42 Example of alpha-hydrogen abstraction.

α-abstraction is a typical process for making alkylidene and alkylidyne compounds with high oxidation states. α-abstraction, unlike α-elimination, does not require an open coordination site on the metal complex. α-abstraction can be caused by steric crowding of the metal, as shown below; the abstraction can happen spontaneously or be stimulated by the addition of a donor ligand, such as phosphine (Scheme 11.43).

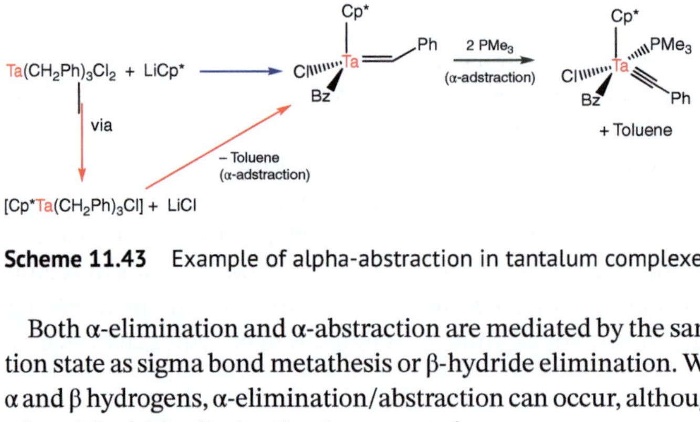

Scheme 11.43 Example of alpha-abstraction in tantalum complexes.

Both α-elimination and α-abstraction are mediated by the same four-center transition state as sigma bond metathesis or β-hydride elimination. When a metal contains α and β hydrogens, α-elimination/abstraction can occur, although it is more common when β-hydride elimination is prevented.

11.4.3.1 Nucleophilic Abstraction

It is a kind of organometallic reaction in which a nucleophilic attack on a ligand results in the removal of a part or all of the original ligands from the metal along with the nucleophile [75, 76].

11.4.3.1.1 Alkyl Abstraction

Nucleophilic abstraction of an alkyl group is relatively unusual, but there are examples of nucleophilic alkyl abstraction. Because reduced metals are frequently poor leaving groups, this process requires the metal to be oxidized first. The M—C bond weakens as the metal oxidizes, allowing nucleophilic abstraction to occur. In the following example of nucleophilic abstraction, G. M. Whitesides and D. J. Boschetto use the halogens Br_2 and I_2 as M–C cleaving agents (Scheme 11.44) [77].

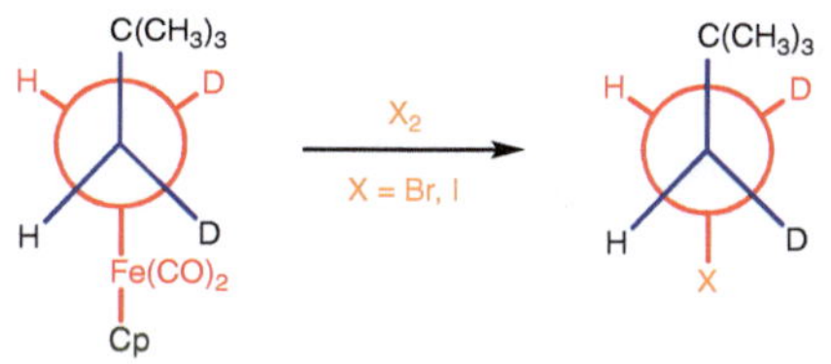

Scheme 11.44 Example of alkyl abstraction.

It is worth noting that the product of this reaction is reversed with respect to the metal's stereochemical center. The mechanism of this reaction can be described in numerous ways, as indicated in the diagram below (Scheme 11.45) [75].

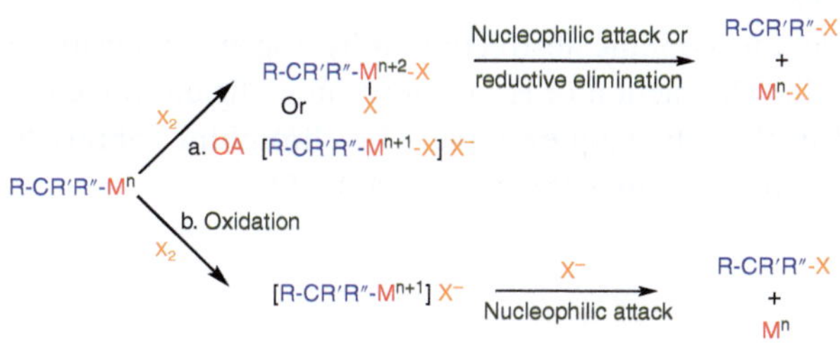

Scheme 11.45 Mechanism of alkyl abstraction.

The OA of the halogen to the metal complex is the initial step in path a. This process produces the oxidized metal center, which is required to weaken the M—C bond. The second step can involve either the halide ion's nucleophilic attack on the alkyl group's α-carbon or RE, both of which result in stereochemistry inversion. In path b, the metal is oxidized first before the halide is added. The second step involves a nucleophilic assault on the α-carbon, which leads to stereochemistry inversion once more.

11.4.3.1.2 Carbonyl Abstraction

In the nucleophilic abstraction of carbonyl, trimethylamine N-oxide (Me_3NO) can be utilized. Me_3NO attacks the carbon of the carbonyl group, pushing electrons onto the metal. After that, the reaction produces CO_2 and NMe_3 (Scheme 11.46) [78, 79].

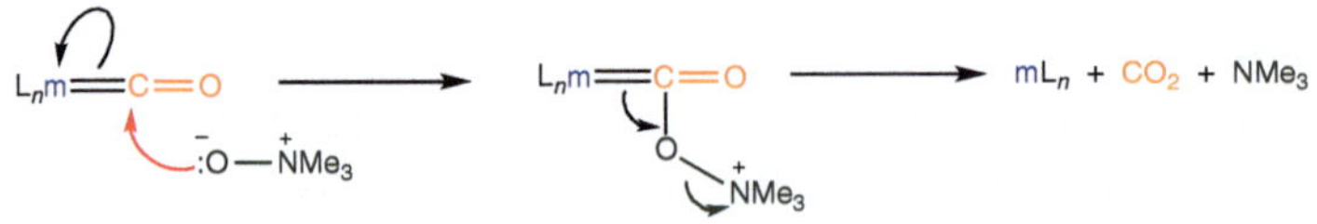

Scheme 11.46 Schematic representation of Carbonyl abstraction.

C. S. Chin and group discovered that one iridium compound undergoes carbonyl abstraction, while another performs hydride extraction (Scheme 11.47) [80].

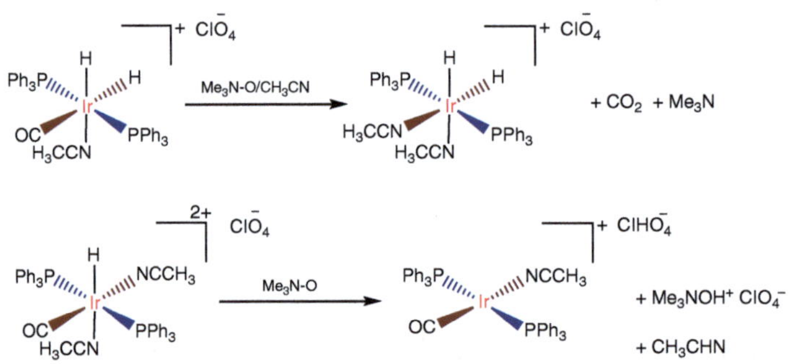

Scheme 11.47 Carbonyl abstraction in iridium compound.

11.4.3.1.3 Hydrogen Abstraction

If the conditions are favorable, nucleophilic abstraction of hydrogen can occur on a metal ligand. The nucleophilic abstraction of H^+ from an arene ligand linked to chromium, as demonstrated in the following example. The chromium's ability to withdraw electrons makes the process simple (Scheme 11.48) [75].

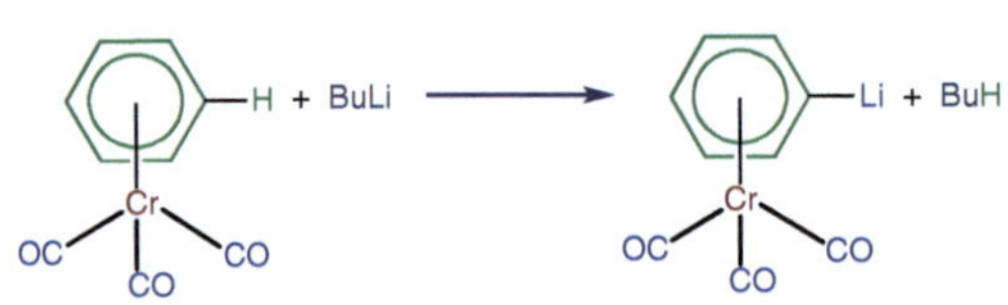

Scheme 11.48 Hydrogen abstraction in chromium complex.

11.4.3.1.4 Methyl Abstraction

A methyl group can be removed from a Fischer carbene through nucleophilic abstraction. The abstracting agent would generally add to the carbene carbon with the addition of a minor abstracting agent. However, in this scenario, the steric bulk of the abstracting agent leads the methyl group to be abstracted. When the methyl group is replaced with ethyl, the reaction slows down by 70 times, as expected with an SN^2 displacement mechanism (Scheme 11.49) [81].

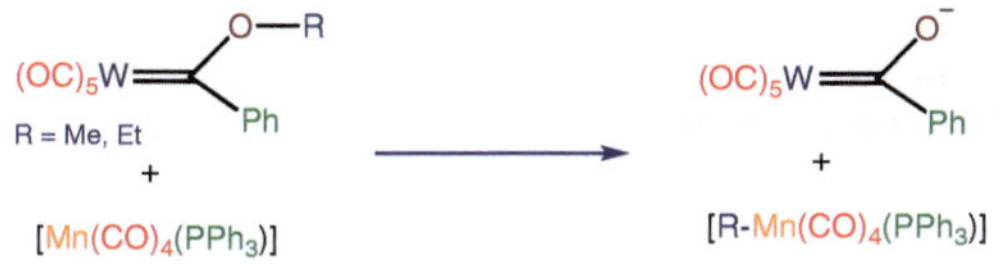

Scheme 11.49 Example of Methyl abstraction.

11.4.3.1.5 Silylium Abstraction

A silicon cation with only three bonds and a positive charge is known as a silylium ion. The ruthenium complex below demonstrates the abstraction of the silylium ion (Scheme 11.50) [82].

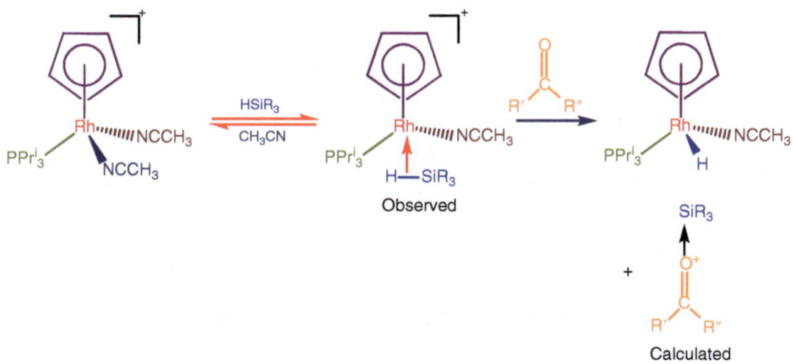

Scheme 11.50 Silylium abstraction in rhodium complex.

One of the acetonitrile groups is replaced by a silicon molecule in the first stage of this process, with the silicon–hydrogen bond coordinating to the ruthenium. A ketone is added in the second stage for nucleophilic abstraction of the silylium ion, while the hydrogen is left on the metal.

11.4.3.1.6 α-acyl Abstraction

An example of nucleophilic abstraction of an α-acyl group is observed when MeOH is introduced to the following palladium complex. The mechanism uses a tetrahedral intermediate to produce the methyl ester and reduced palladium complex displayed (Scheme 11.51) [83].

A similar mechanism was described by R. F. Heck in which OA of an aryl halide is followed by migratory CO insertion and nucleophilic abstraction of the α-acyl by MeOH. The generation of linear acyl derivatives is one of the advantages of this intermolecular nucleophilic abstraction. Lactone and lactum are cyclic compounds formed by the intramolecular attack of these linear acyl derivatives (Scheme 11.52) [84].

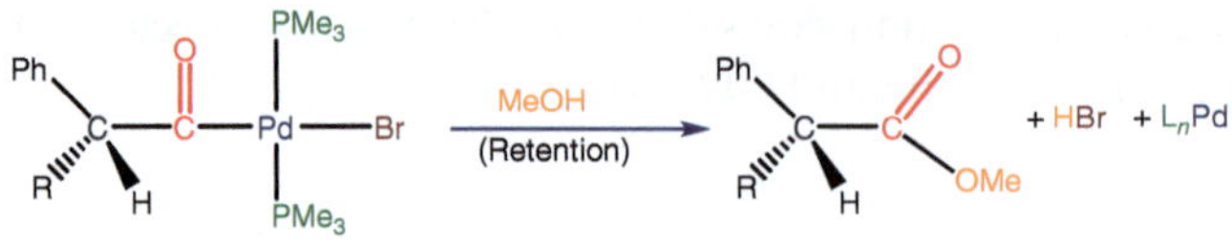

Scheme 11.51 α-Acyl abstraction in palladium complex.

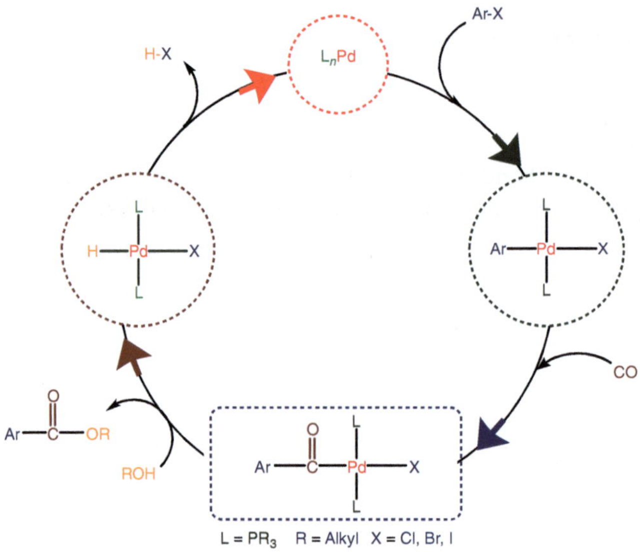

Scheme 11.52 Mechanism of α-Acyl abstraction in palladium complex.

11.5 Conclusion

Organometallic chemistry has always been an interesting field for study and research. This chapter has dealt with different reactions of organometallic compounds. Further efforts need to be made in order to explore the scope of this class of chemicals for application in the fields of synthesis of various compounds through their specific reactions.

References

1 Tolman, C.A. (1970). *J. Am. Chem. Soc.* 92: 2956.
2 (1977). *Chem. Rev.* 77: 313.
3 Bunten, K.A., Chen, L., Fernandez, A.L., and Poë, A.J. (2002). *Coord. Chem. Rev.* 41: 233–234.
4 Love, J.A., Sanford, M.S., Day, M.W., and Grubbs, R.H. (2003). *J. Am. Chem. Soc.* 125: 10103.
5 Darensbourg, D.J. and Graves, A.H. (1979). *Inorg. Chem.* 18: 1257.

6 Kubas, G.J. (2001). *Metal Dihydrogen and σ-Bond Complexes: Structure, Theory, and Reactivity*. Kluwer ISBN 0-306-46465-9.

7 Johnson, C. and Eisenberg, R. (1985). *J. Am. Chem. Soc.* 107 (11): 3148–3160.

8 Buchanan, J.M., Stryker, J.M., and Bergman, R.G. (1986). *J. Am. Chem. Soc.* 108: 1537.

9 Schenck, T.G., Milne, C.R.C., Sawyer, J.F., and Bosnich, B. (1985). *Inorg. Chem.* 24: 2338–2344.

10 Okrasinski, S.J. and Nortom, J.R. (1977). *J. Am. Chem. Soc.* 99: 295–297.

11 Cross, J.L., Garrett, A.D., Crane, T.W. et al. (2004). *Polyhedron* 23: 2831–2840.

12 Cavell, K.J. (1996). *Coord. Chem. Rev.* 155: 209–243.

13 Soderberg, B.C. (1995). *Comprehensive Organometallic Chemistry II*, vol. 12, Chapter 3.5 (ed. E.W. Abel, F.G.A. Stone and G. Wilkinson), 241–297. Oxford: Elsevier.

14 Berke, H. and Hoffmann, R. (1978). *J. Am. Chem. Soc.* 100: 7224–7236.

15 Yamamoto, A.J. (1999). *Chem*, 1027–1037. Soc.: Dalton Trans.

16 Parkin, G., Bunel, E., Burger, B.J. et al. (1987). *J. Mol. Catal.* 41: 21–39.

17 Kayaki, Y. and Yamamoto, A. (2003). *Curr. Meth. Inorg. Chem.* 3: 373–409.

18 Flood, T.C., Jensen, J.E., and Statler, J.A. (1981). *J. Am. Chem. Soc.* 103: 4410.

19 Cossee, P. (1964). *J. Catal.* 3: 80; Arlman, E. J. and Cossee, P. *J. Catal.*, 1964, 3, 99.

20 Chen, E.Y.-X. and Marks, T.J. (2000). *Chem. Rev.* 100: 1391–1434.

21 Coates, G.W. and Waymouth, R.M. (1995). *Comprehensive Organometallic Chemistry II*, vol. 12, Chapter 12.1 (ed. E.W. Abel, F.G.A. Stone and G. Wilkinson), 1193–1208. Oxford: Elsevier.

22 Collman, J.P., Hegedus, L.S., Norton, J.R., and Finke, R.G. (1987; Chapter 11). *Principles and Applications of Organotransition Metal Chemistry*, 577–617. Mill Valley: University Science Books.

23 Coates, G.W., Hustad, P.D., and Reinartz, S. (2002). *Angew. Chem., Int. Ed.* 41: 2236–2257.

24 Mecking, S. (2001). *Angew. Chem., Int. Ed.* 40: 534–540.

25 Gibson, V.C. and Spitzmesser, S.K. (2003). *Chem. Rev.* 103: 283–315.

26 Kaminsky, W. (ed.) (1999). *Metalorganic Catalysts for Synthesis and Polymerisation*. Berlin: Springer-Verlag.

27 Janiak, C. Metallocene catalysts for olefin polymerization. In: *Metallocenes: Synthesis Reactivity Applications*, vol. 2, Chapter 9 (ed. A. Togni and R.L. Halterman), 547–623. Wiley.

28 Silanes, I. and Ugalde, J.M. (2005). *Organometallics* 24: 3233–3246.

29 Rappé, A.K., Skiff, W.M., and Casewit, C.J. (2000). *Chem. Rev.* 100: 1435–1456.

30 Cossee, P. (1964). *J. Catal.* 3: 80–88.

31 Brookhart, M. and Green, M.L.H. (1983). *J. Organomet. Chem.* 250: 395–408.

32 Brookhart, M., Green, M.L.H., and Wong, L.-L. (1988). *Prog. Inorg. Chem.* 36: 1–124.

33 Grubbs, R.H. and Coates, G.W. (1996). *Acc. Chem. Res.* 29: 85–93.

34 Schwartz, J. and Labinger, J.A. (1976). *Angew. Chem., Int. Ed. Engl.* 15: 333–340.

35 Wipf, P., Takahashi, H., and Zhuang, N. (1998). *Pure Appl. Chem.* 70: 1077–1082.

36 Wipf, P. and Kendall, C. (2005). *Top. Organomet. Chem.* 8: 1–25.

37 Wipf, P. and Jahn, H. (1996). *Tetrahedron* 52: 12853–12910.

38 Vaultier, M. and Carboni, B. (1995). *Comprehensive Organometallic Chemistry II*, vol. 11, Chapter 5 (ed. E.W. Abel, F.G.A. Stone and G. Wilkinson), 191–276. Oxford: Elsevier.

39 Brown, H.C. (1975). *Organic Synthesis Via Boranes.* New York: Wiley.

40 Ramachandran, P.V. and Brown, H.C. (ed.) (2001). *Organoboranes for Syntheses.* Washington, DC: American Chemical Society.

41 Beletskaya, I. and Pelter, A. (1997). *Tetrahedron* 53: 4957–5026.

42 Omae, I. (1998; Chapter 6). *Applications of Organometallic Compounds*, 91–106. New York: Wiley.

43 Ledford, J., Shultz, C.S., Gates, D.P. et al. (2001). *Organometallics* 20: 5266–5276.

44 Groen, J.H., Delis, J.G.P., van Leeuwen, P.W.N.M., and Vrieze, K. (1997). *Organometallics* 16: 68–77.

45 Drent, E., van Broekhoven, J.A.M., and Budzelaar, P.H.M. (2002). The alternating copolymerization of alkenes and carbon monoxide. In: *Applied Homogeneous Catalysis with Organometallic Compounds*, 2e, vol. 1, Chapter 2.3.4 (ed. B. Cornils and W.A. Herrmann), 344–361. Weinheim: Wiley-VCH.

46 Bianchini, C. and Meli, A. (2002). *Coord. Chem. Rev.* 225: 35–66.

47 Nozaki, K. and Hiyama, T. (1999). *J. Organomet. Chem.* 576: 248–253.

48 Drent, E. and Budzelaar, P.H.M. (1996). *Chem. Rev.* 96: 663–681.

49 Sen, A. (1993). *Acc. Chem. Res.* 26: 303–310.

50 Cotton, F.A., Wilkinson, G., Murillo, C.A., and Bochmann, M. (1999). *Advanced Inorganic Chemistry*, 6e, 1207–1228. New York: Wiley.

51 Crabtree, R.H. (2005; Chapter 7). Insertion and elimination. In: *The Organometallic Chemistry of the Transition Metals*, 4e, 183–204. Hoboken: Wiley.

52 Sebastian, A., Royo, P., Gómez-Sal, P., and Ramírez de Arellano, C. (2004). *Eur. J. Inorg. Chem.* 3814–3821.

53 Martins, A.M., Ascenso, J.R., de Azevedo, C.G. et al. (2003). *Organometallics* 22: 4218–4228.

54 Thorn, M.G., Lee, J., Fanwick, P.E., and Rothwell, I.P.J. (2002). *Chem. Soc., Dalton Trans.* 3398–3405.

55 Ong, T.G., Wood, D., Yap, G.P.A., and Richeson, D.S. (2002). *Organometallics* 21: 1–3.

56 Gibson, D.H. (1996). *Chem. Rev.* 96: 2063–2095.

57 Leitner, W. (1996). *Coord. Chem. Rev.* 153: 257–284.

58 Yin, X. and Moss, J.R. (1999). *Coord. Chem. Rev.* 181: 27–59.

59 Pandey, K.K. (1995). *Coord. Chem. Rev.* 140: 37–114.

60 Darensbourg, D.J. and Kudaroski, R.A. (1983). *Adv. Organomet. Chem.* 22: 129–168.

61 Gibson, D.H. (2004; Chapter 1.30). *Comprehensive Coordination Chemistry II* (ed. J.A. McCleverty and T.J. Meyer), 595–602. San Diego: Elsevier.

62 Duda, D.M. and McKenna, R. (2004). *Handbook of Metalloproteins*, vol. 3 (ed. A. Messerschmidt, M. Cygler and W. Bode), 249–263. Hoboken: Wiley.

63 Tripp, B.C., Smith, K., and Ferry, J.G. (2001). *J. Biol. Chem.* 276: 48615–48618.

64 Darensbourg, D.J. and Holtcamp, M.W. (1996). *Coord. Chem. Rev.* 153: 155–174.

65 Coates, G.W. and Moore, D.R. (2004). *Angew. Chem., Int. Ed.* 43: 6618–6639.

66 Dell'Amico, D.B., Calderazzo, F., Labella, L. et al. (2003). *Chem. Rev.* 103: 3857–3897.

67 Hogarth, G. (2005). *Prog. Inorg. Chem.* 53: 71–561.

68 Wojcicki, A. (1974). *Adv. Organomet. Chem.* 12: 31–81.

69 Ryan, R.R., Kubas, G.J., Moody, D.C., and Eller, P.G. (1981). *Struct. Bond.* 46: 47–100.

70 Mews, R., Lork, E., Watson, P.G., and Görtler, B. (2000). *Coord. Chem. Rev.* 197: 277–320.

71 Wick, D.D. and Goldberg, K.I. (1999). *J. Am. Chem. Soc.* 121: 11900–11901.

72 Denney, M.C., Smythe, N.A., Cetto, K.L. et al. (2006). *J. Am. Chem. Soc.* 128: 2508–2509.

73 Crabtree, R.H. (2005). *The organometallic chemistry of the transition metals*, 4e, 58. Hoboken, N.J: Wiley ISBN 0-471-66256-9. OCLC 61520528.

74 J. D. Fellmann, R. R. Schrock, D. D. Traficante, Organometallics, 1982, 1, 481; Hartwig, J. F. (2010). Organotransition Metal Chemistry, From Bonding to Catalysis, University Science Books, Mill Valley, CA, 2010.

75 Spessard, G. and Miessler, G. (2010). *Organometallic Chemistry*, 2e, 285–289. ISBN 978-0-19-533099-1.

76 Xu, R., Pang, W., and Huo, Q. (2011). *Modern Inorganic Synthetic Chemistry*, 275–278. ISBN 978-0-444-53599-3.

77 Whitesides, G.M. and Boschetto, D.J. (1971). *J. Am. Chem. Soc.* 93: 1529.

78 Yang, K., Bott, S.G., and Richmond, M.G. (1994). *Organometallics* 13: 3788.

79 Albers, M.O. and Coville, N. (1984). *J. Coord. Chem. Rev.* 53: 227.

80 Chin, C.S., Oh, M., Won, G. et al. (1999). *Bull. Korean Chem. Soc.* 20: 85.

81 Toomey, L.M. and Atwood, J.D. (1997). *Organometallics* 16: 490.

82 Gutsulyak, D.V., Vyboishchikov, S.F., and Nikonov, G.I. (2010). *J. Am. Chem. Soc.* 132: 5950.

83 Stille, J.K. and Lau, K.S.Y. (1977). *Acc. Chem. Res.* 10: 434.

84 Heck, R.F. (1978). *Pure Appl. Chem.* 50: 691.

12

Characterization Techniques of Organometallic Compounds

Javad B. M. Parambath and Ahmed A. Mohamed

University of Sharjah, College of Sciences, Department of Chemistry, Sharjah 27272, UAE

12.1 Introduction

Organometallic compounds are easily characterized using a wide range of instruments for basic and advanced analysis based on the functional groups. Evidence of the presence of functional groups requires basic instruments, such as nuclear magnetic resonance (NMR) spectroscopy, Fourier transform infrared (FTIR), Raman, ultraviolet–visible (UV-Vis) spectroscopy, mass spectrometry, and elemental analysis (EA) [1–9]. Advanced instruments such as neutron diffraction analysis can be used for hydride detection [10], X-ray photoelectron spectroscopy (XPS) to detect the presence of the various types of elements in the compounds [11, 12], and Mössbauer to specifically probe the environment around iron atoms [13–16]. Additional instruments such as single-crystal X-ray diffractometers are used for measuring the bond distances and angles [17–19]. Electron paramagnetic resonance (EPR) has been integrated for the detection of free radicals in clusters of classical compounds and enzyme models [20]. The integration of new instruments for the characterization of organometallic compounds has evolved because of their rapid applications in industry, medicine, and drug delivery. The merge of organometallic compounds with nanotechnology has advanced this field rapidly for novel applications in catalysis and nanomedicine engineering [21].

Some phenomena such as fluxionality require specific tools under low temperatures, such as ^{1}H NMR [22]. The carbon versus nitrogen coordination from isocyanide and cyanide isomers to central metals has been studied using ^{13}C NMR. Similarly, carbonyl versus isocarbonyl coordination was investigated [23]. X-ray crystallography was used to probe oxidative rearrangement in gold organometallics [24]. Gold(I) was oxidized to gold(III) and the ligands rearranged from the trans to the cis isomers. There are many similar examples where the characterization of organometallic compounds targeted a specific phenomenon or probed the final product after basic reactions, such as oxidative addition and reductive elimination. Further, Raman spectroscopy is used to study the stretching frequency of a few functional groups such as isocyanide in a molecular entity

Organometallic Compounds: Synthesis, Reactions, and Applications, First Edition.
Edited by Dakeshwar Kumar Verma and Jeenat Aslam.

or when bonding to metal surfaces for surface enhanced Raman spectroscopy (SERS) studies [3, 25]. One must also mention the need to study the metal–metal interaction using extended X-ray absorption fine structure (EXAFS) in clusters [26]. The unusual double, triple, and possibly quadruple bonds in bimetallic organometallics by Cotton are interesting to mention in this chapter in addition to the many instruments used for their characterization. Isotope labeling has been used to study the migratory insertion reactions of carbonyl functional groups [27].

Several studies conducted for the characterization of organometallic compounds involved basic and sophisticated instruments. In this chapter, we highlight the instruments used for probing the nature of bonding of ligands, phenomena that occur upon bonding certain ligands, and some intriguing interactions between metals in clusters. In our discussion, we categorize the instruments into fundamental and specific instruments. The use of the discussed instruments in specific studies will be highlighted.

12.2 Conventional Methods

The characterization of an organometallic compound implies explicating a complete understanding of its identification, purity, and even its stereochemistry to perceive its structure–property paradigm. The physical parameters, such as refractive indices, boiling and melting points, and dipole moments, are invariable characteristics to ensure the homogeneity of compounds. Although, the successful identification of organometallic compounds necessitates the need for various conventional methods including spectroscopic techniques, some of the important ones are discussed below.

12.2.1 Nuclear Magnetic Resonance (NMR) Spectroscopy

NMR spectroscopic method has evolved significantly for the characterization of organometallic compounds, of particular interest is the structural dependence of metal NMR parameters [1, 2, 9]. NMR is a physicochemical spectroscopic technique where an externally applied radiofrequency interacts with atomic nuclei (with $I \neq 0$) under a magnetization process (Figure 12.1A). For example, ^{1}H, ^{13}C, ^{15}N, ^{19}F, and ^{31}P have a half-integer spin (i.e. spin quantum number = 1/2, 3/2, 5/2, 7/2, 9/2) and are NMR active [9]. Few other atoms can produce broad NMR signals or weak signals due to quadrupolar nucleus asymmetric interaction with the electric field. These low-spin and short longitudinal relaxation (T_1) atoms are also investigated in terms of organometallic compounds through transition metal NMR spectroscopy, such nuclei include ^{57}Fe, ^{59}Co, and ^{103}Rh [1]. Experimental challenges arising due to small magnetic moments, long or short relaxation times, and low natural abundance for these transition metal NMR spectroscopy leads researchers across the world to focus on the most common elements (i.e. ^{1}H, ^{13}C, ^{15}N, and ^{31}P) with a spin −1/2 and these nuclei use to characterize organometallic compounds chemical environment.

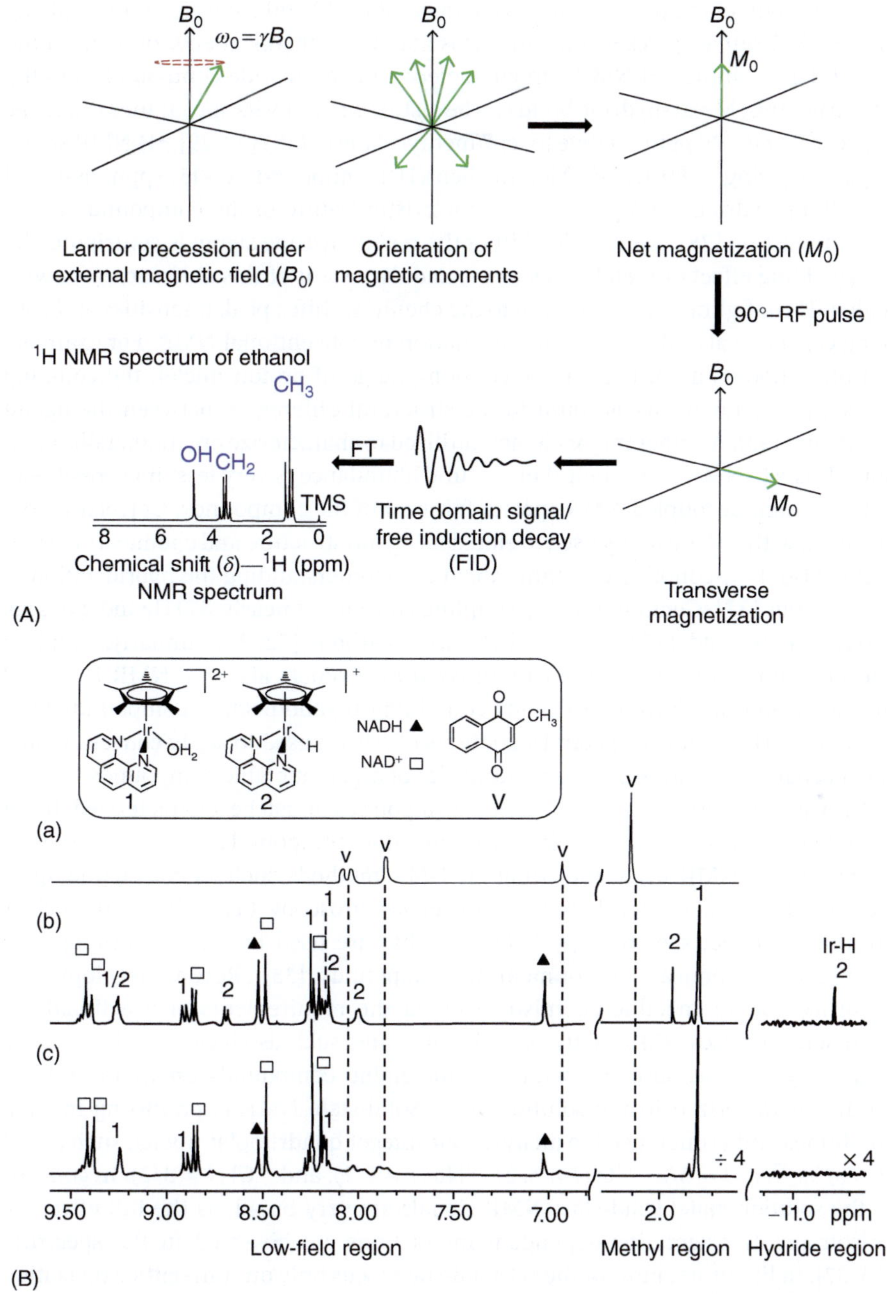

Figure 12.1 (A) Schematic representation of nuclear spin behavior under the influence of an external magnetic field (left) and a magnetization process occurring upon a magnetic pulse (right). (B) ^1H NMR spectra of [(η^5-C_5Me_5)Ir(phen)(H)] and its reaction with menadione. (a) Menadione, (b) formation of hydride compound, and (c) hydride signal disappearance. Source: Reproduced with permission from Liu et al. [28]/John Wiley and Sons.

The position of the peak known as chemical shift (δ) and splitting, or multiplicity of the NMR signals gives significant ideas about the chemical environment of the nuclei. For example, ^{1}H NMR can characterize metal hydride compounds and ligand structures. Metal–hydride bond can be distinguished by its appearance at a range between 0 and −40 ppm relative to tetramethylsilane (TMS) [28, 29]. ^{1}H NMR showing a sharp singlet for [(η^5-C_5Me_5)Ir(phen)(H)] compound at −11.3 ppm assigned to iridium–hydride bond gives the characteristic feature of the compound and its further reaction [28]. An upfield shift of the metal–hydride signal is contributed by the shielding effect of metal d-electrons, and the extent of this shift increases with higher d^n configuration. In addition to the chemical shifts, peak intensities and coupling constants also share critical information in conventional NMR. For example, if a phosphorous atom is near the coupling range of proton nuclei, the coupling constant (J_{P-H}) will vary account to the structural differences between the ligand backbones [30]. Another major element utilized to characterize organometallic compounds is ^{13}C-NMR. Although their natural abundance is very less, it is possible to obtain proton-decoupled NMR spectra for most of the compounds. Correlating the ^{1}H signal with ^{13}C is usually used to elucidate ligand structure and chemical environment. Also, J_{C-H}-coupling constants are vital to understanding the hybridization of C—H bonds [31]. Generally, a J_{C-H}-coupling constant of nearly 250 Hz indicates sp, 160 Hz for sp^2, and 125 Hz for sp^3-hybridized carbons [32, 33] Similarly, coupling constant shifts are valid in the case of *cis–trans* isomers also. ^{31}P NMR is crucial in understanding phosphine ligands containing organometallic compounds [34]. Moreover, ^{1}H decoupled spectral signatures of ^{31}P are used to study conventionally the mechanistic pathways on catalytic cycles of organometallic compounds.

As a general criterion, diamagnetic compounds must be characterized by at least two NMR-active nuclei in organometallic compounds. There are several techniques in NMR itself, including 2D-NMR methods such as correlation spectroscopy (COSY), total coherence transfer spectroscopy (TOCSY), and nuclear Overhauser effect spectroscopy (NOESY), that are used to detect couplings and differentiate chemical environment in compounds [35]. Recent developments in surface organometallic chemistry suggest the requirements of functionalized solid-state surfaces NMR analysis [32]. Substrates such as silica and alumina are frequently used as supports with organometallic compounds on them and are hardly characterized in the solution state. Solid-state NMR is promising in such conditions and is often used to analyze half-integer quadrupolar nuclei, such as ^{11}B ($I = 3/2$), ^{27}Al ($I = 5/2$), ^{17}O ($I = 5/2$), ^{93}Nb ($I = 9/2$), and ^{35}Cl ($I = 3/2$). In contrast to the solution state, solid-state NMR signals are very broad, as the full effects of anisotropic or orientation-dependent interactions are observed in the spectrum [36, 37]. In literature, most of the ^{1}H solid-state focus only on spin–lattice relaxation times concerning temperature to study methyl group rotations or movements in solid polymer chains.

12.2.2 Infrared and Raman Spectroscopy

Infrared (IR) and Raman spectroscopic techniques are widely utilized for the primary characterization and structural unit quantifications in organometallic

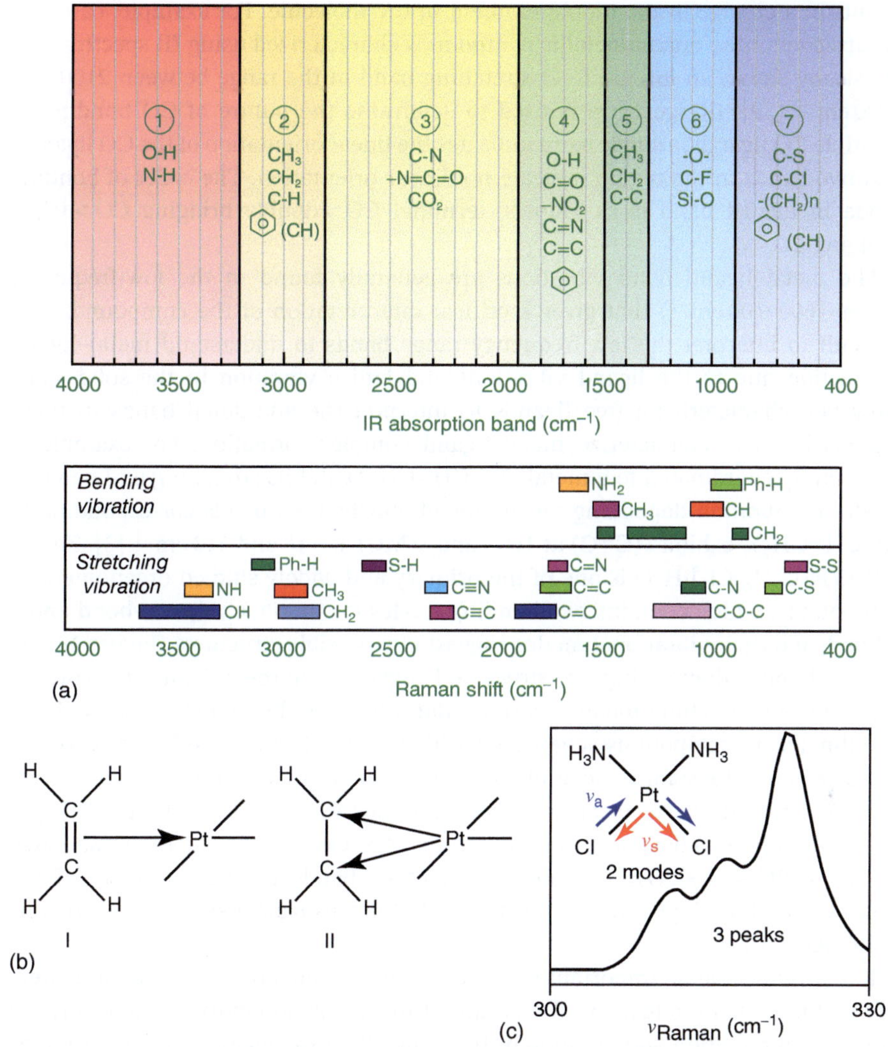

Figure 12.2 (a) IR a band (upper) and Raman shifts (lower) for different functional groups, (b) Pt–C_2H_4 bonding parameters in Zeise's salt. Source: Adapted from Nakamoto [3]. and (c) Raman spectra in the Pt–Cl stretching region for trans-[Pt$(NH_3)_2Cl_2$]. Source: Reproduced with permission from Bélanger-Desmarais et al. [38]/American Chemical Society.

compounds. Among the broad range of IR regions in the electromagnetic spectrum, the characterization of organometallic compounds utilizes the mid-IR range, which covers from 200 to 4000 cm^{-1} [6]. This technique involves the processing of vibrational motions of atomic bonds in metal compounds and ligands. In general, molecules interacting with IR radiation absorb specific wavelengths uniquely for each functional group present in the compound (Figure 12.2a). The position of the IR band indicates additional information, such as the function of a ligand (e.g. terminal or bridging), the nature of the ligand either π-acceptor/donor, and the electronic environment of the metal [3]. However, IR is active only if absorption results in a change in the dipole moment of the molecule. For example, carbonyl groups containing organometallic compounds characterized using IR spectroscopy classically shows an intense C–O stretching band in the range between 2100 and 1700 cm^{-1} [39]. IR spectrum is used to determine the nature of CO bonding in multiple CO ligands: a single band indicates the linear orientation of the CO ligands, and two bands in the region indicate nonlinear orientation. The order of bonding mode in the IR band is as follows: terminal CO > doubly bridging CO > triply bridging CO.

The metal–ligand bond vibrations are generally found in the low-frequency region (600–100 cm^{-1}) that gives a critical interpretation of the compound. It is difficult to interpret the low-frequency range bands to understand metal–ligand interaction due to the ligand vibrations and lattice vibration in the solid state. However, characterizing free ligands to interpret the additional bands in these regions helps to characterize metal–ligand complex formation. For example, a vinyl group coordinated to a metal ($M{-}CH{=}CH_2$) exhibits stretching and bending modes of vibration depending on the metal attached. IR bands corresponding to $Hg(CH{=}CH_2)_2$ exhibit $\nu(C{=}C)$ at 1603 and $\nu(MC)$ at 541 and 513 cm^{-1} [3]. Zeise's salt $K[Pt(C_2H_4)Cl_3]{\cdot}H_2O$ is one of the primary and widely studied organometallic compounds, which contains two bonding modes for the $Pt{-}C_2H_4$: a σ-bond and a π-bond, and their vibrational modes suggest two possible structures (Figure 12.2b). If the σ-bond is dominating, the structure I suggests that the 407 cm^{-1} IR band for Pt(II)-alkene stretching mode is correct. But other possible interpretations of five coordinated Pt-compounds(scheme II) with the two bands at 491 and 403 cm^{-1} corresponding to symmetric and antisymmetric Pt-C vibrational modes were ruled out [3]. Similarly, electron-rich compounds such as $[Fe(C{=}CRR')(L_2)Cp]^+$ (L = phosphine, phosphite) and relatively less electron-rich compounds such as $[Fe(\eta^2\text{-}RC{\equiv}CR')(CO)_2Cp]^+$ show distinct IR bands corresponding to of the η^2-alkyne and η^1-vinylidene ligands that help to investigate alkyne and vinylidene tautomers [40].

Raman effect has an important role in determining the molecular structure, identifying functional groups, and the nature of metal–ligand bonds. A small fraction of scattered radiation with a different frequency than the incident monochromatic light is responsible for the Raman effect. These scattered radiations are processed to identify Raman shifts depending on the energy spacing of the chemical bonds. Raman activity depends on the ability of the molecule to get polarized in an electromagnetic field, i.e. deformation of the electron cloud, which is mandatory for a molecule to show the Raman effect. Although IR and Raman techniques are related to vibrational spectroscopy, a Raman spectrum arises with polarizability and each line has a characteristic polarization that provides additional information about the compound. Raman active vibrations may be IR inactive, and vice versa depending on the symmetry of compounds. Additionally, Raman spectra are possible in aqueous solutions, a medium where IR transmission is limited. Raman analysis was successfully utilized to resolve Pt-Cl stretching modes symmetric and antisymmetric in the 312–333 cm^{-1} regions [38] (Figure 12.2c). Multi-metal compounds are more often characterized using the Raman technique, $\nu(M{-}M)$ band appears strong since large changes in polarizabilities arise due to long stretching and covalent M—M bonds. Raman active vibrations give direct information about the structure of the metal–carbon skeleton, such as strong C=C bands in the spectra of $Hg(CH=CH_2)_2$ [3].

12.2.3 UV–Visible Spectroscopy

Organometallic compounds containing electrons have interactions with the photons. Different possible energy transfer processes occur when a compound interacts with these lights through electronic transitions [41]. These transitions may be within the metal ions, or between the ligand molecules [5]. Many compounds absorb specific wavelengths of UV or visible light, using this spectroscopic technique; the absorption is directly measured to the concentration of the sample and the wavelengths associated can be categorized as electronic excitation. This categorization helps to identify functional groups of organic ligands, oxidation state, and metal–ligand electronic interactions [4]. As shown in Figure 12.3a, the imidazolium ligand has three absorption bands at 336, 250, and 220 nm assigned to $n \rightarrow \pi^*$ and $\pi \rightarrow \pi^*$ transitions [42].

One of the most common electronic transitions occurring in metal compounds is charge transfer. It arises due to the excitation of an electron from the ligand to the metal, or vice versa termed a ligand-to-metal charge transfer (LMCT) and a metal-to-ligand charge transfer (MLCT), respectively. Electronic excitation within the metal ions is known as d–d transitions. Valence electrons present in transition metal compounds are in the d sub-shell; those electronic excitations from one d-orbital to the other require relatively less energy [5, 41]. But the transition is quite inefficient and shows very weak absorbances in the spectrum, whereas MLCT and LMCT require high energy but occur frequently to give strong absorption bands. The extinction coefficient is often known as the absorption coefficient, or the molar absorptivity coefficient is a significant parameter that tells us how strongly a compound absorbs light at a particular wavelength. A different application of UV–Vis spectroscopy in organometallic compounds includes catalytic activity and initiation monitoring. For example, ruthenium species used in alkene metathesis have an intense red–green color with molar absorptivity of $10^3\,l\,mol^{-1}\,cm^{-1}$ [43]. Precatalyst initiation is monitored using this technique as the decrease in the absorbance over time and by obtaining rate constants for precatalyst initiation of compounds, substrates, or different solvents. Grubbs et al., Plenio et al., and Percy et al. have extensively utilized UV–Vis technique to conduct kinetic studies of Grubbs-type and Hoveyda-type catalyst for metathesis [43]. Characterizing metal porphyrins is primarily achieved using the UV–Vis technique. An intense band near 400 nm corresponding to a porphyrin ring known as the Soret band identifies the ligand structure in metal–porphyrin compounds. UV–Vis absorption spectra of (5,10,15,20-tetraphenyl-21-oxaporphyrin [OTPP])$Fe^{III}Cl_2$ and (OTPP)$Fe^{II}Cl$ are shown in Figure 12.3b with a split Soret band and two (iron(III)) or three (iron(II)) broad Q-bands, respectively [4].

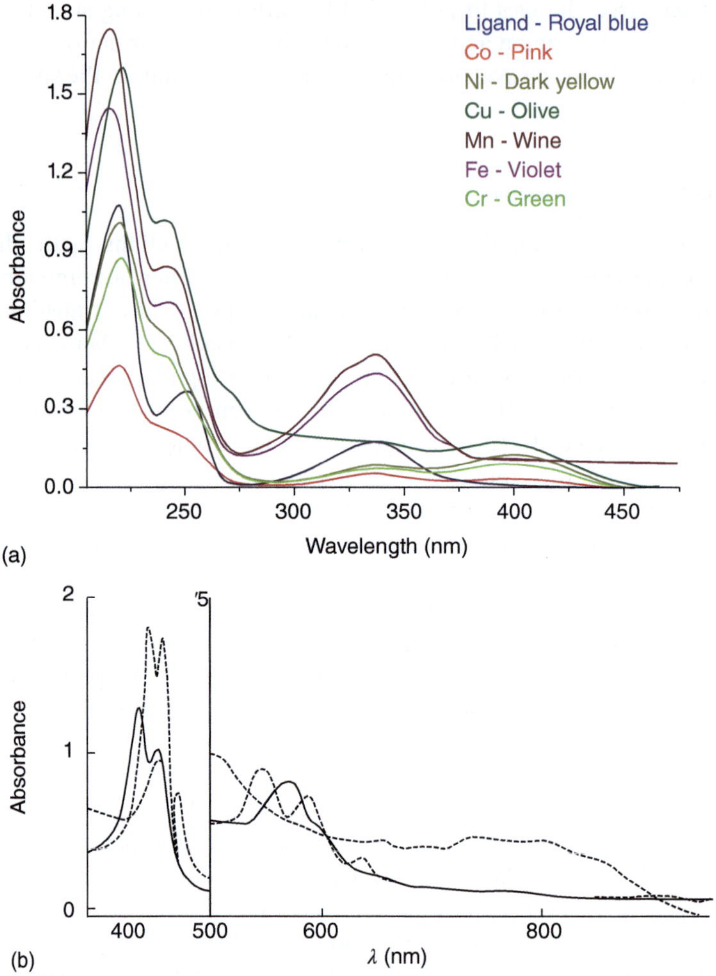

Figure 12.3 (a) UV–visible spectra of the Schiff base metal compounds. Source: Reproduced with permission from Sinha et al. [42]/Springer Nature and (b) UV–Vis spectra of (OTPP)$Fe^{III}Cl_2$ and (OTPP)$Fe^{II}Cl$ in DCM. Source: Adapted from Stute et al. [4].

12.2.4 X-Ray Crystallography

The single-crystal X-ray crystallography technique (SC-XRD) is a standard tool for solving structures of organometallic compounds. X-rays possess a shorter wavelength than visible light with higher energy to penetrate solid structures. In recent years, the development of suitable radiation sources, detectors, and high-speed data processing units has significantly improved the crystallographic structural determination technique [44]. Bragg utilized the X-ray diffraction technique for the first time to characterize single-crystal structures of simple salts [45]. Since the atoms in a crystal are orderly arranged, the incident X-rays reflect from the atomic planes and a constructive interference will occur [46]. According to Braggs law, when the X-ray hits on a crystal plane, its angle of incidence, θ, will reflect with the same angle of scattering, θ. In single-crystal XRD, integrated intensities of the diffraction peaks are utilized to map the electron density within the unit cell of the crystal. A proposed model is then developed into the experimental electron density, refined against the data, and the result will be an accurate molecular structure (Figure 12.4a). The SC-XRD can elucidate the precise location of atoms in the crystal, bond length–bond angle, supramolecular interactions, etc. for determining physicochemical properties of organometallic compounds [48, 49].

Metal ions and ligand donor atoms usually possess finite coordination numbers and geometries. A quantitative analysis using SC-XRD is conducted to describe the impact of the position and nature of the ligand on the orientation of the axial aryloxy substituents in silicon phthalocyanines. Understanding their crystal structures identifies their structural isomers and impact on aromaticity and planarity. Also, the π–π stacking of molecules identified in $(PhO)_2$-SiPc and $(Pyr)_2$-SiPc suggests their 2D arrangement and a 3D geometry for $(2MP)_2$-SiPc [47]. This suggests the effect of structural changes in organometallic compounds in solid-state properties (Figure 12.4b). SC-XRD technique is extensively used to investigate structural information of porous coordination polymers or metal–organic frameworks (MOFs) at different states [18]. This technique helps to analyze such solid structures critically by comparing atomic positions, pore size, shape, and surface chemistry to understand the adsorption properties. Organometallic nonlinear optical crystals are another class of compounds that recently utilizes the SC-XRD technique to characterize their physicochemical parameters. The crystals of the compounds formed with thiocyanate (SCN) ligands have shown interesting nonlinear optical (NLO) properties [50]. The existence of a perfect crystal never happens, since crystal growth and post-synthetic processes always cause defects and damaged areas. Therefore, single-crystal to single-crystal (SCSC) transformation, a rarely occurring phenomenon, is of particular interest and importance. Recently, there are several articles discussing the significance of SCSC in organometallic compounds, solid-state frameworks, structural transformations, and reactivity of coordination polymers [19]. Wei et al. reported reversible SCSC transformation of a tetranuclear iron compound with TPA (tris(2-pyridylmethyl)amine), $[Fe(tpa)\{N(CN)_2\}]_4\cdot(BF_4)_4(H_2O)$ upon guest molecule adsorption/desorption [51]. X-ray structures of di and trinuclear gold(I) and dinuclear gold(II) formamidinate compounds were reported, showing differences in Au—Au bond lengths [52]. Also, X-ray crystallographic technique were utilized to predict the molecular structure of trinuclear mixed-metal gold–silver compounds, such as $[Au(carb)Ag_2(\mu\text{-}3,5\text{-}Ph_2pz)_2]$ and $[Au_2(carb)_2Ag(\mu\text{-}3,5\text{-}Ph_2pz)]\cdot CH_2Cl_2$. Critical information on bond angle and bond length for these systems, including Ag⋯Ag, Au⋯Au, and Au⋯Ag distances and space group (monoclinic) identification, were also determined [53].

12.2.5 Mass Spectroscopy

Mass spectroscopy (MS) is another significant tool to characterize organometallic compounds. The primary focus of MS characterization is the determination of molecular mass and elemental composition of the compounds. Moreover,

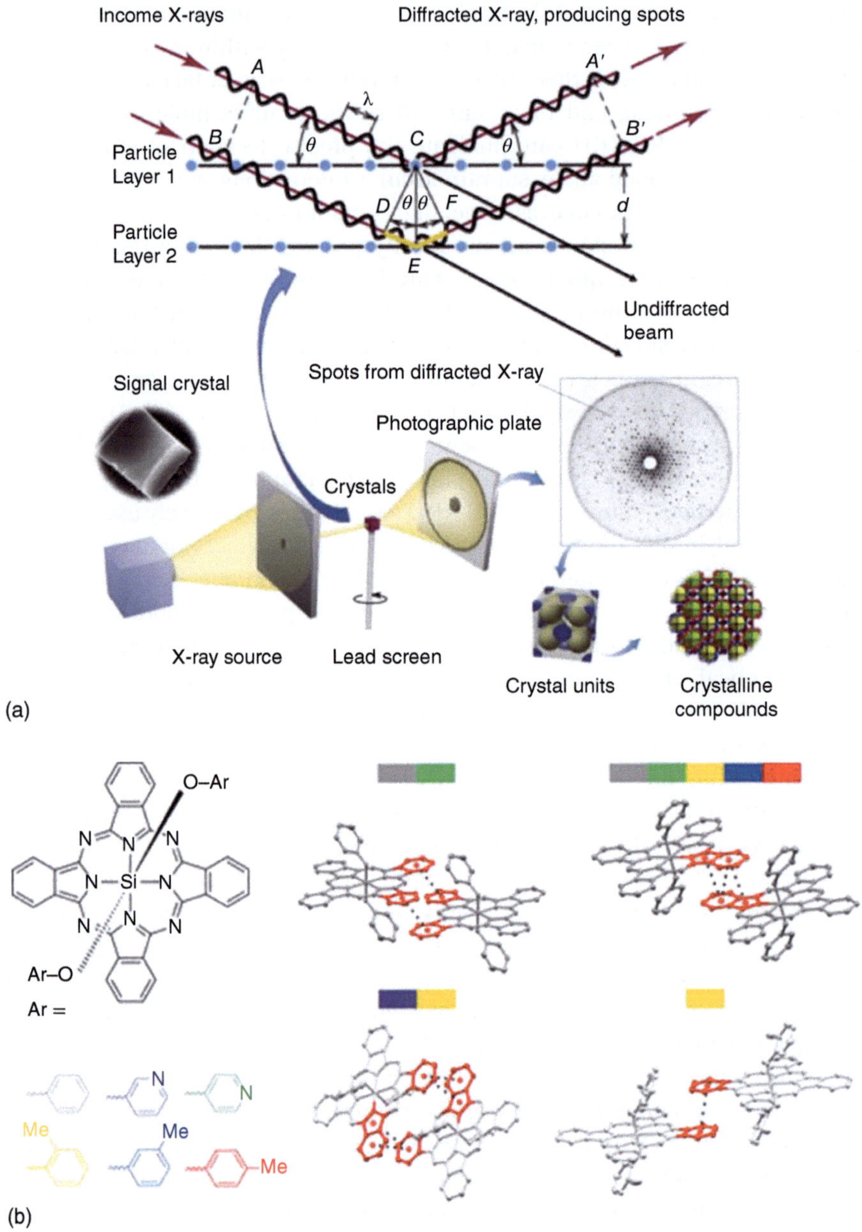

Figure 12.4 (a) Schematic representation showing SC-XRD processes such as X-ray diffraction, data refinement, and crystalline structure analysis. Source: Reproduced with permission from Bashir et. al [45]/Elsevier, (b) SC-XRD determined changes in the chemical structure of aryloxy silicon phthalocyanine derivatives and π–π intermolecular interactions. Source: Reprinted with permission from Raboui et al. [47].

this technique provides basic thermochemical characteristics of organometallic compounds, such as ionization energies, proton affinities, electron affinities, and metal–ligand bond dissociation energies [7, 54]. Various ionization techniques are used to generate molecular ions or charged species in compounds that are activated toward sensitive molecular mass detectors and processing units [55]. The obtained mass loss helps to deduct the elemental composition of the ligand. Also, a gradual loss of metal–ligand bonds can be obtained by activation of the ion.

The identification of atom connectivity in compounds is not always an easy method, and a constructive understanding of other characterization techniques and experimental data is critical. Structural identification of ligands involves the reactions caused by mass loss due to radical generation, neutral molecule, and migration of groups to the metal center. MS technique is often used to identify isomers of compounds, where the electronegativity or nature of bonds such as unsaturation in the substituent helps to classify them. For example, intramolecular interaction helping to distinguish ortho-from meta- and *p*-ferrocenyl benzenes, showing dissociation of molecular ions of ortho isomers gave rise to a strong peak due to the loss of the unsubstituted cyclopentadienyl ring, but their meta and para isomers give minimal intensity of $[P\text{-}C_5H_5]^+$ ions [7]. Chemical ionization in compounds gives high selectivity and sensitivity toward some analytes, such as unsaturated and functionalized hydrocarbons, peptides, crown ethers, and polymers, and can be monitored using gas chromatography–MS experiments (GC–MS). There are two methods to interpret the mass spectra of organometallic compounds, which are as follows:

(i) Valance change concept that accounts for predicting the type of fragmentation for a variety of s compounds and metal chelates.
(ii) Charge localization concept is dominant in the interpretation of π-compounds of transition metals.

12.2.6 Elemental Analysis

Elemental analysis (EA) had an eminent importance in the twentieth century when techniques such as NMR, MS, and SC-X-ray crystallography were less routinely available. Still, EA is crucial in determining the elemental composition of organometallic compounds, it is one of the best available indicators of purity, specifically for compounds or ligands that are not susceptible to high-performance liquid chromatography (HPLC) analysis [56]. In organometallic chemistry, the high sensitivity of metal–carbon bonds often challenges EA. EA, in general, refers to CHNX analysis, which means the determination of common elements in compounds such as carbon, hydrogen, nitrogen, and heteroatoms such as halogens and sulfur. This is simultaneously a qualitative and a quantitative method with the destruction of analyses due to combustion in the process. EA is conducted independent of structural features or functional group characteristics, for example, the determination of carbon content does not provide information regarding their hybridization or other structural parameters. Nevertheless, EA allows conclusions about the elements present and the purity of compounds. Since this technique has major challenges and limitations, reports have shown about 5–10% of the publications show EA data values deviate ≤0.10% from the theoretical calculation [57]. However, EA is still an essential method to guarantee sufficient purity of novel compounds according to most of all journals.

12.3 Unconventional Methods

Unconventional methods of characterizing organometallic compounds involve specific or additional techniques to identify physicochemical parameters of compound structures, on the other hand, these techniques are less widely used until they are relevant to a specific class of compounds or measurement.

12.3.1 Neutron Activation Analysis

Neutron activation analysis (NAA) technique characterizes the gamma rays emitted from organometallic compounds irradiated with neutrons. This nondestructive technique can quantify elements by comparing the rate of gamma rays emitted because it is directly proportional to the concentration of that element [58]. This technique has very high sensitivity and can analyze around 70 elements with very low detection limits, 0.03 ng to 4 μg. Gamma rays emitted are distinct for a specific nuclide, and this can be detected using a high-resolution semiconductor detector [59, 60].

NAA analysis is utilized to understand the fate of the reactions, state of the metal–ligand bond after reaction, hydride on the metals after hydrogenation, and detection of elements [10, 61, 62]. Holmium-166 (^{166}Ho), technetium-99m (^{99m}Tc), platinum-193m (^{193m}Pt), platinum-195m (^{195m}Pt), and indium-111 (^{111}In) are often used to analyze using NAA. The cytotoxicity study on In(III) compounds with *N*(4)-*ortho*-chlorophenyl-2-acetylpyridine thiosemicarbazone and *N*(4)-*para*-fluorophenyl-2-acetylpyridine thiosemicarbazone on Michigan Cancer Foundation (MCF-7) (human breast carcinoma) cells generated by neutron activation helps to identify the therapeutic action of ^{114m}In compounds [62]. Reports have also shown that the cold neutron prompt gamma activation analysis (PGAA) technique is used to determination of the mass of hydrogen in Mg hydride thin films [10]. NAA analysis on a series of tungsten trifluorophosphine carbonyls, ^{187}W, was conducted to understand their chemical behavior in gas and condensed phase [61].

12.3.2 X-Ray Photoelectron Spectroscopy

XPS is a photoelectric technique used to characterize the atomic composition and chemical environment of surfaces. In principle, X-rays are used to excite the core electrons; the kinetic energy of their photoemission will be characteristic of the electron binding energy and are determined using the photoelectric effect equation (Figure12.5a). Elements except hydrogen and helium can be characterized using

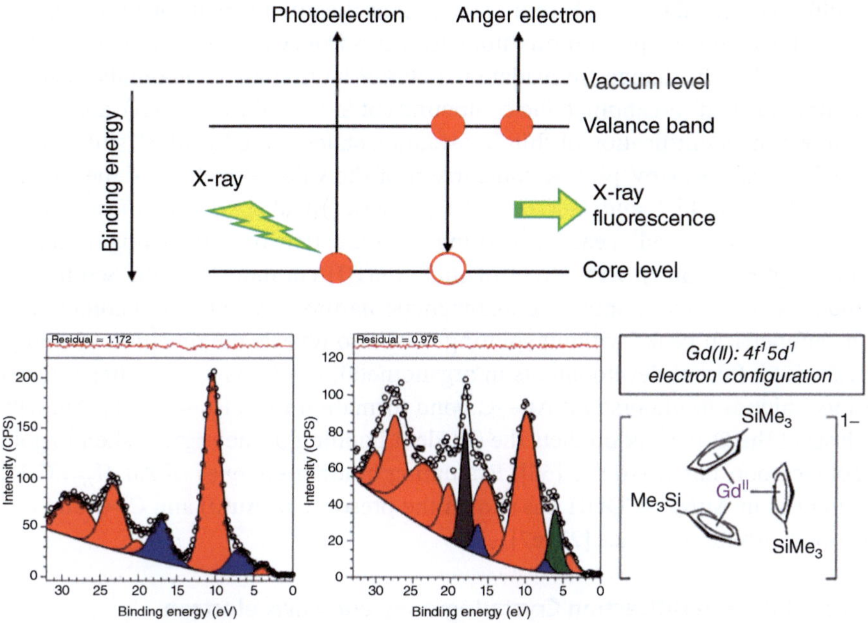

Figure 12.5 (a) Schematic representation of photoemission and relaxation process, (b, c) Valence XPS region of $(C_5H_4SiMe_3)_3Gd(III)$ (c) and [K(crypt)][$(C_5H_4SiMe_3)_3Gd(II)$] (d), red peaks indicate valence orbitals associated with Gd, the blue peaks indicate orbitals of Cp ligand, gray peak indicates the K 2p orbital, and the green peak is associated to the Gd 5d orbital. Source: Reproduced with permission from Huh et al. [63]/American Chemical Society.

XPS. A normal XPS spectrum is a plot of electrons detected to its binding energy. Elements have their own characteristic XPS peaks corresponding to their electronic configuration.

XPS can give highly sensitive chemical state information in organometallic compounds, such as the bonding environment, formal oxidation states, hybridizations, and neighboring atoms [11]. Various XPS studies of organometallic compounds have given considerable insight into the origins of different metal binding energies and their shifts upon environmental changes [64]. For instance, carbon atoms present in the ligand can be distinguished according to their neighboring atoms, such as oxygen. So, a C–O will give a different BE shift than a C–C or O–C–O in C1s spectra. Identification of elements of a compound is in most cases routine and easy unless peaks are overlapping. This characterization allowed by XPS is to understand or differentiate structural similarities between homo/hetero-metallic compounds or homothallic with metal in different oxidation states. A study on different Cu compounds, such as copper(II) phthalocyanine and [Cu(S)-3-aminohexahydroazepine$)_2$]$(BF_4)_2$, reveals that the Cu2p satellite peak intensity increases with the covalent character of the Cu-ligand bond [11]. This arises due to the ligand–metal charge transfer effects that change the electron configuration of exited state giving rise to satellite peaks. Studies of different metal phosphides revealed that the P 2p BE increases as the difference in electronegativity between the metal and phosphorus atom decreases from Cr–P to Co–P [12]. XPS can be used to identify changes in oxidation states of lanthanide compounds. A study on lanthanide (Ln) organometallic compounds of Sm(II), Eu(II), Gd(II), and Tb(II) has shown the identification of shift in oxidation states using Ln 4d BE shifts in the range from 0.3 to 1.0 eV [63]. A comparison of the valence regions of the Gd(III) in $(C_5H_4SiMe_3)_3Gd^{III}$ and Gd(II) in $[(C_5H_4SiMe_3)_3Gd^{II}]^{1-}$ compounds allowed the assignment of a $5d^1$ peak concerning the reduction of a $4f^7$ configuration to $4f^7 5d^1$ (Figure 12.5b,c). XPS study of thallium(III) compounds with Schiff base ligands was reported to show the diamagnetic nature of all prepared compounds with symmetrical peaks without splitting. XPS also reveals the direction of charge transfer [65]. Recent developments in organometallic-gold nanostructures utilized XPS techniques to understand Au—C bond formation in C_8F_{17}–4–C_6H_4–Au. XPS analysis of the C 1s peak predicts the covalent bonding of the organic shell to gold in gold–carbon nanoparticles [66]. Evaluating binding energies of Au $4f_{7/2}$, C 1s, and in O 1s in AuNPs–COOH has shown the presence of Au(0) and C=O and OH functional groups existence [21, 67].

12.3.3 Electron Diffraction Crystallography and Microelectron Diffraction

Electron crystallography is a similar technique to X-ray crystallography where the macromolecular crystal scattering beams are utilized to interpret structural features. However, the electron source used in ED interacts with the atoms in the crystal more strongly than X-rays and gives the following advantages [44, 68, 69]:

(i) It can overcome the need for large crystals for successful structural analysis.
(ii) Compounds undergoing nuclear destruction such as boron compounds can be analyzed.
(iii) 2D and 3D imaging are possible.

Earlier, electron crystallographic techniques were used for protein structural analysis. Widom and coworker and Hovmöller et al. introduced the ED technique to analyze inorganic crystals [70, 71]. High-resolution tunneling electron microscope (HRTEM) images of crystal structures were captured and characterized. Reports have shown several studies using a variety of electron diffraction techniques conducted to characterize organometallic compounds, especially for highly reactive and unstable compounds with smaller crystal sizes. MicroED is a 3D microelectron diffraction technique where microcrystals can be analyzed to determine structural features [72]. In this technique, diffraction patterns are collected from submicron-thick crystals using a low-intensity electron beam under cryogenic temperatures. Looking at a crystal at multiple orientations reveals its reciprocal lattice and permits indexing of its structural features. A recent report shows MicroED characterization of several organometallic compounds using a 300 keV TEM with continuous rotation technique and processed using standard crystallographic software. This technique revealed the crystal structures of zirconocene hydride (Schwartz's reagent), dimeric Pd(II) compound, and different

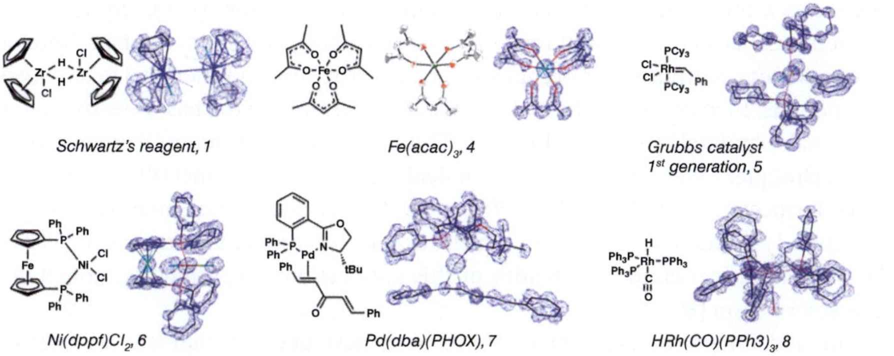

Figure 12.6 Molecular structure and orientations obtained by ambient temperature MicroED method for some organometallic compounds. Source: Reproduced with permission from Jones et al. [73]/American Chemical Society.

paramagnetic and diamagnetic organometallic compounds [73] (Figure 12.6), which cannot be easily obtained using routine NMR or X-ray crystallographic techniques due to insolubility, paramagnetic nature, and instability.

12.3.4 Mössbauer Spectroscopy

Mössbauer spectroscopy is considered nuclear spectroscopy. It was named after Rudolf Mössbauer, a German physicist, who was awarded the Nobel Prize in physics in 1961 [13]. It is mainly used to probe the atomic environment in the solid state. ^{57}Fe Mössbauer spectroscopy technique has been used for the characterization of iron-containing compounds [14–16]. Magnetic properties, electronic structures, and structural investigations have been the most common targeted studies using such spectroscopy techniques. Specifically, the Mössbauer technique can be used to probe the oxidation state of the central iron metal as Fe(0), Fe(II), or Fe(III), spin state, hyperfine magnetic fields, and coordination environment such as octahedral, tetragonal, or trigonal. Mössbauer spectra of iron(II) and iron(III) compounds in the octahedral arrangement were studied as the early classical models of using such technique.

Typical Mössbauer spectra can be interpreted using three parameters: (i) isomer shift (IS), which is the difference in s electron density between the source and the absorber; (ii) quadrupole splitting (QS), which is a shift in nuclear energy levels; and (iii) hyperfine magnetic splitting. Different coordinating ligands were used, such as cyanide, isocyanide, carbonyl, and nitrogen donors [74]. In addition, organometallic iron compounds in the tetrahedral arrangement were investigated [75]. The use of Mössbauer in probing the environment around multinuclear iron compounds such as [8Fe-7S] and nitrogenase models was reported [76, 77].

12.3.5 Electroanalytical Methods

Electrochemical processes are used as an analytical technique to provide qualitative and quantitative information on many organometallic compounds [78]. The responses from cyclic voltammetry (CV), chronoamperometry (CA), chronopotentiometry (CP), electrochemical impedance spectroscopy (EIS), etc. can be calibrated in terms of concentrations or to characterize the structural and chemical reactivity of compounds. Electrostatic potential, charge, and current are the basic electrochemical signals obtained by these techniques. This is also often considered a green analytical technique. Evolution of electrochemical assays on organometallic compounds since ferrocene was published by William E. Geiger and a deep understanding of structure [79] and reactions including electron transfer mechanisms in catalysis by Didier Astruc increased the breadth of this application in specific organometallic characterization [80–82].

Sole and integrated experiments have been developed to characterize organometallic compounds involving redox reactions using electrochemistry. CV of dimeric $[M(DPPMHQ)Cl]_2$ (M = Cu(II), Co(II), and Zn(II)) compounds revealed the existence of two on-electron quasi-reversible peaks corresponding to M(II) → M(III) and M(II) → M(I). Understanding their redox potential differences and distortion in geometry were identified using this experiment [83]. A dirhodium(II) compound, $Rh_2(O_2CCH_3)_2[(C_6H_5)_2P(C_6H_4)]_2$-$2CH_3CO_2H$, was investigated using CV and differential pulse voltammetry in different ligand donor organic solvents such as pyridine, *p*-dithiane, and PMe_2Ph. In all solutions, one-electron oxidation was observed at potentials between +0.25 and +1.04 V vs. Ag–AgCl [84]. CV study for $LAu(SC_6H_4CH_3)$ and $LL[Au(SC_6H_4CH_3)]_2$ [L = PPh_3; LL = dppe, dppp, dppb, dpppn] show two irreversible anodic processes in a potential widow of +0.8 V and +1.6 V vs. saturated calomel electrode (SCE) and a constant potential electrolysis study of $Ph_3PAu(SC_6H_4CH_3)$ and comparing their number of electrons have suggested the reactivity by first oxidation and potential oxidation species were identified [85]. The electronic structure of the radical cation formed by one-electron oxidation of a cymantrene compound of N-heterocyclic carbene (NHC) ligand is characterized using CV. Also, monitoring the $E_{1/2}$ shifts for a series of $MnCp(CO)_2L$ compounds with different ligands provide ligand electronic parameters [86]. Recently, MOFs are examined using this technique. MOF-modified electrodes are studied for their conductivity and spin polarization of the metal ions to understand the characteristics of different metal ions of the MOFs [87]. Reports have also shown an electrochemical method coupled with any other characterization technique to identify molecular structure, reactivity, and mechanism. For example, the spectroelectrochemical method coupled with electron paramagnetic resonance (SEC–EPR) was used to monitor *in situ* electrooxidation of pyrazole-based pincer compounds of nickel $[(PCN)Ni^{II}X]$, X = F, Cl, Br, and I [88].

12.4 Conclusion

Characterization of organometallic compounds plays a crucial role in its applications to shed light on their performance and correlate it with physicochemical, morphological, structural, and optical properties. Due to the different types of characterization techniques, it is impossible to cover all of them in a single chapter.

For this reason, we focused to classify these techniques into conventional methods and unconventional methods. In conventional methods, we have discussed routine analysis techniques, NMR characterization of diamagnetic compounds, key functional group identification techniques such as IR and Raman spectroscopy, analysis of electronic transitions through UV–visible spectroscopy, the precise location of atoms in the structure through single-crystal XRD, and molecular mass and EA identification by MS and CHN analysis. Some organometallic compounds exhibit an unprecedented structural feature or may undergo unexpected reactions, such compounds require a detailed characterization to confirm their validity. Here we have discussed unconventional methods of characterizing organometallic compounds involving specific or additional techniques to identify physicochemical parameters. These techniques include NAA, XPS, electron crystallography, Mössbauer spectroscopy, and electrochemical analytical methods. Thus in this chapter, we focused on providing basic information and classical examples of organometallic compound characterizations required to provide sufficient information to establish the identity of new compounds and purity.

References

1 von Philipsborn, W. (1986). Transition metal NMR spectroscopy-a probe into organometallic structure and catalysis. *Pure Appl. Chem.* 58 (4): 513–528.

2 Gudat, D. (2003). Applications of heteronuclear X/Y-correlation spectroscopy in organometallic and organoelement chemistry: recent developments. *Annu. Rep. NMR Spectrosc.* 51: 60–61.

3 Nakamoto, K. (ed.) (2009). *Infrared and Raman Spectra of Inorganic and Coordination Compounds, Part B: Applications in Coordination, Organometallic, and Bioinorganic Chemistry*, 6e. Wiley.

4 Stute, S., Götzke, L., Meyer, D. et al. (2013). Molecular structure, UV/Vis spectra, and cyclic voltammograms of Mn (II), Co (II), and Zn (II) 5, 10, 15, 20-tetraphenyl-21-oxaporphyrins. *Inorg. Chem.* 52 (3): 1515–1524.

5 Lundquist, R.T. and Cais, M. (1962). Ultraviolet spectra of organometallic compounds[1,2]. *J. Org. Chem.* 27 (4): 1167–1172.

6 Xie, M., Yang, X., Wang, Q., and Zhi, J. (1996). Studies on organometallic compounds by FT-IR. *Spectrosc. Lett.* 29 (1): 53–60.

7 Zagorevskii, D.V. (1999). Organometallics Studied Using Mass Spectrometry.

8 Culmo, R.F. and Shelton, C.T. (2013). The elemental analysis of various classes of chemical compounds using CHN. *Appl. note. Waltham, MA Perkin Elmer*.

9 Pregosin, P.S. (2013). *NMR in Organometallic Chemistry*. Wiley.

10 Cao, L.R., Hattrick-Simpers, J.R., Bindel, R. et al. (2010). Combinatorial study of thin film metal hydride by prompt gamma activation analysis. *J. Radioanal. Nucl. Chem.* 283 (1): 63–68.

11 Wilson, K. and Adam, L. (2010). Applications of XPS to the study of inorganic compounds. *Spectrosc. Prop. Inorg. Organomet. Compd. RSC Publ.* London: 72–86.

12 Grosvenor, A.P., Wik, S.D., Cavell, R.G., and Mar, A. (2005). Examination of the bonding in binary transition-metal monophosphides MP (M= Cr, Mn, Fe, Co) by X-ray photoelectron spectroscopy. *Inorg. Chem.* 44 (24): 8988–8998.

13 Mössbauer, R.L. (1962). Recoilless nuclear resonance absorption of gamma radiation: a new principle yields gamma lines of extreme narrowness for measurements of unprecedented accuracy. *Science* 137 (3532): 731–738.

14 Collins, R.L. and Travis, J.C. (1967). The electric field gradient tensor. In: *Mössbauer Effect Methodology* (ed. I.J. Gruverman), 123–161. Springer.

15 Fujita, F.E. (1975). *Mössbauer Spectroscopy*. Springer.

16 Long, G.J. (1984). Basic concepts of Mössbauer spectroscopy. In: *Mössbauer Spectroscopy Applied to Inorganic Chemistry* (ed. G.J. Long), 7–26. Springer.

17 Anbuselvi, D. (2020). The growth and characterization techniques of organometallic nonlinear optical crystal–manganese mercury thiocyanate–Bis (N–methyl formamide). *Mater. Today Proc.* 33: 4101–4104.

18 Zhang, J.-P., Liao, P.-Q., Zhou, H.-L. et al. (2014). Single-crystal X-ray diffraction studies on structural transformations of porous coordination polymers. *Chem. Soc. Rev.* 43 (16): 5789–5814.

19 Chaudhary, A., Mohammad, A., and Mobin, S.M. (2017). Recent advances in single-crystal-to-single-crystal transformation at the discrete molecular level. *Cryst. Growth Des.* 17 (5): 2893–2910.

20 Jeschke, G. (2005). EPR techniques for studying radical enzymes. *Biochim. Biophys. Acta* 1707 (1): 91–102.

21 Ahmad, A.A.L., Marutheri Parambath, J.B., Postnikov, P.S. et al. (2021). Conceptual developments of aryldiazonium salts as modifiers for gold colloids and surfaces. *Langmuir* 37 (30): 8897–8907.

22 Nikitin, K. and O'Gara, R. (2019). Mechanisms and beyond: elucidation of fluxional dynamics by exchange NMR spectroscopy. *Chem. Eur. J.* 25 (18): 4551–4589.

23 Plečnik, C.E., Liu, S., Chen, X. et al. (2004). Lanthanide–transition-metal carbonyl complexes: new $[Co_4 (CO)_{11}]$ 2-clusters and lanthanide (II) isocarbonyl polymeric arrays. *J. Am. Chem. Soc.* 126 (1): 204–213.

24 Overton, A.T., López-de-Luzuriaga, J.M., Olmos, M.E., and Mohamed, A.A. (2012). Oxidative rearrangement in gold organometallics. *Organometallics* 31 (9): 3460–3462.

25 Koy, M., Bellotti, P., Das, M., and Glorius, F. (2021). N-Heterocyclic carbenes as tunable ligands for catalytic metal surfaces. *Nat. Catal.* 4 (5): 352–363.

26 Müller, A., Jostes, R., and Cotton, F.A. (1980). Trinuclear clusters of the early transition elements. *Angew. Chem. Int. Ed. Engl.* 19 (11): 875–882.

27 Lin, L., Romano, C., and Mazet, C. (2016). Palladium-catalyzed long-range deconjugative isomerization of highly substituted α, β-unsaturated carbonyl compounds. *J. Am. Chem. Soc.* 138 (32): 10344–10350.

28 Liu, Z., Deeth, R.J., Butler, J.S. et al. (2013). Reduction of quinones by NADH catalyzed by organoiridium complexes. *Angew. Chem. Int. Ed.* 52 (15): 4194–4197.

29 Castro, A.C., Balcells, D., Repisky, M. et al. (2020). First-principles calculation of 1H NMR chemical shifts of complex metal polyhydrides: the essential inclusion of relativity and dynamics. *Inorg. Chem.* 59 (23): 17509–17518.

30 Hierso, J.-C., Armspach, D., and Matt, D. (2009). "Through-space" nuclear spin–spin couplings in ferrocenyl polyphosphanes and diphosphino cavitands: A new way of gathering structural information in constrained P (III) ligands by NMR. *C.R. Chim.* 12 (9): 1002–1013.

31 Mosher, M.D. and Ojha, S. (1998). Hybridization and structural properties: a physical organic chemistry experiment. *J. Chem. Educ.* 75 (7): 888.

32 Blanc, F., Copéret, C., Lesage, A., and Emsley, L. (2008). High resolution solid state NMR spectroscopy in surface organometallic chemistry: access to molecular understanding of active sites of well-defined heterogeneous catalysts. *Chem. Soc. Rev.* 37 (3): 518–526.

33 Choi, Y.S. (1994) Studies in fluorine chemistry: 13C NMR investigation of SF5/SO2F fluorinated systems.

34 Meek, D.W. and Mazanec, T.J. (1981). Determination of structural and dynamic aspects of organometallic and coordination chemistry by phosphorus-31 NMR spectroscopy. *Acc. Chem. Res.* 14 (9): 266–274.

35 Topcu, G. and Ulubelen, A. (2007). Structure elucidation of organic compounds from natural sources using 1D and 2D NMR techniques. *J. Mol. Struct.* 834: 57–73.

36 Reven, L. (1994). Solid-state NMR studies of supported organometallics. *J. Mol. Catal.* 86 (1–3): 447–477.

37 Grekov, D., Vancompernolle, T., Taoufik, M. et al. (2018). Solid-state NMR of quadrupolar nuclei for investigations into supported organometallic catalysts: scope and frontiers. *Chem. Soc. Rev.* 47 (8): 2572–2590.

38 Bélanger-Desmarais, N., Schütz, M., and Reber, C. (2019). Remarkably intricate raman spectra of platinum (II)–ligand skeletal modes in diamminedihalido complexes. *J. Phys. Chem. A* 123 (26): 5574–5579.

39 Patmore, D.J. and Graham, W.A.G. (1967). Organometallic compounds with metal-metal bonds. VII. Preparation and a study of the infrared spectra of mono-and bis (tetracarbonylcobalt) derivatives of germanium and tin. *Inorg. Chem.* 6 (5): 981–988.

40 Low, P.J. and Bock, S. (2013). Spectroelectrochemistry: a valuable tool for the study of organometallic-alkyne,-vinylidene,-cumulene,-alkynyl and related complexes. *Electrochim. Acta* 110: 681–692.

41 Förster, H. (2004). UV/vis spectroscopy. *Charact. I* 337–426.

42 Sinha, B., Bhattacharya, M., and Saha, S. (2019). Transition metal complexes obtained from an ionic liquid-supported Schiff base: synthesis, physicochemical characterization and exploration of antimicrobial activities. *J. Chem. Sci.* 131 (3): 1–10.

43 Nelson, D.J. and Percy, J.M. (2014). The influence of structure on reactivity in alkene metathesis. In: *Advances in Physical Organic Chemistry*, vol. 48 (ed. I.H. Williams and N.H. Williams), 81–188. Elsevier.

44 Li, J. and Sun, J. (2017). Application of X-ray diffraction and electron crystallography for solving complex structure problems. *Acc. Chem. Res.* 50 (11): 2737–2745.

45 Bragg, W.L. (1913). The structure of some crystals as indicated by their diffraction of X-rays. *Proc. R. Soc. London. Ser. A* 89 (610): 248–277.

46 Bashir, S., and Liu, J. (2015) Nanocharacterization, advanced nanomaterial and their applications in renewable energy.

47 Raboui, H., Lough, A.J., Plint, T., and Bender, T.P. (2018). Position of methyl and nitrogen on axial aryloxy substituents determines the crystal structure of silicon phthalocyanines. *Cryst. Growth Des.* 18 (5): 3193–3201.

48 Tang, J. and Zhao, L. (2022). Structural control and chiroptical response in intrinsically tetra-and pentanuclear chiral gold clusters. *Inorg. Chem.* 61 (11): 4541–4549.

49 Wijesinghe, K.H., Sakthivel, N.A., Jones, T., and Dass, A. (2020). Crystal structure of Au_{30-x} Ag x (S-t Bu) 18 and effect of the ligand on Ag alloying in gold nanomolecules. *J. Phys. Chem. Lett.* 11 (15): 6312–6319.

50 Subashini, A., Rajarajan, K., Sagadevan, S. et al. (2020). Enhanced properties of cadmium mercury thiocyanate bis (N-methyl formamide): a promising non-linear optical crystal. *Chin. J. Phys.* 67: 52–62.

51 Wei, R., Huo, Q., Tao, J. et al. (2011). Spin-crossover FeII4 squares: two-step complete spin transition and reversible single-crystal-to-single-crystal transformation. *Angew. Chem. Int. Ed.* 50 (38): 8940–8943.

52 Abdou, H.E., Mohamed, A.A., and Fackler, J.P. (2005). Synthesis and X-ray structures of dinuclear and trinuclear gold (I) and dinuclear gold (II) amidinate complexes. *Inorg. Chem.* 44 (2): 166–168.

53 Mohamed, A.A., Galassi, R., Papa, F. et al. (2006). Gold (I) and silver (I) mixed-metal trinuclear complexes: dimeric products from the reaction of gold (I) carbeniates or benzylimidazolates with silver (I) 3, 5-diphenylpyrazolate. *Inorg. Chem.* 45 (19): 7770–7776.

54 Traeger, J.C. (2000). Electrospray mass spectrometry of organometallic compounds. *Int. J. Mass spectrom.* 200 (1–3): 387–401.

55 Farrer, N.J. and McIndoe, J.S. (2007) MS of organometallics (for Volume 6 of EMS), 903–915.

56 Pregl, F. (1917). *Quantitative Micro-analysis of Organic Substances*. Berlin: Springer.

57 Kandioller, W., Theiner, J., Keppler, B.K., and Kowol, C.R. (2022). Elemental analysis: an important purity control but prone to manipulations. *Inorg. Chem. Front.* 9: 412–416.

58 Witkowska, E., Szczepaniak, K., and Biziuk, M. (2005). Some applications of neutron activation analysis. *J. Radioanal. Nucl. Chem.* 265 (1): 141–150.

59 Ehmann, W.D. and Vance, D.E. (1989). Advances in neutron activation analysis. *Crit. Rev. Anal. Chem.* 20 (6): 405–443.

60 Yakovlev, Y.V. and Kolotov, V.P. (1981). Neutron-activation determination of phosphorus using extraction separation of phosphate-ion with dialkylti dinitrates. *Zhurnal. Anal. Khimii* 36 (8): 1534–1539.

61 Bottomley, L.D. and Clark, R.J. (1988). Chemical behavior of tungsten trifluorophosphines following neutron activation. *Radiochim. Acta* 43 (4): 197–204.

62 Oliveira, A.A., Franco, L.L., dos Santos, R.G. et al. (2017). Neutron activation of In (III) complexes with thiosemicarbazones leads to the production of potential radiopharmaceuticals for the treatment of breast cancer. *New J. Chem.* 41 (17): 9041–9050.

63 Huh, D.N., Bruce, J.P., Ganesh Balasubramani, S. et al. (2021). High-Resolution X-ray Photoelectron Spectroscopy of Organometallic (C5H4SiMe3) 3LnIII and [(C5H4SiMe3) 3LnII] 1–Complexes (Ln= Sm, Eu, Gd, Tb). *J. Am. Chem. Soc.* 143 (40): 16610–16620.

64 Feltham, R.D. and Brant, P. (1982). XPS studies of core binding energies in transition metal complexes. 2. Ligand group shifts. *J. Am. Chem. Soc.* 104 (3): 641–645.

65 Gaur, G.K. and Srivastava, S. (2012). Characterization, XPS and Toxicological Study of Organothallium (III) Compounds with Schiff Base Ligands. *Cryst. Struct. Theory Appl.* 1 (3): 97–99.

66 Orefuwa, S.A., Ravanbakhsh, M., Neal, S.N. et al. (2014). Robust organometallic gold nanoparticles. *Organometallics* 33 (2): 439–442.

67 Ahmad, A.A.L., Panicker, S., Chehimi, M.M. et al. (2019). Synthesis of water-soluble gold-aryl nanoparticles with distinct catalytic performance in the reduction of the environmental pollutant 4-nitrophenol. *Catal. Sci. Technol.* 9 (21): 6059–6071.

68 Wan, W., Sun, J., Su, J. et al. (2013). Three-dimensional rotation electron diffraction: software RED for automated data collection and data processing. *J. Appl. Crystallogr.* 46 (6): 1863–1873.

69 Shi, D., Nannenga, B.L., Iadanza, M.G. et al. (2013). Three-dimensional electron crystallography of protein microcrystals. *Elife* 2: e01345.

70 Widom, J. and Klug, A. (1985). Structure of the 3000Å chromatin filament: X-ray diffraction from oriented samples. *Cell* 43 (1): 207–213.

71 Hovmöller, S., Sjögren, A., Farrants, G. et al. (1984). Accurate atomic positions from electron microscopy. *Nature* 311 (5983): 238–241.

72 Hattne, J., Reyes, F.E., Nannenga, B.L. et al. (2015). MicroED data collection and processing. *Acta Crystallogr. Sect. A Found. Adv.* 71 (4): 353–360.

73 Jones, C.G., Asay, M., Kim, L.J. et al. (2019). Characterization of reactive organometallic species via MicroED. *ACS Cent. Sci.* 5 (9): 1507–1513.

74 Collins, R.L., Pettit, R., and Baker, W.A. Jr., (1966). Mössbauer studies of iron organometallic complexes—III: Octahedral complexes. *J. Inorg. Nucl. Chem.* 28 (4): 1001–1010.

75 Mazak, R.A. and Collins, R.L. (1969). Mössbauer Studies of Iron Organometallic Complexes. VII. The Iron (− II) Tetrahedral Compounds. *J. Chem. Phys.* 51 (8): 3220–3225.

76 Shoji, M., Saito, T., Takeda, R. et al. (2007). Assignments of the Mössbauer spectra of an inorganic [8Fe–7S] complex based on the first-principle calculations. *Chem. Phys. Lett.* 446 (1–3): 228–232.

77 Van Stappen, C., Decamps, L., Cutsail, G.E. III, et al. (2020). The spectroscopy of nitrogenases. *Chem. Rev.* 120 (12): 5005–5081.

78 Roullier, L., Lucas, D., Mugnier, Y. et al. (1991). Electrochemical studies on organometallic compounds: XXXIX. Electrochemical bahavior of bis-silylated cyclopentadienyl niobium trihydride. ESR spectroscopic characterization of the dimeric hydride niobium (IV) complex [Nb (η5-$Me_3SiC_5H_4$) 2H]$_2$ (μ-H)$_2$. *J. Organomet. Chem.* 412 (3): 353–362.

79 Geiger, W.E. (2007). Organometallic electrochemistry: origins, development, and future. *Organometallics* 26 (24): 5738–5765.

80 Albach, R.W., Küsthardt, U., Behm, J. et al. (1993). Electrochemical and chemical generation of a mixed-valent organorhenium oxide and its subsequent aggregation. *J. Organomet. Chem.* 450 (1–2): 165–170.

81 Gatard, S., Deraedt, C., Rapakousiou, A. et al. (2015). New polysilyl dendritic precursors of triazolylferrocenyl and triazolylcobalticenium dendrimers–comparative electrochemical study and stabilization of small, active Pd nanoparticles. *Organometallics* 34 (9): 1643–1650.

82 Ruiz, J., Roman, E., and Astruc, D. (1987). Photochemical synthesis, reactivity, and electrochemical properties of [CpFeP2L]+ cations;(Cp= C5H5, C5Me5; L= Co vs. CH3CN; P= phosphine or phosphite). *J. Organomet. Chem.* 322 (1): C13–C15.

83 Ammar, R.A., Alaghaz, A.M.A., and Alturiqi, A.S. (2018). New dimeric Schiff base quinoline complexes: synthesis, spectral characterization, electrochemistry and cytotoxicity. *Appl. Organomet. Chem.* 32 (6): e4361.

84 Chakravarty, A.R., Cotton, F.A., Tocher, D.A., and Tocher, J.H. (1985). Structural and electrochemical characterization of the novel ortho-metalated dirhodium (II) compounds Rh2 (O_2CMe)$_2$ [Ph_2P (C_6H_4)]$_2$. 2L. *Organometallics* 4 (1): 8–13.

85 Chen, J., Jiang, T., Wei, G. et al. (1999). Electrochemical and chemical oxidation of gold (I) thiolate phosphine complexes: formation of gold clusters and disulfide. *J. Am. Chem. Soc.* 121 (39): 9225–9226.

86 Wu, K., Conger, M.A., Waterman, R. et al. (2019). Electrochemical and structural characterization of a radical cation formed by one-electron oxidation of a cymantrene complex containing an N-heterocyclic carbene ligand. *Polyhedron* 157: 442–448.

87 Hsieh, Y.-T., Huang, S.-C., Lu, S.-I. et al. (2021). Electrochemical characterization of and theoretical insight into a series of 2D MOFs, [M (bipy)(C_4O_4)(H_2O)$_2$]· $3H_2O$ (M= Mn (1), Fe (2), Co (3) and Zn (4)), for chemical sensing applications. *RSC Adv.* 11 (43): 26516–26522.

88 Gafurov, Z.N., Kantyukov, A.O., Kagilev, A.A. et al. (2021). Electrochemical methods for synthesis and in situ generation of organometallic compounds. *Coord. Chem. Rev.* 442: 213986.

13

Organometallic Reagents

Ranjan K. Basak[1] *and Ashish K. Asatkar*[2]

[1] *Siddharth University, Shivharsh Kisan P.G. College, Basti, Department of Chemistry, Kapilavastu, Siddharth Nagar, Basti, Uttar Pradesh 272001, India*
[2] *Shaheed Mahendra Karma Vishwavidyalaya, Govt. Gundadhur P.G. College, Kondagaon, Department of Chemistry, Bastar, Jagdalpur, Kondagaon, Chhattisgarh 494226, India*

13.1 Organoboron Reagents

13.1.1 Introduction

Boron, being the fifth element of the periodic table, belongs to the second period and 13th group. Its outermost shell has three electrons, so it needs five electrons more in order to complete its octet and achieve a noble gas configuration. Alternatively, it can lose three electrons to achieve a noble gas configuration. Being very small in size, its ionization potential is very high. So it's easier for boron to gain electrons to fulfill its octet. In order to do so, it forms covalent bonds with three of its electrons to form three bonds. Even though in its bonded state boron does not achieve the noble gas configuration and thus the trivalent compounds of boron behave as electron deficient species, i.e. electrophilic species. So much so that the lowest borane, i.e. BH_3 in a gaseous state is dimeric in nature. The electrophilic boron compounds may gain two electron from a nucleophile and achieve the noble gas configuration. But in this tetravalent state, it bears a negative charge. The size of the boron being so small that the negative charge does not get stabilized at the boron and the tetravalent boron species behave as nucleophilic entities. Thus, the electrophilicity of the trivalent boron compounds and the nucleophilicity of the tetravalent boron species defines and determines the chemistry of boron compounds.

13.1.2 Main Organoboron Reagents and Their Reactions

13.1.2.1 Hydroboration and Subsequent Reactions

Hydroboration of an unsaturated bond (i.e double bond and triple bond) is followed by oxidation, hydrogenation, halogenation, sulfidation, carbonylation and amination to furnish alcohols, aldehydes, ketones, alkenes, alkynes, halogenated compounds, thioethers, carbonyl compounds, and amines, respectively. Hydroboration is the addition of hydroborane and organohydroborane to a double

Organometallic Compounds: Synthesis, Reactions, and Applications, First Edition.
Edited by Dakeshwar Kumar Verma and Jeenat Aslam.

or triple bond yielding more substituted organoborane (Scheme 13.1).

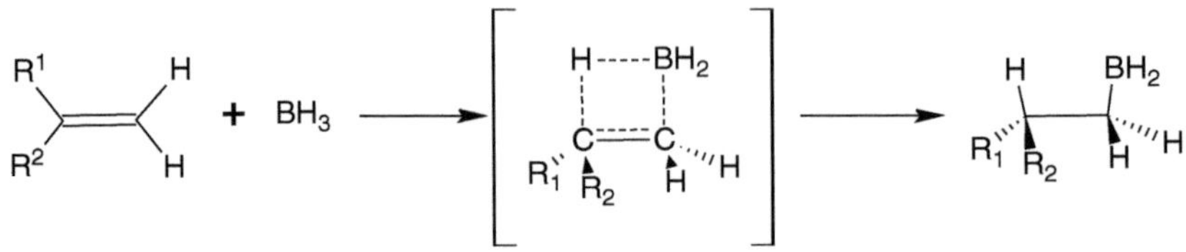

Scheme 13.1 Representative hydroboration reaction.

Hydroboration is a highly regiospecific reaction where the addition of boron moiety occurs on the less substituted carbon while the hydrogen gets added on to the more substituted carbon. The reaction is highly stereospecific as the only syn-addition of hydroborane occurs on either a double or a triple bond. The regioselectivity increases with an increase in the steric demand of the borane reagent and also at the unsaturated bond.

13.1.2.1.1 Hydroboration–Oxidation

Hydroboration–oxidation yields alcohols and aldehydes/ketones from unsaturated compound. The organoborane resulting from the hydroboration reaction undergoes a number of reactions yielding a wide variety of products [1]. The alkylboron furnishes alcohols on treatment with oxidizing agents. The reaction of alkylboranes with H_2O_2 followed by treatment with an alkaline solution of NaOH (Scheme 13.2),

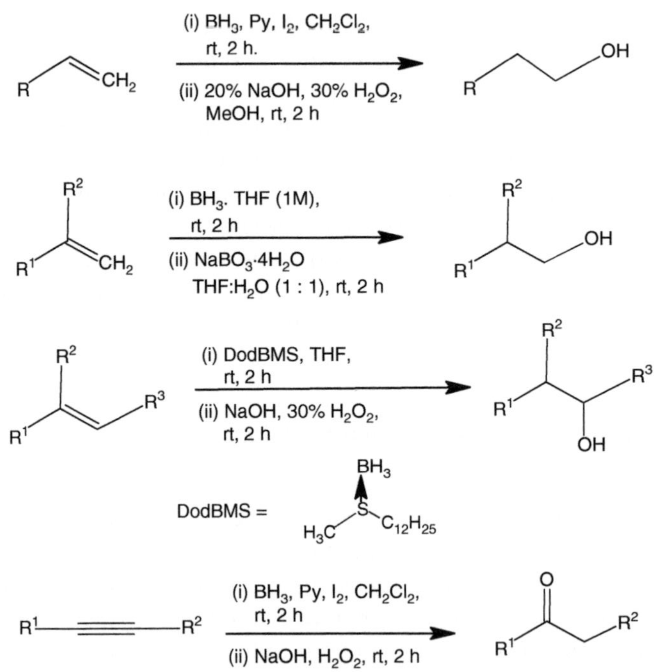

Scheme 13.2 Hydroboration–oxidation.

or NaOAc, or NaH_2PO_4/K_2HPO_4 yields alcohols [2, 3]. The hydroboration–oxidation of a terminal double bond yields primary alcohols, while internal double bond yields secondary alcohols [4].

As the hydroboration occurs at the less substituted carbon, the resulting alcohols obtained are anti-Markovnikov products. Reaction of organoboranes with 2 mol of O_2 yields alcohols (Scheme 13.3). Treatment with 1.5 mol of O_2 per mole of organoboranes followed by the addition of an alkaline solution of H_2O_2 yields corresponding alcohols in very good yields [4d,4e].

1. $R_3B + O_2 \longrightarrow R_2BO_2R$
2. $R_2BO_2R + O_2 \longrightarrow RB(O_2R)_2$
3. $RB(O_2R)_2 + H_2O_2 + 2H_2O \longrightarrow ROH + 2ROOH + B(OH)_3$

Scheme 13.3 Oxidation of organoboranes by O_2.

Hydroboration followed by oxidation of a triple bond leads to aldehydes and ketones. Alkenylborane resulting from addition of hydroborane to terminal alkynyl yields aldehyde, while hydroboration of internal alkyne yields ketone upon treatment with H_2O_2/OH^-. The formation of all the alcohols, ketones, and aldehydes resulting from hydroboration–oxidation reactions occurs with complete retention of configuration. Treatment of alkenes and alkynes with boranes results in hydroboration, yielding organoborane (Scheme 13.4). The organoboranes on

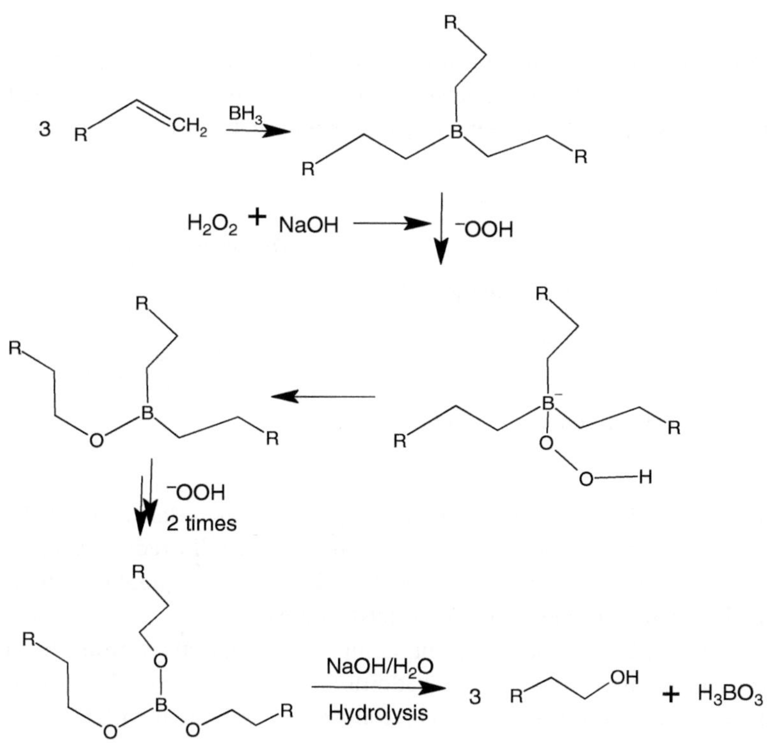

Scheme 13.4 Mechanism of oxidation reaction.

treatment with H_2O_2 in alkaline media (HOO^-) attacks the electrophilic boron center. The resulting borate anion transfers one of the alkyl groups to the oxygen directly attached to the boron. The step repeats twice to yield trialkoxyborane. The nucleophilic attack of the hydroxide ion on the alkoxyborane generates trialkoxyborate anion. The borate anion on hydrolysis gives three moles of alcohol per mole of alkoxyborane.

13.1.2.1.2 Carbonylation

The organoboranes upon treatment with carbon monoxide followed by reduction with trimethoxymethoxy lithium aluminum hydride followed by hydrolysis in and hydrolyzed in alkaline media yields homologated primary alcohols (Scheme 13.5). Reaction of organoborane with H_2O_2/NaOH in a carbon monoxide atmosphere yields homologated tertiary alcohols.

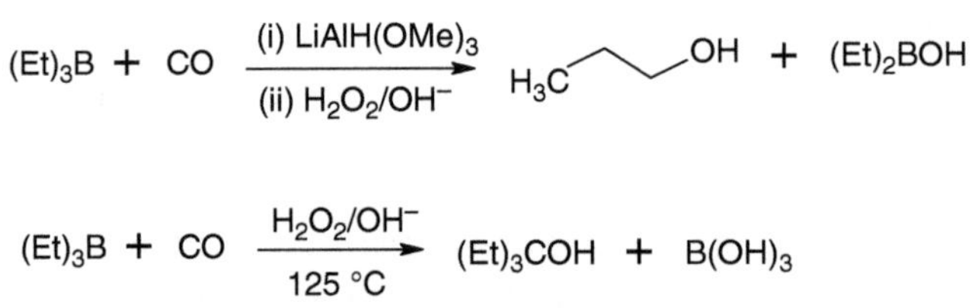

Scheme 13.5 Carbonylation of organoboranes.

Trialkylboranes on reaction with CO in presence of a mild reducing agent followed by oxidation yields aldehyde. However, when the same reaction is hydrolyzed in alkaline media instead of oxidation, it yields primary homologated alcohol (Scheme 13.6).

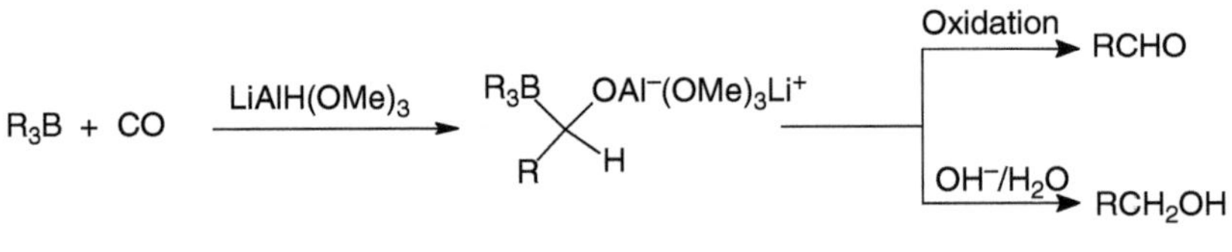

Scheme 13.6 Homologated alcohols and aldehydes.

Similarly, 1-methylcyclopentenone (Scheme 13.7) on hydroboration with 9-BBN followed by carbonylation and subsequent reduction with a mild reducing agent lithium triacetoxyaluminium hydride yields primary alcohol. Subsequent treatment with H_2O_2 in alkaline medium led to tautomerized aldehyde.

In the above carbonylation reactions, migration of only one alkyl group from organoborane occurred after carbonyl group addition. However, if two alkyl groups migrate from the organoborane (Scheme 13.8), followed by treatment with H_2O_2 in alkaline media, results in ketone.

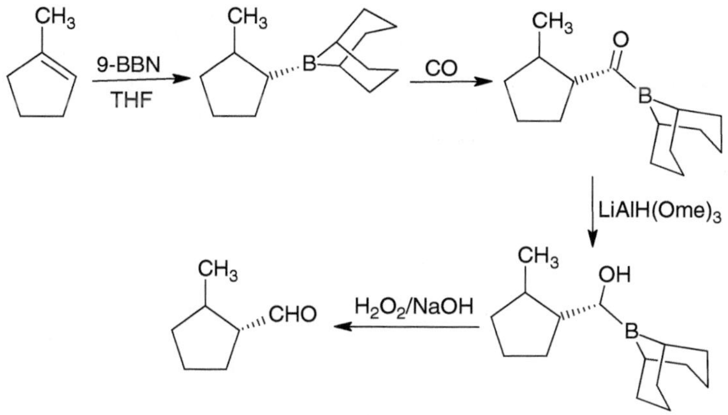

Scheme 13.7 Carbonylation oxidation of cyclic alkene.

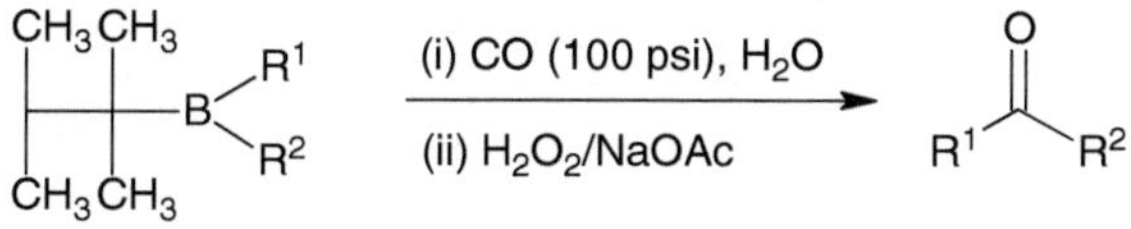

Scheme 13.8 Formation of ketone from carbonylation.

13.1.2.1.3 Hydrogenolysis

Alkylboranes and alkenylboranes are quite resistant to hydrolysis. However, in presence of acetic acid or other carboxylic acids, alkenylboranes and alkylboranes yield alkenes and alkanes, respectively (Scheme 13.9).

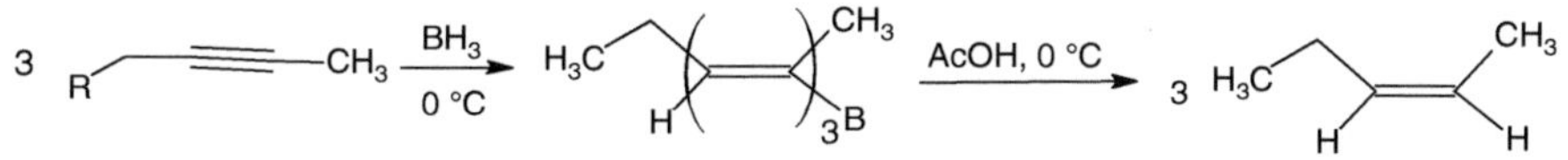

Scheme 13.9 Reduction of organoboranes with carboxylic acid.

The alkenyl and **alkynylboranes** undergo rearrangement if electrophilic reagents are available in order to stabilize the carbanion generated on the double bond due to attack of a nucleophile on the boron atom. Intramolecular transfer of an alkyl group from the tetracoordinated borate anion occurs on the adjacent double bond (Scheme 13.10) [5] because of the available alkyl halide, which stabilizes the anion of the sp^2 carbon of the resulting alkene. The transfer of the alkyl group from the borate anion to the adjacent triple bond occurs due to available R″ electrophile. The transfer of the alkyl group and addition of the electrophile occurs in syn- and anti-fashion on the double bond. However, it has been observed that syn-addition predominates over anti-addition. The migration of the alkyl group is followed by reduction with carboxylic acid, yielding the alkenyls in 90 : 10 ratio. Similarly, in presence of the electrophile α-bromoester (Scheme 13.11), the migration of the alkyl group from the borate ester to the unsaturated bond and addition of the bromoester occurs in anti-fashion rather than syn, as was observed in alkyl halide electrophile case. Further replacement of the boron moiety with hydrogen occurs upon treatment with carboxylic acid.

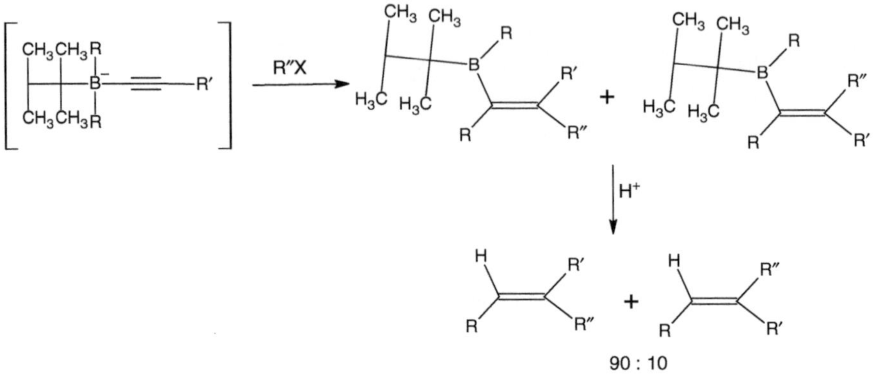

Scheme 13.10 Rearrangement followed by reduction.

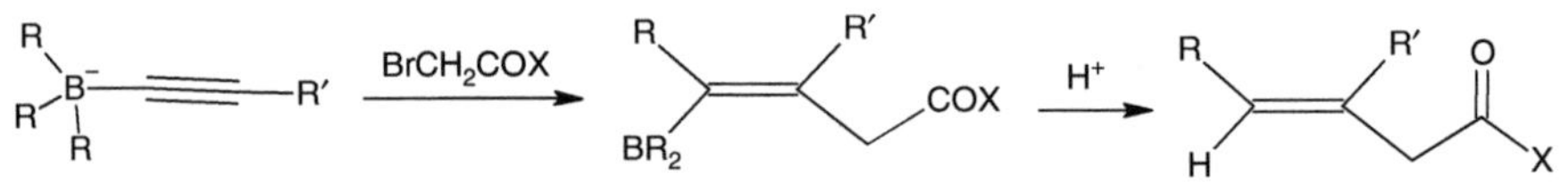

Scheme 13.11 Rearrangement followed by reduction.

13.1.2.1.4 Sulfidation

Thioethers are also obtained from organoboranes (Scheme 13.12) on treatment with an alkene followed by treatment with dimethyl disulfide. The reaction with dimethyl disulfide follows a free radical mechanism.

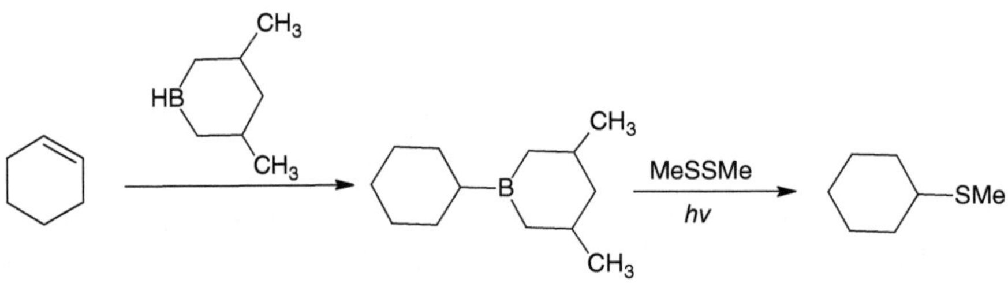

Scheme 13.12 Thioether formation.

13.1.2.1.5 Amination

Alkylamines are prepared from organoboranes by treatment with amines containing good leaving groups (Scheme 13.13) like NH_2Cl, NH_2OSO_3H, 2,4,6-trimethylbenzenesulfonamide, and dimethylchloroamine. The amines develop positive charge upon the treatment with organoborane. One of the alkyl groups migrates from the borate to the nitrogen of the amine with concomitant loss of the leaving group. Hydrolysis of such an adduct yields amine. Treatment of α-pinene with dimethylborane yields trisubstituted organoborane. Treatment of amine containing hydrogen **sulfate** produces an borate amine adduct. Migration of the α-pinene from the borate anion with subsequent hydrolysis yields

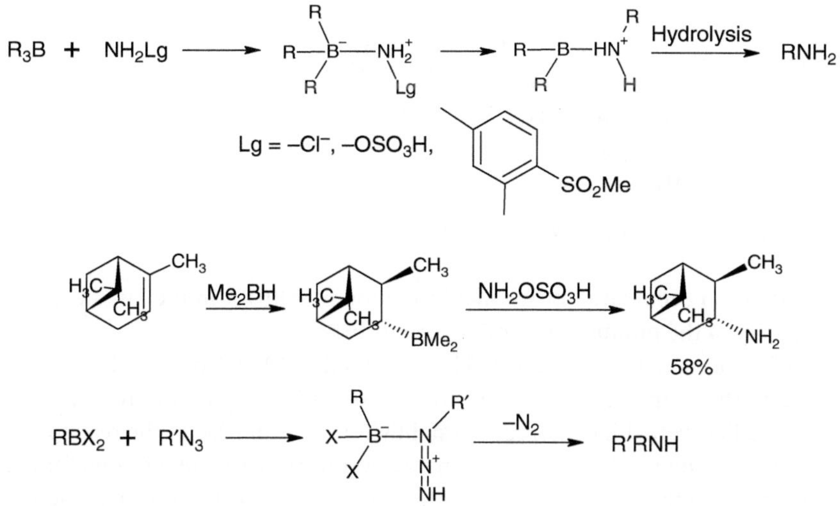

Scheme 13.13 Amination.

3-amino-derivative of the α-pinene. Secondary amines are obtained from organoboranes by treatment with alkyl azide. The amine borane adduct formation is followed by the concomitant migration of the alkyl group to the nitrogen and loss of the N_2 molecule. The rearranged product obtained is a secondary amine. The use of β-iodoazide instead of simple alkyl azide yields aziridine.

13.1.2.1.6 Alkylation of α-Haloesters

Reaction of organoboranes with α-haloesters yields corresponding α-alkylated esters (Scheme 13.14). The nucleophile of the salt of α-haloesters attacks on the electrophilic center of trivalent organoborane. The resulting borate becomes nucleophilic and the molecule loses negative charge by transfer of one of the organic ligands. Hydrolysis of the resulting organoborane yields alkyl substituted ester.

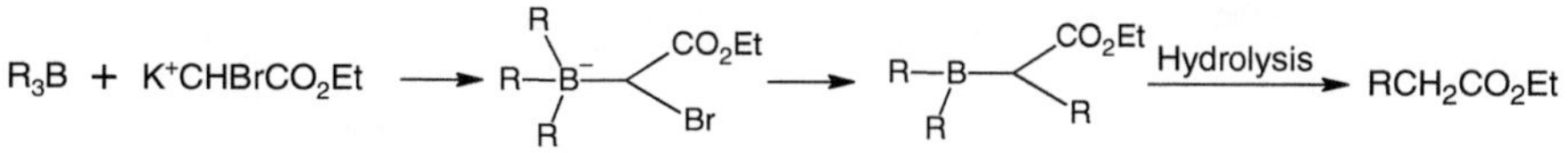

Scheme 13.14 Reaction of organoborane with α-haloester.

13.1.2.2 Suzuki Coupling

C—C bond-forming reaction between aryl, vinyl, and alkyl boronate esters with that of aryl, vinyl, and alkyl halides and pseudohalides using Pd(0) as catalyst is classified as Suzuki coupling [6–8] or Suzuki–Miyaura coupling (Scheme 13.15). The reaction is highly regioselective and stereospecific. The substituted alkene is obtained as a trans product with a new substitution at the less substituted end of the double bond. The regioselectivity is due to the Pd(0) addition at the more substituted end of the double bond and the organic group at the less substituted end. Stereoselectively trans

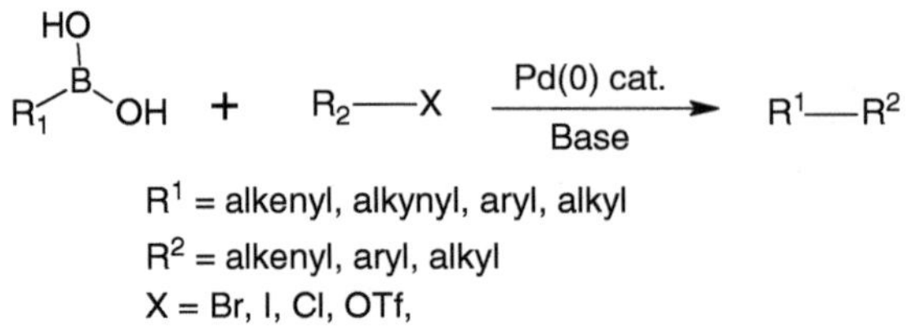

Scheme 13.15 Representative Suzuki reaction.

product formation occurs due to anti-β-elimination of $Pd(0)L_2$ in the last step. The isomeric purity of the products generally exceeds 98%.

The first Suzuki reaction was reported by A. Suzuki in 1979. However, in the subsequent years the scope of the reaction was enlarged by varying the reactants. A number of studies were done in order to establish the role of base in the reaction.

Initially, it was thought that tetracoordinated boron species were required for the transfer of the organic group on the boron to occur. The base used in the reaction gets coordinated to the tricoordinated boron and gave tetracoordinated boronate ion. The boronate ion would deliver the organic group to the palladium species. Finally, reductive elimination of the Pd(0) would deliver the C—C bond. In order to verify the hypothesis, a tetracoordinate boron species was used to perform the reaction (Scheme 13.16), but a poor yield of 9% of the product was obtained. However, no reaction occurred when the tetracoordinated palladium(II) complex was treated with a tricoordinated boronate ester in the absence of a base. However, the reaction of tetracoordinated methoxy palladium(II) complex with tricoordinated boronate ester in the absence of a base furnished product in 79% yield. Thus, it was concluded that the methoxy palladium species was required for the successful reaction to occur and the previous hypothesis of the base attacking the boron species and giving rise to tetracoordinated boronate anion was discarded. The mechanism of the Suzuki reaction and catalytic cycle of palladium accepted presently is given

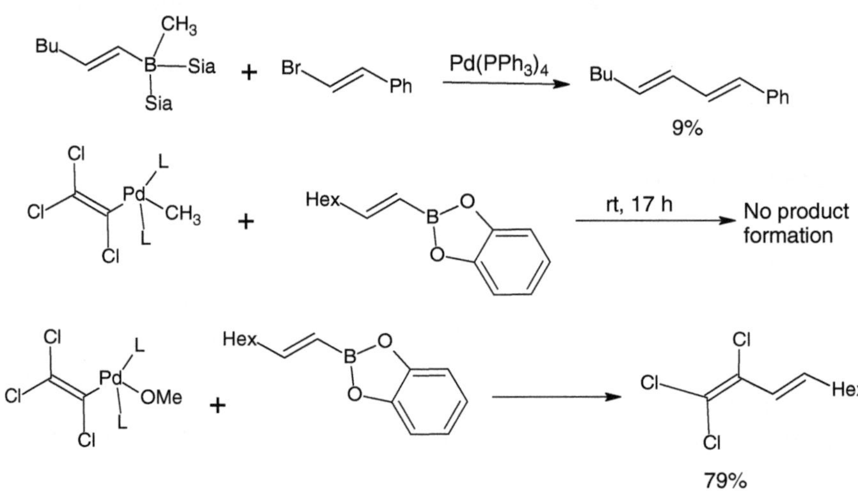

Scheme 13.16 Study of mechanism.

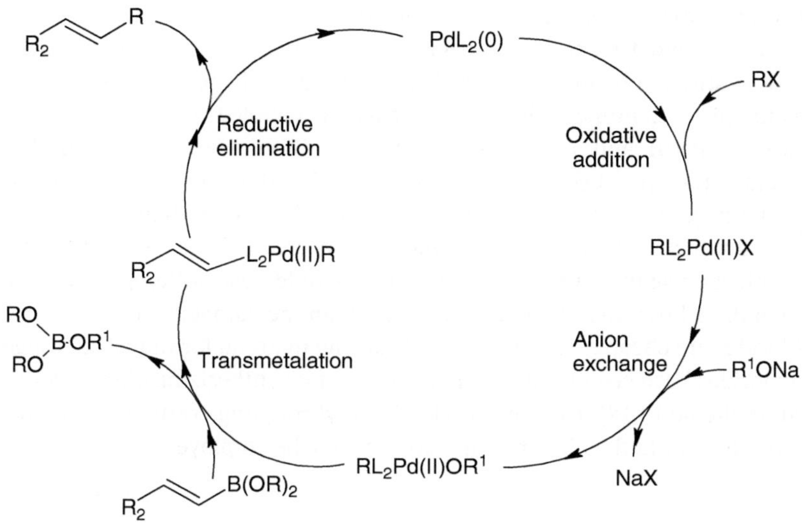

Figure 13.1 Catalytic cycle of Suzuki reaction.

in Figure 13.1. Accordingly, the reaction starts with the oxidative addition of the $L_2Pd(0)$ species to the organic halide to give tetracoordinated $L_2RPd(II)X$. In the second step, the tetracoordinated Pd(II) species undergoes exchange of anion X, with that of alkoxy anion of the alkoxide base, giving rise to $RL_2Pd(II)OR^1$. In the transmetallation step, the alkoxy palladium(II) species undergoes alkoxy anion exchange with the organic group of the tricoordinated borane. Lastly, reductive elimination of the $L_2Pd(0)$ from the tetracoordinated Pd(II) species generates a coupling product and the Pd(0) species continues with the next cycle of reaction.

Coupling reactions of iodides and bromides in the Suzuki reaction occur with relative ease, while the chlorides do not undergo facile reactions. However, coupling reactions of allylic and benzylic chlorides occur with comparative ease. Suzuki reactions have been used for synthesizing biaryls (Scheme 13.17).

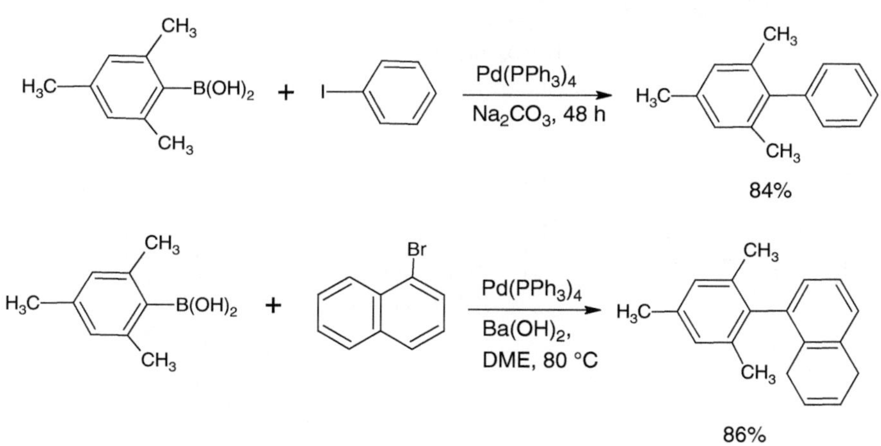

Scheme 13.17 Synthesis of biaryls.

The first such biaryl was synthesized using Suzuki coupling and was reported in 1981. Even hindered biaryls have been obtained in good yields. Synthesis of biaryls gave synthetic chemists an excellent alternative of the Ullman reaction. 2,4,6-trimethylphenylboronic acid is very feasibly coupled with iodobenzene in presence of Pd(0) and a very mild base, to afford the coupled product in 84% yield. However, use of an strong base $Ba(OH)_2$ in the coupling of 2,4,6-trimethylphenylboronic acid with that of 1-bromonapthalene afforded the coupled product in 86% yield. A remarkable decrease in reaction time was observed with the use of a strong base instead of a mild base, although the steric hindrance in the 1-bromonapthalene was greater than the iodobenzene.

Aryl chlorides are cheap and thus are good starting material for industrial usage. The aryl chlorides have been used with relative ease for synthesizing biaryls in very good yields (Scheme 13.18). However, cycloalkyl-aryl coupling occured in presence of potassium fluoride and a higher temperature had to be employed.

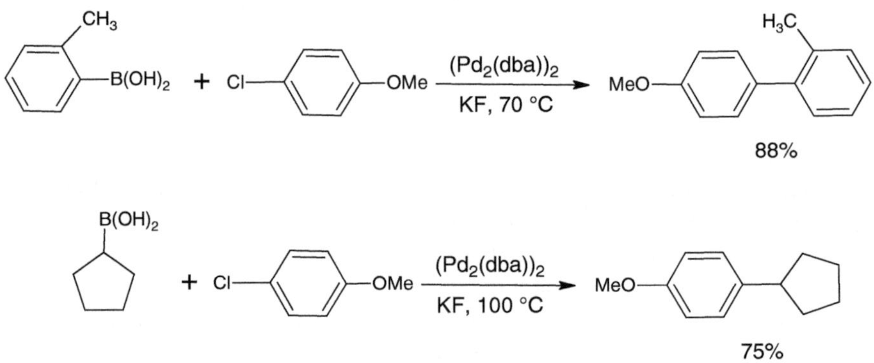

Scheme 13.18 Synthesis of biaryls from aryl chlorides.

Carbonylative Suzuki coupling reactions have been used to introduce carbonyl functionality between the coupling partners (Scheme 13.19). 1-Iodocyclopentene in

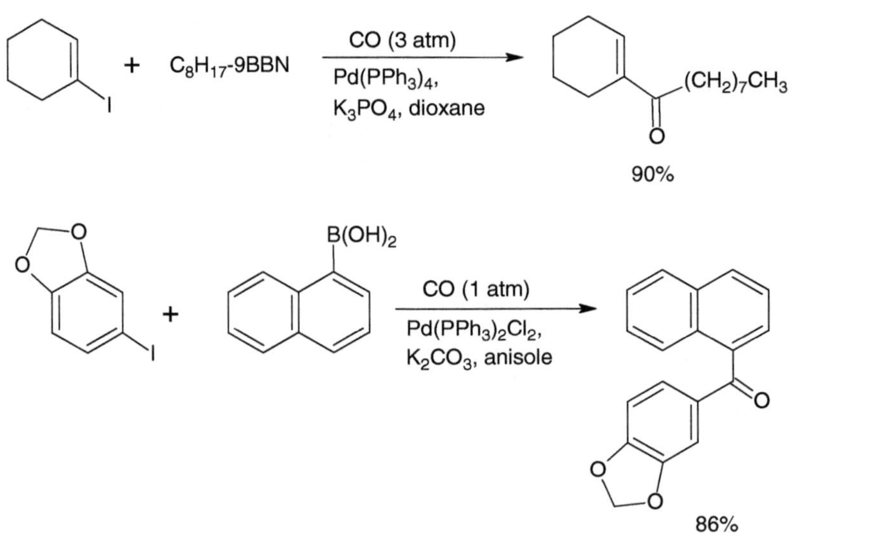

Scheme 13.19 Suzuki reaction.

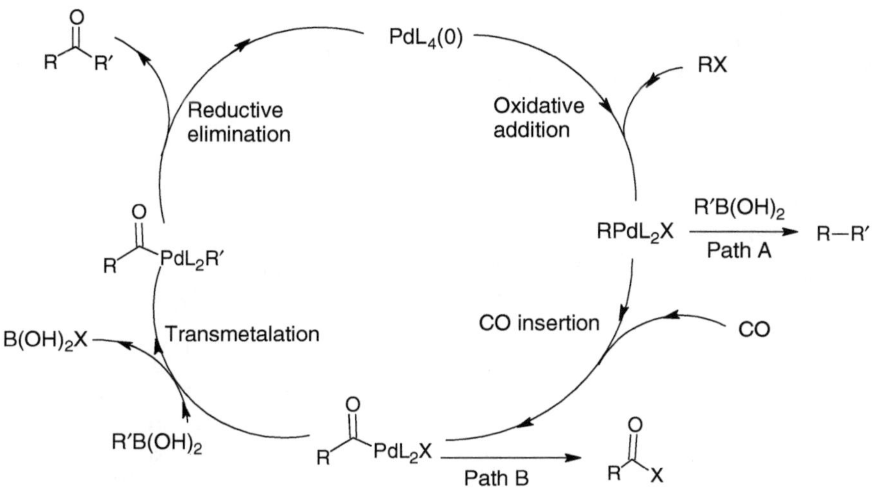

Figure 13.2 Carbonylation catalytic cycle.

CO atmosphere undergoes Suzuki coupling with *n*-octyl-9BBN in presence of Pd(0) and a potassium phosphate base to afford the coupled product in 90% yield. Similarly, 4-iodobenzene reacts with 1-napthaylboronic acid in CO atmosphere to afford a coupled product in 86% yield. The insertion of CO moiety works well when electron donating groups are substituted in organic halides. However, electron deficient substituents on the organic halides retards insertion of CO and the C–C formation reaction occurs without carbonylation (Figure 13.2). The catalytic cycle of CO insertion occurs as usual Suzuki coupling catalytic cycle only with a minor modification of CO insertion between the R and Pd after the oxidative addition of the organic halide. If the organic halide has electron deficient substituents, then the reaction directly proceeds to transmetallation without undergoing CO insertion (path A).

13.1.2.3 Acyltrifluoroborates

Acyltrifluoroborates have been reported to yield amide on reaction with alkyl azides, *O*-benzoylhydroxyl amines, and *O*-carbamoylhydroxyl amines (Scheme 13.20) [9]. The acylborates are difficult to generate because of its instability and tendency toward rearrangement. However, the trifluoroacylboranes have been prepared and used in the synthesis of amides. Treatment of acyltrifluoroborate with azides yields amides. As per the mechanism, trifluoroacylborane gives off potassium fluoride and the resulting acylborane makes an adduct with the nitrogen of the azide. The resulting borate anion transfers the acyl group to the nitrogen with concomitant elimination of the N_2, yielding difluoroborane amide. Subsequent hydrolysis produces amide. Similarly, amides have been prepared from *O*-benzoylhydroxylamine (Scheme 13.21) by treatment with acyltrifluoroborane. The loss of KF leads to acyldifluoroborane, which forms an adduct with nitrogen of the *O*-benzoylhydroxylamine. The migration of the acyl moiety from the borate anion to the amine is aided by the loss of *O*-benzoyl moiety, which is a very good leaving group. The resulting product is a secondary amide, which is obtained in good yields. Similarly, treatment of the *O*-carbomoylhydroxyl amine with trifluoroacyl borate leads to the formation of an adduct with nitrogen of the

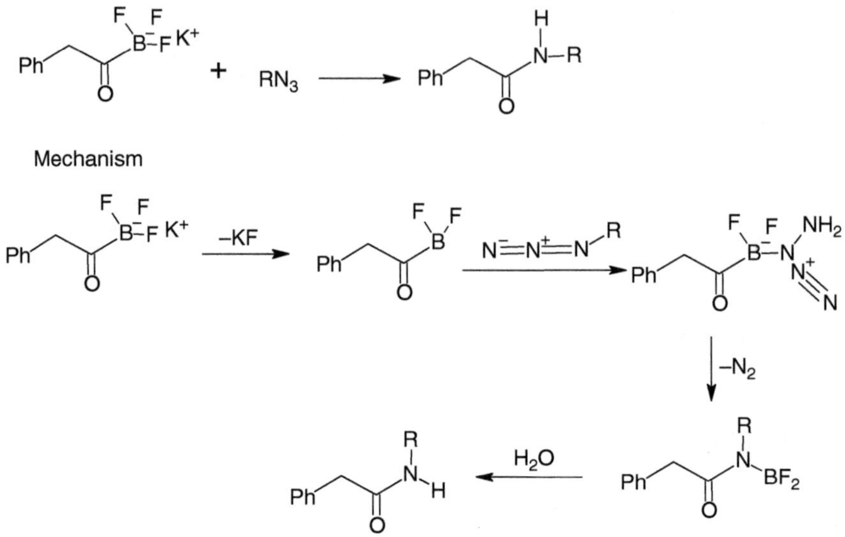

Scheme 13.20 Amide formation from azide.

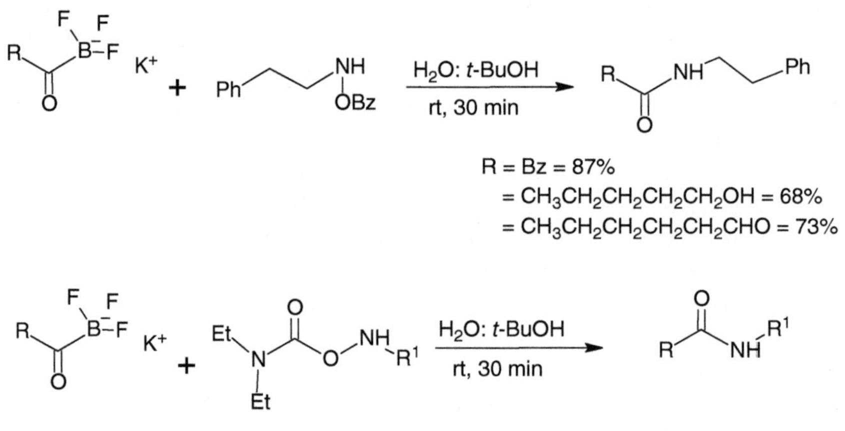

Scheme 13.21 Synthesis of amides from *O*-benzoylhydroxylamine and *O*-carbamoylhydroxylamine.

O-carbomoylhydroxyl amine. The difluoroacyl borate adduct transfers acyl group to the nitrogen with concomitant loss of carbamoyl group. The resulting product is the desired secondary amide.

13.1.2.4 Allylation

Allylation reactions on aldehydes are effected by trialkyl boron compounds (Scheme 13.22) having at least one allyl group attached to the boron [10, 11]. The trialkyl boron being electrophilic coordinates to the oxygen of the carbonyl thus increasing the electrophilicity of the carbonyl carbon. The migration of the allyl group from the boron occurs through a six-membered transition state. The allylated aldehyde upon hydrolysis affords the homoallyl secondary alcohol in 71% yield.

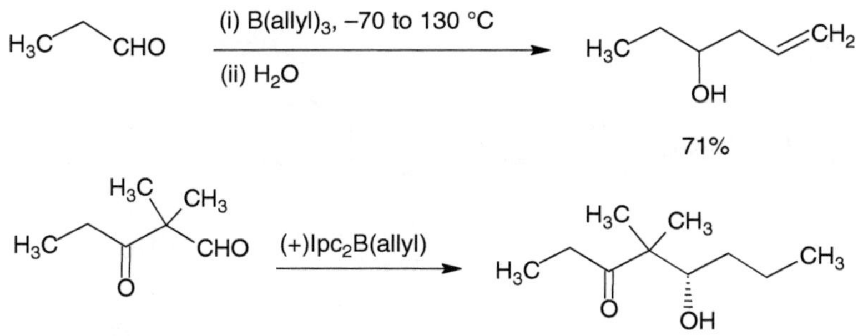

Scheme 13.22 Allylation.

There are a number of asymmetric versions of the allylation reaction on aldehydes. The chiral ligands attached to boron make the allylation reaction asymmetric. Allylation occurs by a cyclic six-membered chair transition state and thus the reaction is stereocontrolled. These tricoordinated boron reagents are prone to air oxidation and easily gets converted to boronic acids.

13.1.2.5 Chan–Lam coupling

Arylboronic acids are coupled with amine and hydroxyl-containing compounds, e.g. phenol, aniline, allylamine, amides, urea, carbamates, sulfonamides, and other heterocycles in presence of copper catalyst. This coupling is known as Chan–Lam coupling (Scheme 13.23) [12–14]. Cyclohexylamine on treatment with toluene boronic acid in presence of $Cu(OAc)_2$ and a base yields *N*-cyclohexylaniline. Similarly, *o*-iodophenol coupled with 4-methoxyphenyl boronic acid in presence of $Cu(OAc)_2$ and a mixture of bases pyridine and triethylamine to yield the product in 81%.

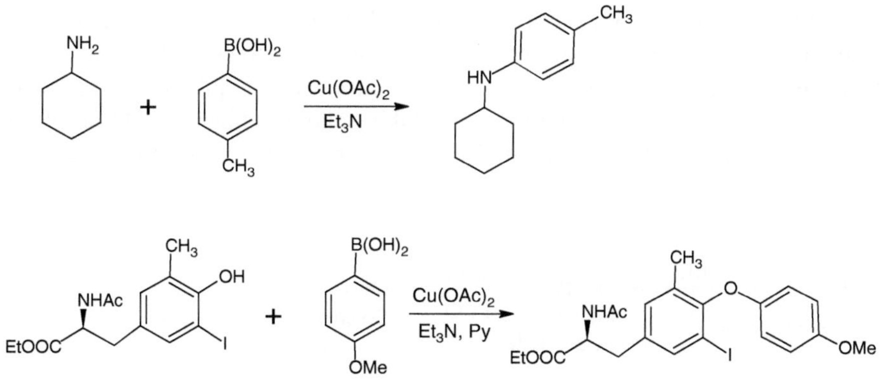

Scheme 13.23 Chan–Lam coupling.

13.1.2.6 Petasis Reaction

Petasis reaction is boronic acid-mediated Mannich reaction (Scheme 13.24) [15]. The reaction follows the mechanism of Mannich reaction. The imine formation is followed by the transfer of the organic ligand from the boronic acid. The reaction has been used to synthesize amino acids stereoselectively [16].

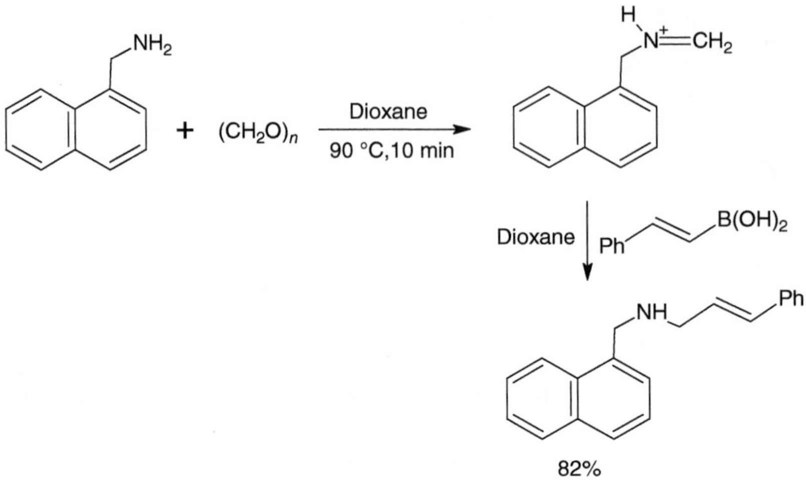

Scheme 13.24 Petasis reaction.

13.2 Organocopper Reagents

13.2.1 Introduction

Organocopper reagents can in general be represented as R_2CuM or RCuXM and R_2CuXM_2, where R can be alkyl, aryl, or alkenyl groups, while M can be Li, Mg, or Zn, and X can be any nontransferable anion or other nontransferable moiety [17]. The chemistry of organocuprates came to light in 1941 through a report by Kharash and Tawney regarding the conjugate addition of Grignard reagent in presence of Cu(I) salt on α,β-unsaturated ketone. Gilman et al. in 1952 reported the formation of an organocuprate reagent, which is also known as Gilman reagent (Scheme 13.25) [18]. The reagent formation occurs when 1 equiv of MeLi is added to CuI to yield a yellow-colored ether-insoluble solid. On further addition of one more equivalent of MeLi yields a soluble, colorless solution of organocuprate reagent R_2CuLi.

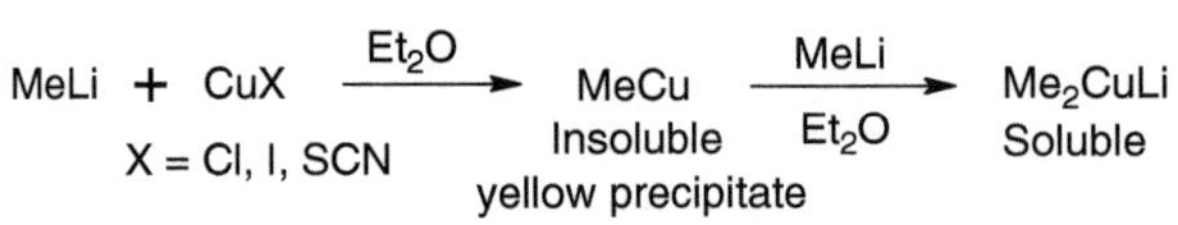

Scheme 13.25 Gilman reagent formation.

Organocuprate reagents offer an excellent way of making C—C bonds through conjugate additions on activated alkenes, substitution of organic halides, tosylates, triflates, epoxide opening, nucleophilic attack on alkynyls, addition on aldehydes, ketones, and imines, and nucleophilic attack on alkynyls to generate allenes (Scheme 13.26).

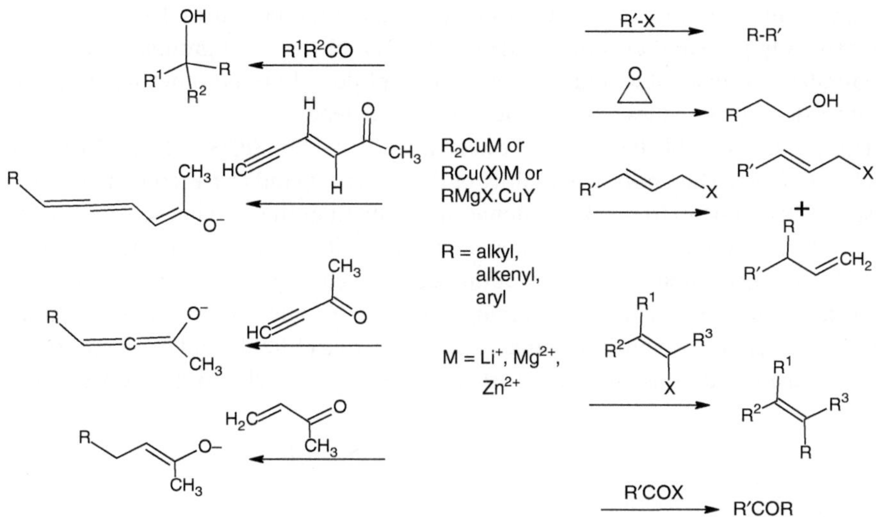

Scheme 13.26 Types of reactions of organocuprates.

13.2.2 Types of Organocuprates Reagents

Although organocuprates have been introduced by Gilman in 1952, the reagent has evolved through time and at present there are following variants of the original reagent.

13.2.2.1 Gilman Reagent (R_2CuM)

A copper-based organometallic reagent with two organic ligands as R and a main-group metal (Li^+, Mg^{2+}, Zn^{2+}) as a cation. The organic ligands may be either same where it is called homocuprates or may be different where it is known as heterocuprates. The representative preparations of few homo and heterocuprates have been shown (Scheme 13.27). Thus, heterocuprates may also be represented as RCu(X)M. Homocuprates are unstable and prone to β-elimination at higher temperatures and thus are prepared at low temperatures. However, heterocuprates are

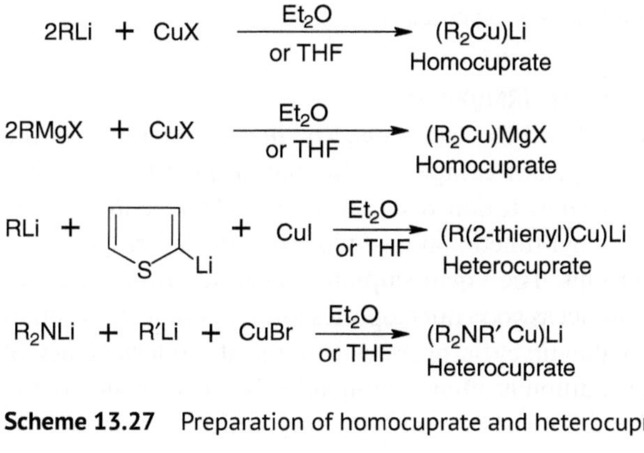

Scheme 13.27 Preparation of homocuprate and heterocuprate.

less prone to β-elimination and thus stable at higher temperatures. Heterocuprates are generally prepared at room temperature. One of the alkyl groups (R) in heterocuprates is transferable and acts as a nucleophile, while the other ligand (X) is nontransferable and does not play the role of a nucleophile.

This nontransferable ligand may be acetynyl, Ph_2P–, Me_3SiCH_2–, PhS–, 2-thienyl, R_2N–, –CN, etc. The nontransferability of the second ligand (X) as compared to the first one (R) is due to its weaker σ-donation ability than the former (Scheme 13.28). The preferential transfer of the desired nucleophilic R group yields the desired product. The weaker σ-donation ability of the second ligand (X) results in the preferred bond formation of R with the incoming electrophile. The nontransferable ligands may be chosen to be enantioselective and thus act as a chiral ligand in enantioselective conjugate additions and also as other nucleophilic attacks on the electrophile.

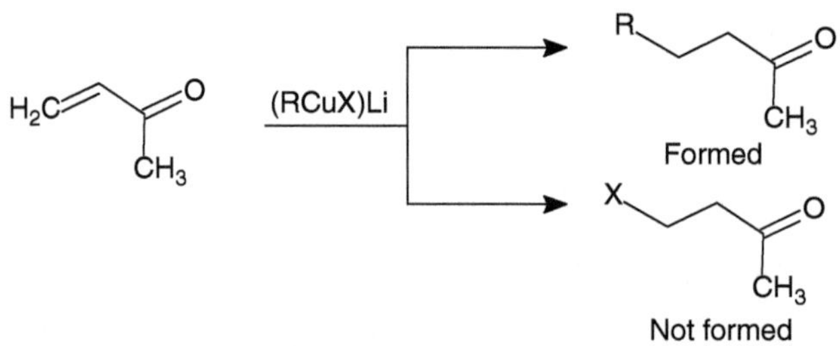

Scheme 13.28 Reaction of heterocuprate.

13.2.2.2 Higher Order Organocuprates ($R_2CuCNLi_2$)

Also called as Lipshutz reagents, were introduced in 1981. These come under the category of cyanocuprates and are readily obtained by the reaction of 2 equiv of RLi with CuCN at low temperature (Scheme 13.29). These reagents are generally employed for substitution in secondary alkyl halides and epoxides where usual organocuprates do not afford the required products in the desired yields.

$$2RLi + CuCN \xrightarrow[-78\text{ to }0\,°C]{Et_2O\text{ or THF}} (R_2CuCN)Li_2$$

Scheme 13.29 Preparation of higher order cyanocuprate.

13.2.2.3 Grignard Copper(I) Reagents (RMgX·CuY)

These are Grignard reagents catalyzed by Cu(I) salts. The *in situ* generation of the organocuprate reagent (Scheme 13.30) competes with the Grignard reagent for the conjugate addition or other substitution reaction. Grignard's reaction readily undergoes 1,2-addition on carbonyl functionalities while organozinc reagents are unreactive toward carbonyl groups. The organocuprates generated from Grignard reagent and organozinc reagents act as good nucleophiles toward conjugate addition and allylic substitution. The catalytic cycle operates through the intermediacy of Cu(I)/Cu(III) ions. Conjugate addition is often accompanied by the inversion of the geometry of the double bond and is consistent with the mechanism of reaction.

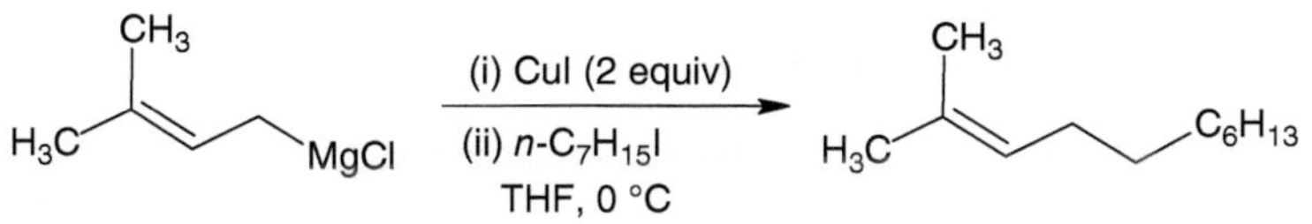

Scheme 13.30 Substitution reaction by Grignard copper reagent.

13.2.3 Reactions of Organocuprates

13.2.3.1 Increasing Chain Length of Alkyl halides

The increase in chain length of an alkyl halide, triflate, or tosylate can be effected by nucleophilic substitution of the leaving group by an organocuprate (Scheme 13.31). Homocuprates being unstable at higher temperatures the reaction is carried out at low temperatures while the heterocuprates being stable, the reaction is carried out at room temperature. So the use of heterocuprates is more economical than homocuprates. However, homocuprates undergo substitution reaction readily with methyl halides and primary alkyl halides while showing frequent elimination reaction with secondary and tertiary alkyl halides. The cyanocuprates afford C—C bond formation products (Scheme 13.32) in secondary alkyl halides. Mechanism of the substitution reaction may proceed via two possible pathways. In one pathway, the reaction may occur by simple nucleophilic S_N2 substitution reaction, while in the second pathway, the oxidative addition of the organocuprate reagent across the carbon–halide linkage is followed by reductive elimination of the copper reagent with concomitant transfer of the R group of the organocopper reagent (Scheme 13.33).

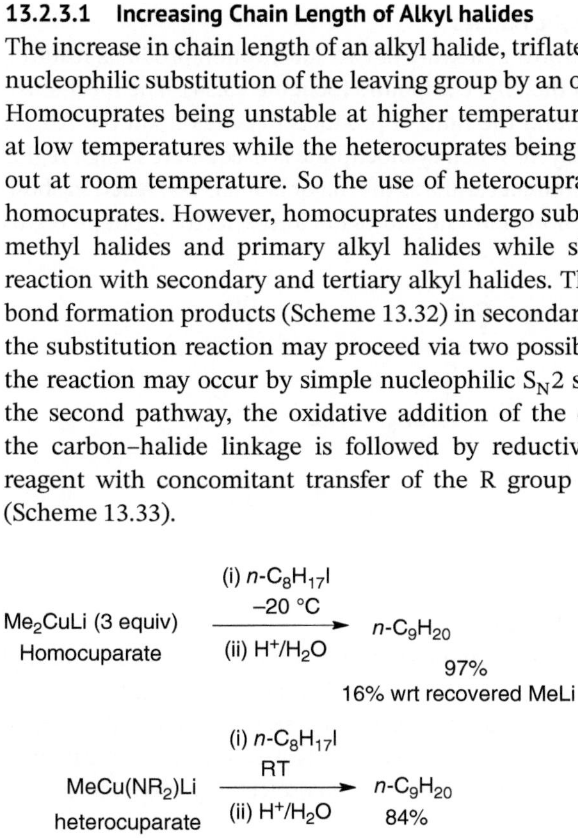

Scheme 13.31 Increasing chain length by nucleophilic substitution reaction.

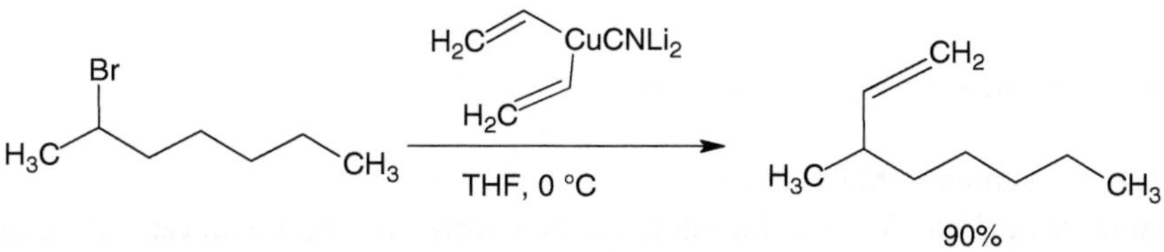

Scheme 13.32 Alkylation at the secondary alkyl halide.

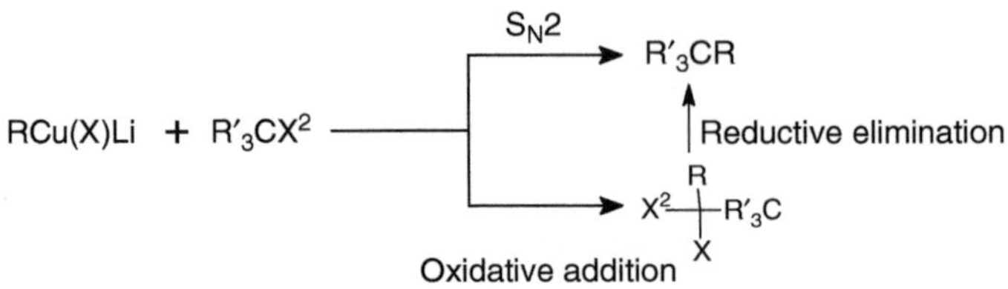

Scheme 13.33 Mechanism of substitution reaction.

13.2.3.2 Alkylation of the Allylic Halides

Alkylation of allylic acetate affords a mixture of two substitution products resulting from S_N2 and S_N2' attacks on the allyl compound (Scheme 13.34). The mechanism of the reaction is not simple, and the ratio of products depends upon the reagent used and the conditions employed. When cyanocuprate is used, there is high regioselectivity and γ-substitution product results as the major isomer. The reaction shows full anti-selectivity. However, homocuprate shows full anti-selectivity but no regioselectivity was observed.

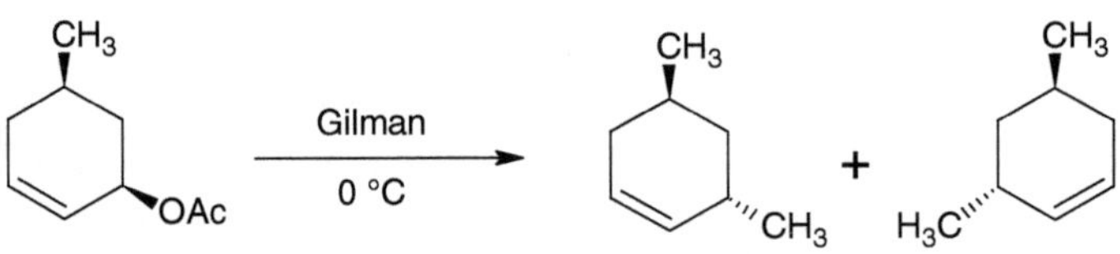

MeCu(CN)Li	4	:	96
Me_2CuLi	50	:	50

Scheme 13.34 Alkylation of allyl acetate.

13.2.3.3 Reaction with Vinyl Halides

The substitution reaction of organocuprates with vinyl halides proceeds with high stereoselectivity (Scheme 13.35). The reaction mechanism as viewed traditionally is oxidative addition of the cuprate at the alkenyl halide bond followed by reductive elimination has changed. The new mechanism is based on theoretical study. However, the old mechanism still explains the stereoselectivity observed in such reactions.

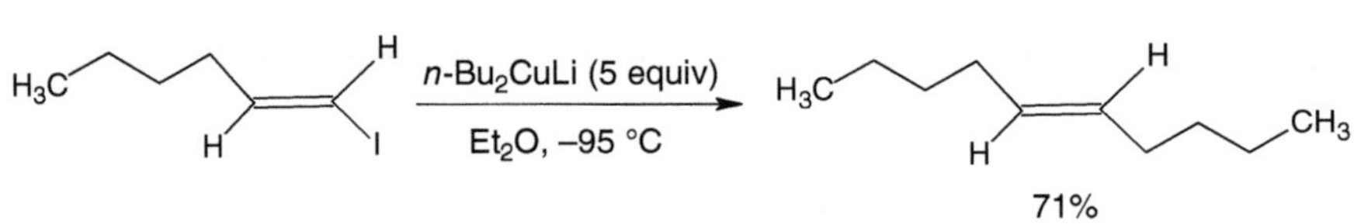

Scheme 13.35 Reaction with alkenyl halide.

13.2.3.4 Alkylation of Acyl Halides

Alkylation of acyl halides and thioesters can be effected by organocuprates yielding ketones (Scheme 13.36). Even α,β-unsaturated acyl halides also afford ketone instead of conjugate addition. Although the mechanism of reaction with acyl halide and thioesters is different from one another, both of them yield ketone.

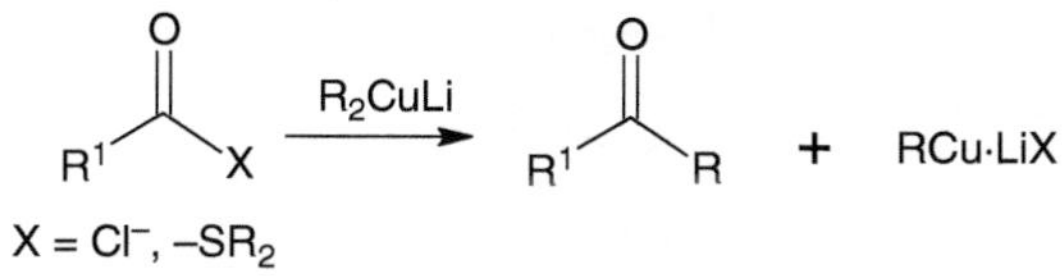

Scheme 13.36 Reaction of organocuprate with acyl halide and thioester.

13.2.3.5 Epoxide Opening

Epoxide ring opening is effected in good yields by cyanocuprates. Although simple organocuprates also result in ring opening of the epoxide but cyanocuprates and higher order cuprates yield better results. The ring opening occurs from the less hindered side (Scheme 13.37). The reaction occurs in a stereoselective manner i.e. anti-attack of the nucleophile. α,β-Unsaturated epoxides undergo ring opening through conjugate addition at the unsaturated bond to yield α,β-unsaturated hydroxy compounds.

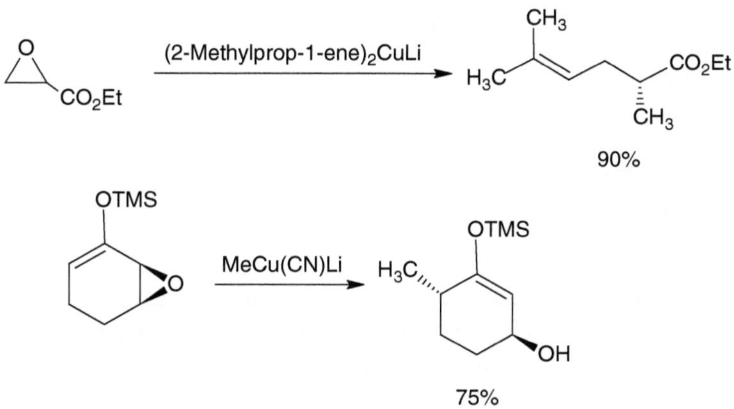

Scheme 13.37 Epoxide opening.

13.2.3.6 Conjugate Additions

While organolithium and Grignard reagents are used for 1,2-addition across conjugate carbonyl compounds, the organocuprates being soft nucleophiles effect regioselectively 1,4-additions. Copper being less electropositive than the main group metals Li and Mg, generates softer nucleophiles. In contrast to the main-group metals, being more electropositive, generates hard nucleophiles. All types of organocuprates, i.e. homocuprates, heterocuprates, cyanocuprates, and higher order cuprates yield conjugate addition products rather than 1,2-addition products. In cases where two sites are available, conjugate addition occurs stereoselectively from the less hindered side (Scheme 13.38) [19].

Mechanistically, addition to activated alkenes occurs by oxidative addition of organocuprate at the conjugate position followed by transfer of alkyl group at the 4-position with concomitant reductive elimination of the copper (Figure 13.3).

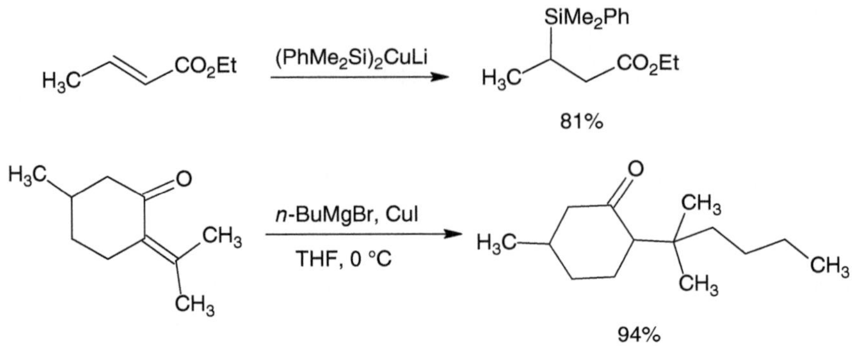

Scheme 13.38 Conjugate addition.

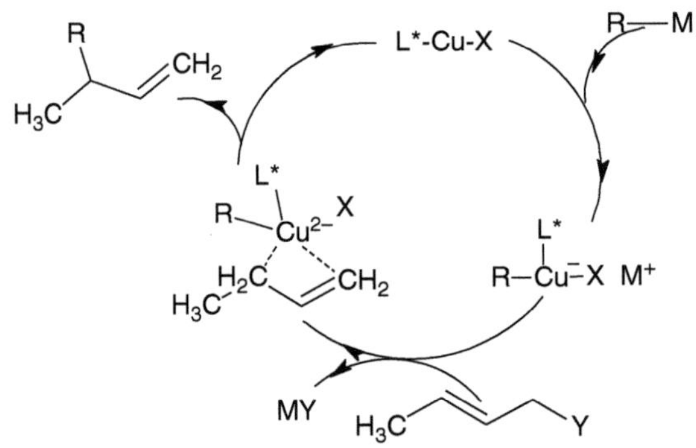

Figure 13.3 Catalytic cycle for conjugate addition.

13.2.3.7 Conjugate Addition in Presence of Lewis Acid and Electrophiles

β,β-Disubstituted enones do not undergo conjugate addition with organocuprate smoothly at low temperatures. However, in presence of a Lewis acid, e.g. BF_3 the reaction proceeds smoothly to afford a conjugate addition product at low temperature (Scheme 13.39). Similarly, ethyl-2,3-dimethylbutenoate reacts with organocuprate to afford a product with less than 1% yield. However, when the same reaction is performed in presence of BF_3, the yield of the product is found to be significantly enhanced.

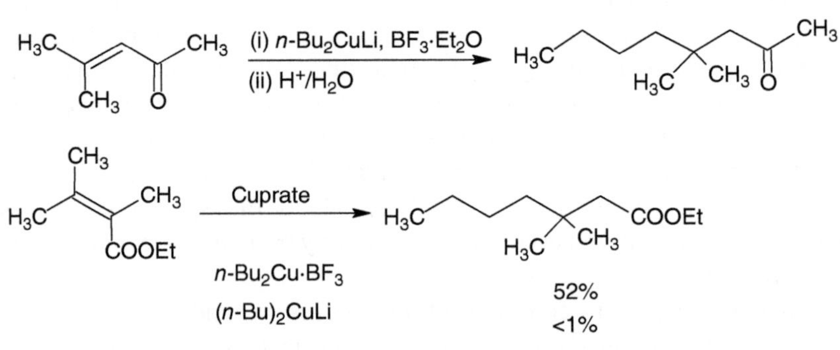

Scheme 13.39 Conjugate addition in presence of Lewis acid.

Similarly, treatment of organocuprate reagent with an activated alkene affords 28% of the conjugate addition product [20] (Scheme 13.40), while in presence of an electrophile TMSCl, the conjugate addition proceeds smoothly to yield 1,4-addition product in 99%.

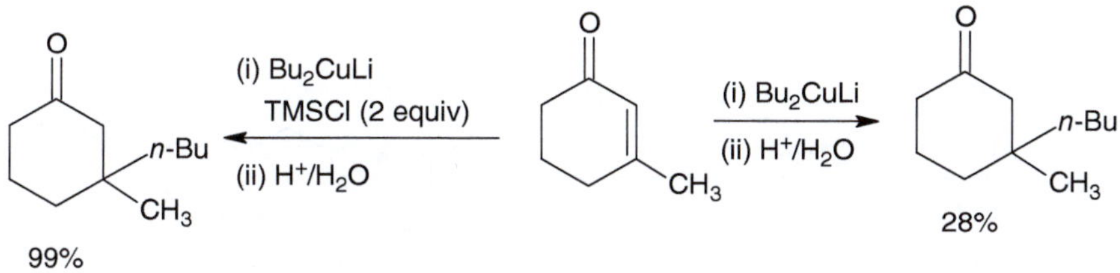

Scheme 13.40 Conjugate addition in presence of electrophile TMSCl.

13.2.3.8 O-Trapping and C-Trapping Reaction

Conjugate addition across an activated double bond leads to addition of the nucleophile at the 4-position to the carbonyl. Now, the addition of an electrophile in order to complete the reaction may occur at the oxygen of the carbonyl or at the 3-position of the conjugate system (Scheme 13.41). O-trapping reagents are R_3SiCl, $(RO)_2P(O)Cl$, while C-trapping reagents are RX, RCHO, X_2, etc.

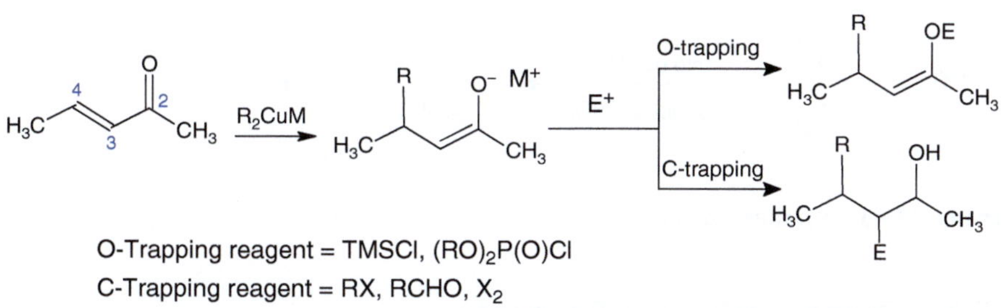

Scheme 13.41 Conjugate addition followed by trapping by electrophiles.

The C-trapping reagent addition occurs in trans fashion or the trans product is the major isomer (Scheme 13.42). C-trapping electrophile, when MeI is used in the reaction led to *trans*-2,3-substituted cyclohexanone as the major isomer [21]. Similarly, reaction of 2-cyclopentenone with an organocuprate reagent, followed by trapping of the nucleophile at 2-position led to trans-2,3-substituted product.

13.2.3.9 Synthesis of Allenes

Organocuprates have been used successfully in generating allenes by nucleophilic attack on the activated alkynes (Scheme 13.43). The Grignard reagent in presence of a catalytic amount of CuBr generates an organocuprate, which competes with the Grignard reagent in the medium for 1,6-conjugate addition to the activated alkyne [22]. The allene is obtained with 79% yield. In another such reaction, the leaving group at the γ-position was knocked out by the conjugate nucleophilic attack at the alkyne to give allene with 70% yield [23].

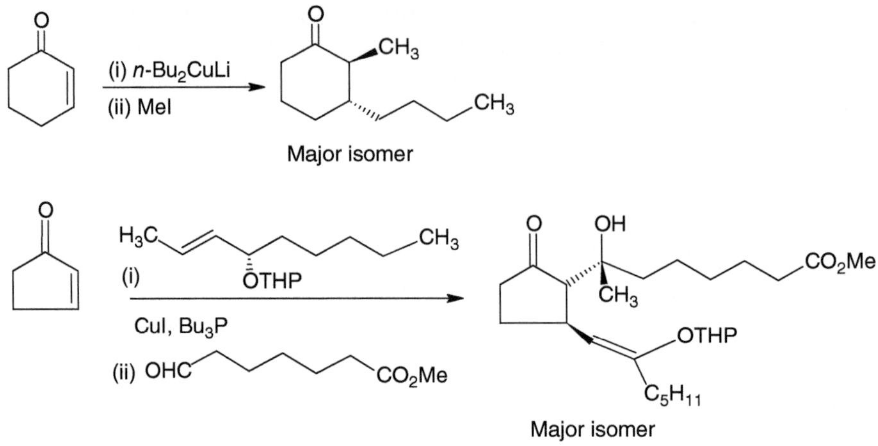

Scheme 13.42 C-trapping of reactions.

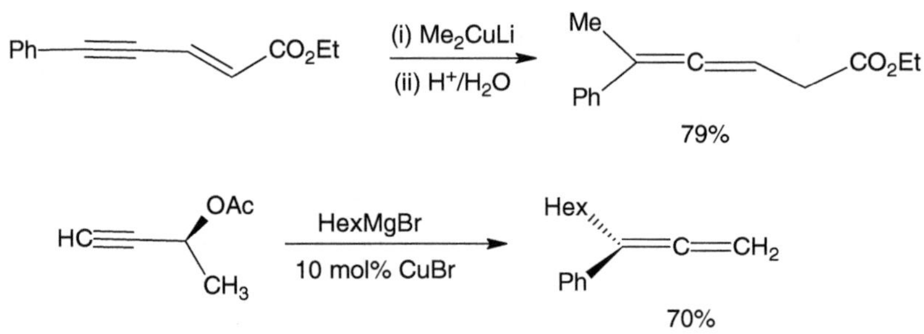

Scheme 13.43 Synthesis of allenes.

13.3 Organopalladium Reagents

13.3.1 Introduction

Pd(0)-catalyzed reactions have revolutionized the C—C bond making process, which is the very core of organic synthesis. The immense contribution of the palladium-catalyzed reactions have been rewarded with 2010 Noble Prize to Professor Richard F. Heck, Professor Ei Ichi Negishi, and Professor Akira Suzuki for the development of palladium-catalyzed reactions. Palladium-catalyzed cross-coupling reaction as it is popularly known, is in general a reaction of an organic halide with that of an organometallic compound or substituted alkene in presence of palladium metal in zero oxidation state in catalytic amounts (Scheme 13.44). The alkyl halide works as an electrophilic donor of organic moiety, and the organometallic compound or the alkene works as a nucleophilic donor of an organic moiety. The two organic moieties are combined to form a C—C bond over the palladium metal in zero oxidation state. However, the palladium catalyst could be added as a Pd(II) or as Pd(0) states. The reaction has been extensively used in the pharmaceutical

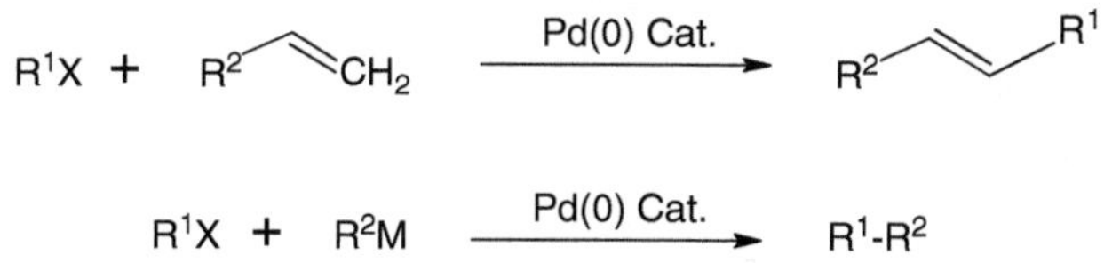

Scheme 13.44 Schematic representation of cross-coupling reaction of palladium.

industry and in other chemical industries for the easy and versatile synthesis of C—C bonds.

Transition metals in organic synthesis have been used since the middle of the twentieth century. The class of metals due to their variable oxidation state has revolutionized chemistry and especially the synthetic approach in organic synthesis. The special reference and significance of palladium in organic chemistry can be traced back to Wacker's process of air oxidation of ethene in presence of palladium to acetaldehyde (Scheme 13.45) [24]. Thereafter, research in organopalladium chemistry picked up pace.

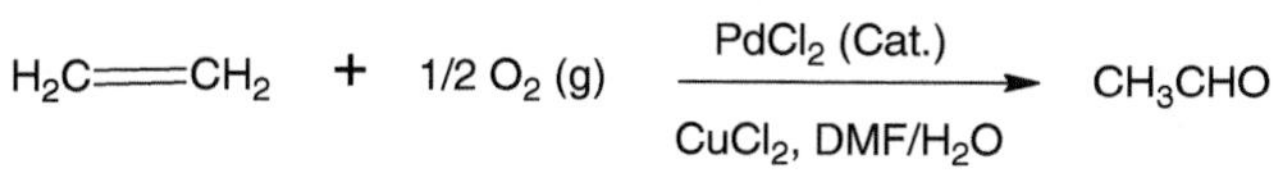

Scheme 13.45 Wacker's process.

13.3.2 Types of C–C Cross-Coupling Reaction Catalyzed by Palladium Metal

13.3.2.1 Heck Reaction

In 1968, Heck reported the arylation and methylation of an alkene [25–29]. The reaction of ArPdX and RPdX with that of an alkene led to the formation of arylated alkene and alkylated alkene (Scheme 13.46). This reaction was conceptually a new method for the formation of C—C bonds and thus was awarded Noble Prize in 2010.

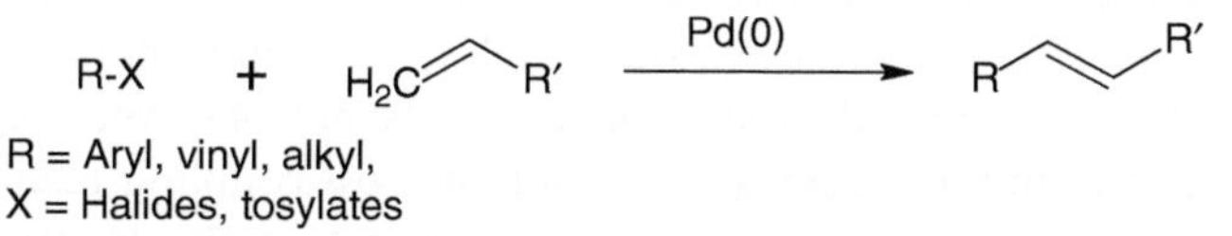

Scheme 13.46 Representative Heck reaction.

The RX can be an aryl halide, vinyl halide, or an alkyl halide. The addition of the organopalladium to the alkene occurs in a syn fashion with palladium on the more substituted end of the double bond (Scheme 13.47). The mechanism and palladium catalytic cycle of the cross-coupling reaction of the Heck reaction (Figure 13.4) begins with Pd(0) catalyst (PdL_2), which adds oxidatively to the organic halide RX and leads to a $RPdL_2X$ with palladium in an oxidation state of +2. Further addition of the Pd(II) on the alkene occurs regioselectively with Pd attaching to the more substituted side and R on the less hindered side of the double bond. Further β-hydride

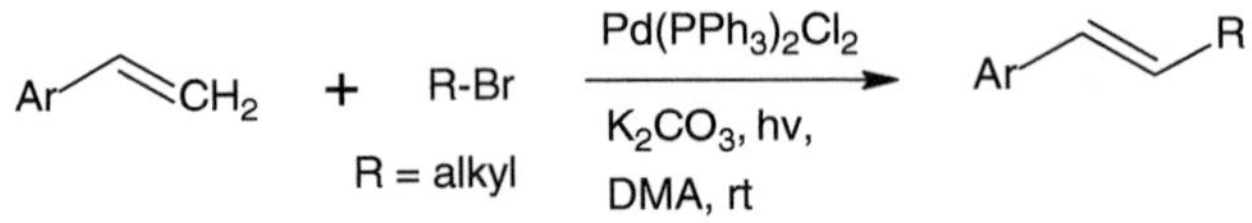

Scheme 13.47 Heck reaction.

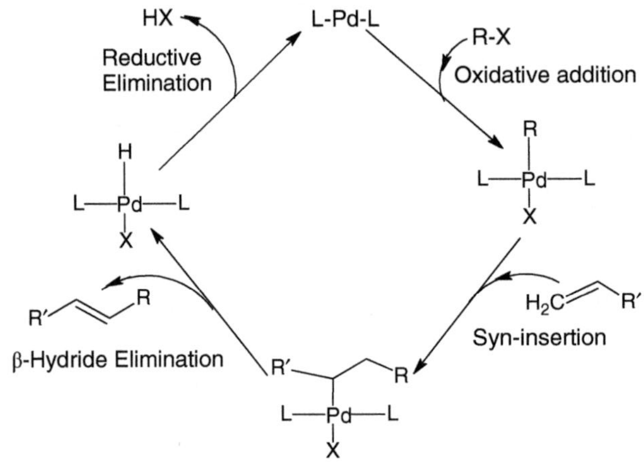

Figure 13.4 Palladium catalytic cycle of Heck reaction.

elimination yields the arylated/alkylated alkene and the $HPdL_2X$ remains. Reductive elimination of HX regenerates to PdL_2 and the catalyst is ready to go for the next cycle of reaction. The Pd(II) catalyst added to the reaction mixture gets reduced to Pd(0) *in situ* and catalyzes the reaction.

13.3.2.2 Negishi Coupling

In 1977, Negishi developed the cross-coupling reaction of organozinc compound with that of organic halide in presence of palladium(0) catalyst (Scheme 13.48) [30, 31]. Zinc, being a mildly electropositive element, generates weak organic nucleophile and thus the functional group tolerance is quite high for the cross-coupling reaction. The same extent of functional group tolerance was not observed for Grignard reagent and organolithium reagents in palladium cross-coupling reactions (Scheme 13.49). The organozinc reagent can be prepared *in situ* by the

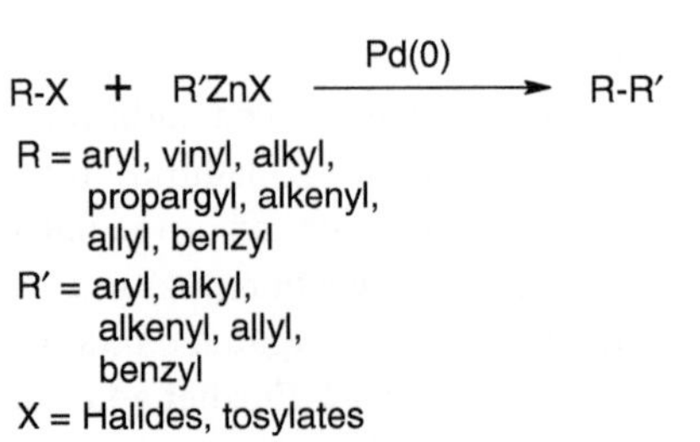

Scheme 13.48 Representative Negishi coupling.

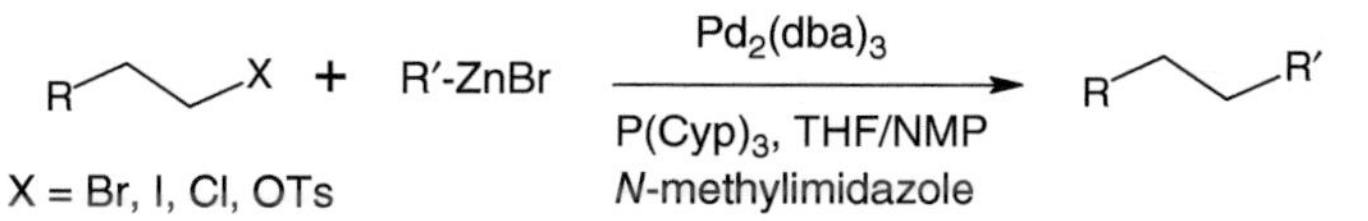

Scheme 13.49 Negishi coupling.

Figure 13.5 General catalytic cycle for Pd(0)-catalyzed cross-coupling reactions.

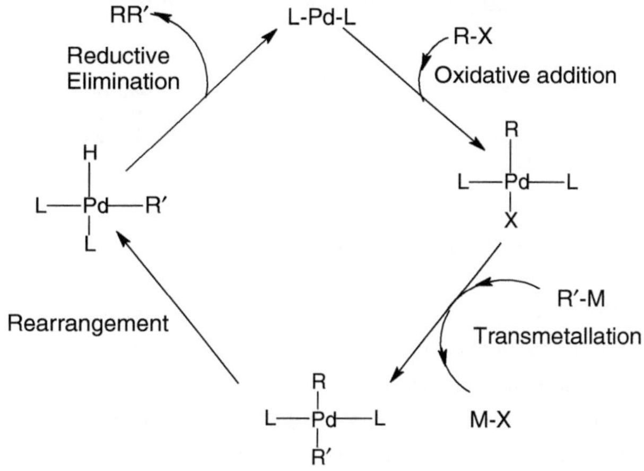

transmetallation reaction of the corresponding Grignard reagent or the organolithium reagent [32]. The organozinc reagent can also be prepared by oxidative addition of the activated Zn(0) to an organic halide. Even unactivated alkyl halides undergo Negishi coupling smoothly (Scheme 13.49) [33].

The general catalytic cycle of the Negishi coupling and other organometallic reagents coupling with the help of Pd(0) in catalytic amounts has been shown in Figure 13.5. The catalytic cycle begins with oxidative addition of the Pd(0) to the organic halide to yield $RPd(II)L_2X$. Transmetallation of R′M leads to exchange of the halide ion of the Pd complex with that of R′ of the organometal. Once R-group enters the Pd complex, the complex rearranges to give cis orientation of the ligands L and cis orientation of the organic groups R·R′. Coupling of the two R and R′ leads to a coupling product and is accompanied by the loss of Pd(0). The catalytic cycle ensures high regioselectivity and stereoselectivity.

13.3.2.3 Suzuki Coupling

This reagent was reported by Professor Akira Suzuki in 1979. The topic has already been adequately discussed under organoboron reagents.

13.3.2.4 Stille Coupling

Cross-coupling reaction of an organic halide with tetraorganotin in presence of Pd(0) catalyst (Scheme 13.50). As the metal carries four organic groups attached to the metal, the relative rates of transmetallation must be kept into consideration while designing the organometallic species. The relative rates of transmetallation of various organic groups are in the order of alkynyl > vinyl > aryl > allyl ≈ benzyl ≫ alkyl. In the given reaction, the vinyl group gets transferred [34] in preference to the other

$$R\text{-}X + R'\text{-}SnR'' \xrightarrow{Pd(0)} R\text{-}R'$$

R = aryl, vinyl, alkyl, acyl
R′ = aryl, alkyl, alkenyl
X = halides

Scheme 13.50 Representative Stille coupling.

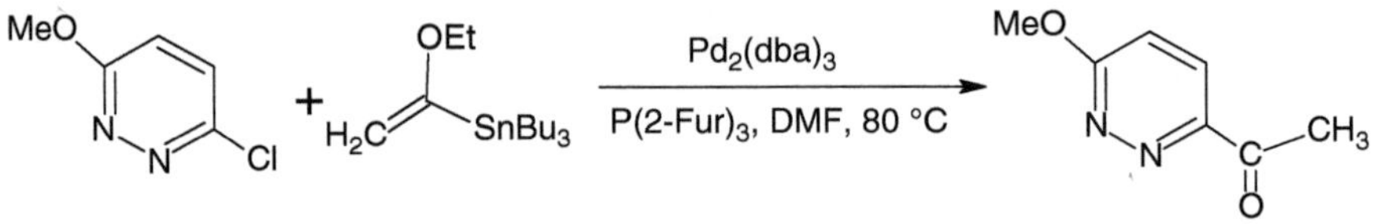

Scheme 13.51 Examples of Stille coupling.

three butyl groups (Scheme 13.51) of the organotin reagent. The tributyltin enol ethyl ether [35] gets coupled with aryl chloride at the vinylic position. The hydrolysis of the coupled product yields the methyl heteroaryl ketone. The tributylphenyltin reacts with 4-methoxy-chlorobenzene [36] to yield biaryl. The phenyl group gets transferred in preference to the three alkyl groups. As the reaction is carried out under neutral conditions, thus rendering it more tolerant to various functional groups than the Suzuki reaction. The only drawback of the organotin reagents is their solubility in various solvents, especially water.

13.3.2.5 Kumada Coupling

Coupling of the organic halide with the Grignard reagent in presence of Pd(0) catalyst is called Kumada coupling (Scheme 13.52). As the Grignard reagent is highly basic, the substrate tolerance of the reaction is significantly reduced, i.e. a very limited number of functional groups can be used in this coupling reaction.

$$R\text{-}X + R'MgX \xrightarrow{Pd(0)} R\text{-}R'$$

R = aryl, vinyl
R′ = aryl, alkyl, alkenyl
X = halides

Scheme 13.52 Representative Kumada coupling.

Phenylmagnesium Chloride [37] reacts with the 2-chloropyridine-5-methylester to yield coupled product at a very low temperature (Scheme 13.53).

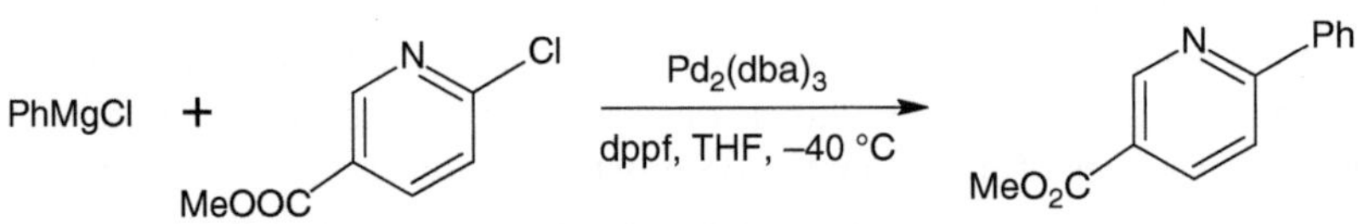

Scheme 13.53 Kumada coupling.

13.3.2.6 Hiyama Coupling

Coupling reaction of organic halide with that of tetraorganosilicon reagent in presence of Pd(0) catalyst is referred to as Hiyama coupling (Scheme 13.54). Like the Suzuki reaction, the reaction requires the use of a base or fluoride reagent for the successful transfer of the organic ligand from the silane reagent [38]. The R′ group is generally taken as F^- or some alkoxy group in order to enhance the rate of reaction (Scheme 13.55). Another approach to enhance the rate of transfer of the R″ group is using silacyclobutane derivatives. The cyclobutane ring aids the transfer of the R″ group from Si to the R group of alkyl halide when the F^- ions attack the silane. An alkyl halide underwent Hiyama coupling when treated with arylsiloxane at room temperature to yield alkylaryl compound (Scheme 13.55) [39].

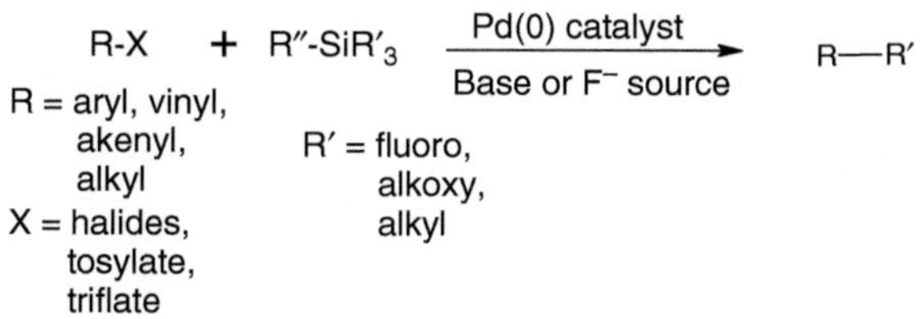

Scheme 13.54 Representative Hiyama coupling.

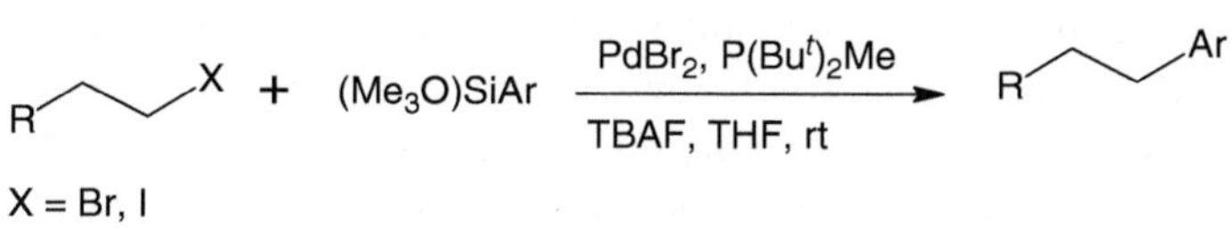

Scheme 13.55 Hiyama coupling.

13.3.2.7 Sonogashira Coupling

Cross-coupling reaction of an organic halide with that of an terminal alkyne in presence of Pd(0) catalyst and Cu(I) salts (CuI, CuCN) [40] is termed as Sonogashira coupling (Scheme 13.56).

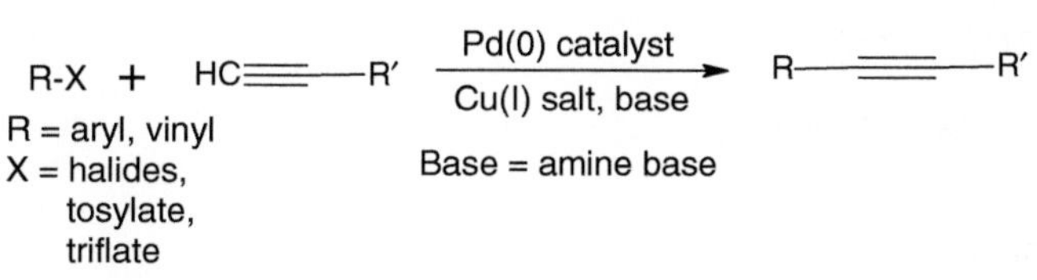

Scheme 13.56 Representative Sonogashira reaction.

The coupling of 4-bromobenzaldehyde occurs with silylated terminal alkyne (Scheme 13.57) in presence of Pd(0) catalyst, CuI, and a base Et_3N to yield arylalkyne.

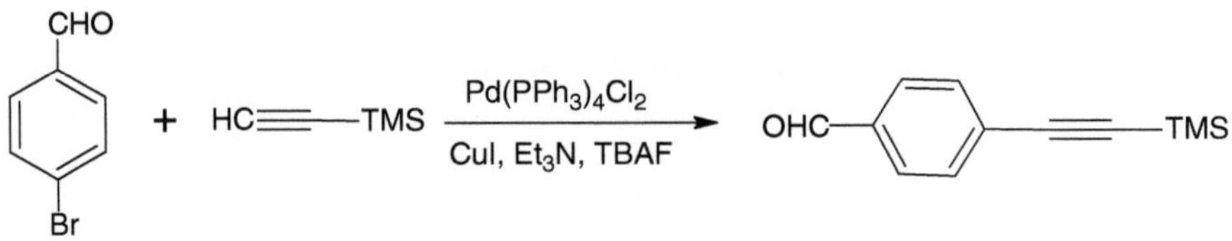

Scheme 13.57 Sonogashira reaction.

13.3.2.8 Butchwald–Hartwig Coupling

While the previous reactions discussed have been essentially C–C coupling, this reaction is about the formation of C—N bond through Pd(0) catalysis (Scheme 13.58).

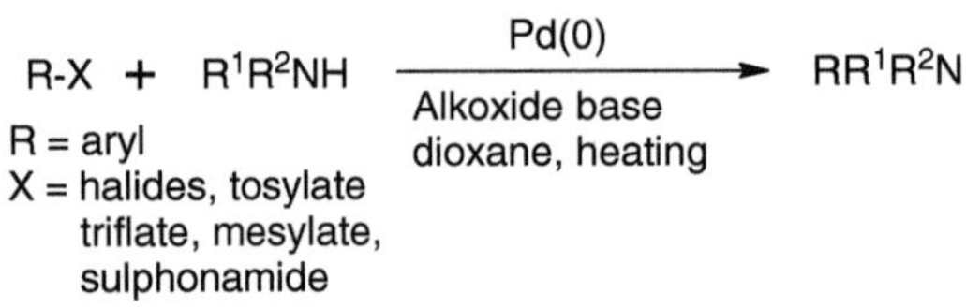

Scheme 13.58 Representative Butchwald–Hartwig reaction.

An organic halide reacts with an amine in presence of a base to yield coupled amine. Buchwald [41] and Hartwig [42] independently reported the coupling reaction in 1995. Aryl halides are conveniently coupled with secondary amine (Scheme 13.59) in presence of a catalytic amount of Pd(0) and a base [43]. The mechanism and the catalytic cycle of the reaction are almost similar to the catalytic cycle of the Suzuki coupling reaction. The catalytic cycle starts with Pd(0) oxidatively adding across the R—X bond. The halide ion of the complex gets exchanged with -OR′ of the alkoxide base. The alkoxide gets exchanged with the organic amine, followed by reductive cleavage of the metal, which yields the coupled amine.

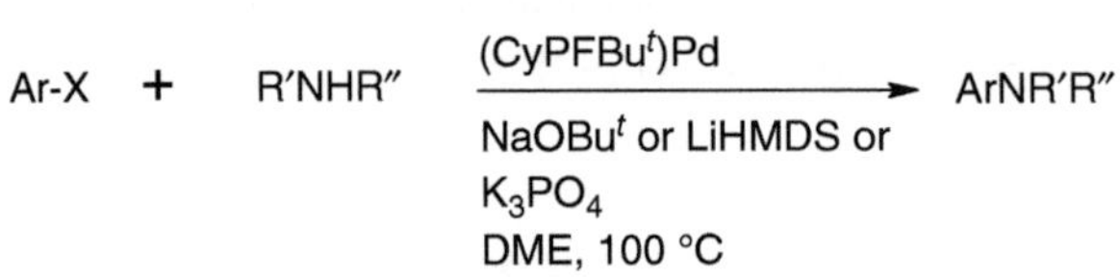

Scheme 13.59 Butchwald–Hartwig coupling.

13.3.2.9 Cyanation Reaction

Pd(0) catalyst is used to substitute the halide group of an organic halide with the nitrile group (Scheme 13.60). The Pd-catalyzed cyanation is a very versatile alternative to cyanation of an aryl halide by the Sandmeyer reaction and the Rosenmund von Braun reaction. In this reaction cyanide ion comes from an inorganic salt (e.g. $Zn(CN)_2$, $K_4[Fe(CN)_6]$). The cyanide ion may act as a catalytic poison for the palladium catalyst at a higher concentration. In order to avoid the catalytic poisoning low concentrations of cyanide ion is provided by $Zn(CN)_2$ instead of using alkali

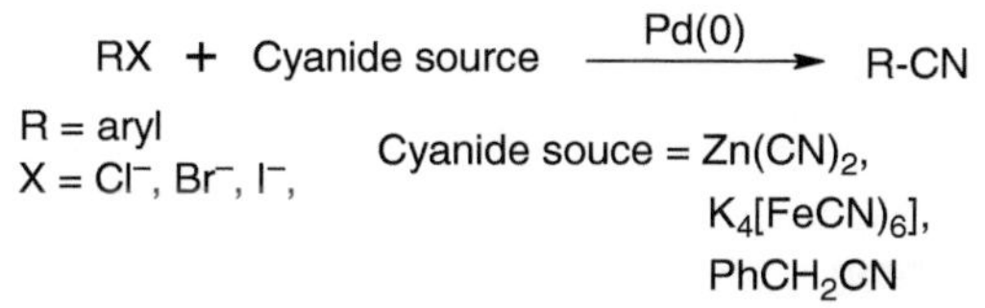

Scheme 13.60 Representative palladium-catalyzed cyanation reaction.

metal cyanide salts. The reaction is performed in DMF solvent as the solubility of $Zn(CN)_2$ in the solvent is significantly low to effect catalytic poisoning. Other sources of CN^- have been used [44] (Scheme 13.61), which are either nontoxic or far less toxic. The mechanism and the catalytic cycle of the reaction are almost identical to other palladium-catalyzed reactions. The reaction begins with the oxidative addition of the Pd(0) catalyst to the aryl halide bond. The halide ion attached to the palladium is exchanged with CN^- by reaction with a halide source. The reductive elimination of the palladium metal furnishes the cyanation product.

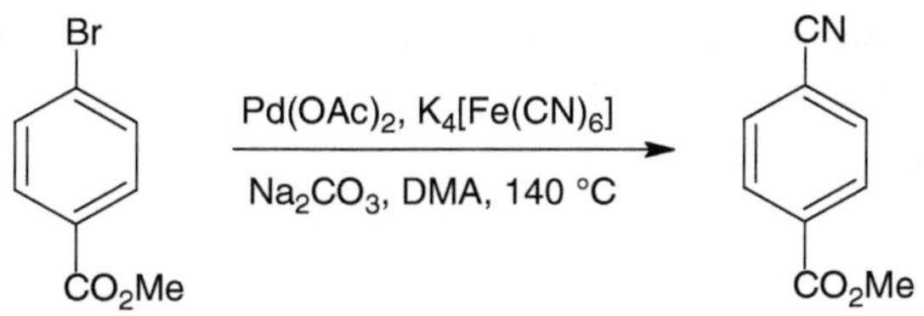

Scheme 13.61 Cyanation using nontoxic sources.

13.3.2.10 Carbonylation Reaction

Carbonylation of organic halides by carbon monoxide atmosphere using Pd(0) catalyst. A nucleophilic reagent is also used in the reaction, which determines whether the product will be an acid, ketone, aldehyde, ester, or amide (Scheme 13.62) [45].

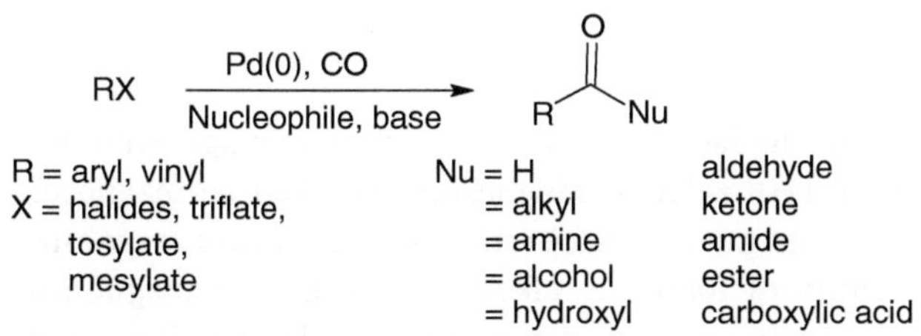

Scheme 13.62 Representative carbonylation reaction.

The reaction has tolerance for a large variety of functional groups. When the reaction is carried out in presence of alcohol, the end product obtained is an ester. The reaction mechanism and the catalytic cycle of the reaction are similar to earlier addition reactions of palladium viz. oxidative addition of the Pd(0) to the alkyl or the vinyl halide followed by insertion of CO. The nucleophilic attack removes the halide ion from the Pd complex. The reductive elimination detaches the metal from the

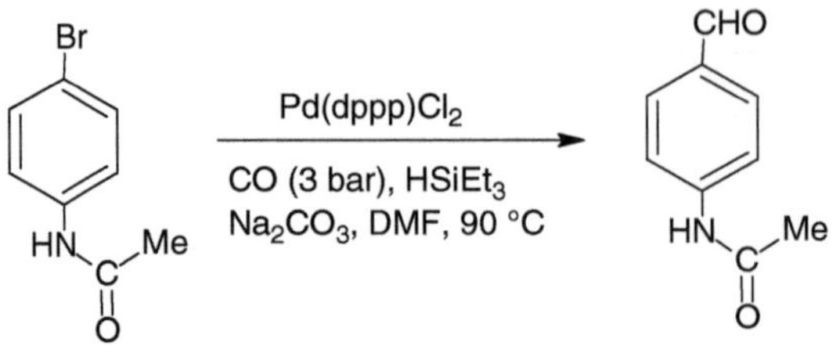

Scheme 13.63 Aldehyde generation from carbonylation reaction .

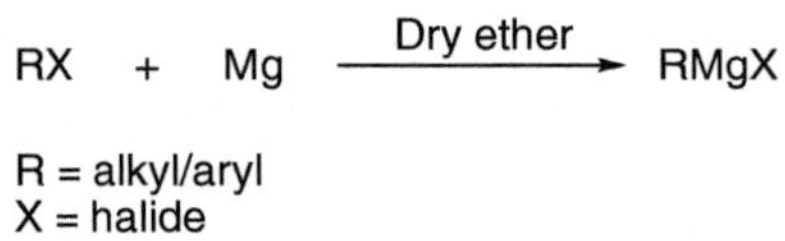

Scheme 13.64 Synthesis of Grignard reagents.

complex and the carbonylated product is obtained. 4-Bromophenylacetamide upon treatment with CO (3 bar) in presence of Pd(0) and a base along with triethylsilane yields 4-substituted benzaldehyde (Schemes 13.63).

13.4 Grignard Reagents

13.4.1 Introduction

Alkyl/aryl magnesium halides (R/Ar–Mg–X) are known as Grignard reagents. The French chemist Victor Grignard received the Nobel Prize in Chemistry in the year 1912 for its development. The Grignard reagent is one of the most popular reagents used in organic synthesis. It has a carbon–magnesium bond and acts as an efficient source of nucleophile (carbanion) (Scheme 13.64 & 13.65).

13.4.2 Synthesis

Grignard reagent is prepared in the laboratory by the reaction of magnesium with alkyl/aryl halide in dry ether/THF solvent. Magnesium is taken as magnesium turnings, but generally magnesium metal surface remains covered with magnesium oxide film, so a small quantity of iodine is taken to activate the magnesium. Magnesium oxide reacts with iodine to form magnesium iodide, which is readily removed from the magnesium turning's surface and exposes the magnesium metal for further reaction with alkyl/aryl halide. The reaction mixture is then refluxed using a water condenser.

Since it can react with the variety of functional groups the solvents like water, ethanol, acetone, benzaldehyde, diethyl ether, acetonitrile, chloroform, carbon tetrachloride, etc. cannot be taken as solvent. Also, it can react with oxygen, carbon dioxide, moisture, etc. present in the air, so an inert atmosphere (Ar or N_2) is maintained during the reaction and dry (anhydrous) solvents (Scheme 13.66) are taken

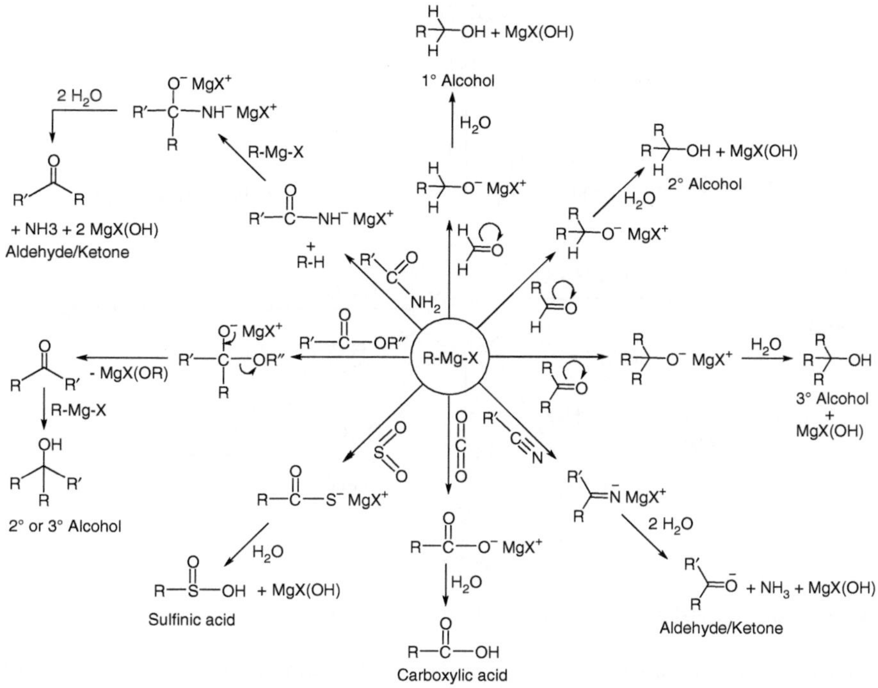

Scheme 13.65 Nucleophilic attack of Grignard reagent on carbon–heteroatom multiple bond.

for the reaction. To avoid the moisture, a $CaCl_2$ guard tube is also used on the top of the condenser, if an inert atmosphere is not provided.

Alky halides react with magnesium easily in diethyl ether, but aryl halides and vinyl halides react slowly and thus need vigorous reaction conditions. Tetrahydrofuran solvent is preferred over diethyl ether due to its higher boiling point and better solvation of Grignard reagents which helps in increasing the rate of reaction. Ether forms a complex with magnesium through the lone pair of electrons present on oxygen atom and vacant orbitals on magnesium.

For any alkyl group, the formation of Grignard reagent depends on the size of the halide. As the size of the halide increases, the formation of Grignard reagent becomes easier. Also, for a particular halide, as the chain length of the alkyl group increases, the formation of Grignard reagent becomes difficult.

13.4.3 Reactivity

Since Grignard reagents provide a nucleophile (carbanion), they give nucleophilic addition reaction on carbon–heteroatom multiple bonds. Carbon is electron deficient in carbon–heteroatom multiple bonds due to the higher electronegativity of heteroatoms and thus the electrophile can easily attack on such carbon.

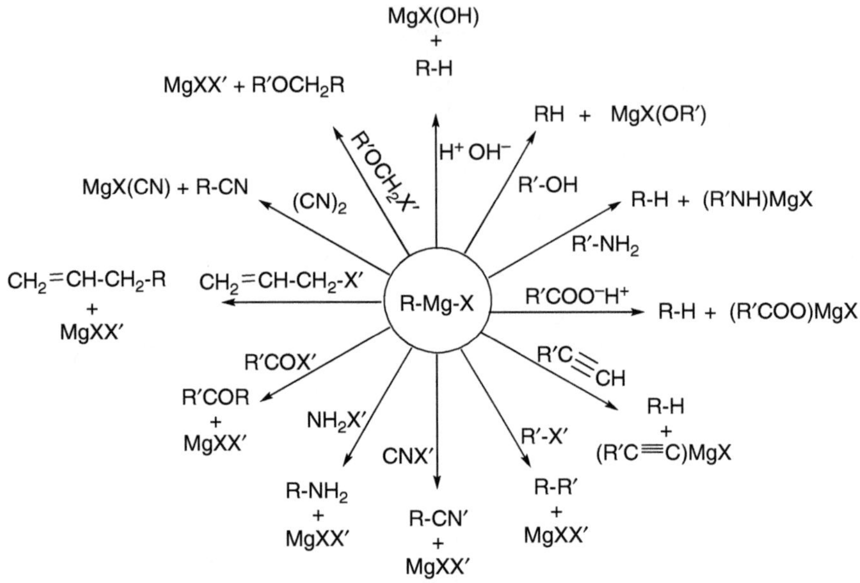

Scheme 13.66 Miscellaneous reactions of Grignard reagent.

Grignard reagents also give nucleophilic substitution reaction. The R^- part of Grignard reagent combines with the positive part of the reactant and the MgX^+ part of Grignard reagent combines with the negative part of the reactant (Scheme 13.66).

In case of α,β-unsaturated carbonyl compounds (Scheme 13.67), the Grignard reagent gives a 1,2-addition product. A similar 1,2-addition is observed with the organolithium reagents, while a 1,4-addition is observed with lithium dialkyl copper (Gilman reagent).

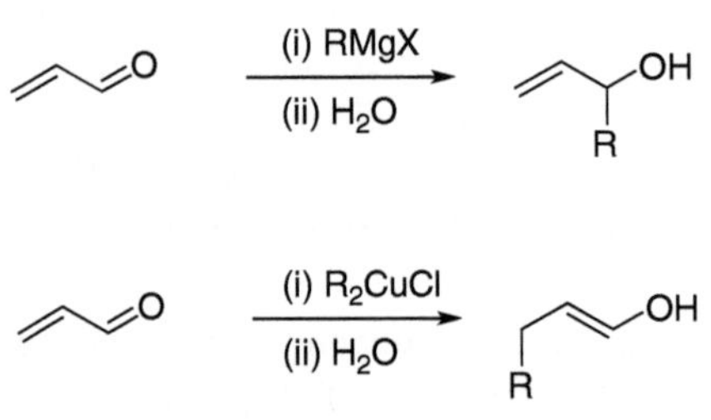

Scheme 13.67 Addition of Grignard reagent on α,β-unsaturated carbonyl compound.

References

1 Negishi, E. (1982). Aliphatic organoboron compounds. In: *Comprehensive Organometallic Chemistry* (ed. G. Wilkinson), 265–301. Elsevier.

2 Clay, J.M. and Vedejs, E. (2005). *J. Am. Chem. Soc.* 127: 5766–5767.

3 Kabalka, G.W., Shoup, T.M., and Goudgon, N.M. (1989). *J. Org. Chem.* 5930–5933.

4 (a) Patra, P.K., Nishide, K., Fuji, K., and Node, M. (2004). *Synthesis* 1003–1006. (b) Brown, H.C., Coleman, R.A., and Rathke, M.W. (1968). *J. Am. Chem. Soc.* 90: 499. (c) Brown, H.C. and Negishi, E. (1967). *J. Am. Chem. Soc.* 89: 5285. (d) Davies, A.G. and Roberts, B.P. (1972). *Acc. Chem. Res.* 5: 387. (e) Davies, A.G. and Roberts, B.P. (1969). *J. Chem. Soc. (B)*, 311.

5 Negishi, E.-I. (1976). *J. Organomet. Chem.* 108 (3): 281–324.

6 Suzuki, A. (2011). *Angew. Chem. Int. Ed.* 50: 6723–6737.

7 Miyaura, N., Yamada, K., Suginome, H., and Suzuki, A. (1985). *J. Am. Chem. Soc.* 107: 972–980.

8 Miyaura, N. and Suzuki, A. (1981). *J. Organomet. Chem.* 213: C53–C56.

9 Scharnagl, F.K., Bose, S.K., and Marder, T.B. (2017). *Org. Biomol. Chem.* 15: 1738–1752; and references cited therein.

10 Mikhailov, B.M. and Bubnov, Y.N. (1964). *Izv. Akad. Nauk SSSR* 10: 1874–1876.

11 Brown, H.C. and Ramachandran, P.V. (1995). *J. Organomet. Chem.* 500: 1–19.

12 Chan, D.M.T., Monaco, K.L., Wang, R.P., and Winters, M.P. (1998). *Tetrahedron Lett.* 39: 2933–2936.

13 Lam, P.Y.S., Clark, C.G., Saubern, S. et al. (1998). *Tetrahedron Lett.* 39: 2941–2944.

14 Evans, D.A., Katz, J.L., and West, T.R. (1998). *Tetrahedron Lett.* 39: 2937–2940.

15 Frauenlob, R., Carcia, C., Bradshaw, G.A. et al. (2012). *J. Org. Chem.* 77: 4445–4449.

16 Diehl, A.M., Ouadoudi, O., Andreadon, E., and Manolikakes, G. (2018). *Synthesis* 50: 3936–3946.

17 Yoshikai, N. and Nakamura, E. (2012). *Chem. Rev.* 112: 2339–2372.

18 Gilman, H., Jones, R.G., and Woods, L.A. (1952). *J. Org. Chem.* 17: 1630.

19 Flemming, I., Roberts, R.S., and Smith, S.C. (1998). *Perkin Trans.* 1: 1209–1214.

20 Nakamura, E., Matsuzawa, S., Horiguchi, Y., and Kuwajima, I. (1986). *Tet. Lett.* 27: 4029–4032.

21 Sato, F. (1988). *J. Org. Chem.* 53: 5590.

22 Krause, N. and Thoranol, S. (1999). *Inorg. Chim. Acta* 296: 1–11.

23 Alexakis, A. (1992). *Pure Appl. Chem.* 64: 387.

24 Smidt, J., Hafner, W., Jira, R. et al. (1959). *Angew. Chem.* 71: 176.

25 Heck, R.F. (1968). *J. Am. Chem. Soc.* 90: 5518.

26 Heck, R.F. (1968). *J. Am. Chem. Soc.* 90: 5526.

27 Heck, R.F. (1968). *J. Am. Chem. Soc.* 90: 5531.

28 Heck, R.F. (1968). *J. Am. Chem. Soc.* 90: 5538.

29 Heck, R.F. (1968). *J. Am. Chem. Soc.* 90: 5542.

30 Negishi, E.-I., King, A.O., and Okukado, N. (1977). *J. Org. Chem.* 42: 1821.

31 King, A.O., Okukado, N., and Negishi, E.-I. (1977). *Chem. Commun.* 683.

32 Krasovskiy, A., Krasovskaya, V., and Knochel, P. (2006). *Angew. Chem. Int. Ed.* 45: 2958–2961.

33 Zhou, J. and Fu, G.C. (2003). *J. Am. Chem. Soc.* 125: 12527–12530.

34 Nunez, A., Abarca, B., Cuadro, A.M. et al. (2009). *J. Org. Chem.* 74: 4166–4176.

35 Zheng, G.Z., Mao, Y., Lee, C.-H. et al. (2003). *Bioorg. Med. Chem. Lett.* 18: 3041–3044.

36 Littke, A.F., Schwartz, L., and Fu, G.C. (2002). *J. Am. Chem. Soc.* 124: 6343–6348.

37 Bonnet, V., Mongin, F., Trecourt, F. et al. (2002). *Tetrahedron* 4429–4438.

38 Li, J.-H., Deng, W.-J., and Liu, Y.-X. (2005). *Synthesis* 3039–3044.

39 Lee, J.-Y. and Fu, G.C. (2003). *J. Am. Chem. Soc.* 125: 5616–5617.

40 Thorand, S. and Krause, N. (1998). *J. Org. Chem.* 63: 8551–8553.

41 Guram, A.S., Rennels, R.A., and Buchwald, S.L. (1995). *Angew. Chem. Int. Ed.* 34: 1348–1350.

42 Hartwig, J.F. (1995). *Tetrahedron Lett.* 36: 3609–3612.

43 Sheng, Q. and Hartwig, J.F. (2008). *Org. Lett.* 10: 4109–4112.

44 Neetha, M., Afsina, C.M.A., Aneeja, T., and Anilkumar, G. (2020). *RSC Adv.* 10: 33683–33699.

45 Barnard, C.F.J. (2008). *Organometallics* 27: 5402–5422.

14

Homogeneous and Heterogeneous Catalysis by Organometallic Complexes

Ratnakar D. Shukla[1], *Bhawna Jain*[2], *Kuleshwar Patel*[3], *Priyanka Singh*[3*], *Dakeshwar Kumar Verma*[3], *Reema Sahu*[3], and *Raghvendra K. Mishra*[4*]

[1] *Pt. Deen Dayal Upadhyay Government College, Department of Chemistry, Bahua, Fatehpur, Uttar Pradesh, India*
[2] *Siddhachalam Laboratory, Raipur, Chhattisgarh, 493221, India*
[3] *Govt. Digvijay PG Autonomous College, Department of Chemistry, Rajnandgaon, Chhattisgarh, 491441, India*
[4] *IMDEA Materials, Madrid, Spain*

14.1 Introduction

The concept of catalysis is very important in the field of chemistry and it is frequently used as a strategic tool in numerous areas such as life sciences, environmental sciences, energy, and industry [1]. Basically, catalysis, a molecular phenomenon, promotes chemical transformation. In fact, catalysis may be divided into two types heterogeneous or homogeneous, depending on the physical states of reactants and catalysts. In the current scenario, the connections between homogeneous catalysis and molecular organometallic chemistry are observed, whereas heterogeneous catalysis shows proximity towards solid-state chemistry and surface science [2]. In this chapter, the applications of organometallic compounds in homogeneous catalysis and several important chemical reactions catalyzed by them have been covered.

Homogeneous catalysis is regarded as one of the most successful applications of organometallic chemistry – a branch of chemistry that has combined organic and inorganic chemistry – to a remarkable degree [3, 4].

Here, the salient features of homogeneous catalysts are as follows:

- Dispersion of homogeneous catalysts "in the same phase" at molecular level.
- Simple, smooth, and reproducible synthesis and manufacturing.
- Easy characterization with the help of chemical and spectroscopic techniques.
- Substrate specific designing is based upon known principles of catalysis.

On the basis of divergent operating systems, previously heterogeneous and homogeneous catalysis were associated with different areas. Subsequently it was thought that both followed different mechanisms of catalysis in elementary steps.

*Corresponding authors.

Organometallic Compounds: Synthesis, Reactions, and Applications, First Edition.
Edited by Dakeshwar Kumar Verma and Jeenat Aslam.

Later, it was observed that both were approaching each other, and this connection was well studied in the field of organometallic surface chemistry.

The applications of organometallic compounds in the area of homogeneous catalysis have gone beyond the frontiers of industry to provide the small-scale synthesis of daily needs in a laboratory [5, 6]. The prominent example of such an application is alkene isomerization using transition metal organometallic complexes as a homogeneous catalyst. The conduction of hydrogenation reactions of substrates like alkene, arene, imine, and ketone is done using various types of transition metal organometallic catalysts. They also proceed via diverse mechanisms involving oxidative addition, homolytic, and heterolytic cleavages of the H—H bond.

In the current scenario, the study and exploration of organometallic compounds as homogeneous catalysts is highly demanding in the field of catalysis. The development of many novel organometallic catalysts accelerates the optimization and subsequent application of organometallic chemistry in homogenous catalysis [7, 8]. The benefits of homogeneous organometallic catalysts have currently spread to all conceivable areas of chemical sciences, from the confines of industry for synthetic processes used on a daily basis to academics for organic synthesis. Presently, several developmental processes related to improving our daily lives have achieved their goals due to the incorporation of organometallic compounds in the field of homogeneous catalysis. Alkene metathesis is regarded as one of the important examples of the aforementioned target, which is executed with the help of homogeneous organometallic catalysts. The production of many novel chemical moieties has been achieved with the help of various homogeneous organometallic catalysts. Consequently, the library of novel synthesized chemical compounds is being explored in various fields of chemical sciences, including pharmaceutical chemistry.

14.2 Organometallic Compounds and Homogeneous Catalysis

When a metal or metalloid possesses at least one bond to a carbon atom directly, the resulting compound is regarded as an organometallic compound. Notably, the carbon metal bond may have either ionic or covalent characteristics. To expand the field of homogeneous transition-metal catalysis, exploitation of a diverse range of transition metal complexes has been done to check their catalytic efficacies. After the successful outcome, employment of homogeneous transition-metal catalysts has been examined for the synthesis of organic molecules [9].

In terms of selectivity of the reaction, it is observed that homogeneous catalysis works better in comparison to heterogeneous catalysis because homogeneous catalysis can be monitored with the incorporation of suitable ligands on the catalytic metal center [10]. One of the noticeable advantages of organometallic catalysis is the stepwise economic synthesis of organic compounds. In addition, conduction of enantioselective reactions employing homogeneous catalysts is done; consequently, high optical yields are obtained and this application has great importance in industry [11]. Furthermore, in homogeneous catalysis, smooth reaction conditions such as low temperature and pressure are frequently sufficient.

Due to the emergence of green and sustainable chemistry in the field of catalysis, the recovery of catalysts is highly desirable. In this context, numerous methods such as precipitation, immobilization on carrier materials, reverse osmosis, and reaction in two-phase liquid–liquid systems are available for catalyst recovery after the reaction [12–14]. Here, we will discuss some important organometallic homogeneous catalysis-driven chemical transformations one by one.

14.3 Catalytic Elementary Reactions

Basically, stepwise categorization of a catalytic cycle for a homogeneous catalyst can be done as follows [15–17]:

I. Generation of the active species,
II. coordination of metal center and the reactants,
III. production of the product, and
IV. elimination of the product with regeneration of the initial catalytically active species

The pictorial representation of the four aforementioned steps is well depicted below.

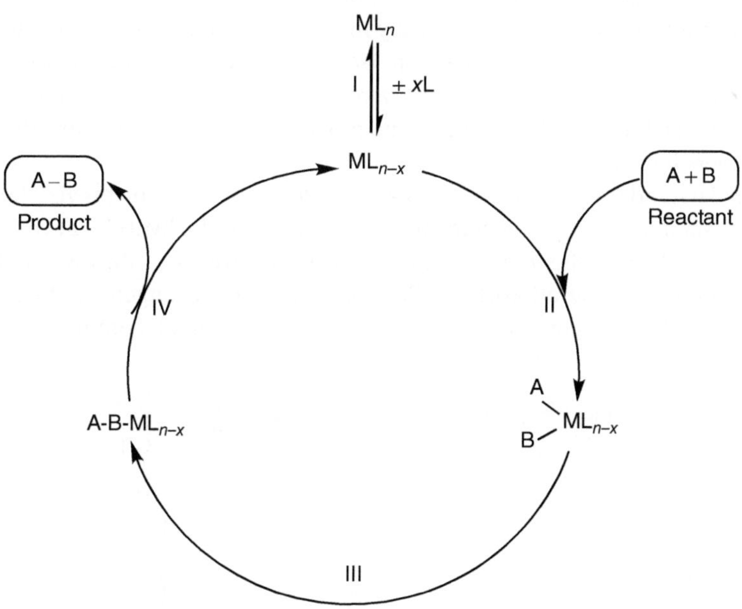

The first step is completed with the *in situ* conversion of starting complex ML_n to ML_{n-x} as a catalytically active species. The presence of vacant coordination sites in catalytically active species is the basic requirement for coordination of the reactants A and B. Often, the generation of active species is done after the dissociation of the ligand, i.e. by the elimination of Lewis bases like phosphines or Lewis acids such as protons.

The second step of the catalytic cycle is initiated with the binding of one or more reactants to the metal center. This step is completed either by oxidative addition, in which molecules like H–H, R–X, or H–X are cleaved and then the two fragments attach to the metal center via σ-bonds or by ligand association, i.e. addition of ligands such as carbon monoxide, alkenes or alkynes, to the metal center. Notably, in oxidative addition, the central metal observes an increase of two in both the oxidation state and coordination number.

The reactions of bond formation between the reactants associated with the metal center are observed in the third step. In addition, insertion reactions are often observed in this step.

In the fourth step, the catalytic cycle is completed with the elimination of the product. Usually, the product is released via the mechanism of reductive elimination. Reductive elimination is just the reverse of oxidative addition.

14.3.1 Isomerization

Mostly transition-metal complexes, particularly transition-metal hydride complexes having the general formula L_nMH (may also be generated *in situ* from ML_n and hydrogen), are explored as homogeneous catalysts for isomerization. In addition to the skeletal rearrangements of cyclic aliphatic compounds, the aforementioned homogeneous catalysts can also catalyze the isomerization of double-bond systems including alkenes, dienes, and allyl compounds [18–20].

The generation of thermodynamically more stable isomers is considered the driving force for the isomerization reaction. Isomerization strongly depends on reaction temperature, and this reaction is an example of an equilibrium reaction. At higher temperatures, it is also observed that the proportion of thermodynamically less stable isomers increases.

To understand the story of alkene isomerization, two mechanisms viz. the hydride mechanism and the allyl mechanism, are available. In the hydride mechanism, the addition of alkene to the transition-metal hydride to give an η^2-alkene complex takes place followed by generation of a σ-alkyl complex after rearrangement. Regeneration of the initial hydride complex occurs after the elimination of a product i.e. olefin.

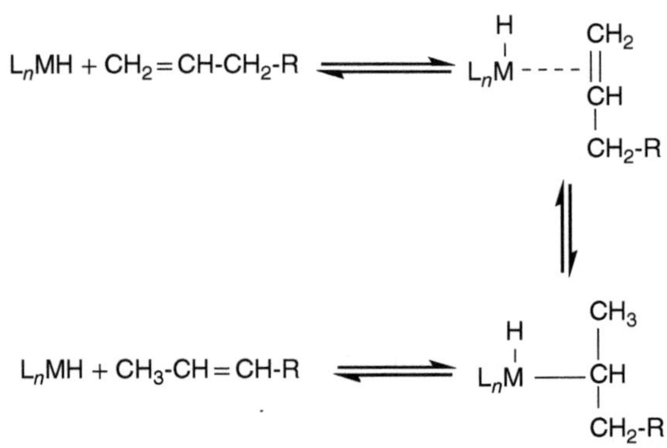

In the allyl mechanism, η^3-allyl complex is formed as an isomerization intermediate via the removal of a proton adjacent to the double bond in the presence of a transition metal.

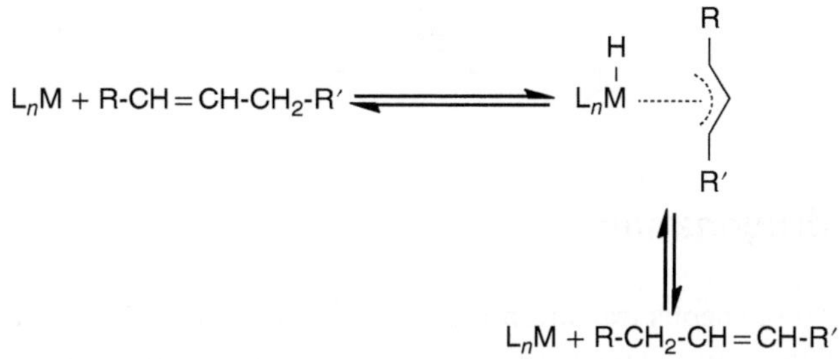

Several metal complexes of groups 6, 8, 9, and 10 can be exemplified as isomerization catalysts. Examples of the most common complexes which are frequently used as catalysts are iron carbonyls, rhodium chloride, rhodium chloro complexes, titanium, and palladium complexes.

14.3.2 Alkene Isomerization

The conversion of 1-hexene to 2- and 3-hexene can be achieved using iron carbonyls as a catalyst.

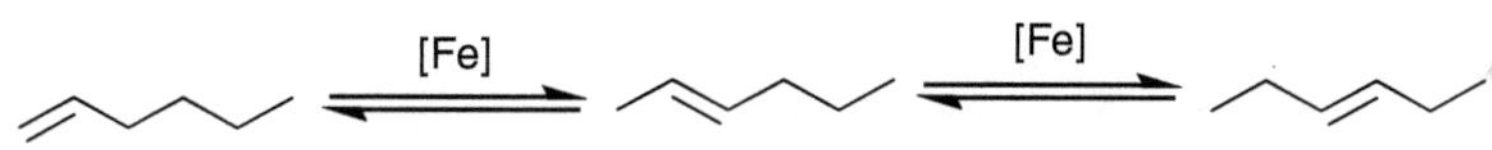

The transformation of exocyclic enones to endocyclic enones via isomerization reaction is done using rhodium(III) trichloride [$RhCl_3$], which is a very suitable catalyst for this reaction. Subsequently 2-substituted cyclopentenones are synthesized.

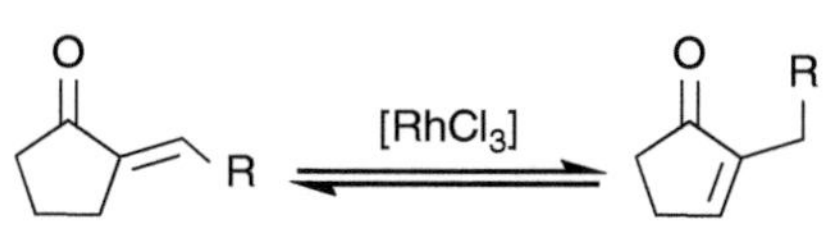

14.3.3 Alkyne Isomerization

Manganese complexes can conduct the conversion of alkynes to their corresponding allenes.

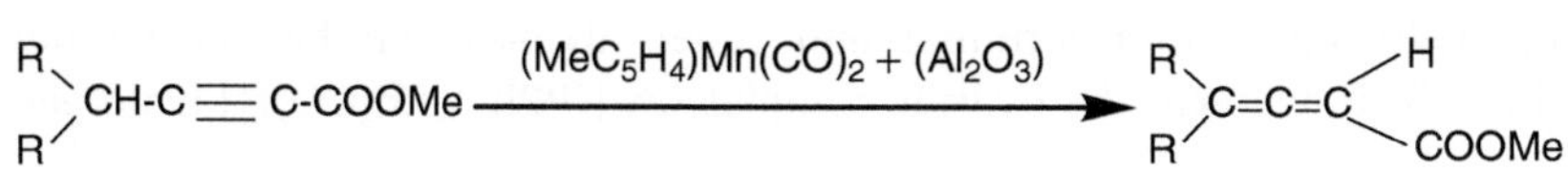

14.3.4 Diene Isomerization

The conversion of thermodynamically more stable 1,3-isomer of cyclooctadiene from 1,5- is afforded in the presence of iron(0) pentacarbonyl [$Fe(CO)_5$]:

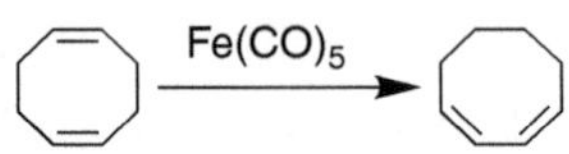

14.4 Hydrogenation

Numerous homogeneous transition-metal catalysts are used for the hydrogenation of alkenes, dienes, and alkynes [21–26]. H_2IrCl_6, [$(PPh_3)_2IrCl(CO)$], [$HCo(CO)_4$], and [$(PPh_3)_2(CO)_2RuCl_2$] are some typical examples of catalysts for hydrogenation. Wilkinson's catalyst [$(PPh_3)_3RhCl$] is one of the most commonly employed catalysts. In the presence of this catalyst, the hydrogenation is carried out under mild conditions. A simple reaction mechanism for hydrogenation reaction in the presence of Wilkinson's catalyst is clearly demonstrated below.

A dihydridorhodium(III) complex is obtained via oxidative addition of hydrogen, which after dissociating a phosphine ligand, creates a vacant coordination site for bonding of an alkene. A hydrido alkyl rhodium complex is obtained after the step of insertion. Then the desired product, an alkane, is obtained after elimination, and the starting complex is regenerated after complexation of the phosphine.

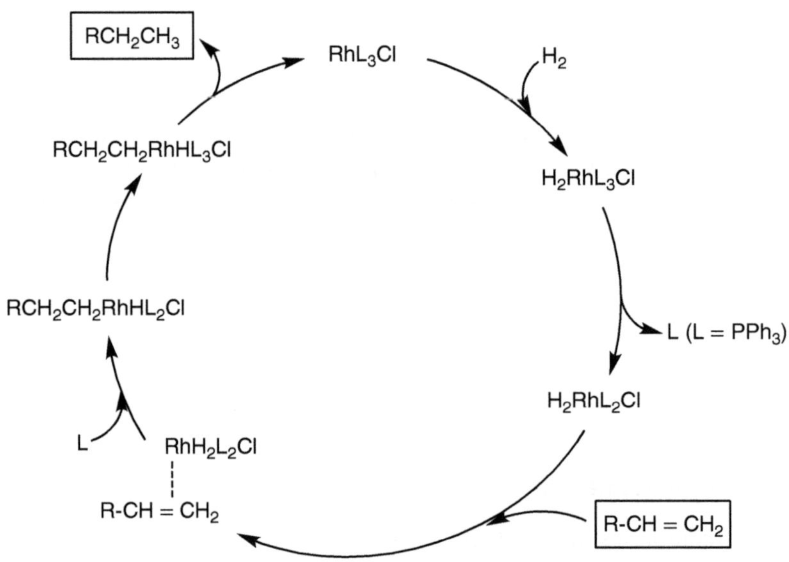

The hydrogenation of terminal and inner alkenes can be carried out by [$(PPh_3)_3RhCl$], whereas the hydride complex i.e. [$(PPh_3)_3Rh(CO)H$] hydrogenates only terminal double bonds. It is also observed that this hydride complex is less reactive in comparison to Wilkinson's catalyst. The hydrogenation of ethene and propene is carried out by the carbonyl chloro complex *trans*-$[(PPh_3)_3Rh(CO)Cl]$, but this catalyst hydrogenates in slow rates only. Furthermore, the hydrogenation reaction becomes fast in the presence of corresponding iridium complex *trans*-$[(PPh_3)_3Ir(CO)Cl]$. Notably, several cationic catalysts such as complexes of the type $[L_2Rh(THF)_2]^+$ shows high catalytic activity.

14.5 Carbon–Carbon Bond Formation

The processes of linear and cyclic oligomerization, telomerization, polymerization, and vinylogation are studied in this section [27–32]. The oligomerization of alkenes results in linear or branched high molecular mass alkenes. In the production of fine chemicals, plasticizers, and detergents, the linear or branched high molecular mass alkenes are used frequently. To understand the oligomerization of alkenes, two common mechanisms viz. the hydride mechanism and the metallacycle mechanism, are reported in the literature. In hydride mechanism, a metal hydride complex is considered as a catalytically active species and into the M—H bond of a metal hydride complex, the insertion of an alkene is observed. Next, due to the insertions of alkene molecules into the resulting metal–carbon (M—C) bond, the long-chain metal alkyls are generated. The β-hydrogen elimination terminates the reaction; consequently, the desired oligomer is liberated, followed by regeneration of the metal hydride.

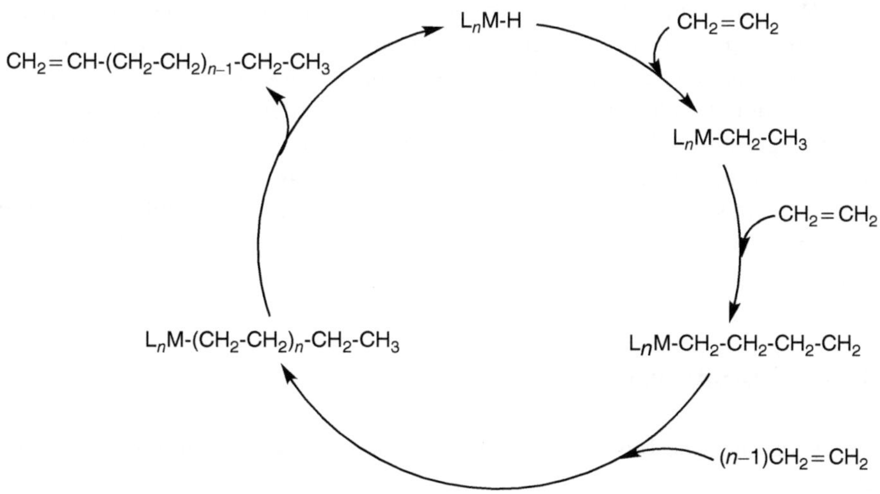

Dimerization of ethene into butane is an example of the metallacycle mechanism. A bis(ethene) complex produces metallacyclopentane, which is then rearranged and hydrogen shifted to form a butene complex.

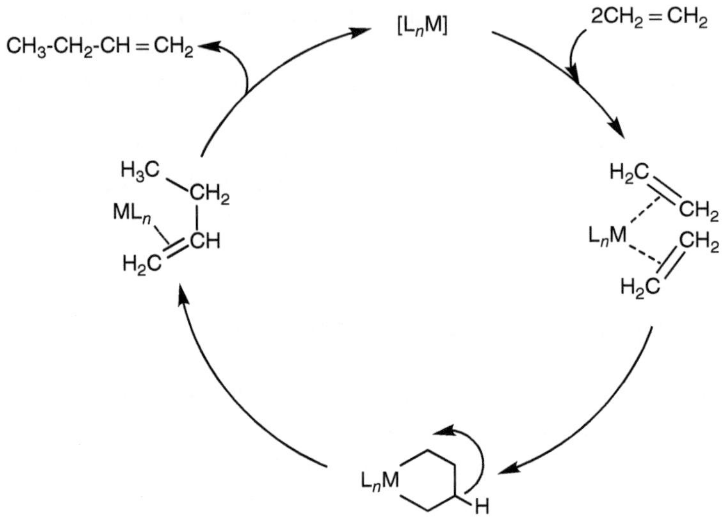

Development of several catalysts has been reported for ethene oligomerization. Ziegler–Natta catalyst is a prominent example of the aforementioned group. Ziegler–Natta catalysts are generated from an aluminum alkyl and a transition-metal component. Titanium, cobalt, nickel, and zirconium are the most common transition metals for these catalysts. Generally, titanium and zirconium are employed in the form of their halides or alkoxides, whereas nickel and cobalt are used as acetyl acetonates or allyl complexes. Adducts of the transition-metal complex and aluminum component are considered catalytically active species. When titanium tetrachloride and ethyl aluminum dichloride react, a complex is formed that has a free coordination site available for an olefin to bond to.

In addition to Ziegler–Natta catalysts, several other metal compounds have catalytic properties for the oligomerization of alkenes, in these cases; no activation by aluminum alkyls is required. Some of the important example of these type of catalysts are cobalt dinitrogen complex $[CoH(N_2)(PPh_3)_3]$, rhodium trichloride $[RhCl_3]$, ruthenium trichloride $[RuCl_3]$, and palladium dichloride $[PdCl_2]$.

14.6 Metathesis

In general, metathesis is defined as the exchange of alkylidene groups between two alkenes [28–32].

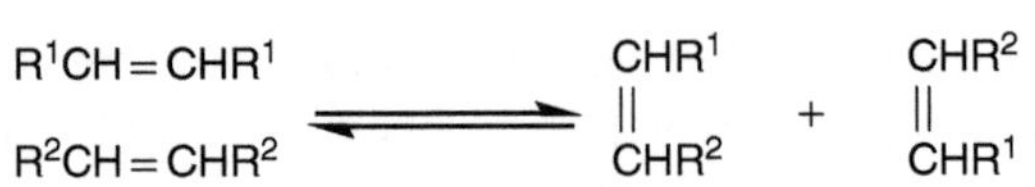

WCl_6-EtOH-R_2AlCl is regarded as one of the most common homogeneous catalytic systems for metathesis. In fact, the reactive intermediates are commonly observed in the form of carbene complexes.

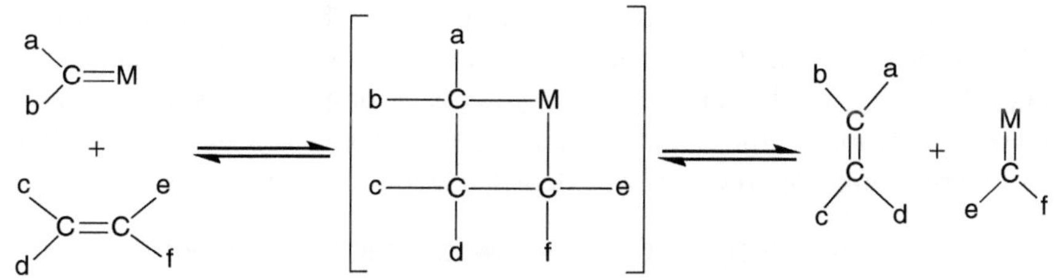

A detailed mechanism for the generation of metal from the catalytic system WCl_6-Me_2AlCl is depicted below.

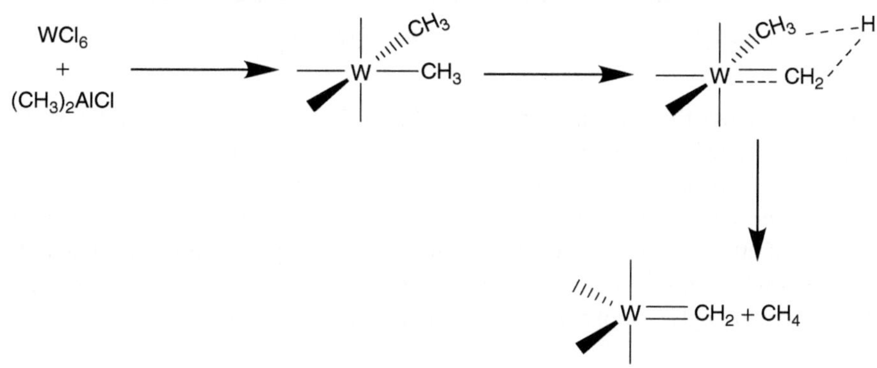

Huls (1980) reported the generation of polyoctenamer from cyclooctene via metathesis. The trade name for the synthesized polyoctenamer is Vestenamer 8012.

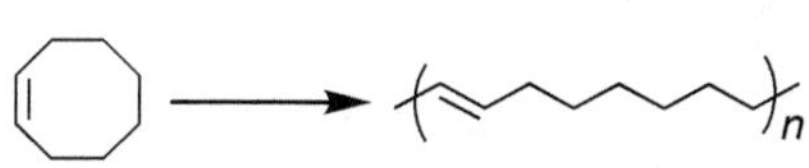

The conversion of α,ω-dienes from cycloolefins and ethane is studied as the Feast process. 1,5-cyclooctadiene, generated via cyclodimerization of butadiene, is explored as the starting material for the Feast process.

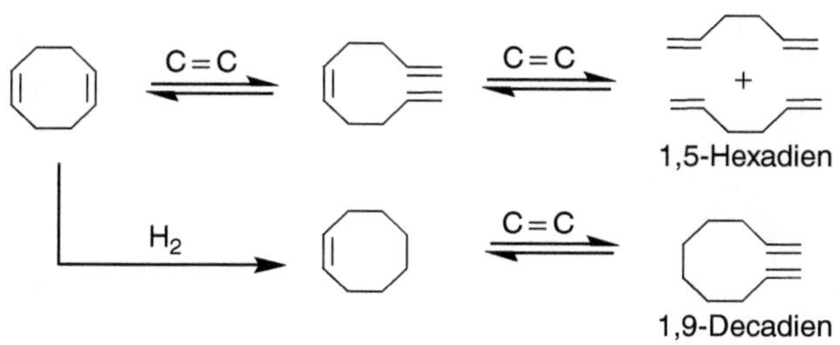

14.7 Oxidation

The Wacker–Hoechst process is a well-known process for oxidation. In this process, the decomposition of ethene chloro complexes of palladium(II) into acetaldehyde and palladium(0) occurred in the presence of water. The use of air environment facilitates the oxidation of Pd(0) back to Pd(II), and the use of a Cu(I)/Cu(II) couple allows the reaction to be catalyzed. This process is completed in several steps, but the overall reaction of ethene and oxygen to give acetaldehyde can be written in one step, and the complete process is well depicted below [33–35].

$$1.\ C_2H_4 + PdCl_2 + H_2O \longrightarrow CH_3CHO + Pd(0) + 2HCl$$

$$2.\ Pd(0) + 2CuCl_2 \longrightarrow PdCl_2 + 2CuCl$$

$$3.\ 2CuCl + 2HCl + 1/2O_2 \longrightarrow 2CuCl_2 + H_2O$$

$$S\ C_2H_4 + 1/2O_2 \longrightarrow CH_3CHO$$

Notably, when the terminal or inner alkenes is explored as substrates in this reaction, ketones are generated as ketones instead of aldehydes. For example, acetone is obtained from propene. The detailed mechanism of acetaldehyde generation via the Wacker–Hoechst process is being demonstrated. In the first step, the addition of ethene and water to the metal center takes place. In the next step, the hydroxyethyl complex is generated via oxypalladation. Finally, after rearranging into a hydridovinyl alcohol complex, acetaldehyde is obtained.

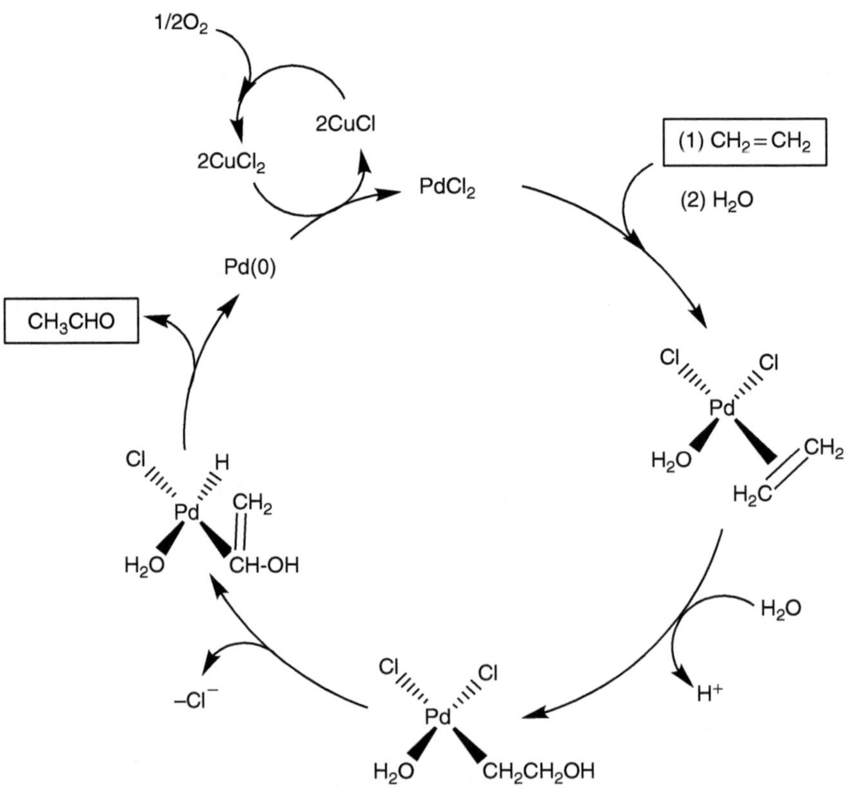

14.8 Reactions with Carbon Monoxide

In this section, transition-metal catalyzed reactions of carbon monoxide with hydrogen, alkenes, alkynes, and alcohols are discussed [36–39].

The substrate for the Fischer–Tropsch reaction is synthesis gas, which is generated from the reaction of carbon monoxide with hydrogen. Recently, Fischer–Tropsch chemistry research has primarily focused on controlling the formation of side products such as alcohols, short-chain olefins, and long-chain α-olefins. Extensive investigations into the reactions of hydrogen and carbon monoxide in the presence of a homogeneous catalyst are highly explored. In this context, synthesis of ethylene glycol is afforded in presence of a rhodium catalyst [40, 41].

Roelenat Ruhrchemie reported the hydroformylation reaction, which is considered one of the most important reactions of carbon monoxide and alkenes [42].

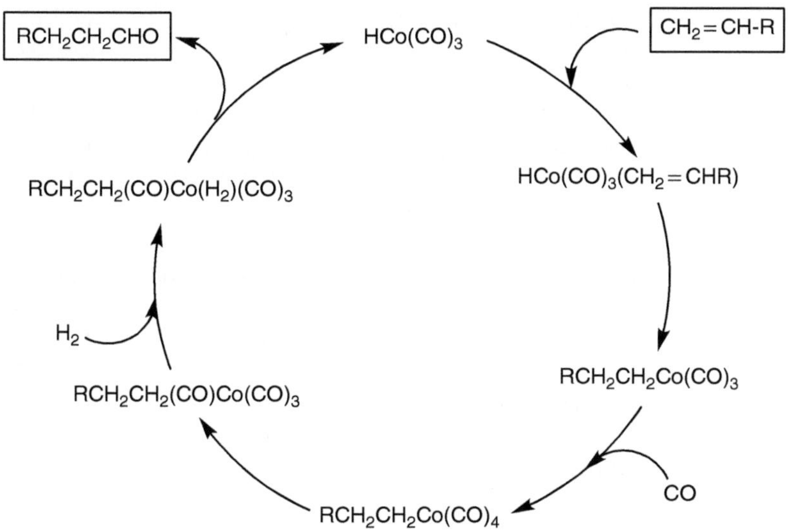

In the hydroformylation reaction, propionaldehyde is obtained from ethane and synthesis gas using a cobalt carbonyl catalyst. In fact, aldehydes are the primary products of hydroformylation reactions, which are often transformed into other products, mostly alcohols. The mechanism of cobalt-catalyzed hydroformylation involves going through the formation of hydridocobalt carbonyl, which is regarded as a catalytically active species. An acyl complex is produced via the insertion of an olefin and carbon monoxide. Finally, aldehyde is produced followed by hydrogenation [43, 44].

14.9 Heterogenous Catalysis

Heterogeneous catalysis involves the use of solid or metallic catalysts. The reactants and/or products have different phases as compared to those of the catalyst. Phase difference stands for solid, liquid, or gaseous states along with immiscible mixtures as well as the presence of interfaces [45–47]. Table 14.1 enlists different

Table 14.1 List of different heterogenous catalysis-based reactions.

S. No.	Reaction	Substrate	Catalyst	Product	References
1	Polymerization	Propene	$MgCl_2/TiCl_4$	Polypropylene	Cavallo and coworkers [48]
2	Hydrogenation	N_2	Hydride-based complexes of Co, Fe, Ru, Rh, Ir, Mn, Mo, Ni, Ti, Ta	NH_3	Kageyama and coworkers [49]
3	Hydrogenation	N_2	Ru	NH_3	Marakatti and Gaigneaux [50]
4	Hydrogenation	N_2	Fe	NH_3	Li and coworkers [51]
5	Dehydro-genation	Polyethylene glycol	Ni/ZrO_2	H_2	Wei and coworkers [52]
6	Oxidative coupling	Methane	Li/MgO	Ethane + ethylene	Tarr and coworkers [53]
7	Oxidation	Methane	Pt, Ni	$CO + H_2$ + formaldehyde	Tarr and coworkers [53]
8	Oxychloration	Ethylene + Cl_2	CuPdCl	Vinyl chloride	Tarr and coworkers [53]
9	Oxychloration	Ethane + Cl_2	AgMnCoO	Vinyl chloride	Tarr and coworkers [53]
10	Oxidation	Ethane	MoVNbO	Acetic acid	Tarr and coworkers [53]
11	Oxidation	*n*-Butane	$(VO)_2P_2O_7$	Maleic anhydride	Tarr and coworkers [53]
12	Oxidation	Ethylene	Ag/Al_2O_3	Ethyl oxide, acetaldehyde, acetic acid	Tarr and coworkers [53]

heterogeneous catalysis-based reactions. The steps involved in heterogeneous catalysis (Figure 14.1) are:

(i) *Diffusion of reactants toward the surface*: Concentration of the reactant and thickness of the surface affects the rate of diffusion.
(ii) *Adsorption of reactants*: The adsorption of reactants on a surface can be through chemisorption or physisorption. In the case of chemisorption, bonds are formed between the reactant and the surface of the catalyst. The binding efficiency of a molecule or atom is known as the sticking coefficient, which is simply the percentage or proportion of molecules that remain bound to the surface.
(iii) *Reaction*: Atoms and/or molecules bound to the catalyst surface react with each other, resulting in bond formation.
(iv) *Desorption of product(s)*: Products are desorbed from the catalyst surface.
(v) *Diffusion of product(s)*: The products diffuse away from the catalyst surface.

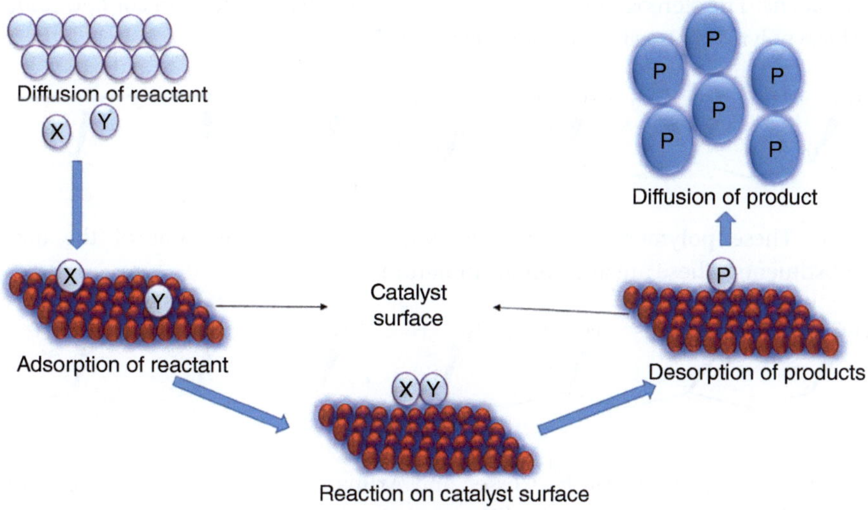

Figure 14.1 Steps involved in heterogeneous catalysis.

14.9.1 Ziegler–Natta Catalyst

Alkenes are polymerized in the presence of transition-metal complexes and aluminum alkyls. Ziegler et al. (1955) proposed the Ziegler–Natta polymerization of ethylene to give polyethylene in presence of $TiCl_4$ and $Al(C_2H_5)_3$ [46, 47, 49]. The noble prize (1963) in chemistry was shared by German chemist Karl Ziegler for the discovery of first Ti-based catalyst and Giulio Natta for preparation of stereoregular polymers by using this catalyst. Various polyolefins have been prepared commercially using the Ziegler–Natta catalyst; some of these include polyethylene, polypropylene, polycycloolefins, polyisoprene, polyacetylene, copolymers of ethylene, alkenes, etc.

14.9.1.1 Stereoregularity

Natta found that the polymers have a crystalline structure owing to stereoregularity. Polymers can have different tacticities depending on the orientation of alkyl groups in polymer chains having $(-CH_2-CH_2-)_n$ units. The stereoregularity of polymers is dependent upon the type of catalyst, which is used to prepare it.

Isotactic: All the stereogenic (CHR) centers have same configuration. These polymers are crystalline.

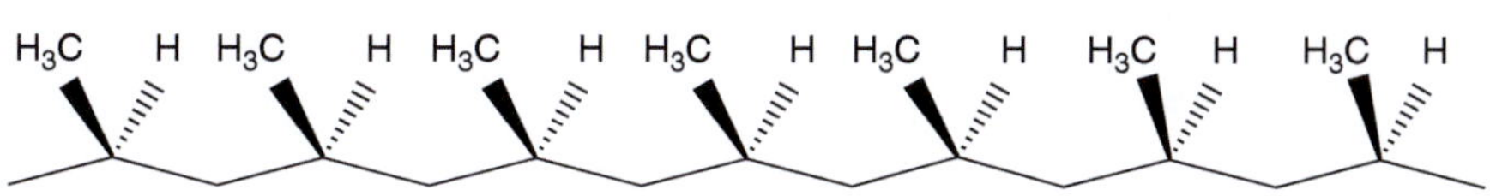

Syndiotactic: The stereogenic centers (CHR) have relatively alternate configuration. These polymers also show crystalline nature.

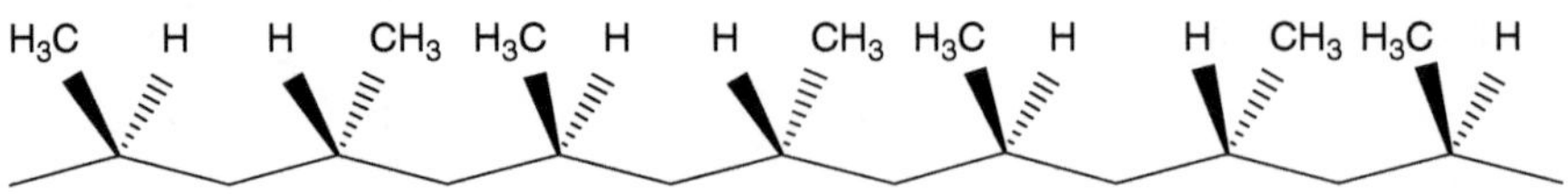

Atactic: These polymers do not have any regular arrangement of the alkyl substituents. These are amorphous in nature.

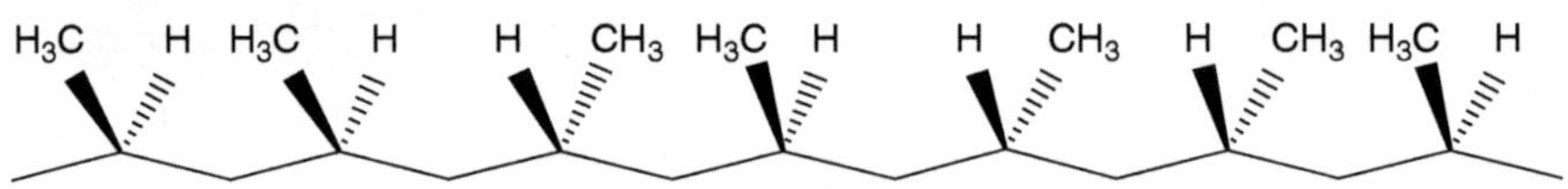

The mechanism proposed by Cossec and Arlman will be discussed here. $TiCl_4$ reacts with $Al(C_2H_5)_3$ to give $TiCl_3$ [45, 46]. A titanium alkyl complex is obtained by the reaction between $TiCl_3$ and $Al(C_2H_5)_3$. The alkene gets inserted into the Ti—C bond, resulting in the formation of a large alkyl complex. The chain length increases by further insertion of alkene.

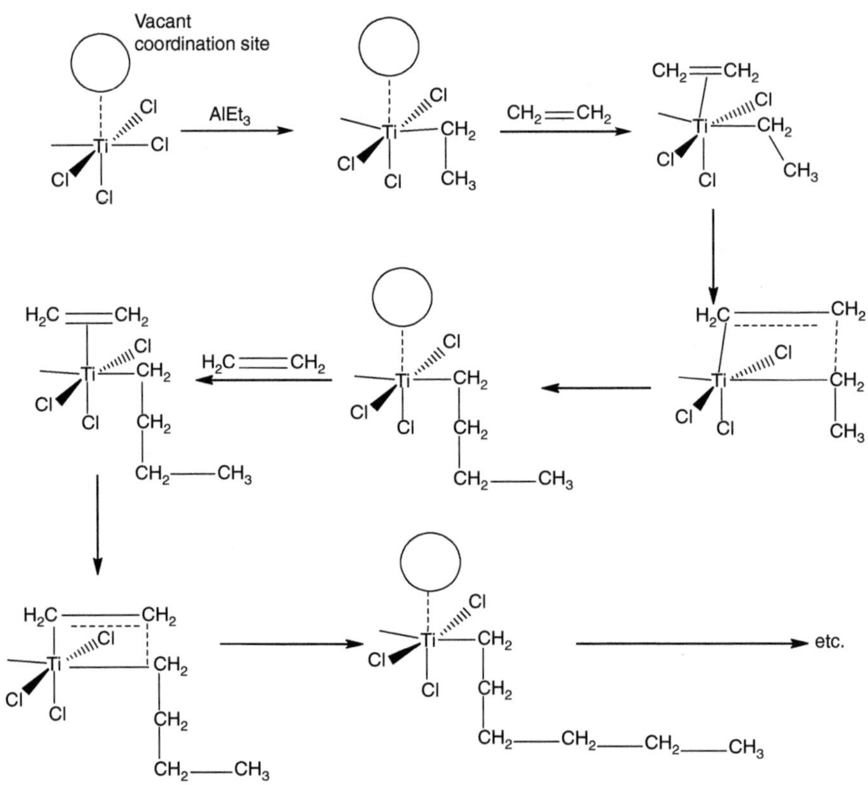

Gabe and coworkers (1985) have discussed the formation of a metallacylbutane intermediate. According to this mechanism, the steps involved are

1. Alkylidene is obtained from metal alkyl complex.
2. Ethylene gets added to form metallcyclobutane.
3. Ethylene then gets added to the metal–carbon bond.

It's difficult to distinguish the mechanism being followed. But Cossec–Arlman mechanism is the commonly observed one for polymerization reactions. Green and coworkers have presented strong evidence in support of this mechanism [46].

14.9.2 Water Gas Reaction

The reaction between steam (water) and a carbon source such as coke or coal at high temperature and pressure is known as the water–gas reaction [47].

$$\underset{\text{Red hot coke}}{C} + \underset{\text{Steam}}{H_2O} \longrightarrow \underbrace{CO + H_2}_{\text{Water gas}}$$

$$CH_4 + \underset{\text{Steam}}{H_2O} \xrightarrow{700\text{–}1000\,°C} CO + 3H_2$$

$$CO + H_2O \longrightarrow CO_2 + H_2$$

This reaction yields an equimolar mixture of hydrogen and carbon monoxide gases, also known as synthetic or syn gas. Some amounts of carbon dioxide gas might also be used in various other reactions. The Fischer–Tropsch process utilizes transition metals as catalysts to produce hydrocarbons, alcohols, alkanes, etc., from synthetic gas.

$$H_2 + CO \xrightarrow{Co} \text{Alkanes}$$

$$3H_2 + CO \xrightarrow{Ni} CH_4 + H_2O$$

$$2H_2 + CO \xrightarrow{\text{Co or Zn/Cu}} CH_3OH$$

This process was used for the production of synthetic fuels in various countries during the Second World War. However, due to the high cost of obtaining H_2 and CO gases from coal or petroleum, all of these initiatives proved to be ineffective. Steam reforming involves the reaction between natural gas (methane) and steam at elevated temperatures and pressures to yield carbon monoxide and hydrogen.

$$CH_4 + \underset{\text{Steam}}{H_2O} \xrightarrow{700\text{–}1000\,°C} CO + 3H_2$$

This process is used for producing hydrogen gas on an industrial scale. Carbon monoxide obtained as a product can be recycled again to produce hydrogen gas.

$$CO + H_2O \xrightarrow[400\,^{\circ}C]{\text{Fe-Cr or Zn-Cu}} CO_2 + H_2$$

This reaction is thermodynamically favorable and can yield 99% pure hydrogen gas, if CO_2 is removed chemically.

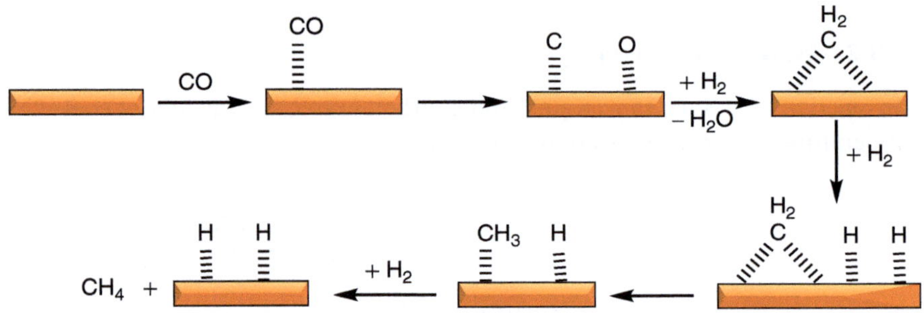

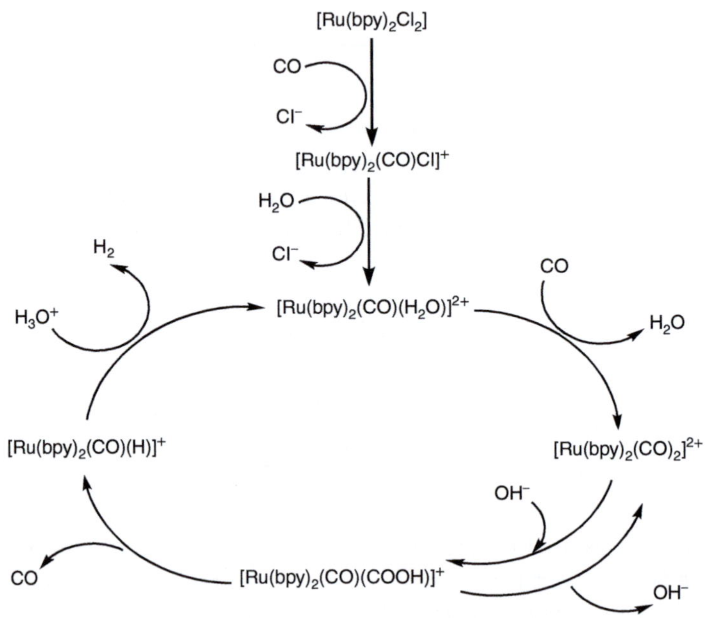

14.9.3 Zeolites

Zeolite is an aluminosilicate mineral that has a crystalline structure, silicon, aluminum, and oxygen. Its chemical formula is $M_xAl_xSi_{1-x}O_2{\cdot}yH_2O$, where M is a metal ion. AlO_4 and SiO_4 have tetrahedral structures and form covalent networks of zeolite. It consists of a frame of cavities and channels inside which cations, water and small molecules can exist due to the way zeolite is used as a molecular sieve. It is generally used as an adsorbent and catalyst. The term zeolite was coined by the mineralogist Axel Fredrik Kronstadt (Swedish) in 1756. Zeolites are produced industrially on a large scale but are also present in nature. More than 40 zeolites are known in nature.

Due to its porous nature, it can accommodate many cations such as Na^+, K^+, Ca^{2+}, Mg^{2+}, and others. All these cations are easily replaced by another contact solution. Analcime, phillipsite, clinoptilolite, and heulandite are examples of common zeolites. $Na_2Al_2Si_3O_{10}{\cdot}2H_2O$ (natrolite) is a simple example of the mineral form of zeolites. Volcanic pillars and ash sheets treated with basic ground water are the natural sources of zeolites. Natural zeolites are frequently impure and polluted with other minerals, metals, or other zeolites. Its distinct porous assembly and adaptable acidity make it dynamic in various reactions. They are used for separation and trapping of molecules. They are extensively applied as ion-exchange coats in domestic as well as viable water softening and purification processes. Mesoporous zeolite materials are broadly applied as catalysts in the petrochemical industry for hydrocracking and liquid catalytic cracking. The acidic zeolites can ease many acid-catalyzed reaction like isomerization and alkylation. Zeolites have micro-porous nature to bind some ions of fission products from nuclear processes and to be detached from the waste permanently.

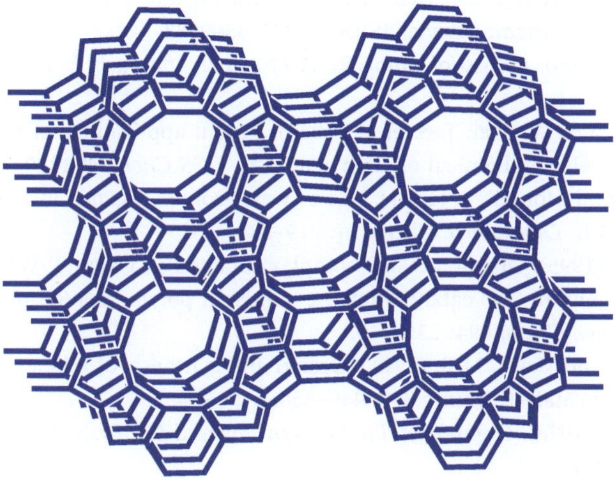

14.10 Conclusion

The use of oganometallic chemistry in the field of catalysis has always been fascinating to researchers. Heterogeneous and homogeneous catalysis have been discussed in this chapter. Continuous research work is going on in this field. However, further research needs to be done in this field to explore the applicability of oganometallic compounds in various reactions.

References

1 Roduner, E. (2014). Understanding catalysis. *Chem. Soc. Rev.* 43: 8226–8239.

2 Speight, J.G. (2017). Heterogeneous catalysis and homogeneous catalysis. In *Rules of Thumb for Petroleum Engineers*, J.G. Speight (Ed.). https://doi.org/10.1002/9781119403647.ch192, Scrivener Publishing LLC.

3 Buckingham, J. (ed.) (1984). *Dictionary of Organometallic Chemistry*, vol. 1–3. London: Chapman and Hall.

4 Haiduc, I. and Zuckerman, J.J. (1985). *Basic Organometallic Chemistry*. Berlin: Walter de Gruyter.

5 Behr, A. (2005). *Organometallic Compounds and Homogeneous Catalysis*. Weinheim: Wiley-VCH Verlag GmbH & Co. KGaA.

6 Herrmann, W. (2022). Organometallic compounds for homogeneous catalysis. In: *Catalysis from A to Z* (ed. W. Herrmann, B. Cornils, H. Zanthoff and J.-H. Xu). https://doi.org/10.1002/9783527809080.cataz11989.

7 Blaser, H.-U., Indolese, A., and Schnyder, A. (2000). Applied homogeneous catalysis by organometallic complexes. *Curr. Sci.* 78 (11): 1336–1344.

8 Cadierno, V. (2021). Recent advances in organometallic chemistry and catalysis. *Catalysts* 11 (5): 646.

9 Denisov, E.T., Sarkisov, O.M., and Likhtenshtein, G.I. (2003). Catalysis by metal complexes, Chapter 17. In: *Chemical Kinetics: Fundamentals and New Developments*, 1e (ed. E.T. Denisov, O.M. Sarkisov and G.I. Likhtenshtein), 472–501. Amsterdam: Elsevier ISBN: 9780444509383.

10 Hagen, J. (2006). *Industrial Catalysis: A Practical Approach*, 1–14. Weinheim: Wiley-VCH.

11 Cheng, X., Lei, A., Mei, T.-S. et al. (2022). Recent applications of homogeneous catalysis in electrochemical organic synthesis. *CCS Chem.* 4: 1120–1152.

12 Kalck, P. and Monteil, F. (1992). Use of water-soluble ligands in homogeneous catalysis. *Adv. Organomet. Chem.* 34: 219–284.

13 Wache, S. (1995). Organometallic complex catalysis in water II. Water soluble organoruthenium (IV) catalysts for the emulsion polymerization of norbornene. *J. Organomet. Chem.* 494: 235–240.

14 Wan, K.T. and Davis, M.E. (1994). Design and synthesis of a heterogeneous asymmetric catalyst. *Nature* 370: 449–450.

15 Kochi, J.K. (1978). *Organometallic Mechanisms and Catalysis*. New York: Academic Press.

16 Parshall, G.W. (1978). Industrial applications of homogeneous catalysis. *J. Mol. Catal.* 4: 243–270.

17 Cornils, B. and Herrmann, W.A. (1996). *Applied Homogeneous Catalysis with Organometallic Compounds*, vol. 1, 2. Weinheim: Wiley-VCH Verlag.

18 Biellmann, J.F. and Jung, M.J. (1968). The mechanism of isomerization of an olefin and its possible relation to the mechanism of the catalytic hydrogenation with tris (triphenylphosphine) rhodium chloride. *J. Am. Chem. Soc.* 90: 1673.

19 Tani, K., Yamagata, T., Akutagawa, S. et al. (1984, 1984). Metal-assisted terpenoid synthesis. 7. Highly enantioselective isomerization of prochiral allylamines catalyzed by chiral diphosphine rhodium (I) complexes. Preparation of optically active enamines. *J. Am. Chem. Soc.* 106: 5208–5217.

20 Ma, W., Liu, B., Wang, D. et al. (2021). Catalytic isomerization of olefins and their derivatives: a brief overview. In: . https://doi.org/10.5772/intechopen.99076.

21 Vaska, L. and Rhodes, R.E. (1965). Homogeneous catalytic hydrogenation of ethylene and acetylene with four-coordinated iridium and rhodium complexes. *J. Am. Chem. Soc.* 87: 4970.

22 James, B.R. (1973). *Homogeneous Hydrogenation*. New York: Wiley-Interscience.

23 McQuillin, F.J. (1976). *Homogeneous Hydrogenation in Organic Chemistry*. Dordrecht: Reidel Publ.

24 Haack, K.-J. et al. (1997). Catalyst precursor, catalyst, and intermediate in the Ruii-catalyzed asymmetric hydrogen transfer between alcohols and ketones. *Angew. Chem.* 109: 297–300.

25 Struijk, J., Moene, R., van der Kamp, T., and Scholten, J.J.F. (1992). Partial liquid-phase hydrogenation of benzene to cyclohexene over ruthenium catalysts in the presence of an aqueous salt solution: II. Influence of various salts on the performance of the catalyst. *Appl. Catal. A* 89 (1): 77–102.

26 James, B.R. (1997). Synthesis of chiral amines catalyzed homogeneously by metal complexes. *Catal. Today* 37: 209–221.

27 Lefebvre, G. and Chauvin, Y. (1970). Aspects of homogeneous catalysis. In: *Carlo Manfredi (Milano)*, vol. 1 (ed. R. Ugo), 107.

28 Komiya, S. and Yamamoto, A. (1972). Reactions of hydrido complexes of ruthenium and rhodium with carbon dioxide involving reversible insertion. *J. Organomet. Chem.* 46 (2): 58–60.

29 Keim, W., Behr, A., and Röper, M. (1982). *Comprehensive Organometallic Chemistry*, vol. 8 (ed. G. Wilkinson), 371. Oxford: *Pergamon Press*.

30 Strupinska, J. (1991). Oligomerization of α-olefins to higher oligomers. *Chem. Rev.* 91: 613–648.

31 Keim, W. and Maas, H. (1996). Copolymerization of ethylene and carbon monoxide. *J. Organomet. Chem.* 514: 271–276.

32 Bianchi, M., Menchi, G., Matteoli, U., and Piacenti, F. (1996). Homo and co-oligomerization of C2-C5 olefins in the presence of cluster ruthenium complexe. *J. Organomet. Chem.* 526: 33–41.

33 Katz, T.J. (1977). The olefin metathesis reaction. *Adv. Organomet. Chem.* 16: 283.

34 Grubbs, R.H. (1982). *Comprehensive Organometallic Chemistry*, vol. 8 (ed. G. Wilkinson), 499. Oxford: Pergamon Press.

35 Ivin, K.J. (1983). *Olefin Metathesis*. London: Academic Press.

36 Herrmann, W.A., Schattenmann, W.C., Nuyken, O., and Glander, S.C. (1996). Allylruthenium (IV) complexes as highly efficient ROMP catalysts. *Angew. Chem.* 108: 1169–1170.

37 Eleuterio, H.S. (1991). Olefin metathesis: chance favors those minds that are best prepared. *J. Mol. Catal.* 65: 55–61.

38 Bäckvall, J.E., Akermark, B., and Ljunggren, S.O. (1979). Stereochemistry and mechanism for the palladium (II)-catalyzed oxidation of ethene in water (the Wacker process). *J. Am. Chem. Soc.* 101: 2411.

39 Bäckvall, J.-E. (1983). Palladium in some selective oxidation reactions. *Acc. Chem. Res.* 16: 335–342.

40 Romano, A.M. and Ricci, M. (1997). New efficient catalysts for the aerobic oxidation of ethers. *J. Mol. Catal. A: Chem.* 120: 71–74.

41 Wender, J. and Pino, P. (ed.) (1968). *Organic Syntheses via Metal Carbonyls*, vol. 1 (1968), vol. 2 (1977). New York: Wiley-Interscience.

42 Pino, P. (1980). Hydroformylation of olefinic hydrocarbons with Rh and Co catalysts. *J. Organomet. Chem.* 200: 223–242.

43 Orchin, M. (1981). $HCo(CO)_4$, the quintessential catalyst. *Acc. Chem. Res.* 14: 259–266.

44 Sneeden, R.P.A. (1982). Organic syntheses where carbon monoxide is the unique source of carbon. In: *Comprehensive Organometallic Chemistry*, vol. 8 (ed. G. Wilkinson), 19. Oxford: Pergamon Press.

45 Housecroft, C.E. and Sharpe, A.G. (2018). *Inorganic Chemistry* (print and electronic), 5e. Pearson Education Limited.

46 Ivin, K.J., Rooney, J.J., Stewart, C.D., and Green, M.L.H. (1978). Mechanism for the stereospecific polymerization of olefins by Ziegler–Natta catalysts. *Chem. Commun.* 604–606.

47 Lee, J.D. (2021). *Concise Inorganic Chemistry*, 4e. Wiley India Pvt. Ltd.

48 Bahri-Laleh, N., Hanifpour, A., Mirmohammadi, S.A. et al. (2018). Computational modeling of heterogeneous Ziegler–Natta catalysts for olefins polymerization. *Prog. Polym. Sci.* 84: 89–114.

49 Kobayashi, Y., Tang, Y., Kageyama, T. et al. (2017). Titanium-based hydrides as heterogeneous catalysts for ammonia synthesis. *J. Am. Chem. Soc.* 139 (50): 18240–18246.

50 Marakatti, V.S. and Gaigneaux, E.M. (2020). Recent advances in heterogeneous catalysis for ammonia synthesis. *ChemCatChem* 12 (23): 5838–5857.

51 Liu, J.C., Ma, X.L., Li, Y. et al. (2018). Heterogeneous Fe_3 single-cluster catalyst for ammonia synthesis via an associative mechanism. *Nat. Commun.* 9 (1): 1–9.

52 Yan, B., Wu, J., Xie, C. et al. (2009). Supercritical water gasification with Ni/ZrO_2 catalyst for hydrogen production from model wastewater of polyethylene glycol. *J. Supercrit. Fluids* 50 (2): 155–161.

53 Miessler, G.L., Fischer, P.J., and Tarr, D.A. (2014). *Inorganic Chemistry*, 5e. Pearson Education, Inc.

15

Cluster Compounds: Boranes, Heteroboranes, and Metallaboranes

Elyor Berdimurodov[1], Khasan Berdimuradov[2], Abduvali Kholikov[1], Khamdam Akbarov[1], Omar Dagdag[3], Brahim El Ibrahimi[4], and Dakeshwar Kumar Verma[5]

[1] *National University of Uzbekistan, Faculty of Chemistry, Tashkent, 100034, Uzbekistan*
[2] *Shahrisabz branch of Tashkent Institute of Chemical Technology, Faculty of Industrial Viticulture and Food Production Technology, Shahrisabz, 181306, Uzbekistan*
[3] *Centre for Materials Science, College of Science, Engineering and Technology, University of South Africa, Johannesburg 1710, South Africa*
[4] *Ibn Zohr University, Department of Applied Chemistry, Faculty of Applied Sciences, 86153, Morocco*
[5] *Government Digvijay Autonomous Postgraduate College, Department of Chemistry, Rajnandgaon, Chhattisgarh, 491441, India*

15.1 Introduction

15.1.1 Fundamentals of Cluster Compounds: Boranes, Heteroboranes, and Metallaboranes

The boranes, heteroboranes, and metallaboranes-based cluster compounds contained mainly 1–12 boron atoms. The general formulation of borane-based clusters is B_xH_y. These compounds were constructed with only borane and hydrogen atoms. They differ from hydrocarbon compounds in their molecular structure and molecular weight [1]. The heteroboranes-based cluster compounds contained heteroatoms such as nitrogen, phosphorous, and other nonmetal atoms. The heteroatoms influence the molecular properties of cluster compounds. Recently, the transition metal elements have been combined with the boranes-based cluster compounds to form the metallaboranes clusters. Figure 15.1 illustrates the representations of the molecular and electronic structures of B_xH_y [2]. Figure 15.2 shows the crystal structures and numbers of skeletal electron pairs for borane-based clusters [2]. It is clear from these illustrations that the B_xH_y-based clusters have inner and outer spheres. There are two types of B—H bonds: endo B—H bonds and exo B—H bonds. According to the electronic representations, the hybrid sp_z orbitals are the centroid of the cluster compounds while the tangential p_x and p_y orbitals are covered by the cluster centers [2]. Table 15.1 indicates the classifications in borane clusters [1]. The borane-based clusters are classified according to the number of borane atoms (n).

The elements of boranes, heteroboranes, and metallaboranes-based cluster compounds are chemically linked together by the nonclassically bonded compounds. This is due to the fact that in these compounds there are enough electrons to form the

Organometallic Compounds: Synthesis, Reactions, and Applications, First Edition.
Edited by Dakeshwar Kumar Verma and Jeenat Aslam.

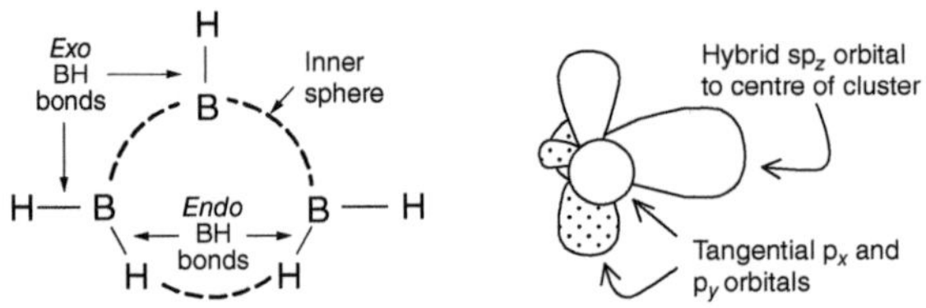

Figure 15.1 Representations of molecular and electronic structures of B_xH_y [2].

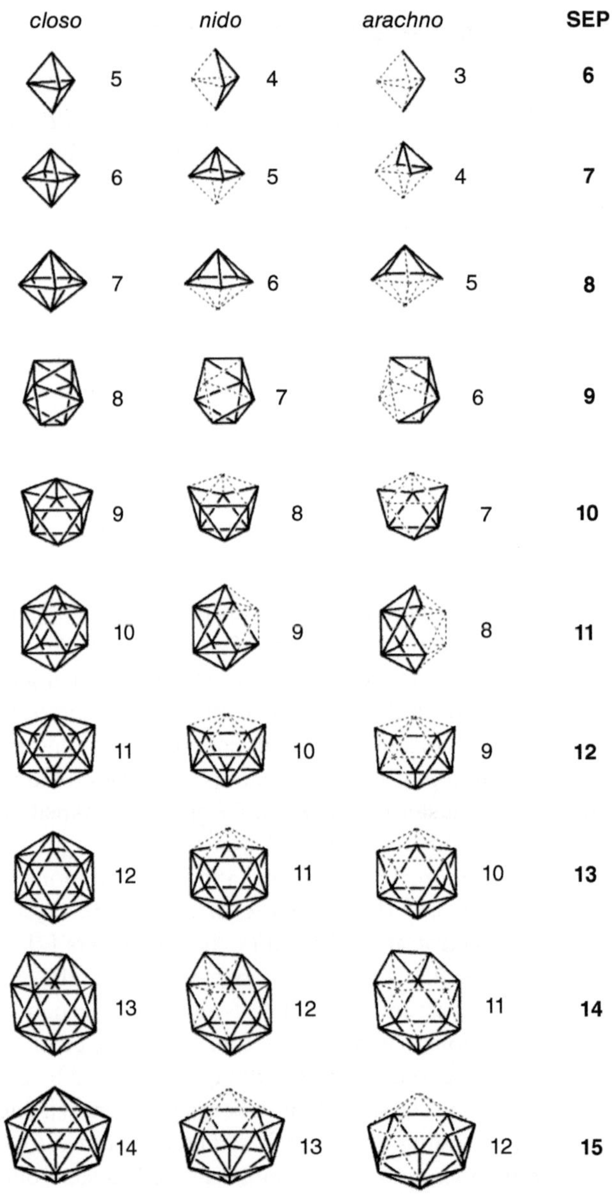

Figure 15.2 Crystal structures and numbers of skeletal electron pairs for borane-based clusters [2].

Table 15.1 Classifications in borane clusters.

Cluster type	Chemical formula
hypercloso-	B_nH_n
closo-	$B_nH_n{}^{2-}$
closo-	$B_nH_n{}^{2-}$
closo-	$B_nH_n{}^{2-}$
closo-	$B_nH_n{}^{2-}$

Table 15.2 Multi-cluster descriptors.

Prefix	Meaning
megalo-	Multiple conjoined clusters
conjuncto-	Conjoined clusters
klado-	Branched clusters

2-electron and 2-center bonds among the elements. William Lipscomb [3] described a new formulation for the bonding in borane-based clusters:

- 2-center 2-electron bonds (in B—B, B—H and BH_2)
- 3-center 2-electron B—B—B bonds
- 3-center 2-electron B—H—B hydrogen bridges

This expression was extended by the lowest unoccupied orbitals and the highest occupied molecular orbital theories. For instance, the electrons of $[B_{12}H_{12}]^{2-}$are equally distributed between 12 boron atoms [3].

The borane-based clusters have some multi-cluster boranes. Their molecular structure is different from the classical boranes. For instance, the BH_3 and B_6H_{10} compounds reacted with the simple borane clusters to form the *conjuncto*-borane clusters (Table 15.2). In these reactions, the borane cluster subunits are joined by the sharing of boron atoms [1].

$$B_6H_{10} + (BH_3) \rightarrow B_7H_{11} + H_2$$

$$B_7H_{11} + B_6H_{10} \rightarrow B_{13}H_{19} + H_2$$

On the other hand, the B—B bond is responsible for the subunit modifications. These clusters were formed by the $PtBr_2$ as a catalyst and ultraviolet irradiation of *nido*-boranes [4].

15.2 Main Part

15.2.1 Cluster Compounds of Boranes and Heteroboranes: Synthesis, Reactions, and Applications

15.2.1.1 Basics

The cluster compounds of heteroboranes were synthesized by the formation of the B—S, B—P, B—N, and B—C bonds. These compounds are named sulfurboranes, phosphoroboranes, nitroboranes, and carboranes. The cluster compounds of heteroboranes differ from each other related to their size, molecular structure, chemical or physical properties, and applications [5]. In another review, the Si, As, Se, Sb, and Te atoms also combined with the borane to form the B—Si, B—As, B—Se, B—Sb, and B—Te bonds [6].

15.2.1.2 Synthesis

Joseph et al. [7] synthesized new S and Se-based heteroborane clusters according to the following reaction schema (Figure 15.3). These reactions were carried out at high temperatures with the S and Se powders. The reaction was carried out in a toluene solvent. In the formed heteroborane cluster structure, the S and Se heteroatoms are directly linked with the borane atoms by the formation of the B—S and B—Se bonds.

Qiu [8] suggested the high thermal reaction for the synthesis of the carboranes (Figure 15.4). The reaction productivity was increased with the addition of Ag metal, in which the Ag acts as catalytic effects. The formed carboranes are stable in acidic and alcoholic solutions. These carboranes reactions were carried out at temperatures ranging from 600 to 700 °C.

Olejniczak et al. [9] synthesized some carboranes clusters (Figure 15.5) via solid support cleavage and deprotection, oxidation and capping, and coupling and detritylation modifications. All reactions were carried out by the B—H and C—H bonds.

15.2.1.3 Reactions

Vrána et al. [10] studied some reactivity properties of cluster compounds of heteroboranes (Figure 15.6): *closo*-1,10-dicarbadecaborane is referred to as para (p, green trace); *closo*-1,6-dicarbadecaborane is referred to as meta (m, blue race);

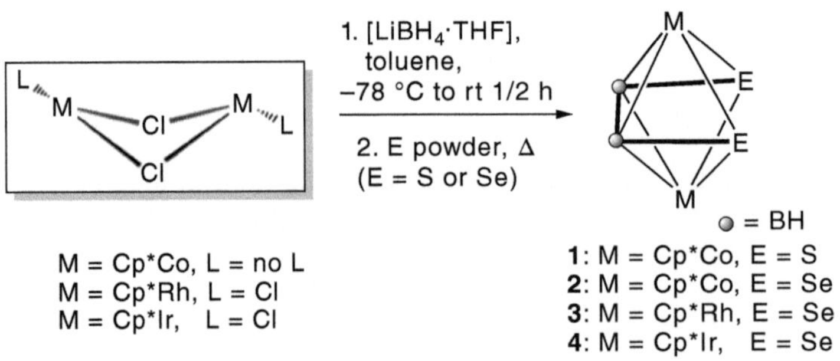

Figure 15.3 Synthetic procedure of S and Se-based heteroborane clusters [7].

Figure 15.4 Synthesis schema of some carboranes. Source: Qiu [8]/With permission of Elsevier.

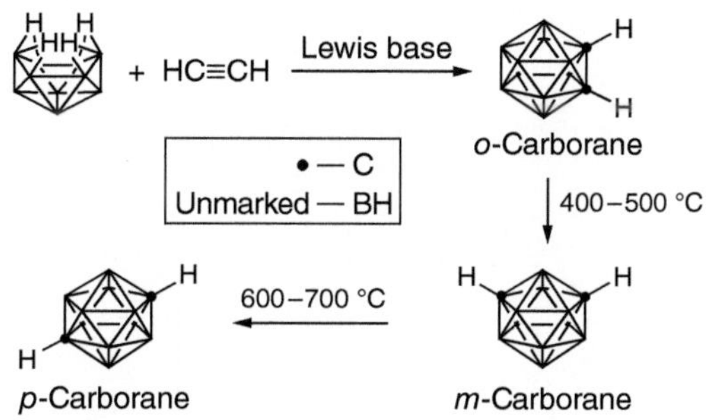

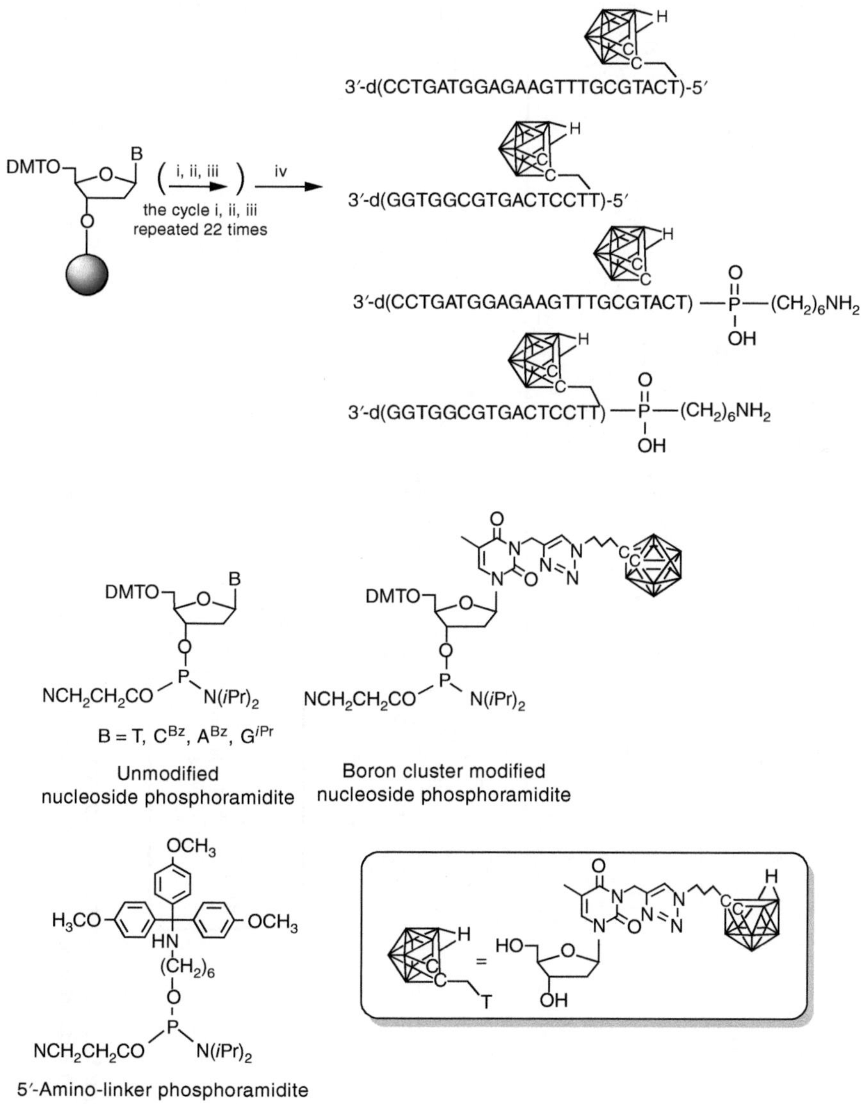

Figure 15.5 Preparation of some carboranes clusters. Source: Olejniczak et al. [9]/MDPI/CC BY 4.0.

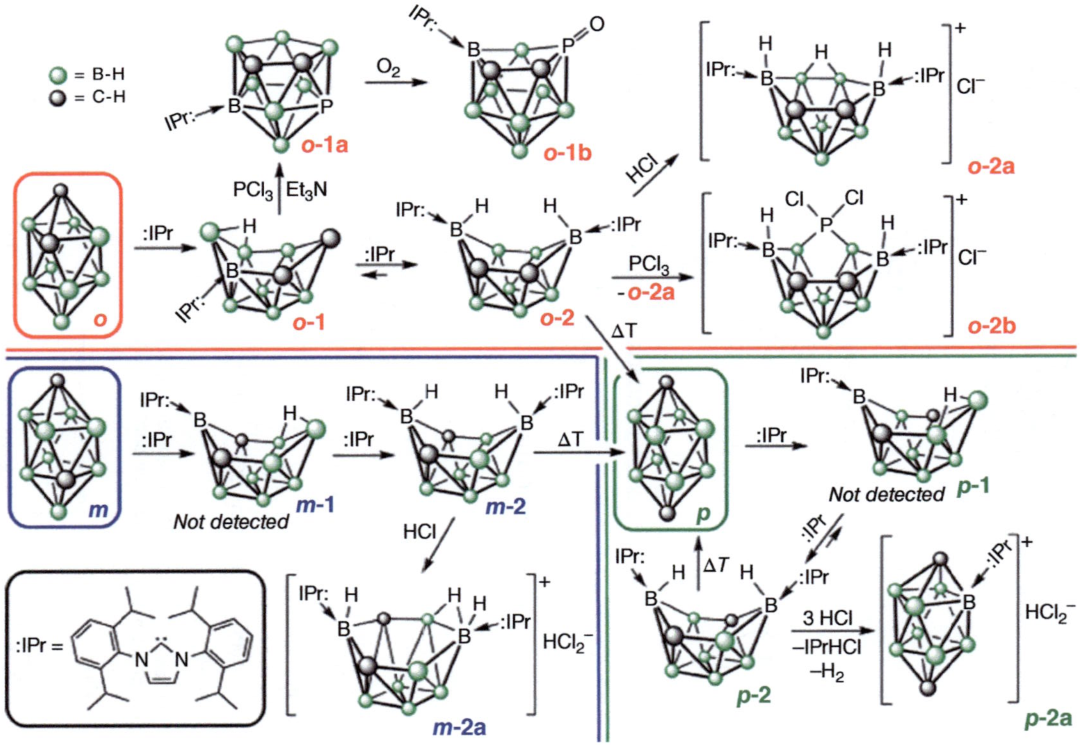

Figure 15.6 Some reactivity properties of cluster compounds of heteroboranes: *closo*-1,10-dicarbadecaborane is referred to as para (p, green trace); *closo*-1,6-dicarbadecaborane is referred to as meta (m, blue race); and *closo*-1,2-dicarbadecaborane is referred to as ortho (o, red trace) [10].

closo-1,2-dicarbadecaborane is referred to as ortho (o, red trace). It is indicated that these compounds reacted with oxygen, hydrochloric acid, and PCl_3 compounds.

Bakardjiev et al. [11] studied some nucleophilic reactions of heteroboranes (Figure 15.7). It was found that these cluster compounds have higher nucleophilic reaction ability. Almost all nucleophilic reactions are done by the B—H bonds. In this research work, the B—O bonds were formed by the nucleophilic reaction mechanisms.

15.2.1.4 Applications

In modern times, carborane clusters are widely used in cancer therapy. They are modifying the cancer therapy drugs to enhance their effectiveness [12–14]. This is due to their excellent chemical and physical properties such as bioavailability, potency, and agent binding affinity. In breast cancer therapy, steroid estrogens play a significant role. The estrogen hormones have sensitive receptors, which support the interactions between the protein and DNA molecules. These hormones easily interacted with the DNA molecules, and as a result, gene transcription was carried out [12]. The heteroborane clusters are widely used to block the receptor part of hormones. Figure 15.8 shows the interactions of the carborane cluster with the estrogen receptors: estrogen receptor antagonist carborane derivatives (**11**, **12**); estrogen receptor agonist carborane derivatives (**10a–g**); hormone and ERα antagonists (**7–9**).

The heteroboranes are modified with amino acids and proteins to create new bioactive peptides and unnatural amino acids. These compounds are used in medicine for various medicinal applications. On the other hand, the heteroboranes clusters are boron neutron capture therapy (BCNT), boron neutron capture synovectomy, and medical imaging [13, 15].

15.2.2 Cluster Compounds of Metallaboranes: Synthesis, Reactions, and Applications

15.2.2.1 Basics

Metallaborane clusters compounds were mostly formed with the transition metals. The researchers devoted the metallaborane clusters into four categories related to the nature of interactions between the metal and borane atoms: transition metal complexes of boron; Me-complexes with boron-containing ligands; metallaboranes; and borides. The bonds between the borane and metal were built by nonclassical interactions: 2-center 2-electron bonds between boron and the metal center [16, 17].

15.2.2.2 Synthesis

Bag et al. suggested a new synthesis methodology for metallaborane clusters (Figure 15.9) [18]. These synthetic reactions were done *in situ* models at low temperatures. Cr, Mo, and W $(CO)_4$ complexes were used. As a result, the previously reported compounds $[\{Cp^*W(CO)_2\}_2B_2H_2M(CO)_4]$, (M = Mo; W) [19], and $[(Cp^*W)_2B_5H_9]$ [20] were obtained, as well as the air and moisture-sensitive brown solids 1, 2, and 3.

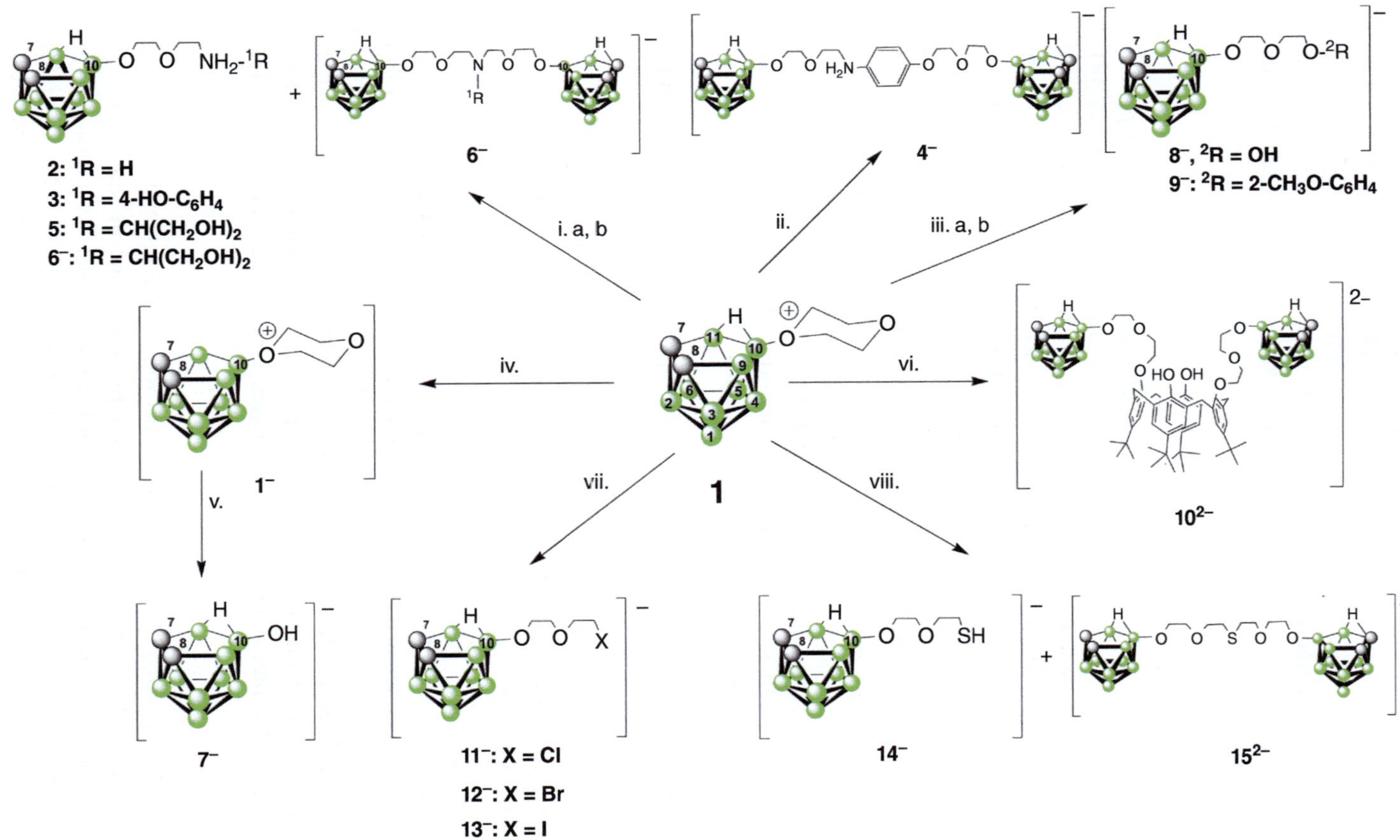

Figure 15.7 Some nucleophilic reactions of heteroboranes. (b) for **9**⁻: K_2CO_3, 1-HO-2-CH_3O-C_6H_4 reflux; (iv) 2.5 M KOH rt; (v) 2.5 M KOH, reflux, (vi) K_2CO_3, CH_3CN, *t*-Bu-calix[4]arene 0.5 equiv; (vii) Bu_4NX, 2 equiv ether, rt; (viii) NaSH·H_2O in excess, THF, rt (b) for **3,** THF rt; (ii) 1-HO-C_6H_4-4-NH_2, K_2CO_3 in CH_3CN, 80 °C; (iii) (a) for **8**⁻: KOH water-ether, rt (a) THF for **2**, ACE® pressure flask. Source: Bakardjiev et al. [11]/MDPI/CC BY 4.0.

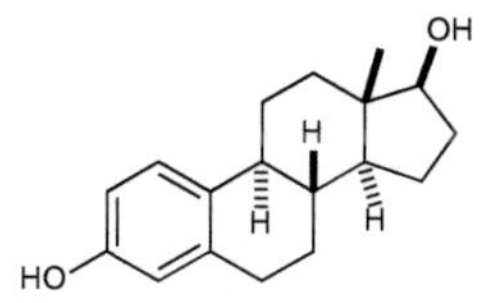

7

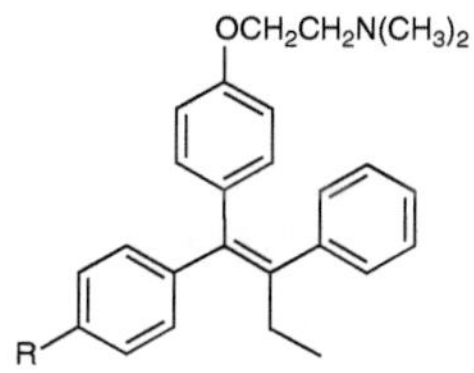

8 R = H
9 R = OH

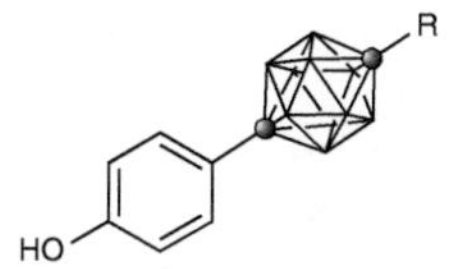

.10

a R = H
b R = OH
c R = CH_2OH
d R = CH_2CH_2OH
e R = $CH_2CH_2CH_2OH$
f R = COOH
g R = NH_2

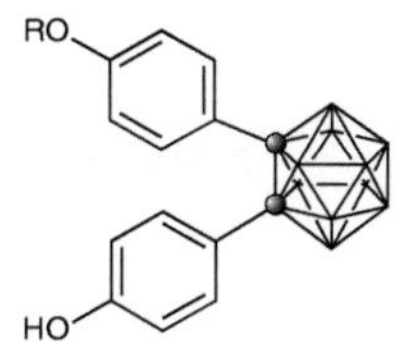

11 R = H
12 R = $CH_2CH_2N(CH_3)_2$

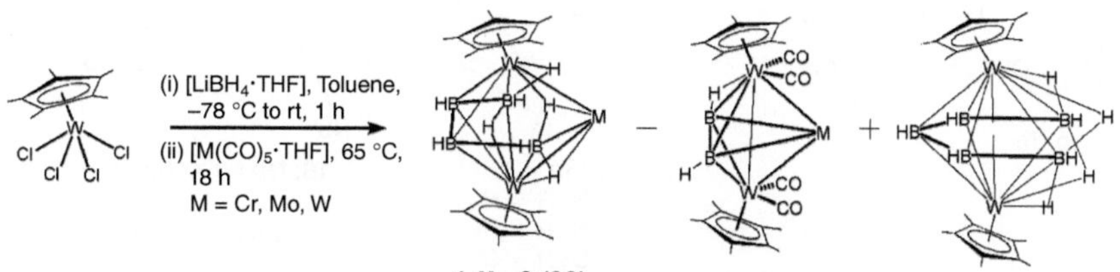

1, M = $Cr(CO)_4$
2, M = $Mo(CO)_4$
3, M = $W(CO)_4$

M = $Mo(CO)_4$
M = $W(CO)_4$

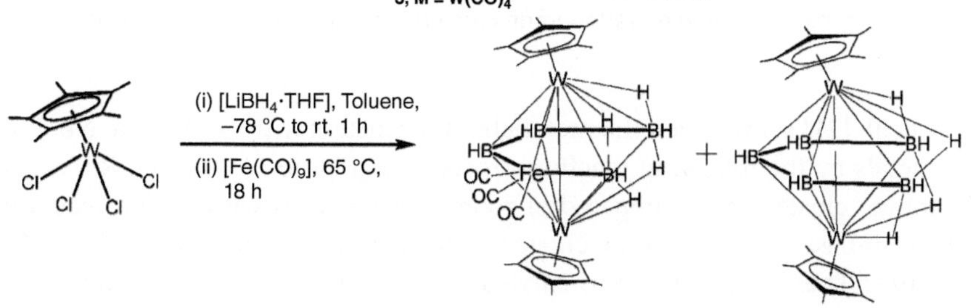

4

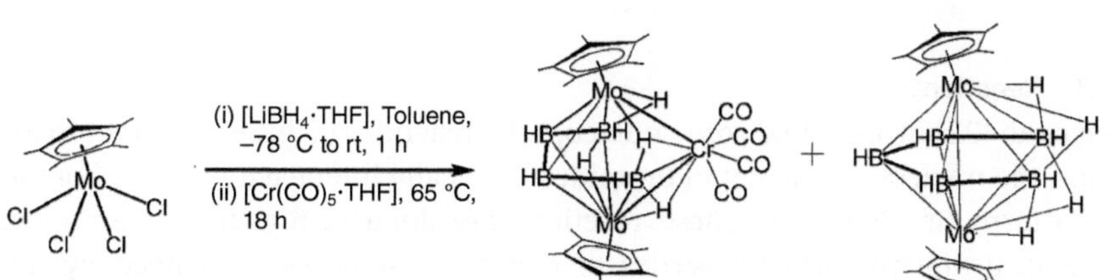

5

Figure 15.8 The interactions of the carborane cluster with the estrogen receptors: estrogen receptor antagonist carborane derivatives (**11, 12**); estrogen receptor agonist carborane derivatives (**10a–g**); and Hormone and ERα antagonists (**7–9**) [12].

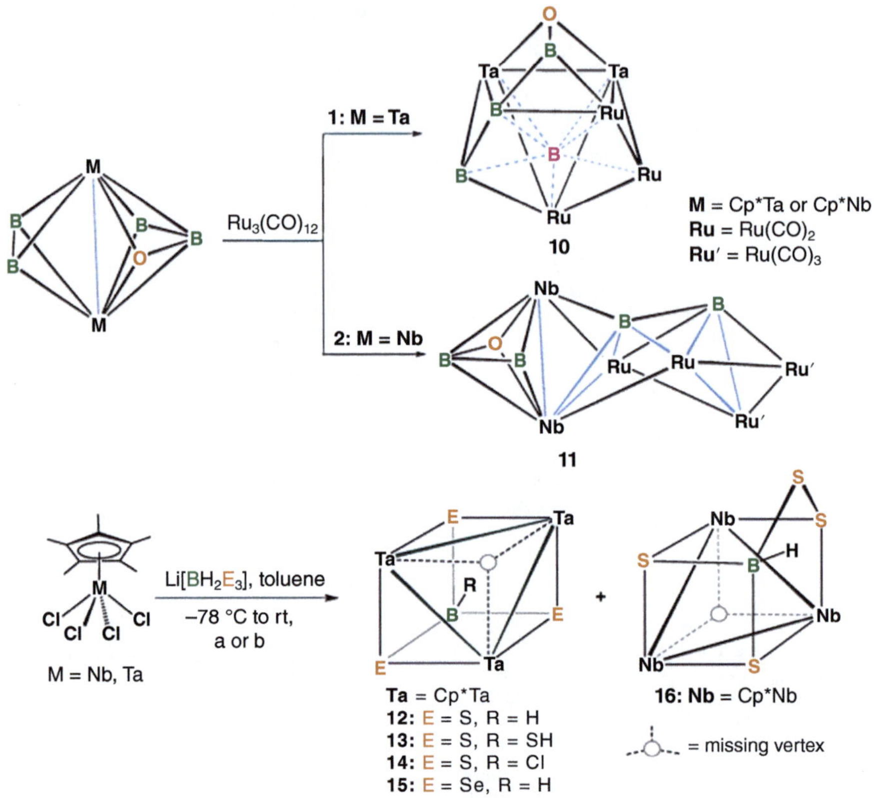

Figure 15.9 Synthesis of some metallaborane clusters. Source: Bag et al. [18]/MDPI/CC BY 4.0.

Pathak et al. [21] synthesized the metallaborane clusters with Fe, Mo, Ta, Nb, Ru, and Co metals by the following reaction procedures (Figure 15.10). The Ru-based metallaborane clusters were prepared at room temperatures with a high reaction yield. The Nb-based metallaborane clusters were synthesized at very low (−78 °C) temperatures. These reactions were carried out with the S and Se heteroborane clusters.

15.2.2.3 Reactions

Yuvaraj et al. [22] investigated some nucleophilic reactions (Figure 15.11). In these reactions, the nucleophilic agent of bromide chemically interacted with the borane atoms to form Br—B bonds. These reactions are done at higher temperatures. The chemical and physical properties of borane clusters were changed by the nucleophilic reactions.

15.2.2.4 Applications

The metallaborane clusters are mostly used in nanomedicine, targeted radiopharmaceuticals, medicinal imaging agents, cancer diagnostic agents, and BNCT of tumors

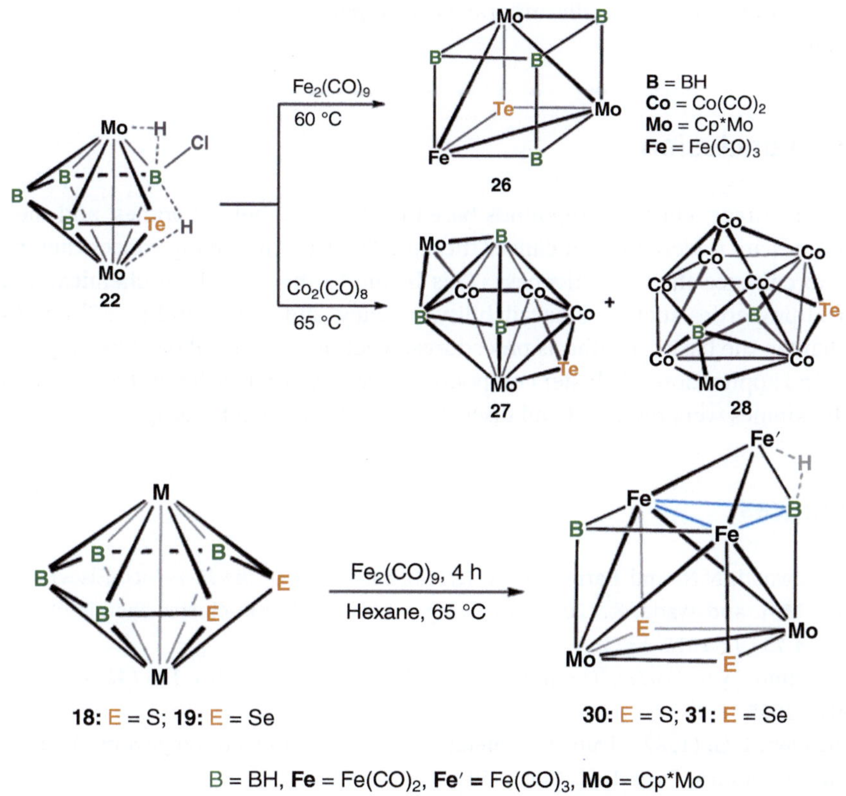

Figure 15.10 Synthesis of some metallaborane clusters (Fe, Mo, Ta, Nb, Ru, and Co) [21].

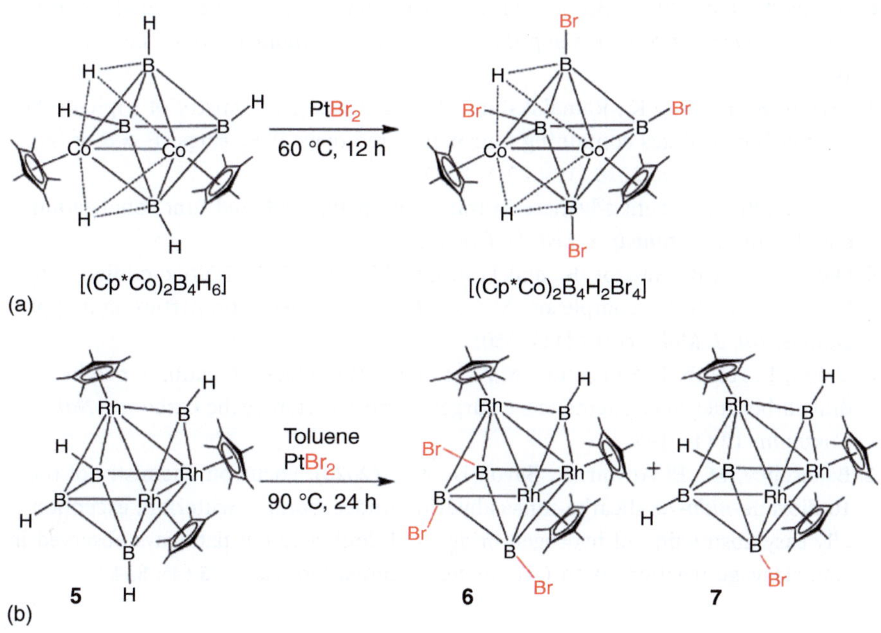

Figure 15.11 Some reactions of metallaboranes [22].

[23–26]. The Rh-based metallacarboranes were performed to enhance molecular imaging.

15.3 Conclusion

In modern times, cluster compounds based on boranes, heteroboranes, and metallaboranes are widely used in cancer therapy. They are modifying cancer therapy drugs to enhance their effectiveness. This is due to their excellent chemical and physical properties such as bioavailability, potency, and agent binding affinity. In this chapter, the basics, synthesis procedures, reactions, chemical or physical properties, and applications of cluster compounds based on boranes, heteroboranes, and metallaboranes were reviewed and discussed with the relevant examples.

References

1 Greenwood, N.N. and Earnshaw, A. (2012). *Chemistry of the Elements*. Elsevier.

2 Fox, M.A. and Wade, K. (2003). Evolving patterns in boron cluster chemistry. *Boron* 75 (9): 1315–1323.

3 Lipscomb, W.N. (1977). The boranes and their relatives. *Science* 196 (4294): 1047–1055.

4 Sneddon, L.G. (1987). Transition metal promoted reactions of polyhedral boranes and carboranes. *Pure Appl. Chem.* 59 (7): 837–846.

5 Olid, D., Nunez, R., Vinas, C., and Teixidor, F. (2013). Methods to produce B—C, B—P, B—N and B—S bonds in boron clusters. *Chem. Soc. Rev.* 42 (8): 3318–3336.

6 Schubert, D.M. (2000). Boron hydrides, heteroboranes, and their metalla derivatives. In: *Kirk-Othmer Encyclopedia of Chemical Technology* (ed. C. Ley), 354. Wiley.

7 Joseph, B., Barik, S.K., Ramalakshmi, R. et al. (2018). Chemistry of triple-decker sandwich complexes containing four-membered open B_2E_2 rings (E = S or Se). *Eur. J. Inorg. Chem.* 2018 (19): 2045–2053.

8 Qiu, Z. (2015). Recent advances in transition metal-mediated functionalization of *o*-carboranes. *Tetrahedron Lett.* 56 (8): 963–971.

9 Olejniczak, A.B., Nawrot, B., and Leśnikowski, Z.J. (2018). DNA modified with boron–metal cluster complexes $[M(C_2B_9H_{11})_2]$—synthesis, properties, and applications. *Int. J. Mol. Sci.* 19 (11): 3501.

10 Vrána, J., Holub, J., Samsonov, M.A. et al. (2021). Access to cationic polyhedral carboranes via dynamic cage surgery with N-heterocyclic carbenes. *Nat. Commun.* 12 (1): 1–7.

11 Bakardjiev, M., El Anwar, S., Bavol, D. et al. (2020). Focus on chemistry of the 10-dioxane-*nido*-7,8-dicarba-undecahydrido undecaborate zwitterion; exceptionally easy abstraction of hydrogen bridge and double-action pathways observed in ring cleavage reactions with OH^- as nucleophile. *Molecules* 25 (4): 814.

12 Zargham, E.O., Mason, C.A., and Lee, M.W. Jr., (2019). The use of carboranes in cancer drug development. *Int. J. Cancer Clin. Res.* 6: 110–113.

13 Armstrong, A.F. and Valliant, J.F. (2007). The bioinorganic and medicinal chemistry of carboranes: from new drug discovery to molecular imaging and therapy. *Dalton Trans.* 38: 4240–4251.

14 Issa, F., Kassiou, M., and Rendina, L.M. (2011). Boron in drug discovery: carboranes as unique pharmacophores in biologically active compounds. *Chem. Rev.* 111 (9): 5701–5722.

15 Valliant, J.F., Guenther, K.J., King, A.S. et al. (2002). The medicinal chemistry of carboranes. *Coord. Chem. Rev.* 232 (1–2): 173–230.

16 Braunschweig, H. and Colling, M. (2001). Transition metal complexes of boron – synthesis, structure and reactivity. *Coord. Chem. Rev.* 223 (1): 1–51.

17 Kennedy, J.D. (2016). Macropolyhedral metallaboranes – aspects of preparation, constitution and structure. *Coord. Chem. Rev.* 323: 71–86.

18 Bag, R., Saha, S., Borthakur, R. et al. (2019). Synthesis, structures and chemistry of the metallaboranes of group 4–9 with M_2B_5 core having a cross cluster M—M bond. *Inorganics* 7 (3): 27.

19 Mondal, B., Bag, R., Ghorai, S. et al. (2018). Synthesis, structure, bonding, and reactivity of metal complexes comprising diborane (4) and diborene (2): $[\{Cp^*Mo(CO)_2\}_2\{\mu\text{-}\eta_2\text{:}\eta_2\text{-}B_2H_4\}]$ and $[\{Cp^*M(CO)_2\}_2B_2H_2M(CO)_4]$, M = Mo, W. *Angew. Chem. Int. Ed.* 57 (27): 8079–8083.

20 Weller, A.S., Shang, M., and Fehlner, T.P. (1999). Synthesis of mono-and ditungstaboranes from reaction of Cp^*WCl_4 and $[Cp^*WCl_2]_2$ with $BH_3 \cdot THF$ or $LiBH_4$ ($Cp^* = \eta_5\text{-}C_5Me_5$). Control of reaction pathway by choice of monoboron reagent and oxidation state of metal center. *Organometallics* 18 (1): 53–64.

21 Pathak, K., Nandi, C., and Ghosh, S. (2022). Metallaheteroboranes with group 16 elements: aspects of synthesis, framework and reactivity. *Coord. Chem. Rev.* 453: 214303.

22 Yuvaraj, K., Roy, D.K., Arivazhagan, C. et al. (2015). Chemistry of early and late transition metallaboranes: synthesis and structural characterization of periodinated dimolybdaborane $[(Cp^*Mo)_2B_4H_3I_5]$. *Pure Appl. Chem.* 87 (2): 195–204.

23 Ma, P., Pellizzeri, T.M.S., Zubieta, J., and Spencer, J.T. (2020). Synthesis and characterization of oxonium functionalized rhenium metallaborane. *J. Chem. Crystallogr.* 50 (1): 14–20.

24 El-Zaria, M.E., Janzen, N., and Valliant, J.F. (2012). Room-temperature synthesis of Re(I) and Tc(I) metallocarboranes. *Organometallics* 31 (16): 5940–5949.

25 Goszczyński, T.M., Kowalski, K., Leśnikowski, Z.J., and Boratyński, J. (2015). Solid state, thermal synthesis of site-specific protein–boron cluster conjugates and their physicochemical and biochemical properties. *Biochim. Biophys. Acta* 1850 (2): 411–418.

26 Laskova, J., Kozlova, A., Białek-Pietras, M. et al. (2016). Reactions of *closo*-dodecaborate amines. Towards novel bis(*closo*-dodecaborates) and *closo*-dodecaborate conjugates with lipids and non-natural nucleosides. *J. Organomet. Chem.* 807: 29–35.

16

Applications of Organometallic Compounds for Carbon Dioxide Fixation, Reduction, Gas Adsorption, and Gas Purification

Zhoveta Yhobu[1], Monica Vijayakumar[1], Nagaraju D. Hanumantharayudu[2], and Srinivasa Budagumpi[1]

[1]*Jain University, Centre for Nano and Material Sciences, Jain Global Campus, Kanakapura, Ramanagaram, Bangalore, 562112, India*
[2]*REVA University, School of Applied Sciences, Department of Chemistry, Kattigenahalli, Yelahanka, Bangalore, 560064, India*

16.1 Organometallic Compounds for Fixation of CO_2

CO_2 is a high-energy reactant and is also thermodynamically and kinetically stable. The synthetic fixation of CO_2 has increased since: (i) CO_2 being a greenhouse gas, its reduction is necessary; (ii) it is cost-effective and can be used as one carbon species for the production of several chemicals; (iii) its fixation in the atmosphere is one of the important biological processes. The chemistry of carbon dioxide with organometallic compounds is an important aspect of converting waste products into useful chemical materials. Certain organometallic compounds have been developed to react with CO_2 for catalytic transformation and CO_2 separation. An example is the bismuth compounds shown in Scheme 16.1 containing Bi—O bonds that can enable irreversible CO_2 fixation at room temperature [1].

Organometallic compounds are designed in such a way that they can be employed in the copolymerization of CO_2 and epoxides from the combination of a hydrogen donor compound and a metal-based system that activates the CO_2 and the ring opening of the epoxide. The hydrogen donor compounds that are commonly used are primary amines, phenols (dihydric or trihydric), water, aromatic dicarboxylic acids, and aromatic hydroxycarboxylic acids [2].

Several organometallic catalysts with well-defined systems have been developed for the copolymerization of CO_2 and epoxides like zinc glutaric derived from zinc oxide and glutaric acid. Substituting the zinc glutarate with organic solvents as the reaction media for copolymerization of CO_2 and epoxides makes the copolymerization process more environmentally friendly and economically feasible [2].

N-heterocyclic carbenes (NHCs), cyclic compounds bearing a divalent carbon atom connected to a nitrogen in a heterocyclic system, have been used as a ligand for the construction of silver-NHCs as a catalyst in the cycloaddition of CO_2 to terminal epoxides. The silver NHC complexes is influenced by the NHC substituents, halides,

Organometallic Compounds: Synthesis, Reactions, and Applications, First Edition.
Edited by Dakeshwar Kumar Verma and Jeenat Aslam.

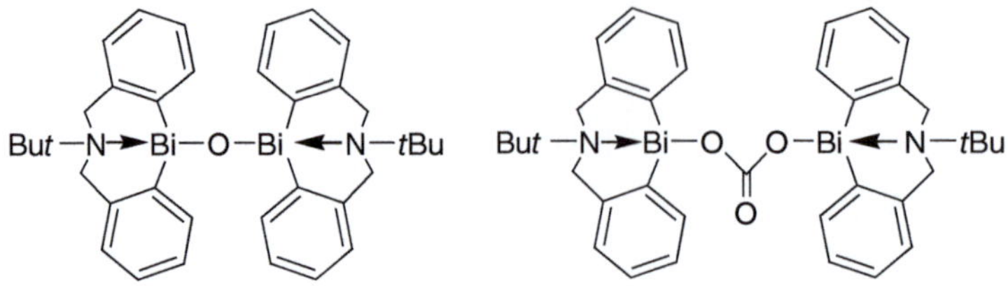

Scheme 16.1 CO_2 fixation by bismuth compounds containing Bi-O bonds.

and additives and have been used as a catalyst for CO_2, exhibiting 77–92% yield of styrene carbonate in 4-dimethylaminopyridine (DMAP) as a co-catalyst in the absence of solvent. It was observed that the catalytic activity was enhanced as the number of alkyl chains increased on the imidazolium ring. These Ag-NHC complexes could be reused several times with no loss in the activity and selectivity with high conversion [3].

16.2 Organometallic Compounds in Reduction of CO_2

As fuel demand has increased around the globe, researchers are examining ways to reduce CO_2 emissions using environment friendly carbon fuel. Carbon dioxide as fuel has a low-potential energy and is also stable. Hydrocarbon fuels have been biased when compared to fuel cells and batteries because of their high energy density, transportation, and ease of storage. CO_2 can be operated by two methods: (i) decreasing CO_2 straight into fuel products such as CH_4, CH_3OH, C_2H_5OH, and other high-grade hydrocarbons (ii) transport it into water gas-CO_2 reduction to CO and H_2O to H_2 in the presence of organic acids or any suitable ones and assist in a Fischer–Tropsch process, which yields the combustible product as outlined in Figure 16.1. The reduction of CO_2 into fuel products is a complex process and requires a catalyst [4]. The reduction of CO_2 electrochemically or photochemically into industrially important compounds is a difficult task. Many efforts have been

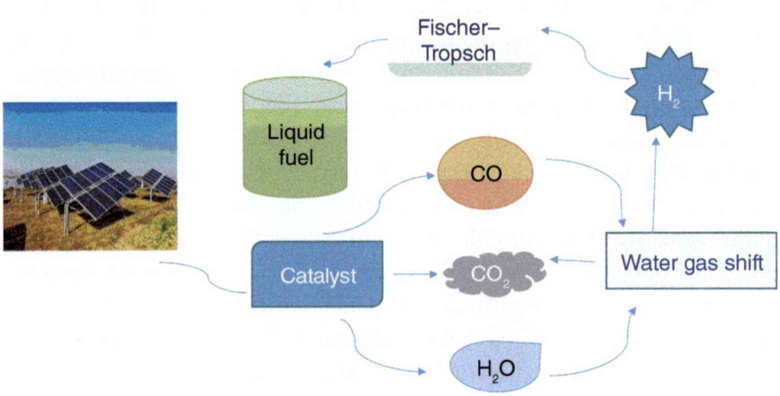

Figure 16.1 Schematic of reduction of CO_2.

made to increase the efficiency of CO_2 reduction reactions (CO_2 RRs) by employing heterogenous and homogenous catalysts, transition metal containing complexes, metal nanostructures, carbon-based nanocomposites, and oxide-derived metals [5].

The largest sustainable energy reaction that takes place on earth is photosynthesis. This is highly desirable and there have been efforts made to mimic this process using organometallic compounds. Sunlight-based photocatalytic CO_2 reduction by artificial photosynthesis for the production of industrially important compounds using solar technology is one of the most promising solutions. When coupled with photoelectrochemical cells or photovoltaic devices, there is an efficient electrochemical reduction of CO_2 [6]. The photoredox reduction of CO_2 is a much more challenging task because of the stable linearity present in the CO_2 molecules. An efficient catalyst is required that enables the capture, activation, and reduction of CO_2 molecules by proton-coupled multielectron transport that prevents the formation of thermodynamically unfavorable high energy intermediates [7].

Reduction of CO_2 directly by one electron to CO_2^- is exceptionally unfavorable because of its inertness, and it requires a formal reduction potential of around −1.90 V vs. Normal Hydrogen Electrode (NHE) in a neutral aqueous solution. The other alternative method to reduce the CO_2 is by proton-assisted multiple electron transfer. Table 16.1 depicts the positive redox potentials where CO_2 is reduced to many diverse products. The increase in the number of electrons and protons that have been intricated in the CO_2 reduction reaction shifts the standard reduction potential from the anode site, thus favoring the CO_2 reduction thermodynamically. As there is an increase in the number of protons and electrons, the kinetic barrier gradually increases. However, accomplishing these reactions with high thermodynamic potentials is difficult, and also it is essential for catalysts to reduce the energy barriers in the activation of CO_2 and allows for the successive proton transfers [8].

One of the first attempts at homogenous photocatalytic CO_2 reduction was based on non-noble metals as catalysts, using $CoCl_2$ as a molecular electrocatalyst. Later, a multitude of non-noble metals with varying ligands were employed for the reduction

Table 16.1 Equilibrium potentials of CO_2 and H^+ reduced products at pH 7.

Half reaction	Potential vs. SCE (V)
$CO_2 + e^- \rightarrow CO_2^{\cdot -}$	$E° = -2.14$
$CO_2 + 2H^+ + 2e^- \rightarrow HCO_2H$	$E° = -0.85$
$CO_2 + 2H^+ + 2e^- \rightarrow CO + H_2O$	$E° = -0.77$
$CO_2 + 4H^+ + 4e^- \rightarrow CH_2O + H_2O$	$E° = -0.72$
$CO_2 + 6H^+ + 6e^- \rightarrow CH_3OH + H_2O$	$E° = -0.62$
$CO_2 + 8H^+ + 8e^- \rightarrow CH_4 + 2H_2O$	$E° = -0.48$
$2H^+ + 2e^- \rightarrow H_2$	$E° = -0.66$

of CO_2 photochemically [4]. Great progress has been attained by making use of these metal complexes as molecular catalysts in CO_2 reduction.

These complexes can be utilized in the conversion of CO_2 electrochemically, thermally, or photochemically to various byproducts like CO, carboxylic acids, formates, acrylates, etc. [4]. The specificity and activity for CO_2 reduction with different metals with varied robustness are influenced by several intrinsic properties like the metal center, the steric, and electronic balance from the appropriate ligand system.

16.2.1 Nickel Complexes

Nickel is a cost-efficient metal that has exhibited a good activity in reducing CO_2. Organometallic compounds have been employed for CO_2 reduction, an instance is the usage dinuclear nickel complexes synthesized from $Ni(COD)_2$ with bis(diphenylphosphine)amine) (dppa). On reducing CO_2 with the complexes the products that were mainly obtained were CO, CO_3^{-2}, and even formate was present in minimal amounts. Based on the theoretical studies and spectroelectrochemical studies, it was observed that ligand-based reduction was consistent where the π^* orbital of the phenyl ring of the isocyanide group was able to shuttle the electrons in and out of the closed shell d^{10}–d^{10} $[Ni_2]$ system. The nickel complex with phosphine substitution exhibited a constant $1e^-$ reduction at −0.99 vs. ferrocene in the corresponding cyclic voltammogram. There was a substantial increase in the current observed for the reduction wave in the cyclic voltammogram in the presence of CO_2. It was observed that the nickel substituted phosphine complex was reduced singly into a radical anionic species that was susceptible toward CO_2 [9].

Nickel complexes with polypyridyl ligands such as phenyls, biphenyls 2,2′:6′,2″-terpyridine have also been utilized as catalysts in CO_2 reduction photocatalytically. The terpyridyl nickel complexes $[Ni(terpyX)_2]^{2+}$ (X = H, COOH, SH) shown in Scheme 16.2 have been applied as catalysts by linking to cadmium sulfide (CdS), which acted as quantum dots and as photosensitizer (PS). It was observed that only CdS-$[Ni^{II}(terpyS)_2]^{2+}$ displayed reasonable catalytic activity in reduction of CO_2 under air mass 1.5G ($\lambda > 400$ nm) irradiation with triethanolamine a reductant in water. Reduction of CO_2 by the photocatalyst releases one "terpyridyl" ligand after $1e^-$ photoreduction and forms a Ni–CO_2 intermediate. This intermediate further undergoes reduction and a reaction with CO_2 as an oxide (O^{2-}), which is an acceptor and forms a Ni–CO intermediate with CO_3^{-2} as a byproduct [10].

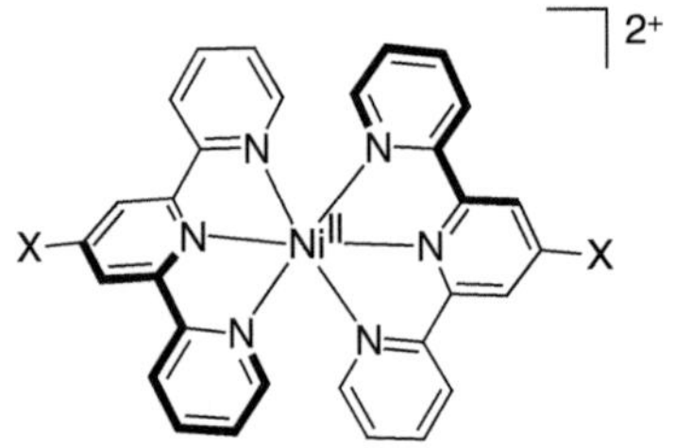

Scheme 16.2 Terpyridyl nickel complexes employed in CO_2^- reduction.

A cyclam ligand, i.e. carboxylate functionalized ligand, was employed in the synthesis of a nickel complex (shown in Scheme 16.3) as an electrocatalyst in the reduction of CO_2. The activity of the carboxylated catalysts has been ascribed to the pendant carboxylic group that acts as a proton transfer agent [11].

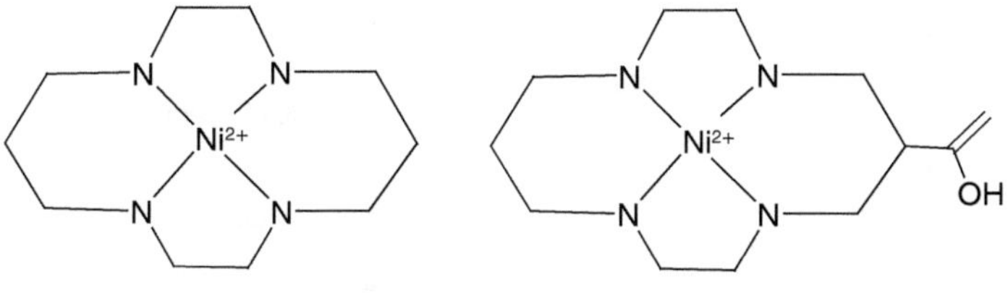

Scheme 16.3 Cyclam based nickel complexes in CO_2 reduction.

Two dinuclear nickel macrocycle complexes and a mononuclear nickel complex that was assisted by three modified macrocyclic ligands shown in Scheme 16.4 has also been used for reducing CO_2. On conducting electrochemical experiments, the complex that was bridged by a single bond showed the best activity in reducing the CO_2 to CO with a 95% faradaic efficiency and a turn over frequency (TOF) value of 1.54 h^{-1}. The other dinuclear complex that was bridged by a phenyl ring displayed Faradaic Efficiency (FE) of 25% and 1.62 TOF and the mononuclear complex exhibited FE of 62%. The nickel dinuclear complex bridged with a single bond showing enhanced reduced catalytic activity is attributed to the appropriate distance between the two Ni(II) centers, enabling the reduction of CO_2 to CO electrochemically through synergistic catalysis [12].

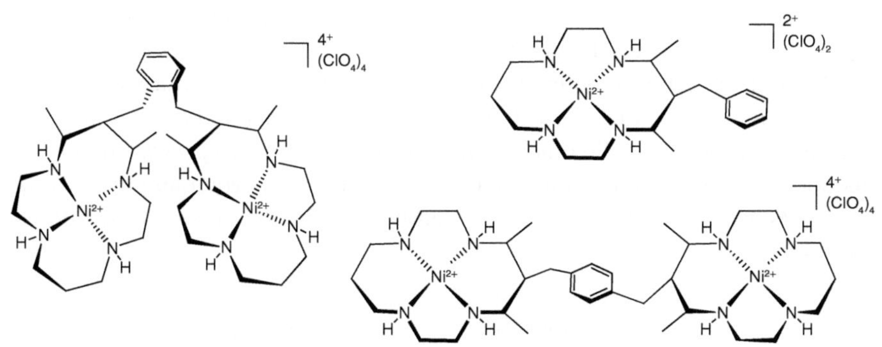

Scheme 16.4 Mononuclear and dinuclear nickel modified macrocyclic ligands in CO_2 reduction.

16.2.2 Iron and Manganese Complexes

Iron complexes with 6,6′-di(3,5-di-*tert*-butyl-2-hydroxybenzene)-2,2′-bipyridne also called as (^{tbu}dhbpy[H]$_2$)-ligand system has been employed as electrocatalysts in the reduction of CO_2 to formate as depicted in Scheme 16.5. From cyclic voltammetry (CV) controlled potential electrolysis and Density Functional Theory (DFT) calculations, it was found that the iron complex exhibited Nernstian behavior

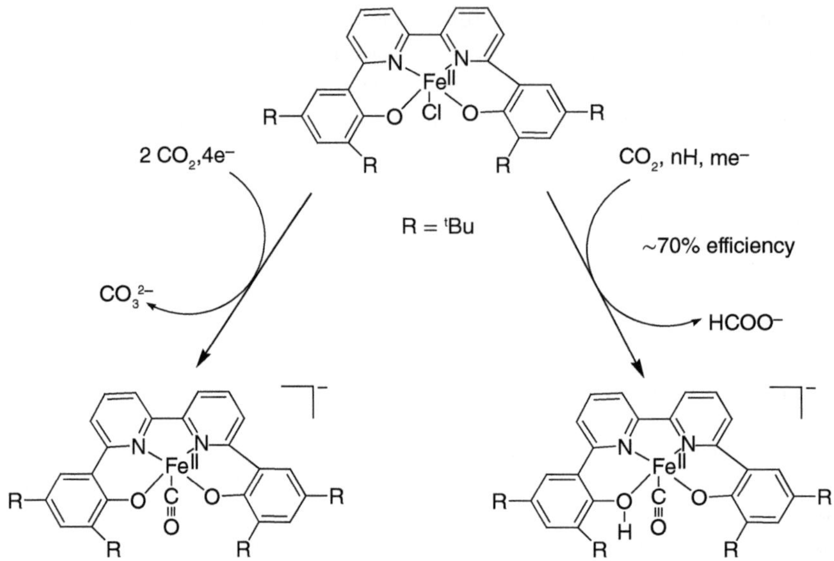

Scheme 16.5 Iron complexes with (^{tbu}dhbpy)-ligand system in reducing CO_2 to formate.

when phenol was added as a proton source and reduced CO_2 to formate and CO with a kinetic inclination for formate. From mechanistic studies, it was suggested that there is a possibility for the protonation and reduction of the overall complex where there is no loss at the active site of the metal center through cooperative metal–ligand behavior [13].

Iron(0) complexes with a cyclopentadienone ligand system shown in Scheme 16.6 were used as efficient electrocatalysts in CO_2 reduction. These complexes were powerful and molecularly defined systems that were able to reduce CO_2 to CO and H_2O. These Fe complexes exhibited cooperative interaction between the metal center and the cyclopentadienone ligand. A proton source was generated after the reduction, which eased the binding of the CO_2 to the metal center and the cleavage of the C—O bond. An intermediate was identified, which was a part of the catalytic cycle [14].

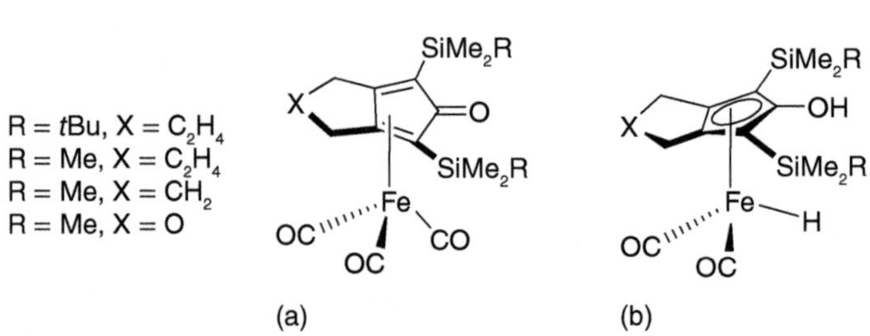

Scheme 16.6 Iron complexes with cyclopentadienone ligand system in reduction of CO_2.

16.2.3 Copper and Cobalt Complexes

The use of earth-abundant metal catalysts is promising for photocatalytic application in the reduction of CO_2. A Cu(II) complex (shown in Scheme 16.7) with a quaterpyridine ligand is a robust and exceedingly selective catalysts to reduce the gas in acetonitrile with $[Ru(bpy)_3]^{2+}$ as photosensitizer. With the addition of water, the yield of CO was significantly reduced and a turn over frequency of 12 400 was achieved [15].

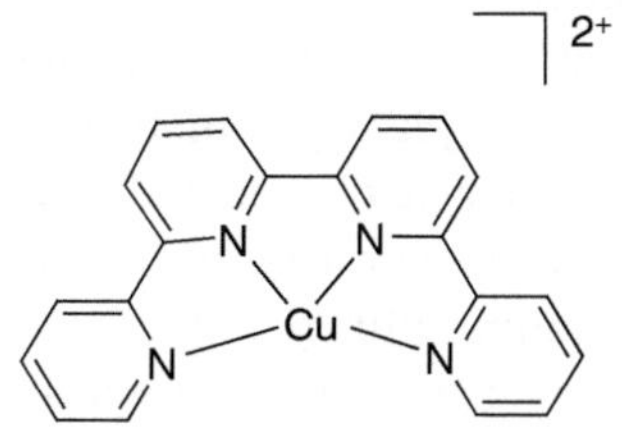

Scheme 16.7 Copper complex employed with quaterpyridine ligand in CO_2 reduction.

Cobalt complexes, which have distorted trigonal bipyramidal geometry, have also been used as catalysts in reducing CO_2 both in electrochemical and photochemical aspects. A cobalt complex with polypyridine ligands i.e., 2,6-bis[2-(2,2′-bipyridin-6′-yl)ethyl]pyridine (L1) and 2,6-bis[2-(4-methoxy-2,2′-bipyridin-6′-yl)-ethyl]pyridine (L2) with pentadentate chelation and a monodentate solvent was employed for CO_2 reduction. The pentadentate ligands in the L2 ligand system exhibited a stronger σ donation from the methoxy groups as shown in Scheme 16.8 making the cobalt center electron rich with higher association constants with CO_2. The redox couples of the cobalt complexes were found to be $E_{1/2} = -1.77$ and 1.85 V vs. $Ag/AgNO_3$ with $[Ru(bpy)_3]^{2+}$ as a quencher, tri-*p*-tolyamine as a reversible quencher and trimethylamine as a sacrificial electron donor. The complex under photochemical conditions, Co(I) gets reduced to Co(0) with photogeneration of $[Ru(bpy)_3]^{2+}$, and liberation of CO and H_2. In the CH_3CN solution, formate product was generated as the major product. From the electrochemistry and time-resolved spectroscopy, it was indicated that the CO binding rate with the $[Co^{I}L2]^{+}$ species was slow and

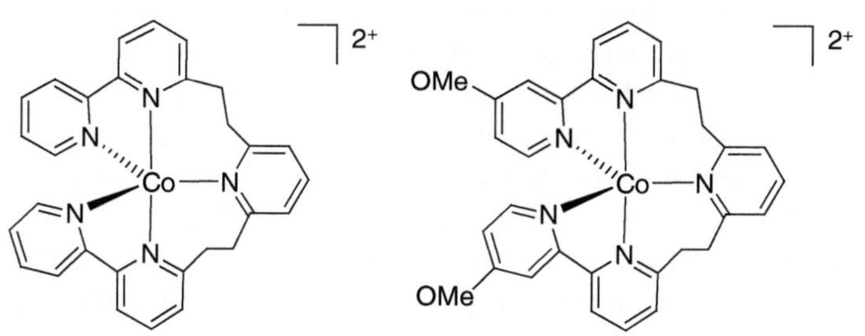

Scheme 16.8 Cobalt complexes with varying ligand systems for CO_2 reduction.

it was found consistent with the DFT calculations. The CoL1 species undergoes large structural changes from trigonal bipyramidal geometry to distorted tetragonal geometry, leading to the slower reduction of CO_2 than the photochemical formation of the primary quencher [16].

Cobalt complexes with pyridylmonoimine as ligand systems that differed in their structural flexibility degrees were investigated as electrocatalysts in the reduction of CO_2. The square planar-based cobalt complex undergoes reductive dimerization. Once the ligand has been reduced, the dimer dissociates resulting in the formation of a monomer species that is active for CO_2 reduction. In order to overcome this reductive dimerization, a coordinating ligand has been added, i.e. triphenylphosphine (PPh_3), which leads to the change in the flexibility of the square planar geometry to a flexible ligand distorted geometry, the complex structure is depicted in Scheme 16.9. This more flexible cobalt complex works with more positive catalytic activity for CO_2 reduction. CO poisoning occurs because of the stronger binding affinity of the CO to the flexible complex. This binding affinity of the CO to the ligand system is because of the [Co(L-R-L)] complex is more reducing than the rigid one that leads to the increase in the stabilization of the CO adducts [17].

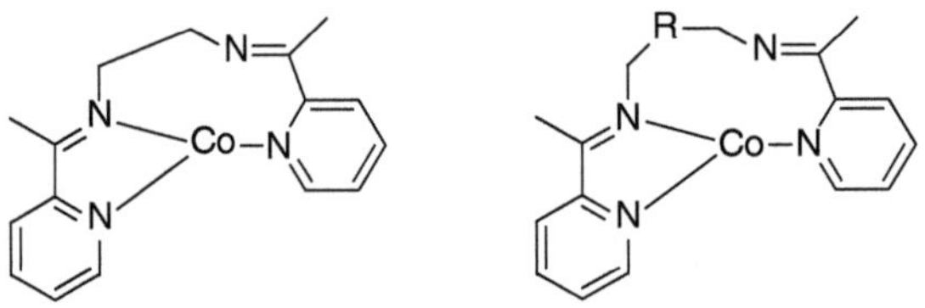

Scheme 16.9 Cobalt complexes bearing pyridylmonoimine ligand with varying ligand systems for CO_2 reduction.

16.2.4 Palladium Complexes

Extensive studies have been done on palladium complexes bonded to phosphine pincers for CO_2 reduction. Among the transition metals that have been discussed, Pd complexes exhibited higher faradic efficiency. But the drawback of this catalyst was its short lifetime because of inert bimetallic species formation. Phosphine-based Pd(II) pincer complexes, binuclear complexes were separated out, which had two square planar palladium centers and with a bridging ligand system, which was hexa-coordinated and the intermediate formed during the electrochemical CO_2 reduction. These complexes exhibited high reaction rates, but after the reduction of both Pd atoms, they decomposed rapidly. In order to retard the decomposition, pincer complexes with different ligand sizes were designed. The bridging was hampered but the decomposition of the catalyst occurred through hydride formation. Replacing the Pd(II) with a Ni(II) in phosphine pincer displayed increased reaction rates [4].

Palladium complexes bearing a tridentate phosphine ligand and a solvent molecule (shown in Scheme 16.10) were found to catalyze the CO_2 reduction to CO electrochemically. The palladium catalysts exhibited high selectivity and activity.

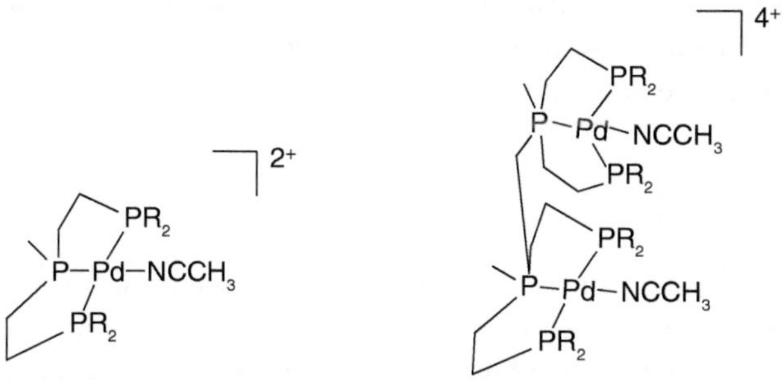

Scheme 16.10 Pincer based palladium complexes for reduction of CO_2.

Catalyst with two palladium atoms showed enhanced catalytic rate when a second palladium atom was incorporated into the catalyst. The increase in the catalytic rate is due to the cooperative binding of palladium atoms to CO_2. A palladium atom in the catalyst binds to the carbon atom of CO_2 and the other palladium atom binds to the oxygen atom [6].

The palladium complexes with the type $[Pd(Bu_3P^+etpE)(CH_3CN)](BF_4)_3$ and $[PdMe_3P^+etpE)(CH_3CN)](BF_4)_3$ showed significant catalytic activity in the reduction of CO_2. Catalyst with butyl group displayed the around 15% current efficiency for production of CO under acidic conditions during the controlled potential electrolysis.

Palladium complexes with a square planar geometry surrounded by a tridentate ligand system of type PXP (X = C, N, O, S, and As) have been prepared by Bryan D. Steffey and group. The Pd complex of the type $[Pd(PCP)\text{-}(CH_3CN)](BF_4)$ where PCP is 2,6 bis((diphenylphosphino)methyl)phenyl) exhibited remarkable catalytic activity in the presence of H_3PO_4 and CO_2 and generated a Pd(I) intermediate upon reduction. This intermediate accounted for the observed CO_2 dependence. The rate kinetics indicated that it was a first-order reaction [18].

16.2.5 Ruthenium, Rhenium, and Rhodium Complexes

Ruthenium, rhodium, and rhenium belong to the noble metal group and are expensive but exhibit significant activity in the reduction of CO_2.

It was reported that the electrocatalytic CO_2 reduction with ruthenium(II) and rhenium(I) bipyridine complexes with pendant amine ligand systems. In both the complexes, CO was the major product obtained [19].

$[Rh\text{-}bpy(TFMS)_2)]^+$ where bpy = 2,2′-bipyridine and TFMS = trifluoromethanesulfonate anion, this complex exhibited a faradic efficiency of 80% in the production of formate in 0.1 M Bu_4NPF_6 in CH_3CN. In this rhodium catalyst, the bpy acts as an electron transfer agent. Reduction of these ligands reduces the complex with CO_2 to form $[Rh\text{-}(bpy)_2(CO_2)H]^-$.

Ruthenium complexes with polypyridyl ligands are well-established in homogenous catalysis. Bipyridine and terpyridine ligands associated with Ru metals like $[Ru(tpy)(bpy)(MeCN)]^{2+}$ are good electrocatalysts in the production of CO with faradic high efficiency of 85% [20]. At negative potentials, reduction of CO_2 leads to the loss of a solvent molecule and results in a metallo-carboxylate intermediate during the CO_2 coordination in the rate limiting step.

16.2.6 N-Heterocyclic Carbenes

NHCs are excellent ligand systems, which can be finely tuned electronically and sterically. These NHCs are excellent sigma donor systems and have been found to be as impressively used as ligands. Once a ligand has been chosen, the coordination of the catalyst is fixed.

Pd(II) catalysts with NHCs as ligands were employed, and these were found to be more efficient than the one that had a pincer ligand. The onium-based pendant group on Pd-NHC enhanced the selectivity for CO that had provided the flexibility and hydrophobicity.

Bis-NHC Pd complexes of pincer and lutidine-based ligands shown in Scheme 16.11 were employed as a catalyst in reducing CO_2 in the presence of Trifluoroacetic Acid (TFA), a strong acid. The lutidine-linked complexes reduced CO_2 to CO at potentials ranging at −1.6 V vs. Ag/$AgNO_3$. The palladium complexes were one electron reduction catalysts and were reversible because of the charge delocalization onto the pyridyl substituents that avoided the decomposition into the inert bimetallic species [21].

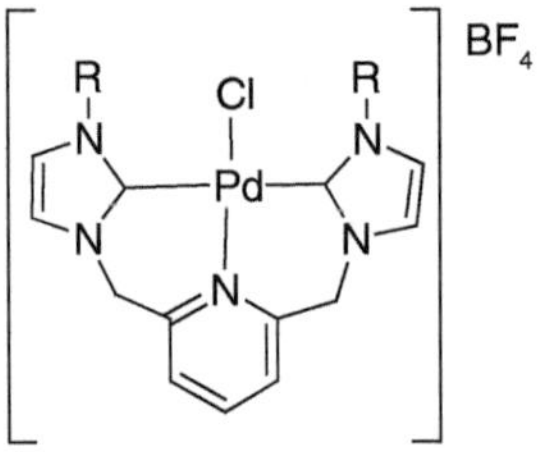

Scheme 16.11 Palladium complexes with varying alkyl chains employed in CO_2 reduction.

Dalton D. Burks et al. combined a NHC and pyridine moiety rings onto a pincer scaffold with a methoxy group as an efficient catalyst for reducing CO_2 to CO. The ruthenium complex substituted with a methoxy group was a vitalized catalyst in contrast to the other ruthenium complex, which is an unsubstituted pincer as shown in Scheme 16.12 and is inactive. It was illustrated that it was the methoxy group that could greatly enhance the photocatalysis [22].

Nickel pincer complexes are known to reduce CO_2 photocatalytically through a photosensitizer (PS). A reducing species has been generated when the first photoexcitation takes place. This reducing species accepts an electron from the sacrificial donor. Then there is a transfer of electron to the Ni catalyst from the reduced PS that interacts with the CO_2 and then accepts the second electron that reduces the CO_2 to CO along with the formation of H_2O or CO_3^{2-} (Scheme 16.13).

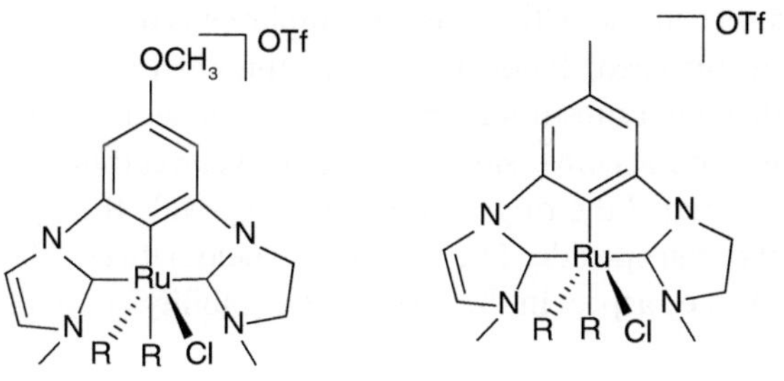

Scheme 16.12 Ruthenium-NHC complexes in CO_2 reduction.

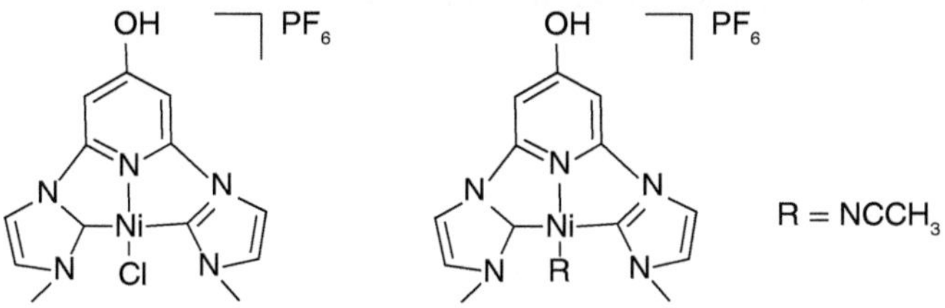

Scheme 16.13 Pincer based nickel-NHCs in CO_2 reduction.

The nickel complex with chlorido as a ligand system was employed as a catalyst in reducing CO_2 electrocatalytically by cyclic voltammetry (CV) technique and compared with other nickel catalysts. A drastic increase in the catalyst reactivity was observed where there was an effect from a single remote atom change from hydrogen in the nickel catalyst to an oxygen anion in chlorido substituted on the pincer ligand. Chlorido substituted complex displayed a catalytic current on the third reduction wave. This catalyst exhibited photocatalytic reactivity in the presence of a photosensitizer. The reactivity of this catalyst exhibited a TOF of $2.2\,h^{-1}$ for the first four hours before there was production of CO and it slowed down gradually and ceased by six hours. By controlling the concentration of the protons in the solution, the pincer-based catalysts can also be termed switchable catalysts.

Based on the lability of the hybrid ligands, six Fe(II) complexes with varying alkyl substituents have been reported. These Fe(II) complexes on amalgamation with PPNCl also called as bis(triphenylphosphoranylidene) ammonium chloride shown in Scheme 16.14 exhibited excellent catalytic potential. Based on UV-vis

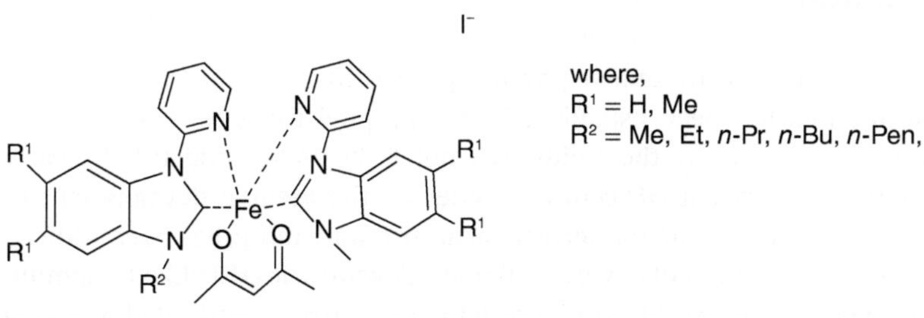

Scheme 16.14 Iron-NHC complexes employed in CO_2 reduction.

and FT-IR studies, they revealed that the activity of Fe(II) complexes is due to the trans-effect between the pyridine nitrogen donor atoms and the corresponding NHC ligand. The Fe complex with a butyl substituent in amalgamation with PPNCl exhibited a good catalytic response for the conversion of CO_2 into cyclic carbonates where the chloride ion of PNNCl initiated the ring opening of the epoxide through nucleophilic attack on the carbon of the epoxide. The trans-effect held between the NHC and the pyridine ring played a key aspect in the activity and stability of the Fe catalysts [23].

Several NHC-based complexes had been investigated for reducing the CO_2 to CO and have displayed moderate efficiency and low stability, such as tricarbonylpyridyl-NHC Mn^I complexes shown in Scheme 16.15.

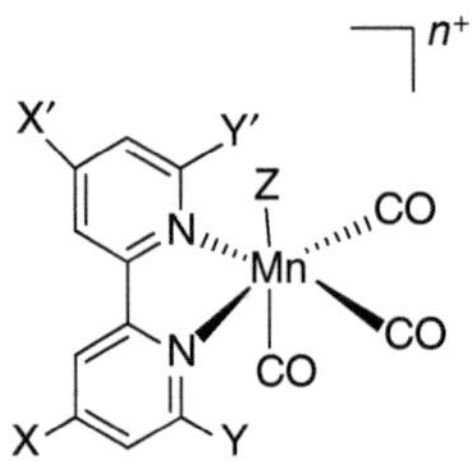

Scheme 16.15 Manganese NHC complexes employed in CO_2 reduction.

NHC-based pure organometallic tricarbonyl Mn^I complexes were prepared and have been used for the electrocatalytic reduction of CO_2. In Mn complex, the pyridine ring was replaced by a NHC that affected the catalytic performance, which increased the turn over frequency and selectivity for the production of CO. In presence of anhydrous aprotic solvent, the bis-NHC catalyst efficiently converted CO_2 to CO, but the complexes with bypyridyl rings were found to be inactive without a definite proton source. On comparing the Mn complexes bearing the polypyridyl motifs, the corresponding Mn complex had shown the turn over frequency in several orders of higher magnitude than the former ones. The Mn atom has a strong nucleophilic character that holds for the positive ligand effect on catalysis [24].

16.3 Organometallic Compounds in Gas Adsorption and Purification

The ability of certain solids to remove physical properties like color from solutions and odor from gases has been established and accepted for ages. However, the phenomenon of such acts by the solids was not fully understood till the early twentieth century. The twentieth century witnessed the gradual development in understanding of this phenomenon termed as adsorption, and it has been utilized in purification and separation of gases and liquids. Several scientists like Langmuir, Brunauer, Emmett and Teller, McBain and Bakr, and Barrer contributed to laying the foundation for understanding adsorption. The results of their contribution evidenced that adsorption is performed optimally in porous solids and it is influenced by the physical attraction that results between the surface of a porous solid and the component molecules of either the gas or liquid from the bulk phase. Adsorption, thus, is defined as the process by which molecules from gas or liquid accumulate on a solid surface, resulting in the formation of a thin film. It is a phenomenon that occurs naturally and plays a crucial role in various industrial applications ranging from purification and separation to energy storage and catalysis. In academia, adsorption is widely studied for its potential in industrial applications as well as its importance as a characterization tool for understanding pore size distribution and surface area of solid materials [25]. The most common example of gas adsorption is the usage of hydrophilic sorbent in a packed column for the removal of moisture in a gas stream. Another example is the large-scale removal of H_2S gas and mercaptans from natural gas. This application of adsorption for separation and purification of gases holds great promise. Generally, adsorptive gas purification or separation involves the flow of a mixture of gases through an adsorbent packet column to get a product consisting of gases that adhere weakly to the adsorbent. This is followed by the desorption and removal of the gases that are adhered to the adsorbent so that the adsorbent can be used again.

There are several mechanisms that can be attributed to the adsorption of the gas molecules on the adsorbent, they are: (i) size or shape exclusion, certain gases in a gaseous mixture are unable to enter the pores present in a adsorbent, other gases enter the pores and get adsorbed, this is also termed as the molecular sieving effect; (ii) difference in the adsorbate packing and surface interactions causes favorable adsorption for some gases and unfavorable conditions for other gases, this is termed the thermodynamic equilibrium effect; (iii) the difference in the kinetic rates in the gas mixture, certain gas species enter the pores and get adsorbed relatively faster than other gas species, this is termed the kinetic effect; (iv) quantum sieving effect. This arises due to the difference in diffusing rates of the micropores for light molecules, allowing for separation of gas components [26].

16.3.1 Metal Organic Frameworks (MOFs) for Gas Adsorption

Gas adsorption is enabled by porous materials as adsorbents, with zeolites being one of the most popular and used adsorbents due to its uniform pore size, shape, and distribution, which are qualities possessed few materials. There is a quest for the fabrication and development of efficient materials, and in this quest, MOFs have been gaining interest as catalysts for gas adsorption. MOFs are a new class of porous materials that have emerged. They also termed as porous coordination networks, metal–ligand coordination polymers, and porous coordination polymers, among other names, these materials have become a popular and exciting topic of research and its related applications. MOFs are basically hybrid materials that possess both inorganic and organic properties. They are made up of either a single metal or a cluster of polynuclear metals that are linked by organic ligands by coordination bonds. The strong coordination bond present between the metal and the ligand results in structurally well-defined framework structures, both geometrically and

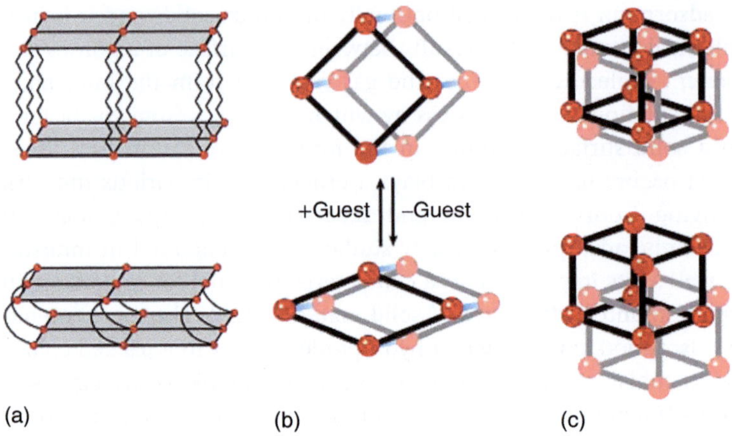

Figure 16.2 Illustration of the flexible behaviors of MOFs on interacting with gas molecules in (a) 2-dimension; (b) 1-dimension; (c) 3-dimension. Red spheres represent the metals; the lines represent the organic linkers. Source: Reproduced with permission from Murdock et al. [27]/American Chemical Society.

crystallographically. MOFs are one of the most versatile materials that can be tuned in numerous ways, like its mode of synthesis; choice of transition metals, which can be made to possess various geometries by virtue of their coordination number; choice of organic ligands ranging from cyanides, halides, organic molecules like 4,4′-bipyridine and benzenedicarboxylic acid. These organic and inorganic parts bind together to form motifs of different dimensions that exhibit flexibility in interaction with guest molecules as exhibited in Figure 16.2 [27]. This tunability and versatility of MOFs allows for it to be designed with various properties like their pore structures, pore sizes, and surface functionalization and thus enables it to be used as selective gas adsorption and separation materials [26, 28].

16.3.2 Classification of MOFs as Adsorbents for Gas Separation

MOFs can be classified into two categories, i.e. rigid and flexible classes. Rigid MOFs, by virtue of their name, are the class of MOFs that are robust, possessing a fixed porosity similar to zeolites. Flexible MOFs, on the other hand, possess soft frameworks that are subject to change due to external forces/influence like guest molecules, pressure, temperature, etc. This phenomenon enables applications of these MOFs in situations and the environment and outshines traditional adsorbents like activated carbon and zeolites.

16.3.2.1 Rigid MOFs as Adsorbents for Selective Gas Separation

Rigid MOFs in most cases functions like typical zeolites and thus its adsorption and separation work on the principles of molecular sieving, adsorption influenced by different strength and interaction between the adsorbent and the guest molecule/adsorbate. Based on these working principles, rigid MOFs can be classified below

16.3.2.1.1 Gas Adsorption Attributed to Shape/Size Exclusion

This category of MOFs as adsorbent relies on the traditional principle that is present in zeolites, termed the molecular sieving effect. Studies have revealed that the employment of a MOF developed from manganese formate possesses 3D structures with 1D channels that are made up of larger cages that are connected via a smaller window. It is evidenced that at lower temperatures i.e. 78 K, the MOF selectively adsorbs H_2 gas over other gases possessing larger sizes like Ar and N_2, at elevated temperature of 195 K, it adsorbs CO_2 over CH_4. At the two different temperature conditions, the adsorption capacity of the larger gas molecules was found to be almost zero. This property of selectivity by the MOF was due to the smaller aperture of the 1D channels translating to the size exclusion principle [29].

16.3.2.1.2 Gas Adsorption Attributed to Adsorbate–Surface Interactions

This category of adsorption by the MOF although relies on the traditional physical factors like pore size and structure, it is majorly influenced by thermodynamics equilibrium and kinetics equilibrium. These kinds of interactions exhibit the selectivity of gases on properties like H-bonding, dipole moment, quadruple moment, and the properties the pores possess. An example of this phenomenon was exhibited by $Cu_2(pzdc)_2(pyz)$, here pzdc is pyrazine-2,3-dicarboxylate and pyz is pyrazine. The application of this complex as an adsorbent evidenced high selectivity of acetylene gas (C_2H_2) over CO_2, which possesses 1D open channel with 4–6 Å cross-section size and O atoms on the channel surface. The O atom acts as the site for adsorption and it favorably adsorbs C_2H_2 over CO_2 at different sorption isotherms at low pressure and temperature. The reason behind the favorable binding of C_2H_2 is the stronger H-bonding that arises between the C_2H_2 and the O atoms on the channel surface as compared to the H-bonding present with CO_2 and the O atoms [30].

16.3.2.1.3 Gas Adsorption Attributed to Shape/Size Exclusion and Adsorbate–Surface Interactions

The third interaction for adsorption of gas in rigid MOF is attributed to both the interactions that have been discussed, i.e. the interactions between the pore walls and the gas molecules and pore size. An illustration of this interaction was evidenced by $Er_2(PDA)_3$ (here PDA is 1,4-phenylendiacetatedianion), a MOF material possessing 3D structural framework with circular 1D channels and Er(III) sites as illustrated in Figure 16.3. The material on exposure to Ar, N_2, and CO_2 displayed selective adsorption of CO_2 due to size and the interaction between gas molecule and active sites on the channel surface. Furthermore, it was evidenced that the presence of Er(III), p-electrons, and polar groups present in the framework resulted in the introduction of an electric field resulting in a dipole in the CO_2 molecule. There was also suggestions that CO_2, which possess a quadrupole moment resulted in further increase in the adsorption and a donor–acceptor affinity between CO_2 molecules and Er(III) ions was suggested [31].

16.3.2.2 Flexible MOFs as Adsorbents for Gas Separation

MOFs are framework supported by coordination bonds as previously discussed and other interactions like π–π bonding, van der Waals interactions, and H-bonding.

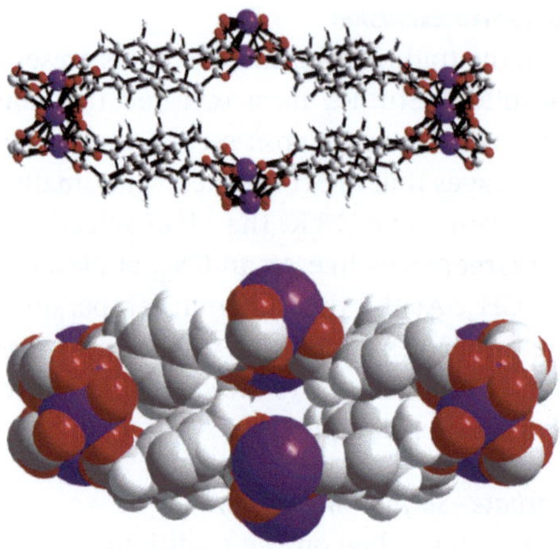

Figure 16.3 Cross section of the open channel present in $[Er(PDA)_{1.5}]$; (Top) ball and stick model, (bottom) space filled model. Color scheme: C, gray; H, white; O, red; Er, blue. Source: Reproduced with permission from Pan et al. [31]/American Chemical Society.

These collective interactions present in MOFs give them flexibility under different conditions. MOFs that possess flexibility properties may show variations in their structure during adsorption and desorption processes. They show a more defined porous structure on interaction with guest molecules. The processes that are involved in flexible MOFs as gas adsorbents differ from rigid MOFs. The adsorption isotherms display hysteretic behaviors because of change in framework structure due to the adsorption and desorption process [26]. These MOFs retain properties that are present in rigid MOFs, however in flexible MOFs the structural rearrangement factor is considered in all the cases. The classification of flexible MOFs based on their working principle is described below:

16.3.2.2.1 Gas Adsorption Attributed to Shape/Size Exclusion with Pore Shape/Size Change

The interaction of this type of flexible MOF is based on the principles of shape/size exclusion effect along the channels, which can also be influenced by pore shape/size change. An example of this type of adsorption is observed in Cd(pzdc)(bpee) where pzdc is pyrazine-2,3-dicarboxylate and bpee is py–CH=CH–py (py is pyridine), Cd(pzdc) is the layer and bpee acts as the pillar ligands and are connected forming 3D structural frameworks with 1D channels. It was observed through the adsorption studies at 298 K that the channel adsorbs H_2O and MeOH and expands; however it does not adsorb gases with larger molecular sizes like EtOH and THF, which is attributed to the shape/size exclusion [32].

16.3.2.2.2 Gas Adsorption Attributed to Adsorbate–Surface Interactions with Pore Shape/Size Change

This adsorption property works similar to the rigid MOFs that are based on the interaction of the surface and adsorbate. The property majorly depends on the surface of the pores that are rendered flexible. This interaction is observed in a flexible MOF i.e. MIL-53, a metal-benzenedicarboxylate, i.e. $M(OH)(O_2C–C_6H_4–CO_2)$; M is either trivalent aluminum or chromium. The MOF is 3D structured, possessing a 1D diamond channel exhibiting breathing phenomenon on hydration and dehydration.

During the hydration, the pores get slightly deformed due to the hydrogen-bonding interaction between the O atom of the carboxylate and hydroxo groups of the MOF and the H_2O molecule. It was also observed that in its dehydrated form, the MOF interacts with CO_2 and CH_4 molecules and exhibits adsorption behaviors different from its hydrated form [33].

16.3.2.2.3 Gas Adsorption Attributed to Structural Rearrangement or Gate-Opening Induced by Adsorbate–Surface Interactions

This type of interaction is observed in dynamic MOFs possessing small or no pores. These MOFs on interaction with certain guest molecule results in the expansion of their pores, and it is termed the gate-opening process. The expansion of the pore is determined by the adsorbate–surface interactions. An example of this interaction is evident in ZIF-20 i.e. $Zn(Pur)_2(DMF)_{0.75}(H_2O)_{1.5}$, here Pur = purinate, it possesses large cages connected through small windows able to selectively adsorb CO_2 over CH_4. The MOF evidenced that CO_2 adsorption was much higher than CH_4 due to stronger interactions arising between the pore surface and the CO_2 molecule. ZIF-20 primarily possesses pores with apertures that are smaller than the diameter of the gas molecules CO_2 and CH_4. This smaller pore size of the ZIF-20 points to the fact that there is a gate opening process involved via the large cages present in the ZIF-20 structure that undergoes a window-widening process due to the ligands swinging and allowing the movement of the guest molecules through the channel. This process is enabled due to the dipole moment, quadruple moment, and the H-bonding possessed by the MOF, which also influences its flexibility [34].

16.3.2.2.4 Gas Adsorption Attributed to Adsorbate-Specific Gate-opening Pressures

This interaction results in the specific gate opening of the pores due to the strength of the molecular interactions that translate to pressures that guest molecules possess by virtue of their properties. It may be noted that each guest molecule possesses a specific pressure and below this pressure the guest molecules are unable to be adsorbed. This property possessed by the adsorbent can thus be utilized for the separation of gases that they possess by the differences in their pressures. This type of interaction was evident in [Cu(dhbc)2(4,4′-bpy)], here dhbc is dihydrobenzoic acid, bpy is bipyridine. This complex is present in hydrated form in a 2D sheet motif. The motif possesses interlocking ridges and hollows assembled in upright fashion by the benzene planes, the complex possesses 1D channels that can accommodate water molecules. The structure is further stabilized by the π–π stacking interaction present between neighboring dhbc ligands. This MOF on exposure to different gases, i.e. N_2, O_2, CO_2, and CH_4 at low pressure, displayed negligible adsorption. The adsorption of the gases increased at specific gate opening pressures for each of the gases, further these gases displayed a decrease in adsorption in pressures that are not attributed to their gate opening pressures. This points to the conclusion that different gases possess distinct gate opening pressures attributed to the differences in molecular interaction forces of different guest molecules [35].

16.4 Gas Purification with MOFs

Section 16.3.2 has discussed the selective adsorption of MOFs and these adsorption properties of MOFs point to the potential of MOFs for gas separation and purification. Recent studies have seen MOFs emerge in gas separation application by means of isotope separation, kinetic separation and membrane-based separation. MOFs due to its adsorbing properties have displayed the adsorption of a specific gas molecule when exposed to a mixture to a mixture of gases. MOFs can exhibit selectivity to different gas molecules on the basis of applied temperature, pressure, adsorbate–surface interactions, and molecular sieving effect. The separation of gas by MOFs based on the process involved is detailed in this section.

16.4.1 Kinetic Separation

This process is governed by the difference in the rates of diffusion of different components. This process is highly rated and plays a crucial role in various industrial applications related to chromatography. The main factors that affect the diffusion rate are the size and shape of the adsorbate molecule, shape and size of the adsorbent pores, adsorption strength, affinity of the adsorbent sites, the adsorbate molecule concentration, temperature, pressure, etc. Kinetic separation is governed and is determined by adsorption that is time-dependent or the discharge of various adsorbate molecules. Examples of kinetic adsorption in real-time applications are fixed-bed adsorption and gas-chromatographic separation. One of the first applications of gas separation and purification by a MOF was performed using HKUST-1/MOF-199, which consists of dimeric metal units that are linked by benzene-1,3,5-tricarboxylate. HKUST-1 on application for gas separation, was able to separate gases mixtures like C_2H_4/C_2H_6, CO_2/CO, and $CO_2/MeOH$. HKUST-1 also evidenced that it is possible to separate and purify air from gases like CO_2, moisture, high-molecular weight hydrocarbons and NO [36].

16.4.2 Quantum Sieving Effect for H_2/D_2 Separation

The molecular sieving principle does not apply in isotope separation due to similar adsorption properties of the isotopes. Quantum sieving process can be applied for the separation of isotope molecules [37]. Quantum sieving works on the preferred adsorption of heavier molecules because of the contrast in quantum energy levels possessed by the isotope that is confined in a narrow space similar to the de Broglie wavelength. Quantum sieving effect has been found to be a property possessed by certain MOFs, an example is $Zn_3(bdc)_3[Cu(Pyen)]\cdot(DMF)_5(H_2O)_5$ where H_2bdc is 1,4-benzenedicarboxylic acid and $PyenH_2$ is 5-methyl-4-oxo-1,4-dihydro-pyridine-3-carbaldehyde. The mixed metal organic 3D framework consists of tessellated sheets of $Zn_3(bdc)_3$ 2D sheets pillared by Cu(Pyen), the MOF on desolvation yields planar open copper centers with curved pores ranging about 5.6×12.0 Å in *c* axis and irregular ultramicropores in *b* axis direction to form 2-D arrangement of pores.

The MOF on investigation with H_2 and D_2 indicated that the adsorption takes place on the two sides of the open Cu(II) centers in the ultramicropores resulting in high enthalpy of adsorption. Virial analysis of isotherm data of H_2 and D_2 suggests greater D_2 surface interactions as compared to H_2 surface interactions, this is due to variation in quantum statistical mass effect affecting the vibrational energy levels of the two gases. The adsorption kinetics also indicated D_2 possessed larger rate constants for and lower activation energy than H_2. This phenomenon is attributed to the quantum effect of H_2 with higher zero-point energy that results in more effective collision in the ultramicropores thereby producing a higher barrier for diffusion.

16.4.3 Membrane-Based Gas Separation

This is one of the most commonly used methods for gas separation in industries and is preferred over other methods due to its operational benefits. The success of this method relies on the membrane materials and its attributed properties. This method has been recently applied for MOFs due to the MOFs' porous structure. A result of this attempt led to the development of a MOF possessing a copper net with benzene-1,3,5-tricarboxylate (btc) linkers. The $Cu_3(btc)_2$ was developed into a thin film and employed as a membrane for the separation of H_2/N_2, H_2/CH_4,and H_2/CO_2. The investigation indicated that the MOF membrane has high H_2 permeation flux and H_2 permeation selectivity [38].

References

1 Yin, S.F. et al. (2008). Efficient fixation of carbon dioxide by hypervalent organobismuth oxide, hydroxide, and alkoxide. *Angew. Chem. Int. Ed.* 47 (35): 6590–6593. https://doi.org/10.1002/anie.200802277.

2 Bahramian, B. and Dehghani, F. (2016). New catalytic systems for fixation of carbon dioxide into valuable poly(alkylene carbonates). In: *Advanced Catalytic Materials – Photocatalysis and Other Current Trends* (ed. N. Luis and J.-A. Wang), 69–93. InTech https://doi.org/10.5772/61969.

3 Taşci, Z. et al. (2012). A new application area for Ag-NHCs: CO_2 fixation catalyst. *ChemCatChem* 4 (6): 831–835. https://doi.org/10.1002/cctc.201100430.

4 Idrees, M.Z. et al. (2021). Efficient palladium(II) electrocatalysts with thiophene anchored pyridinium amidates for CO_2 reduction. *J. CO2 Util.* 44 (August 2020): 101384. https://doi.org/10.1016/j.jcou.2020.101384.

5 Jiang, X. et al. (2018). Carbon dioxide electroreduction over imidazolate ligands coordinated with Zn(II) center in ZIFs. *Nano Energy* 52 (II): 345–350. https://doi.org/10.1016/j.nanoen.2018.07.047.

6 Miedaner, A., Noll, B.C., and DuBois, D.L. (1997). Synthesis and characterization of palladium and nickel complexes with positively charged triphosphine ligands and their use as electrochemical CO_2-reduction catalysts. *Organometallics* 16 (26): 5779–5791. https://doi.org/10.1021/om970519c.

7 Wang, S. and Wang, X. (2016). Imidazolatsysteme zur CO_2-Abscheidung und photochemischen Reduktion. *Angew. Chem.* 128 (7): 2352–2364. https://doi.org/10.1002/ange.201507145.

8 Therrien, J.A., Wolf, M.O., and Patrick, B.O. (2014). Electrocatalytic reduction of CO_2 with palladium bis-N-heterocyclic carbene pincer complexes. *Inorg. Chem.* 53 (24): 12962–12972. https://doi.org/10.1021/ic502056w.

9 Simón-Manso, E. and Kubiak, C.P. (2005). Dinuclear nickel complexes as catalysts for electrochemical reduction of carbon dioxide. *Organometallics* 24 (1): 96–102. https://doi.org/10.1021/om0494723.

10 Kojima, T. (2021). Photocatalytic carbon dioxide reduction using nickel complexes as catalysts. *ChemPhotoChem* 5 (6): 512–520. https://doi.org/10.1002/cptc.202000263.

11 Neri, G. et al. (2014). A functionalised nickel cyclam catalyst for CO_2 reduction: electrocatalysis, semiconductor surface immobilisation and light-driven electron transfer. *Phys. Chem. Chem. Phys.* 17 (3): 1562–1566. https://doi.org/10.1039/C4CP04871G.

12 Cao, L.M. et al. (2018). The synergistic catalysis effect within a dinuclear nickel complex for efficient and selective electrocatalytic reduction of CO_2 to CO. *Green Chem.* 20 (4): 798–803. https://doi.org/10.1039/c7gc03451b.

13 Nichols, A.W. et al. (2018). Electrocatalytic reduction of CO_2 to formate by an iron Schiff base complex. *Inorg. Chem.* 57 (4): 2111–2121. https://doi.org/10.1021/acs.inorgchem.7b02955.

14 Rosas-Hernández, A. et al. (2017). Cyclopentadienone iron complexes as efficient and selective catalysts for the electroreduction of CO_2 to CO. *Catal. Sci. Technol.* 7 (2): 459–465. https://doi.org/10.1039/C6CY02352E.

15 Guo, Z. et al. (2017). Photocatalytic conversion of CO_2 to CO by a copper(II) quaterpyridine complex. *ChemSusChem* 10 (20): 4009–4013. https://doi.org/10.1002/cssc.201701354.

16 Shimoda, T. et al. (2018). Photocatalytic CO_2 reduction by trigonal-bipyramidal cobalt(II) polypyridyl complexes: the nature of cobalt(I) and cobalt(0) complexes upon their reactions with CO_2, CO, or proton. *Inorg. Chem.* 57 (9): 5486–5498. https://doi.org/10.1021/acs.inorgchem.8b00433.

17 Nie, W. et al. (2020). Electrocatalytic CO_2 reduction by cobalt bis (pyridylmonoimine) complexes: effect of ligand flexibility on catalytic activity. *ACS Catal.* 10 (9): 4942–4959. https://doi.org/10.1021/acscatal.9b05513.

18 Steffey, B.D. et al. (1994). Synthesis and characterization of palladium complexes containing tridentate ligands with PXP (X = C, N, O, S, As) donor sets and their evaluation as electrochemical CO_2 reduction catalysts. *Organometallics* 13 (12): 4844–4855. https://doi.org/10.1021/om00024a030.

19 Hazari, N., Iwasawa, N., and Hopmann, K.H. (2020). Organometallic chemistry for enabling carbon dioxide utilization. *Organometallics* 39 (9): 1457–1460. https://doi.org/10.1021/acs.organomet.0c00229.

20 Chen, Z., Chuncheng Chen, C., Weinberg, D.R. et al. (2011). Electrocatalytic reduction of CO_2 to CO by polypyridyl ruthenium complexes. *Chem. Commun.* 47: 12607–12609. https://doi.org/10.1039/C1CC15071E.

21 Deluca, E.E. et al. (2019). Improved electrocatalytic CO_2 reduction with palladium bis(NHC) pincer complexes bearing cationic side chains. *Organometallics* 38 (6): 1330–1343. https://doi.org/10.1021/acs.organomet.8b00649.

22 Burks, D.B. et al. (2018). Nickel(II) pincer complexes demonstrate that the remote substituent controls catalytic carbon dioxide reduction. *Chem. Commun.* 54 (31): 3819–3822. https://doi.org/10.1039/c7cc09507d.

23 Chen, F. et al. (2021). Hemilabile N-heterocyclic carbene and nitrogen ligands on Fe(II) catalyst for utilization of CO_2 into cyclic carbonate. *Appl. Organomet. Chem.* 35 (2): 1–10. https://doi.org/10.1002/aoc.6099.

24 Franco, F. et al. (2018). A highly active N-heterocyclic carbene manganese(I) complex for selective electrocatalytic CO_2 reduction to CO. *Angew. Chem. Int. Ed.* 57 (17): 4603–4606. https://doi.org/10.1002/anie.201800705.

25 Thomas, W.J. and Crittenden, B. (1998). *Adsorption Technology and Design*, 270. Butterworth–Heinemann, Elsevier https://doi.org/10.1016/B978-0-7506-1959-2.X5001-0.

26 Li, J.R., Kuppler, R.J., and Zhou, H.C. (2009). Selective gas adsorption and separation in metal–organic frameworks. *Chem. Soc. Rev.* 38 (5): 1477–1504. https://doi.org/10.1039/b802426j.

27 Murdock, C.R., Lu, Z., and Jenkins, D.M. (2013). Effects of solvation on the framework of a breathing copper MOF employing a semirigid linker. *Inorg. Chem.* 52 (4): 2182–2187. https://doi.org/10.1021/ic3026304.

28 Wang, B. et al. (2018). Applications of metal–organic frameworks for green energy and environment: new advances in adsorptive gas separation, storage and removal. *Green Energy Environ.* 3 (3): 191–228. https://doi.org/10.1016/j.gee.2018.03.001.

29 Dybtsev, D.N. et al. (2004). Microporous manganese formate: a simple metal–organic porous material with high framework stability and highly selective gas sorption properties. *J. Am. Chem. Soc.* 126 (1): 32–33. https://doi.org/10.1021/ja038678c.

30 Matsuda, R. et al. (2005). Highly controlled acetylene accommodation in a metal–organic microporous material. *Nature* 436 (7048): 238–241. https://doi.org/10.1038/nature03852.

31 Pan, L. et al. (2003). Porous lanthanide-organic frameworks: synthesis, characterization, and unprecedented gas adsorption properties. *J. Am. Chem. Soc.* 125 (10): 3062–3067. https://doi.org/10.1021/ja028996w.

32 Maji, T.K. et al. (2004). Expanding and shrinking porous modulation based on pillared-layer coordination polymers showing selective guest adsorption. *Angew. Chem. Int. Ed.* 43 (25): 3269–3272. https://doi.org/10.1002/anie.200453923.

33 Bourrelly, S. et al. (2005). Different adsorption behaviors of methane and carbon dioxide in the isotypic nanoporous metal terephthalates MIL-53 and MIL-47. *J. Am. Chem. Soc.* 127 (39): 13519–13521. https://doi.org/10.1021/ja054668v.

34 Hayashi, H. et al. (2007). Zeolite A imidazolate frameworks. *Nat. Mater.* 6 (7): 501–506. https://doi.org/10.1038/nmat1927.

35 Kitaura, R. et al. (2003). Porous coordination-polymer crystals with gated channels specific for supercritical gases. *Angew. Chem. Int. Ed.* 42 (4): 428–431. https://doi.org/10.1002/anie.200390130.

36 Min Wang, Q. et al. (2002). Metallo-organic molecular sieve for gas separation and purification. *Microporous Mesoporous Mater.* 55 (2): 217–230. https://doi.org/10.1016/S1387-1811(02)00405-5.

37 Beenakker, J.J.M., Borman, V.D., and Krylov, S.Y. (1995). Molecular transport in subnanometer pores: zero-point energy, reduced dimensionality and quantum sieving. *Chem. Phys. Lett.* 232 (4): 379–382. https://doi.org/10.1016/0009-2614(94)01372-3.

38 Guo, H. et al. (2009). "Twin copper source" growth of metal–organic framework membrane: $Cu_3(BTC)_2$ with high permeability and selectivity for recycling H_2. *J. Am. Chem. Soc.* 131 (5): 1646–1647. https://doi.org/10.1021/ja8074874.

17

Emerging Role of Organometallic Compounds for Drug Delivery, Renewable Energy, and Wastewater Treatment

Shokoh Parham[1], Hadi Sarvari[2,3], Seyed-Shirin Parham[4], and Hadi Nur[1,5]

[1] *Universiti Teknologi Malaysia, Ibnu Sina Institute for Scientific and Industrial Research(ISI-SIR), Centre for Sustainable Nanomaterials, Johor Bahru Campus, 81310 Skudai, Johor, Malaysia*
[2] *Islamic Azad University, Department of Civil Engineering, Isfahan (Khorasgan) Branch, Jey Street, 81551-39998, Isfahan, Iran*
[3] *The Hong Kong Polytechnic University, Department of Building and Real Estate, Hung Hom, Kowloon, Hong Kong, China*
[4] *Islamic Azad University, Department of Veterinary, Shahrekord Branch, Rahmatiyeh, 8813733395, Shahrekord, Iran*
[5] *Universitas Negeri Malang, Department of Chemistry, Faculty of Mathematics and Natural Sciences, Jl. Semarang 5, 65145, Malang, Indonesia*

17.1 Introduction

Organometallics, identified by their prefix "organo-," have the highest prevalence in a variety of applications. These compounds normally contain metals and transition metals, semimetals, lanthanides, and actinides. The use of the18-electron rule helps predict the chemical bonding, reactivity, and stability of organometallics. These metal complexes contain a minimum of one carbon–metal bond in the form of a direct carbon–metal (sigma) or a metal complex (pi – bond and lone pair) bond. The organometallics' metal–carbon bond may be ionic, covalent, or in-between, contributing significantly to chemical reactions [1].

Recent studies have revealed that organometallics, characterized as metal compounds that contain a minimum of one direct covalent metal–carbon bond, can contribute as good candidates for future medications such as anticancer drugs [2]. Metal ions play a critical role in chemical diversity when searching for new treatments and diagnostic agents [3]. Organometallic compounds are structurally various (at a rage of linear to octahedral or even beyond), show a higher diversity of stereochemistry compared to their organic counterparts (there are 30 stereoisomers for one octahedral complex with six different ligands), and can control the main kinetic features (including the ligand hydrolysis rate) through rational ligand design. In addition, these compounds show kinetic stability, typically no charge, relative lipophilicity, and a low oxidation state of their metal atoms [2].

Organometallics have found a broader scope of application in technologies and industries compared to their organic counterparts of different metals. For instance, organometallics are applied in catalysis or anti-catalysis, mainly in polymer

Organometallic Compounds: Synthesis, Reactions, and Applications, First Edition.
Edited by Dakeshwar Kumar Verma and Jeenat Aslam.

chemistry, and are important out of all proportion to the tonnage of tin utilized. Therefore, they have important applications in the construction industry [4].

There are many kinds of research on the different applications, including pharmaceuticals and drug delivery, renewable energy, and wastewater treatment [5–13]. Organometallic chemistry became an important industrial field in the 1950s due to its contribution as a basic science focusing on homogeneous catalysis [14–18] as well as coordinating olefin polymerization. Yet, the commercial concentration on large-scale polymers and commodity chemical production decreased from 1975 to 1985 because of the transformation in the chemical industries. Hence, fewer studies have focused on the investigation of the catalytic processes for these materials. The recent attention to the field of organometallic chemistry has resulted from the changes in the business environment toward fine chemicals, specialty polymers, and electronic materials [18]. The usage of organometallic compounds as stoichiometric reagents or catalysts has developed the way organic molecules are being prepared in different industries, including agrochemical, fine chemicals, electronics, and pharmaceuticals [19]. In this regard, this chapter aims to show

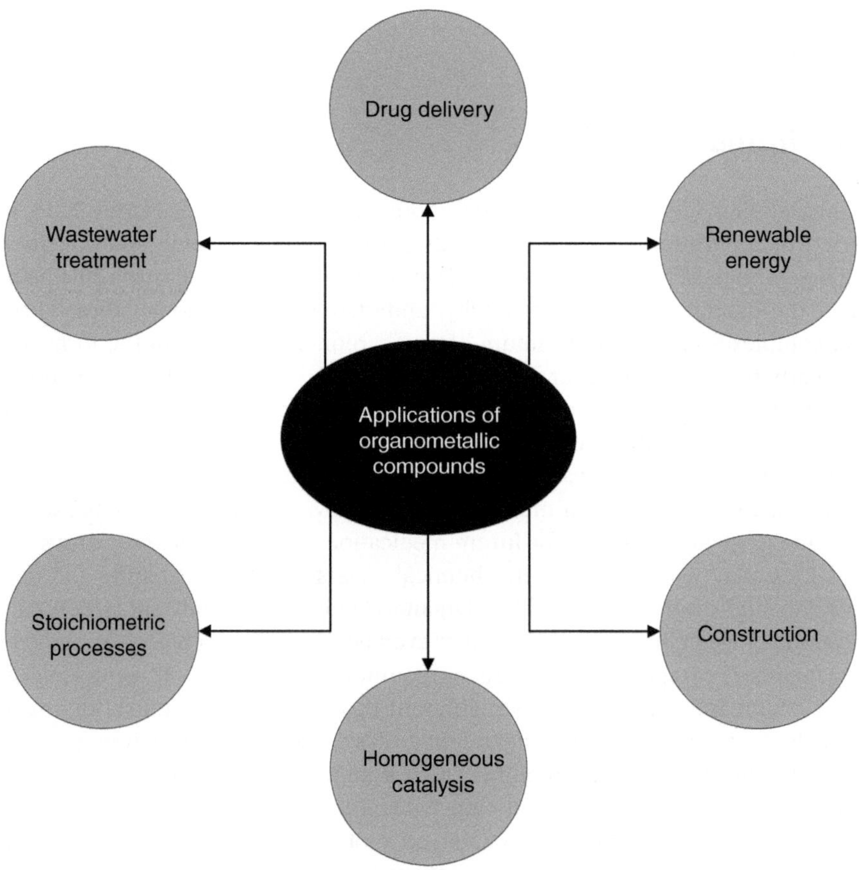

Figure 17.1 The different applications of organometallic compounds.

the applications and properties of organometallic compounds in drug delivery, renewable energy, and wastewater treatment. Figure 17.1 shows the different applications of organometallic compounds.

17.2 Organometallic Compounds

Organometallic chemistry, a combination of organic and inorganic chemistry, includes studying chemical compounds that contain a minimum of one metal–carbon bond, with carbon belonging to an organic molecule. Organometallics have broad applications in homogeneous catalysis and are practically used in stoichiometric processes and medicinal biochemistry [20]. Since ferrocene was discovered in 1951, much attention has been paid to novel organotransition metal chemistry. The distinctive structure, extraordinary reactive properties, and significant contribution of these complexes to large-scale industrial catalysis and fine chemical synthesis have piqued the interest of many scientists, including chemists working in the organic, inorganic, and theoretical fields [21, 22]. Organometallics contain a minimum of one carbon–metal bond in the form of a direct carbon–metal (sigma) or a metal complex (pi – bond) bond. The organometallics' metal–carbon bond may be ionic, covalent, or in-between, contributing significantly to chemical reactions [1, 21, 22].

It has been recently found that organometallic compounds are suitable candidates for designing efficient catalysts, including olefin metathesis and transfer hydrogenation. The thermodynamics and kinetics of ligand substitution and redox reactions of metal ions, particularly Ru^{II}, can be controlled using their electronic and steric impacts. These compounds possibly provide new mechanisms for medications to act by incorporating the target outer-sphere recognition along with controlled activation characteristics that consider ligand substitution and redox processes based on metals and ligands. A study investigating η^6-arene, η^5-cyclopentadienyl sandwich and half-sandwich complexes of Fe^{II}, Ru^{II}, Os^{II}, and Ir^{III} showed their potential effects on several health complications such as cancers, malaria, etc. [23]. Organometallics have been long known because of their substantial contribution to catalysis [24–34]; however, their role in chemical biology still needs to be clarified. These complexes cytotoxicity or instability when subjected to physiological environment is one of their noteworthy characteristics.

Consequently, few studies have considered the potential applications in biological fields. The amazing physiochemical characteristics of these compounds, including their stable chemical features, diverse structures, and distinguished photo- and electrochemical characteristics, have contributed significantly to the field of chemical biology [24]. Also, optoelectronic studies have recently focused on organometallic molecules, proposing the potential for application as flexible functional materials to be applied in energy interconversion [35]. Such compounds have an application in solar cells and other important fields, including wastewater treatment [36–39]. Table 17.1 listed the different organometallic compounds used for pharmaceuticals and drug delivery, renewable energy, and wastewater treatment [40–59]. Figure 17.2 shows the properties of organometallic compounds.

Table 17.1 The different organometallic compounds are used for pharmaceuticals and drug delivery, renewable energy, and wastewater treatment.

Compound	Result	Application	Study condition
Ruthenium-derived compounds (RDCs) [40]	RDC11 led to the growth inhibition of different tumors in mouse models	Anticancer therapy	*In vitro* and *in vivo*
Organometallic-montmorillonite material (K10-APTES-3Gly-Fe) [41]	This compound shows the application in the wastewater purification: pollutant photodegradation and good antibacterial efficiencies	Wastewater treatment	*In vitro*
Hetero-tri-organometallic compounds [42]	Strong effects on gram-positive bacteria consisting of multiresistant *Staphylococcus aureus* (MRSA)	Antibiotics	*In vitro*
Alq_3-5Cl organometallic compound [43]	Theoretical and empirical studies regarding the use of the Alq_3-5Cl organometallic compounds together with layers of graphene oxide in solar cells seem to be effective in optimizing their layering processes	Renewable energy (Solar cell)	*In vitro*
Cr organometallics based on the lead structure [44]	Complexes that contain a (η^6-pentamethylbenzene) $Cr(CO)_3$ moiety show considerably higher activity compared to other analogs that contain metals and the metal-free derivatives while exhibiting antibacterial effects on the gram-positive *Bacillus subtilis* but not on gram-negative *Escherichia coli*	Antibiotic	*In vitro*
Metallated terpolymer donors PM6Irx (x = 1%, 3%, 5%) [45]	The PM6Ir1:Y6-based polymer solar cells (PSCs) have the highest power conversion efficiency (PCE) equal to 16.71% and short-circuit current density (J_{SC}), open-circuit voltage (V_{OC}), and fill factor (FF) of 26.16 mA cm^{-2}, 0.848 V, and 75.33%, respectively	Renewable energy (Solar cell)	*In vitro*
Fe–Mg metal–organic complex adsorbent (Fe–Mg–BDC) [46]	Fe–Mg–BDC was effective in removing Pb(II) through the ion exchange mechanism. Efficient removal of As(V) could be also observed. Therefore, Fe–Mg organometallics showed promising results for the elimination of anions and cations that contain oxygen	Wastewater treatment	*In vitro*
1,4-Dicarboxybenzene modified Fe/Co metal organic [47]	Highly efficient removal of trace lead(II) from wastewater	Wastewater treatment	*In vitro*

(*Continued*)

Table 17.1 (Continued)

Compound	Result	Application	Study condition
Gold(III) 5,10,15-triphenylcorrole (Au–C1), gold(III) 10-(*p*-trifluoromethyl-phenyl)-5,15-diphenylcorrole (Au–C2), and gold(III) 10-(pentafluorophenyl)-5,15-diphenyl-corrole (Au–C3) [48]	These gold(III) corroles display excited state lifetime ≥25 μs and low emission quantum yield <0.15%. In combination with Au–C1, Au–C2, and Au–C3, vacuum-deposited organic solar cells (OSCs), giving respective PCEs of 2.7%, 3.0%, and 1.8%	Renewable energy (Solar cell)	*In vitro*
3-[1-(2-hydroxyphenyl) ethylideamino]-2-phenyl-3,4-dihydroquinazolin-4(3*H*)-one (**LH**) [49]	Notable anti-inflammatory activity was observed	Anti-inflammatory activity	*In vitro*
Cyclopentadienyl–ruthenium(II) and iron(II) [50]	The leading compound 14Ru had considerably higher cytotoxicity compared to oxaliplatin to HCT116 cells, which triggered higher activity and apoptosis levels of caspase-3 and -7 dose-dependently	Anticancer therapy	*In vitro*
AlGaInP [51]	AlGaInP layers bandgap ranges 1.9–2.2 eV. The quantum efficiency of samples having a higher composition of aluminum decreased while increasing their bandgap-voltage offset, remaining at 500 mV or less, to a bandgap of 2.1 eV	Renewable energy (Solar cell)	*In vitro*
Organoiron dendrimers attached to the well-known drug ibuprofen [52]	Higher inhibitory activity against methicillin-resistant *S. aureus*, *vancomycin-resistant Enterococcus faecium*, and *Staphylococcus warneri* compare to the reference drugs and all dendrimers display considerable anti-inflammatory activity	Anti-inflammatory and antimicrobial	*In vitro* and *in vivo*
Series of novel oxicams and coordinated them to Ru^{II}(cym)Cl and Os^{II}(cym)Cl moieties (η^6-*p*-cymene = cym) [53]	The isoxazolyl motif-containing ligand 1 and related complexes with Ru^{II}(cym)Cl **1a** and the Os analog **1b** had anticancer effects with IC50 values ranging similarly to the results of Ru^{III} experimental drug IT-139. Overall, the Os compounds showed equal or somehow higher strength compared to the Ru derivatives	Anticancer therapy	*In vitro*
Rh^{III}(Cp*) and Ir^{III}(Cp*) compound [54]	These materials had *in vitro* anticancer effects better than cisplatin and the experimental drug IT-139 but showed lower levels of toxicity in hemolysis as well as *in vivo* zebrafish models	Anticancer therapy	*In vitro* and *in vivo*

(Continued)

Table 17.1 (Continued)

Compound	Result	Application	Study condition
Cyanovinylhydrazone bridged diferrocenyl compounds [55]	Fabrication of a solar cell device utilizing the diferrocenyl donor–acceptor compounds as a bifunctional system indicated an average open-circuit voltage (V_{OC}) equal to 331 mV	Renewable energy (Solar cell)	*In vitro*
Organometallic osmium compounds [56]	This compound shows the anticancer ability	Anticancer therapy	*In vitro* and *in vivo*
NiO/perovskite/[6,6]-phenyl C61-butyric acid methyl ester (PC61BM) [57]	A 9.51% efficiency in converting power under AM 1.5 G illumination	Renewable energy (Solar cell)	*In vitro*
Organometallic tetranuclear Ru(II) arene [58]	These complexes seemed suitable for their antimetastatic effects, decreasing the cancer cell's ability to invade	Anticancer therapy	*In vivo*
Organometallic vapor phase epitaxy (VPE) growth of GaAs and AlGaAs [59]	The development of a simple theory aimed at relating the development rate and composition of AlGaAs epitaxial layer to the input fluxes. Based on Auger investigations, the GaAs-to-AlGaAs transition was obtained in <40 Å	Renewable energy (Solar cell)	*In vitro*

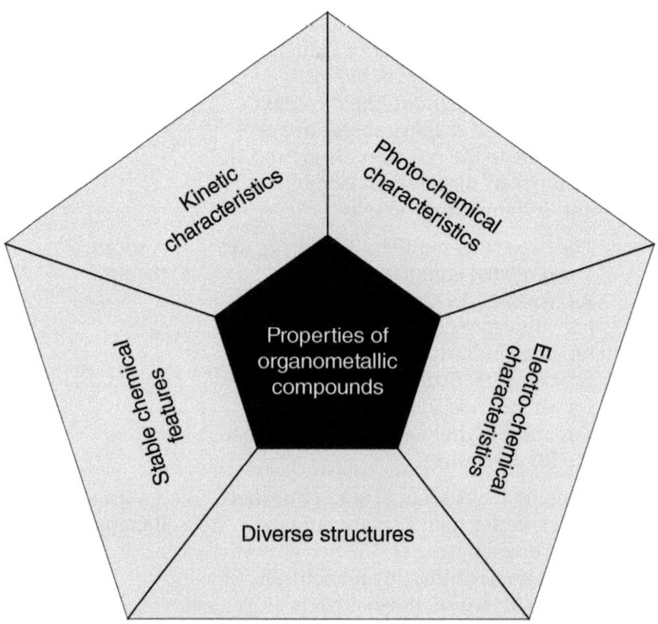

Figure 17.2 The properties of organometallic compounds.

17.3 Organometallic Compounds for Drug Delivery

Organometallic compounds show kinetic stability, typically no charge, relative lipophilicity, and a low oxidation state of metal atoms. Given these basic characteristics of classical coordination metal compounds, they are supposed to be suitable for designing new medicines, possibly with novel metal-specific functions. It is noteworthy that nearly all ordinary organometallic compounds, including metallocenes and half-sandwich with extensive applications in catalysis or biosensing areas, are currently taken into account in the field of medical chemistry [2].

Organometallics have recently contributed as a critical class of anticancer agents [60–71], along with other well-known inorganic coordination compounds, including carboplatin, cisplatin, NAMI-A, FFC14A, etc. A study indicated a series of three Pt(II)–alkynyl organometallic molecules, including a 1,2,3-triazole ring in the molecular structure. The pyrazine and ethynyl units bridge the Pt(II) center and the triazole ring. Given the antiproliferation effects of "Pt–ethynyl" and "1,2,3-triazole" reported separately, the biological effects of molecules containing both have been investigated. These reports on Pt(II) organometallics with 1,2,3-triazole ligands still need more exploration. Research on human osteosarcoma (MG-63) and breast cancer (MDAMB-231) has demonstrated that novel organoplatinum complexes containing 1,2,3-triazole are significantly more effective at inhibiting cell proliferation than cisplatin [72]. Yet, a study focused on the bioactive compounds $[La(L)(H_2O)_3Cl]Cl_2 \cdot 2H_2O$ (i), $[Er(L)(H_2O)_3Cl]Cl_2$ (ii), $[Yb(L)(H_2O)_2Cl_2]Cl \cdot 4H_2O$ (iii), $[Ru(L)(H_2O)_2Cl_2]Cl \cdot H_2O$ (iv), and $[Ta(L)(H_2O)Cl_5]$ (v) from 2-acetylferrocene derivative Schiff base (L). As shown, all compounds had greater anticancer effects compared to free ligands. Besides, the majority of compounds had acceptable antibacterial effects [73]. Many have recently concentrated on developing novel anticancers that contain metals except platinum to overcome the common restrictions of antineoplastic medicines based on this metal. In the meantime, there has been considerable interest in organopalladium complexes as they are highly stable in physiological conditions, with several instances indicating acceptable *in vitro* and *in vivo* antiproliferation effects against different cisplatin-sensitive and cisplatin-resistant tumors and usually exhibit differences in their action mechanisms from platinum-based medications [74].

Organometallics have increasingly attracted attention for their contributions as anticancers and antibiotics. As a transition metal ion, rhenium plays a crucial role in this regard. Its tri- and dicarbonyl compounds are particularly appealing to scientists because they are highly stable, relatively simple to prepare, and have unique photophysical and luminescent properties that allow them to be combined for diagnosis and treatment, allowing the tracking of molecules inside of cells [75]. Another study [76] reported on the historical and contemporary developments in the preparation of tiny organometallics for antibacterial purposes and potential future development for clinical uses. Previous researchers reported synthesizing, characterization, and biological effects of six bioorganometallic conjugates of ciprofloxacin with ferrocenyl, ruthenocenyl, and cymantrenyl entities, investigating the antimicrobial effects of these compounds on gram-positive and gram-negative bacteria [77].

Another study investigated a trimetallic complex (**1a**), containing a Re $(CO)_3$ moiety as the lead structure, along with a systematic structure–activity relationship (SAR) investigation for different gram-positive pathogenic bacteria consisting of methicillin-resistant *Staphylococcus aureus* (MRSA) strains [78]. As a result, these organometallic compounds play a significant role in medical applications or drug delivery systems.

17.4 Organometallic Compounds for Renewable Energy

Organometallics can be alternatively used instead of conventional metal nanoparticle-based approaches to develop electrocatalysts for fuel cells and electrolyzers. Although this area can be promising in overcoming the restrictions of electrocatalysts and preparing the grounds for developing novel catalysts in the energy field, it still needs to be explored more closely [79]. Considering the considerable dependence of human life on access to energy, the challenges brought by the energy crisis can seriously threaten humans' future lives, making the energy crisis one of the main issues worldwide. Hence, finding efficient alternatives such as renewable energy would be of critical importance [35, 80].

Alternative or renewable energies have been the focus of different sciences worldwide recently, leading to more interest in this field from the public to develop renewable energies and improve energy interconversion technologies. Two significant examples, recently attracting a lot of attention, are light to electricity (generating electricity by photovoltaic cells) and electricity to light (generating light by light-emitting diodes) transformations [35]. Investigations into organometallics for each of these transformations have confirmed the potential contribution of these metal compounds and polymers to creating, converting, and using them in a safe and effective manner [35, 81].

Some researchers reported on the application of organometallics to renewable energies such as solar cells [82–92]. Wong et al. used low-bandgap organometallic polyyne polymer, showing power-conversion efficiency (PCE) equal to 4.1%, as an effective alternative [93]. Such powerful absorbing conjugated organometallic congeners are supposed to play a key role in identifying environmentally compatible high-tech solutions for the crisis in the field of energy demands [94]. One researcher reported on the typical production of bulk heterojunction solar cells (BHJSCs) through switching thin photoactive polymer films between two electrodes like indium tin oxide (ITO) and aluminum. The quality of the ITO electrode surface can be improved by the poly(3,4-ethylene-dioxythiophene):poly (styrene sulfonate) (PEDOT:PSS) layer while also facilitating the hole injection and extraction [95]. The organic or organometallic absorbers and electron-accepting methanofullerene can be dissolved in an identical solvent and cast into a blend to make the interpenetrated electron donors and acceptors at the scale of nanometer length [35]. Organometallics of several transition metals, including Re^{I}, Os^{II}, and Ir^{III}, have received a lot of attention recently [35, 84, 96, 97].

17.5 Organometallic Compounds for Wastewater Treatment

Demand management, water savings, and conservation projects have recently been taken into account for future water usage strategies, as different regions worldwide are supposed to experience considerable challenges concerning accessibility, quality, and water usage and disposal regulations. In ultrafiltration, a membrane is used to separate dangerous viruses, bacteria, organics, and different water pollutants [8]. Some research about wastewater treatment [98–100] has discussed the significance of ensuring chemical safety whenever selenium complexes are used to implement environmental protection by guaranteeing an overall safety culture. The study analyzed electrochemical procedures followed to determine the nature of selenium and different organo-selenium substances [101]. Another study reported on the successful fabrication of antiviral ultrafiltration membranes (UF) with the use of amphiphilic polyvinylchloride-*co*-acrylic acid) (PVC-AA) integrated into organometallic manganese acetylacetonate $Mn(acac)_3$. Consequently, the membrane hydrophilicity and antifouling characteristics improved as manganese acetylacetonate $Mn(acac)_3$ load increased to 1% in the polymeric solution. Using bovine serum albumin ($1\,g\,l^{-1}$), sodium alginate ($2\,g\,l^{-1}$), and a genuine wastewater sample, the membrane with 1 wt% $Mn(acac)_3$ (UF3) demonstrated the most excellent antifouling features, which is 99% of flux recovery ratio. Effects against fouling and antiviral were revealed by PVCAA-$Mn(acac)_3$. According to viral experiments with the use of real sewage waste, the membranes UF2 and UF3 eradicated viruses by 100%. These organometallic-based antiviral amphiphilic membranes have removed proteins from wastewater [8].

Some research discussed releasing these materials into the environment [102–105]. On the other hand, in the last decade, many researchers focused on the design and preparation of water pipes by using organometallic compounds [106–109]. One research reported that two types of porous magnetic ferrite nanowires that contained manganese ($MnFe_2O_4$ and Mn-doped Fe_3O_4) underwent successful synthesis using organometallics' thermal decomposition, whereas also utilizing nitrilotriacetic acid (NA) as a chelating agent for coordination with different Fe(II) and Mn(II) ion ratios. These nanowire samples exhibit a high level of ability extract organic and heavy metals ions from wastewater [110]. Therefore, the antimicrobial organometallic compound can be used to prepare antimicrobial sewer and water pipes for water treatment in the construction industry. Figure 17.3 shows the last five years of developments in organometallic compounds for pharmaceuticals and drug delivery, renewable energy and wastewater treatment.

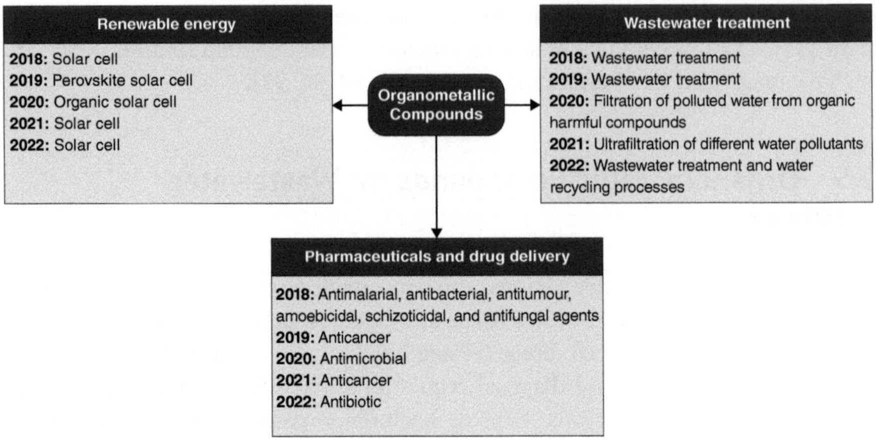

Figure 17.3 The last five years of developments in organometallic compounds for pharmaceuticals and drug delivery, renewable energy, and wastewater treatment.

17.6 Conclusion

This chapter has provided detailed information on the emerging role of organometallic compounds in different fields of industry, including medications such as drug delivery, renewable energy, and wastewater treatment. Numerous examples concerning the practical application of these materials in the fields of medicine, wastewater treatment, and energy were elaborated in detail. The merit of these materials in comparison to different applications, such as drug delivery and renewal energy, was also discussed. The applications of organometallic compounds for the fabrication of solar cells were also highlighted. In addition, the application of these materials to the treatment of cancer was highlighted. Finally, these profound materials could be improved to have commercial applications.

17.7 Outlook

Considering the trends in organometallic compounds, combining these materials with biomaterials, particularly derived from natural and renewable agents with biocompatibility, biodegradability, and nontoxicity, can be extensively explored by further research. In addition, natural nano-capsules, bio-nano composites, natural biopolymers, herbal extracts, and essential oils are also supposed to be the most demanded to be combined with organometallic compounds to improve relative activity, synthesis, and applications in different fields of industry, including medicine, energy, environment, and construction, would be the center of attention in future studies.

Acknowledgment

The authors would like to thank Universiti Teknologi Malaysia (UTM) for providing the platform and research facilities. Furthermore, the authors would like to thank The Hong Kong Polytechnic University for its support.

References

1 Mudi, S.Y., Usman, M.T., and Ibrahim, S. (2015). Clinical and industrial application of organometallic compounds and complexes: a review. *Am. J. Chem. Appl.* 2 (6): 151–158.

2 Gasser, G., Ott, I., and Metzler-Nolte, N. (2011). Organometallic anticancer compounds. *J. Med. Chem.* 54 (1): 3–25.

3 Paunescu, E., McArthur, S., Soudani, M. et al. (2016). Nonsteroidal anti-inflammatory- organometallic anticancer compounds. *Inorg. Chem.* 55 (4): 1788–1808.

4 Ghazi, D., Rasheed, Z., and Yousif, E. (2018). Review of organotin compounds: chemistry and applications. *Development* 3: 4.

5 Tong, K.K., Hanif, M., Lovett, J.H. et al. (2020). Thiourea-derived chelating ligands and their organometallic compounds: investigations into their anti-cancer activity. *Molecules* (16): 3661.

6 Khanvilkar, P., Pulipaka, R., Shirsath, K. et al. (2019). Organometallic binuclear Ru (II) complexes: design, synthesis, DNA/BSA binding interactions and in-vitro cytotoxicity against HeLa cell line. *Inorg. Chem. Commun.* 102: 134–140.

7 Kozieł, S., Komarnicka, U.K., Ziółkowska, A. et al. (2020). Anticancer potency of novel organometallic Ir (iii) complexes with phosphine derivatives of fluoroquinolones encapsulated in polymeric micelles. *Inorg. Chem. Front.* 7 (18): 3386–3401.

8 Shalaby, M., Mansor, E.S., Abdallah, H. et al. (2021). Antiviral amphiphilic membranes based on the organometallic compound for protein removal from wastewater with fouling-resistant. *J. Polym. Res.* 28 (5): 1–5.

9 Tripathi, S., Sharma, P., and Chandra, R. (2021). Degradation of organometallic pollutants of distillery wastewater by autochthonous bacterial community in biostimulation and bioaugmentation process. *Bioresour. Technol.* 338: 125518.

10 Singh, A.K., Kumar, A., Bilal, M., and Chandra, R. (2021). Organometallic pollutants of paper mill wastewater and their toxicity assessment on Stinging catfish and sludge worm. *Environ. Technol. Innov.* 24: 101831.

11 Kozyreva, L.V., Fadeev, O.V., and Yudin, A.O. (2020). The algorithm of development safe methods for depositing metallic coatings by CVD-method of organometallic compounds. *Environment* 1: 2.

12 Sánchez-Bastardo, N., Schlögl, R., and Ruland, H. (2020). Methane pyrolysis for CO_2-free H_2 production: a green process to overcome renewable energies unsteadiness. *Chem. Ing. Tech.* 92 (10): 1596–1609.

13 Madadkhani, S., Aghakhanpour, R.B., Singh, J.P. et al. (2019). A trimetallic organometallic precursor for efficient water oxidation. *Sci. Rep.* 9 (1): 1–8.

14 Gentner, T.X. and Mulvey, R.E. (2021). Alkali-metal mediation: diversity of applications in main-group organometallic chemistry. *Angew. Chem. Int. Ed.* 60 (17): 9247–9262.

15 Colacot, T.J. and Sivakumar, V. (ed.) (2020). *Organometallics in Process Chemistry*. Springer Nature.

16 Hapke, M. and Hilt, G. (ed.) (2020). *Cobalt Catalysis in Organic Synthesis: Methods and Reactions*. Wiley.

17 Fleissner, S., Pittenauer, E., Pecak, J., and Kirchner, K. (2022). Characterization of selected organometallic compounds by electrospray ionization- and matrix-assisted laser desorption/ionization-mass spectrometry using different types of instruments: possibilities and limitations. *RCM* 36 (10): e9281.

18 Parshall, G.W. (1987). Trends and opportunities for organometallic chemistry in industry. *Organometallics* 6 (4): 687–692.

19 Gardner, B.M., Seechurn, C.C., and Colacot, T.J. (2020). Industrial milestones in organometallic chemistry. In: *Organometallic Chemistry in Industry: A Practical Approach*, vol. 30 (ed. T.J. Colacot and C.C.C. Johansson Seechurn), 1–22. John Wiley & Sons, Inc.

20 Elschenbroich, C. (2016). *Organometallics*, 3e. Wiley.

21 Straub, B.F. (ed. J.F. Hartwig) *Organotransition Metal Chemistry. From Bonding to Catalysis*. John Wiley & Sons, Inc.

22 Hartwig, J.F. and Collman, J.P. (2010). *Organotransition Metal Chemistry: From Bonding to Catalysis*. Sausalito, CA: University Science Books.

23 Noffke, A.L., Habtemariam, A., Pizarro, A.M., and Sadler, P.J. (2012). Designing organometallic compounds for catalysis and therapy. *Chem. Commun.* 48 (43): 5219–5246.

24 Patra, M. and Gasser, G. (2012). Organometallic compounds: an opportunity for chemical biology? *ChemBioChem* 13 (9): 1232–1252.

25 Herrmann, W.A. and Cornils, B. (1997). Organometallic homogeneous catalysis—quo vadis? *Angew. Chem. Int. Ed. Engl.* 36 (10): 1048–1067.

26 Sautet, P. and Delbecq, F. (2010). Catalysis and surface organometallic chemistry: a view from theory and simulations. *Chem. Rev.* 110 (3): 1788–1806.

27 Echavarren, A.M., Hashmi, A.S., and Toste, F.D. (2016). Gold catalysis – steadily increasing in importance. *Adv. Synth. Catal.* 358 (9): 1347.

28 Piers, W.E. (2011). Future trends in organometallic chemistry: organometallic approaches to water splitting. *Organometallics* 30 (1): 13–16.

29 Astruc, D. (2007). *Organometallic Chemistry and Catalysis*. Berlin: Springer.

30 Lefebvre, F. and Basset, J.M. (1999). Recent applications in catalysis of surface organometallic chemistry. *J. Mol. Catal. A: Chem.* 146 (1–2): 3–12.

31 Cornils, B., Herrmann, W.A., Beller, M., and Paciello, R. (ed.) (2017). *Applied Homogeneous Catalysis with Organometallic Compounds: A Comprehensive Handbook in Four Volumes*. Wiley.

32 Copéret, C., Chabanas, M., Petroff Saint-Arroman, R., and Basset, J.M. (2003). Homogeneous and heterogeneous catalysis: bridging the gap through surface organometallic chemistry. *Angew. Chem. Int. Ed.* 42 (2): 156–181.

33 Cornils, B. and Herrmann, W.A. (ed.) (2004). *Aqueous-Phase Organometallic Catalysis: Concepts and Applications*. Wiley.

34 Yermakov, Y.I. (1983). Organometallic compounds in the preparation of supported catalysts. *J. Mol. Catal.* 21 (1–3): 35–55.

35 Wong, W.Y. (2009). Challenges in organometallic research – great opportunity for solar cells and OLEDs. *J. Organomet. Chem.* 694 (17): 2644–2647.

36 Yin, X., Song, Z., Li, Z., and Tang, W. (2020). Toward ideal hole transport materials: a review on recent progress in dopant-free hole transport materials for fabricating efficient and stable perovskite solar cells. *Energy Environ. Sci.* 13 (11): 4057–4086.

37 Varadwaj, P.R. (2017). Methylammonium lead trihalide perovskite solar cell semiconductors are not organometallic: a perspective. *Helv. Chim. Acta* 100 (7): e1700090.

38 Marzano, G., Ciasca, C.V., Babudri, F. et al. (2014). Organometallic approaches to conjugated polymers for plastic solar cells: from laboratory synthesis to industrial production. *Eur. J. Org. Chem.* 2014 (30): 6583–6614.

39 Lu, M., Wu, X.J., Zeng, D.C., and Liao, Y. (2012). Distribution of PCDD/Fs and organometallic compounds in sewage sludge of wastewater treatment plants in China. *Environ. Pollut.* 171: 78–84.

40 Meng, X., Leyva, M.L., Jenny, M. et al. (2009). A ruthenium-containing organometallic compound reduces tumor growth through induction of the endoplasmic reticulum stress gene CHOP. *Cancer Res.* 69 (13): 5458–5466.

41 Mekidiche, M., Khaldi, K., Nacer, A. et al. (2021). Organometallic modified montmorillonite application in the wastewater purification: pollutant photodegradation and antibacterial efficiencies. *Appl. Surf. Sci.* 569: 151097.

42 Wenzel, M., Patra, M., Senges, C.H. et al. (2013). Analysis of the mechanism of action of potent antibacterial hetero-tri-organometallic compounds: a structurally new class of antibiotics. *ACS Chem. Biol.* 8 (7): 1442–1450.

43 Ciobotaru, I.C., Polosan, S., and Ciobotaru, C.C. (2018). Organometallic compounds for photovoltaic applications. *Inorg. Chim. Acta* 483: 448–453.

44 Patra, M., Gasser, G., Pinto, A. et al. (2009). Synthesis and biological evaluation of chromium bioorganometallics based on the antibiotic platensimycin lead structure. *ChemMedChem* 4 (11): 1930–1938.

45 Zhang, M., Ma, X., Zhang, H. et al. (2022). Metallated terpolymer donors with strongly absorbing iridium complex enables polymer solar cells with 16.71% efficiency. *Chem. Eng. J.* 430: 132832.

46 Shi, X., Mai, X., Wei, R. et al. (2021). Removing Pb^{2+} and As(V) from polluted water by highly reusable Fe–Mg metal–organic complex adsorbent. *Powder Technol.* 383: 104–114.

47 Hong, J., Kang, L., Shi, X. et al. (2022). Highly efficient removal of trace lead(II) from wastewater by 1,4-dicarboxybenzene modified Fe/Co metal organic nanosheets. *J. Mater. Sci. Technol.* 98: 212–218.

48 Lai, S.L., Wang, L., Yang, C. et al. (2014). Gold(III) corroles for high performance organic solar cells. *Adv. Funct. Mater.* 24 (29): 4655–4665.

49 Hunoor, R.S., Patil, B.R., Badiger, D.S. et al. (2011). Synthesis, characterization, anti-inflammatory and analgesic activity of transition metal complexes of 3-[1-(2-hydroxyphenyl)ethylideamino]-2-phenyl-3, 4-dihydroquinazolin-4(3*H*)-one. *Appl. Organomet. Chem.* 25 (6): 476–483.

50 Florindo, P.R., Pereira, D.M., Borralho, P.M. et al. (2015). Cyclopentadienyl–ruthenium (II) and iron (II) organometallic compounds with carbohydrate derivative ligands as good colorectal anticancer agents. *J. Med. Chem.* 58 (10): 4339–4347.

51 Perl, E.E., Simon, J., Geisz, J.F. et al. (2016). Development of high-bandgap AlGaInP solar cells grown by organometallic vapor-phase epitaxy. *IEEE J. Photovoltaics* 6 (3): 770–776.

52 Abd-El-Aziz, A.S., Abdelghani, A.A., El-Ghezlani, E.G. et al. (2021). Pharmacological evaluation of novel organoiron dendrimers as antimicrobial and anti-inflammatory agents. *Macromol. Biosci.* 21 (1): 2000242.

53 Ashraf, A., Aman, F., Movassaghi, S. et al. (2019). Structural modifications of the antiinflammatory oxicam scaffold and preparation of anticancer organometallic compounds. *Organometallics* 38 (2): 361–374.

54 Parveen, S., Hanif, M., Leung, E. et al. (2019). Anticancer organorhodium and-iridium complexes with low toxicity in vivo but high potency in vitro: DNA damage, reactive oxygen species formation, and haemolytic activity. *Chem. Commun.* 55 (80): 12016–12019.

55 Barik, T., Ghosh, A., Mobin, S.M., and Chatterjee, S. (2021). Solvent free synthesis of vinylcyanoacetohydrazone bridged diferrocenyl organometallic compounds as bifunctional molecule for donor–acceptor and photovoltaic properties. *J. Organomet. Chem.* 933: 121648.

56 Abraham, K. (2018). In vitro and in vivo chronopharmacology of a new generation of an organometallic anticancer drug complex. Doctoral dissertation. University of Warwick.

57 Wang, K.C., Jeng, J.Y., Shen, P.S. et al. (2014). P-type mesoscopic nickel oxide/organometallic perovskite heterojunction solar cells. *Sci. Rep.* 4 (1): 1–8.

58 Subarkhan, M.K., Ren, L., Xie, B. et al. (2019). Novel tetranuclear ruthenium(II) arene complexes showing potent cytotoxic and antimetastatic activity as well as low toxicity in vivo. *Eur. J. Med. Chem.* 179: 246–256.

59 Aebi, V., Cooper, C.B. III, Moon, R.L., and Saxena, R.R. (1981). Characterization of organometallic VPE grown GaAs and AlGaAs for solar cell applications. *J. Cryst. Growth* 55 (3): 517–525.

60 Khan, R.A., Bin Sharfan, I.I., Alterary, S.S. et al. (2022). Organometallic (η^6-*p*-cymene) ruthenium(II) complexes with thiazolyl-based organic twigs: en route towards targeted delivery via human serum albumin of the potential anticancer agents. *Appl. Organomet. Chem.* 36 (3): e6550.

61 El-Bindary, M.A., El-Desouky, M.G., and El-Bindary, A.A. (2022). Metal–organic frameworks encapsulated with an anticancer compound as drug delivery system: synthesis, characterization, antioxidant, anticancer, antibacterial and molecular docking investigation. *Appl. Organomet. Chem.* 36: e6660.

62 Li, Q., Shao, M., Ran, W. et al. (2022). An AIE-featured triphenyltin(IV)-triphenylamine acylhydrazone compound and anticancer application. *Dyes Pigm.* 201: 110231.

63 Peña, L., Jiménez, C., Arancibia, R. et al. (2022). Heterobimetallic complexes containing organometallic acylhydrazone ligands as potential inhibitors of human carbonic anhydrases. *J. Inorg. Biochem.* 232: 111814.

64 Sindhu, M., Kalaivani, P., and Prabhakaran, R. (2022). Enhanced anticancer property of bio-organometallic nano composites: design, characterization, and biological evaluation. *Appl. Organomet. Chem.* 36 (1): e6488.

65 Shekarkhand, M., Zare, K., Monajjemi, M. et al. (2022). Computational study of heterocyclic anticancer compounds through NBO method. *Nexo Revista Científica* 35 (1): 367–381.

66 Mounica, A., Balachandran, C., Gopalakrishnan, D. et al. (2022). Synthesis and antiproliferative activity of novel organometallic cobalt(III) complex encapsulated in polydiacetylene-phospholipid nanoformulation. *Inorg. Chim. Acta* 530: 120701.

67 Han, T., Wu, Y., Han, W. et al. (2021). Antitumor effect of organometallic half-sandwich Ru(II)–arene complexes bearing a glutathione S-transferase inhibitor. *Inorg. Chem.* 60 (17): 13051–13061.

68 Arshad, J., Tong, K.K., Movassaghi, S. et al. (2021). Impact of the metal center and leaving group on the anticancer activity of organometallic complexes of pyridine-2-carbothioamide. *Molecules* 26 (4): 833.

69 Nguyen, C., Toubia, I., Diring, S. et al. (2021). Synergetic anticancer activity of gold porphyrin appended to phenyl tin malonate organometallic complexes. *Dalton Trans.* 50 (13): 4583–4592.

70 Kowalski, K. (2021). Organometallic nucleosides—synthesis, transformations, and applications. *Coord. Chem. Rev.* 432: 213705.

71 Rodriguez-Prieto, T., Michlewska, S., Hołota, M. et al. (2021). Organometallic dendrimers based on ruthenium(II) N-heterocyclic carbenes and their implication as delivery systems of anticancer small interfering RNA. *J. Inorg. Biochem.* 223: 111540.

72 Singh, K., Gangrade, A., Jana, A. et al. (2019). Design, synthesis, characterization, and antiproliferative activity of organoplatinum compounds bearing a 1,2,3-triazole ring. *ACS Omega* 4 (1): 835–841.

73 Deghadi, R.G., Mohamed, G.G., and Mahmoud, N.F. Bioactive La(III), Er(III), Yb(III), Ru(III) and Ta(V) complexes of new organometallic Schiff base: preparation, structural characterization, antibacterial, anticancer activities and MOE studies. *Appl. Organomet. Chem.* 36: e6675.

74 Scattolin, T., Voloshkin, V.A., Visentin, F., and Nolan, S.P. (2021). A critical review of palladium organometallic anticancer agents. *Cell Rep. Phys. Sci.* 2 (6): 100446.

75 Schindler, K. and Zobi, F. (2022). Anticancer and antibiotic rhenium tri-and dicarbonyl complexes: current research and future perspectives. *Molecules* 27 (2): 539.

76 Patra, M., Gasser, G., and Metzler-Nolte, N. (2012). Small organometallic compounds as antibacterial agents. *Dalton Trans.* 41 (21): 6350–6358.

77 Szczupak, Ł., Kowalczyk, A., Trzybiński, D. et al. (2020). Organometallic ciprofloxacin conjugates with dual action: synthesis, characterization, and antimicrobial and cytotoxicity studies. *Dalton Trans.* 49 (5): 1403–1415.

78 Patra, M., Wenzel, M., Prochnow, P. et al. (2015). An organometallic structure-activity relationship study reveals the essential role of a $Re(CO)_3$ moiety in the activity against gram-positive pathogens including MRSA. *Chem. Sci.* 6 (1): 214–224.

79 Bellini, M., Bevilacqua, M., Marchionni, A. et al. (2018). Energy production and storage promoted by organometallic complexes. *Eur. J. Inorg. Chem.* 2018 (40): 4393–4412.

80 Turner, J.A. (1999). A realizable renewable energy future. *Science* 285 (5428): 687–689.

81 Nazeeruddin, M.K. and Grätzel, M. (2007). Transition metal complexes for photovoltaic and light emitting applications. In: *Photofunctional Transition Metal Complexes* (ed. V.W.W. Yam), 113–175. Springer.

82 Liu, L., Ho, C.L., Wong, W.Y. et al. (2008). Effect of oligothienyl chain length on tuning the solar cell performance in fluorene-based polyplatinynes. *Adv. Funct. Mater.* 18 (18): 2824–2833.

83 Wong, W.Y., Wang, X., Zhang, H.L. et al. (2008). Synthesis, characterization and photovoltaic properties of a low-bandgap platinum(II) polyyne functionalized with a 3,4-ethylene dioxythiophene-benzothiadiazole hybrid spacer. *J. Organomet. Chem.* 693 (24): 3603–3612.

84 Wadman, S.H., Kroon, J.M., Bakker, K. et al. (2010). Cyclometalated organoruthenium complexes for application in dye-sensitized solar cells. *Organometallics* 29 (7): 1569–1579.

85 Szlufcik, J., Majewski, J., Buczkowski, A. et al. (1989). Screen-printed titanium dioxide anti-reflection coating for silicon solar cells. *Sol. Energy Mater.* 18 (5): 241–252.

86 Luechai, A., Pootrakulchote, N., Kengthanomma, T. et al. (2014). Photosensitizing triarylamine- and triazine-cored porphyrin dimers for dye-sensitized solar cells. *J. Organomet. Chem.* 753: 27–33.

87 Privalov, T., Boschloo, G., Hagfeldt, A. et al. (2009). A study of the interactions between I^-/I_3^- redox mediators and organometallic sensitizing dyes in solar cells. *J. Phys. Chem. C* 113 (2): 783–790.

88 Anizaim, A.H., Zainuri, D.A., Zaini, M.F. et al. (2020). Comparative analyses of new donor-π-acceptor ferrocenyl-chalcones containing fluoro and methoxy-fluoro acceptor units as synthesized dyes for organic solar cell material. *PLoS One* 15 (11): e0241113.

89 Avila, J., La-Placa, M.G., Longhi, E. et al. (2019). Ruthenium pentamethyl-cyclopentadienyl mesitylene dimer: a sublimable n-dopant and electron buffer layer for efficient n–i–p perovskite solar cells. *J. Mater. Chem. A* 7 (45): 25796–25801.

90 Dai, F.R., Zhan, H.M., Liu, Q. et al. (2012). Platinum(II)–bis (aryleneethynylene) complexes for solution-processible molecular bulk heterojunction solar cells. *Chem. Eur. J.* 18 (5): 1502–1511.

91 Gale, R.P., Fan, J.C., Turner, G.W., and Chapman, R.L. (1985). High-efficiency double-heterostructure AlGaAs/GaAs solar cells. Massachusetts Institute of Technology, Lexington (USA). Lincoln Laboratory.

92 Kashyout, A.B., Soliman, H., Fathy, M. et al. (2012). CdSe quantum dots for solar cell devices. *Int. J. Photoenergy* 2012: 952610.

93 Wong, W.Y., Wang, X.Z., He, Z. et al. (2010). Metallated conjugated polymers as a new avenue towards high-efficiency polymer solar cells. In: *Materials for Sustainable Energy: A Collection of Peer-Reviewed Research and Review Articles from Nature Publishing Group* (ed. V. Dusastre), 51–57. Nature Publishing Group.

94 Wong, W.Y. (2008). Metallopolyyne polymers as new functional materials for photovoltaic and solar cell applications. *Macromol. Chem. Phys.* 209 (1): 14–24.

95 Thompson, B.C. and Fréchet, J.M. (2008). Polymer–fullerene composite solar cells. *Angew. Chem. Int. Ed.* 47 (1): 58–77.

96 Chan, W.K., Hui, C.S., Man, K.Y. et al. (2005). Synthesis and photosensitizing properties of conjugated polymers that contain chlorotricarbonylbis (phenylimino) acenaphthene rhenium(I) complexes. *Coord. Chem. Rev.* 249 (13–14): 1351–1359.

97 Kumaresan, D., Shankar, K., Vaidya, S., and Schmehl, R.H. (2007). Photochemistry and photophysics of coordination compounds: osmium. In: *Photochemistry and Photophysics of Coordination Compounds II* (ed. V. Balzani and S. Campagna), 101–142. Springer.

98 Manikandan, S., Subbaiya, R., Saravanan, M. et al. (2022). A critical review of advanced nanotechnology and hybrid membrane based water recycling, reuse, and wastewater treatment processes. *Chemosphere* 289: 132867.

99 Sarode, S., Upadhyay, P., Khosa, M.A. et al. (2019). Overview of wastewater treatment methods with special focus on biopolymer chitin-chitosan. *Int. J. Biol. Macromol.* 121: 1086–1100.

100 Absalan, Y., Alabada, R., Ryabov, M. et al. (2020). Removing organic harmful compounds from the polluted water by a novel synthesized cobalt(II) and titanium(IV) containing photocatalyst under visible light. *Environ. Nanotechnol. Monit. Manage.* 14: 100304.

101 Popov, V., Pashinin, V., Demidov, A., and Sukhov, P. (2022). Selenium compounds in the waste water. In: *XIV International Scientific Conference "INTERAGROMASH 2021"*, 353–360. Cham: Springer.

102 Vinayaka, K.S. and Kadkol, S. (2022). Advances in bioremediation of organometallic pollutants: strategies and future road map. In: *Biological Approaches to Controlling Pollutants* (ed. S. Kumar and M.Z. Hashmi), 233–239. Woodhead Publishing.

103 Sulivan Jouanneau, A.A., Durand, M.J., and Thouand, G. (2022). Detection and effects of metal and organometallic compounds with microbial bioluminescence and Raman spectroscopy. In: *Handbook of Cell Biosensors* (ed. G. Thouand), 825. Springer.

104 Cavalheiro, J., Zuloaga, O., Prieto, A. et al. (2017). Occurrence and fate of organic and organometallic pollutants in municipal wastewater treatment plants and their impact on receiving waters (Adour Estuary, France). *Arch. Environ. Contam. Toxicol.* 73 (4): 619–630.

105 Craig, P.J. and Jenkins, R.O. (2004). Organometallic compounds in the environment: an overview. In: *Organic Metal and Metalloid Species in the Environment*, 1–5.

106 Forsyth, D.S. and Jay, B. (1997). Organotin leachates in drinking water from chlorinated poly(vinyl chloride) (CPVC) pipe. *Appl. Organomet. Chem.* 11 (7): 551–558.

107 Quevauviller, P., Donard, O.F., and Bruchet, A. (1991). Leaching of organotin compounds from poly(vinyl chloride) (PVC) material. *Appl. Organomet. Chem.* 5 (2): 125–129.

108 Jones-Lepp, T.L., Varner, K.E., and Hilton, B.A. (2001). Speciation and detection of organotins from PVC pipe by micro-liquid chromatography–electrospray–ion trap mass spectrometry. *Appl. Organomet. Chem.* 15 (12): 933–938.

109 Sadiki, A.D. and Williams, D.T. (1999). A study on organotin levels in Canadian drinking water distributed through PVC pipes. *Chemosphere* 38 (7): 1541–1548.

110 Cui, H.J., Shi, J.W., Yuan, B., and Fu, M.L. (2013). Synthesis of porous magnetic ferrite nanowires containing Mn and their application in water treatment. *J. Mater. Chem. A* 1 (19): 5902–5907.

18

Computational Approaches in Some Important Organometallic Catalysis Reaction

Pratibha Gumasta[1], Naresh C. Deshmukh[2], Aseel A. Kadhem[3], Snehlata Katheria[4], Reena Rawat[5], and Bhawana Jain[6]

[1] *Sai Mahavidyalaya, Department of Chemistry, Sector 6, Bhilai, Chhattisgarh, 49006, India*
[2] *Durg University, Kalyan P. G. College, Department of Chemistry, Sector 7, Bhilai, Chhattisgarh, 49006, India*
[3] *Iraqi Ministry of Education/Wasit Education Directorate, Department of Chemistry, Wasit, Kut, Al Khajia District, Iraq*
[4] *University of Lucknow, Department of Chemistry, Babuganj, Lucknow, Uttar Pradesh, 226007, India*
[5] *Echelon Institute of Technology, Department of Chemistry, Faridabad, Haryana, 121101, India*
[6] *Siddhachalam Laboratory, Department of Chemical Science, Ring road 2, Raipur, Chhattisgarh, 493221, India*

18.1 Introduction

Due to its broad application in reaction mechanisms, organometallic catalysis attracts attention in the area of academics as well as industry too. In many catalytic reactions, Pd, Ni, Rh, and Ir catalysts are highly important [1]. In the experimental reaction mechanism approach, chemists had a high work load to design catalysts. To mitigate this huge workload, scientists developed automated computational workflows and generated reaction energy profiles without or with minimal human intervention [2]. Those automated workflows generate suitable structures for reactants, products, intermediates, and transition states with analysis; for example, they determine their action rates and their steps. It also saves time by understanding the reaction mechanism and standardizing the discovery. Fully automated workflows also had new catalysis reaction approaches and an easy way to use advanced computational tools, which transformed experimental group catalysis research. Computational methods are a powerful tool for the design of experimental molecular catalysts that provide important theoretical predictions for catalysis reactions and also decrease the time for screening.

In homogeneous [3–6] and heterogeneous catalysis [7], automated computational workflows are used. However, it untested in real applications and also not possible to investigate all reaction pathways because of complex and highly interconnected nature of the chemical space. Thus, it was necessary to describe chemical space and potential reaction mechanisms. The calculations, like transition state calculations, require troubleshooting processing for accurate results by automated workflow without human intervention.

Organometallic Compounds: Synthesis, Reactions, and Applications, First Edition.
Edited by Dakeshwar Kumar Verma and Jeenat Aslam.

Computational catalysis could use in catalytic reaction mechanisms and help for identification of controlling factors. Giuseppe Cassone et al. described electric fields production by interacting atoms, molecules, and complex matter through modifying chemical reactions via molecular dynamic methods [8]. Ainara Nova et al. described bifunctional catalysts actions onamide hydrogenation, a reaction which conveniently yields amines and alcohols using dihydrogen as reducing agent. And via computational methods, reaction times could be improved [9]. Richard Zare and Christian Chamberlayne described the chemical reactions arising from static electric charges on insulating surfaces (Teflon) in electrochemical cells via a computation method [10, 11]. They also described chemical reactivity of charge carrying species and also voltage imparted through discharging electrostatically charged surface. Tao Cheng, William Goddard, et al. explain the single atom electrocatalyst anchored on MoS_2 through computational methods for nitrogen reduction reactions, which provide an environmentally green alternative and also replace the water gas shift production of H_2 with hydrogen atom which is extracted from water [12]. Samantha Johnson, Simone Raugei, et al. describe a computational study of the reactivity of metalloporphyrins for NH_3 oxidation. They also use density functional theory (DFT) and molecular dynamics simulations to assess the ability of tetraphenylporphyrin (TPP)M complexes (M = Cr, Mn, Fe, Co, Ni, Mo, Ru, W, and Os) to coordinate and weaken the N—H bonds of ammonia, for N—N bond formation and N_2 generation [13]. Julien Panetier and Xiaohui Li used DFT calculations for investigate the electronic structure and reactivity in CO_2 reduction reactions by tungsten electrocatalysts series. Based on the energy-span model and theoretical Tafel plots, designed CO_2 reduction electrocatalysts reaction [14]. Through theoretical and computational methods we understand the mechanism of reaction with numerous aspects of catalysis.

The incredible advancement in computational methods, computing power, software, and other related tools were heavily used for the calculation of rationalize with prediction of organometallic reactions. Computational chemistry becomes an important tool to gain insight knowledge of mechanism of catalytic organometallic reactions, where catalytic species were hard to study which used experimental approaches [15]. It had many challenges for mechanism of catalytic reactions. To use of computational tools in catalytic reactions, there are few approaches available [15]. Here, we described of Pd-, Ni-, Rh-, and Ir-mediated catalytic reactions by computational approaches and DFT [16, 17] use for computational performance in calculation of energy and their geometry.

18.2 Computational Method

These days, computational methods of organometallic reactions, are frequently limited by their precision with time and try to match the accuracy in computational assessment and also complete all reaction-based calculations within a reasonable time [18]. For example, like the coupled cluster method CCSD(T) [19, 20], which is usually considered the gold standard in quantum chemistry but was not possible

Table 18.1 Highly preferable DFT methods are as follows.

GGA	Meta-GGA	Hybrid-meta GGA	Hybrid-GGA	Range separated	References
BP86	M06L	M06	B3LYP	ωB97X-D	[29–33]
B97D	TPSS	M062X	B3PW91 PBE0	DFT-D3	[33–37]

to use in calculations which were involved in relevant chemical complexes of Pd, Ni, Rh, and Ir, especially for optimization of geometry [21]. Thus, for elements in organometallic reaction, DFT method use for compromise between computational precise with speed [22–25].

18.2.1 Geometry Calculation

To obtain chemical information through calculation, exact molecular geometries are necessary for property and energy calculations [26–28]. There were various DFT methods for calculating molecular calculations (Table 18.1).

As Martin and coworkers described [38], for the calculation of species of Pd, Ni, Rh, and Ir had not any method. Except for the hybrid method, which dominates Pd, Ni, Rh, and Ir related organometallic calculations. Here, DFT had ten important methods for Pd, Ni, Rh, and Ir sections. Previously, B3LYP was preferably used for optimization of the geometry of Pd, Ni, Ir, and Rh species [39]. But due to the growth of other methods like ωB97X-D and the Minnesota functionals, application of B3LYP decreased.

Doyle, Norrby, and coworkers described, Pd-catalyzed allylic fluorination via B3LYP method and reproduce X-ray structures of allylic Pd(II) chloride with 5% error (Figure 18.1a) [41]. Pratt et al. described, CCSD(T) reference data and B3LYP apply for structure of reactants and intermediates of alkyl coupling reaction with Ni complexes (Figure 18.1b) [42]. Similarly, Hughes and coworkers used B3LYPfor reproducing X-ray data of alkene–Ir(Cp) complexes (Figure 18.1c) [43]. Except B3LYP, other methods like BP86, B3PW91, and PBE0 also showed better results of regenerated crystallographic structures of organometallic complexes, than B3LYP [44, 45]. Bühl and coworkers described, gas-phase electron diffraction and showed their data via B3PW91, which could be better than B3LYP for structural optimization of various second-row transition-metal complexes like Rh(NO)$(PF_3)_3$and $RhCp(C_2H_4)_2$ (Figure 18.1d) [40].

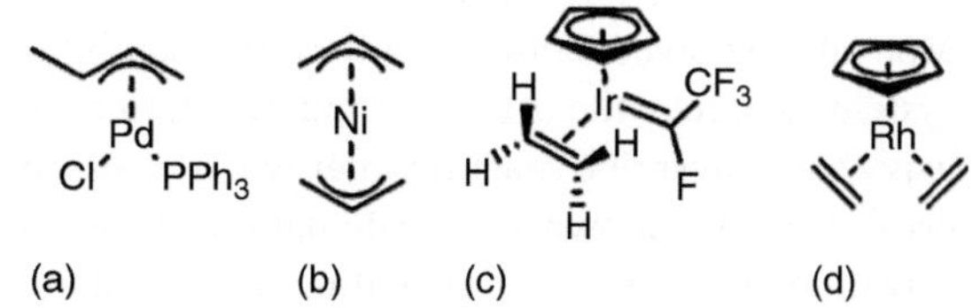

Figure 18.1 Structure of transition metal complexes obtained through DFT methods like B3LYP methods. Source: Adapted from Waller et al. [40].

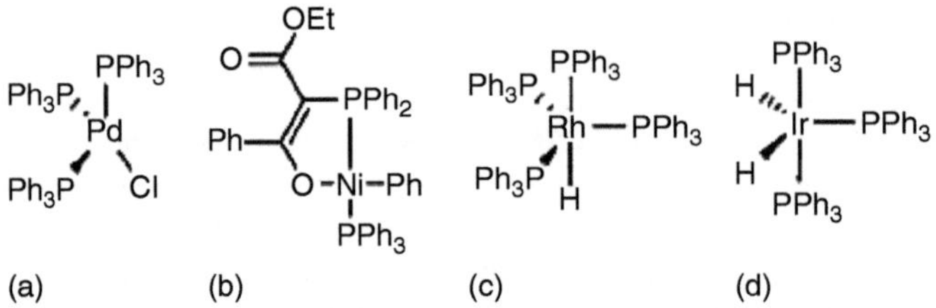

Figure 18.2 Transition metal complexes developed through DFT methods by group of Jensen. Source: Adapted from Minenkov et al. [25].

However, B3LYP and other methods were limited and did not describe well-dispersed interactions [46]. Thus, it cannot be preferred for Pd, Ni, Ir, and Rh catalytic geometry optimization. It also depends on nature, as well as the degree of intramolecular interactions in structures. For example, Jensen and coworkers described, the expansion of the geometry of various DFT methods in organometallic complexes (Figure 18.2) [25]. The organometallic complexes, precursors of catalysts and intermediates in the catalytic reaction, which were compared with computational results of crystallographic data of the DFT method, showed the best results in geometry optimization, were dispersion optimized DFT method ωB97X-D, whereas B3LYP had huge statistical errors.

18.2.2 Energy

For Pd, Ni, Rh, and Ir complex-based catalytic reactions, calculation of energy is also important [27]. Generally, energy of reaction and activation barriers have huge importance and also provide qualitative or quantitative analysis of various organometallic reactions for synthesis of organometallic reactions. Unfortunately, organometallic energy calculations cover only related small reactions [15, 47]. For the last eight years, for energy calculations, the B3LYP method has been fed, in compression with DFT methods. In various cases, it was replaced by dispersion-corrected methods, like Minnesota functionals or DFT-D3, because of the better reproducing of specific chemical parameters like activation barriers and reaction energies [21, 22, 33, 48–50]. These days, the M06 method is preferred for energy calculations, which is followed by B3LYP, DFT-D3, and M06L methods. Whereas, for transition metals, it dominated by SDD and followed by LANL2DZ.

Recently, Grimme and coworkers described different DFT methods like TPSS, PBE0, B3LYP, PW6B95, etc. for the calculation of reaction energies of Pd(II) dimer dissociation [49]. Where enthalpies and Gibbs free energy are measured via calorimetry titration. They used, PW6B95-D3 with BJ damping methods [51, 52]. M06 also showed better results after combining with Grimme's D3-dispersion system, whereas the DFT method did not count properly for dispersion and also was able to describe reaction energy. They also included that the highest source of error could be shown with calculation of the Gibbs free energy of solvation. The solvation energy is calculated at upto 2–3 kcal mol^{-1} via COSMO-RS and SMD.

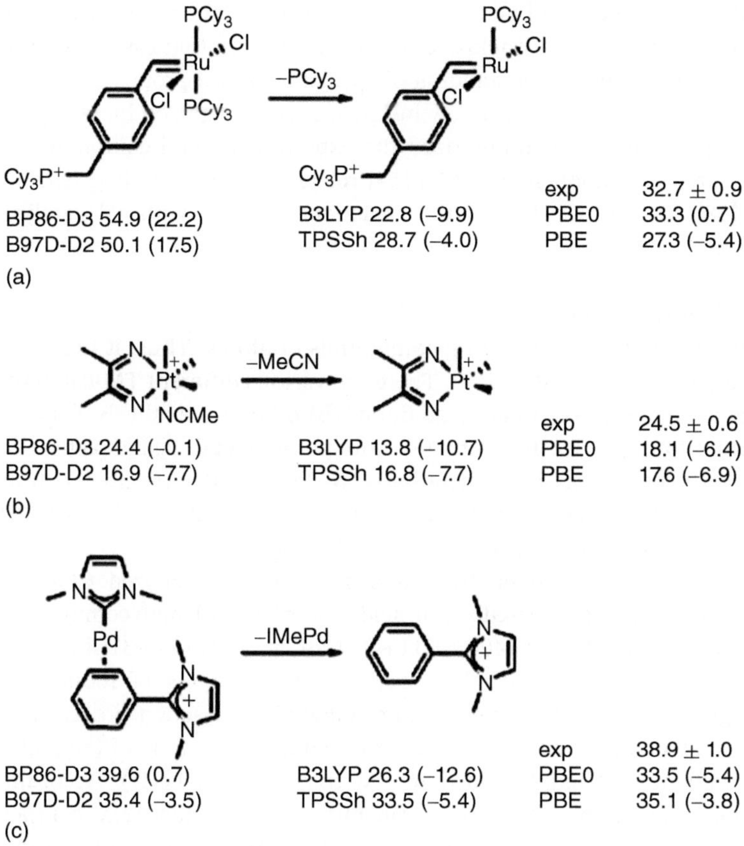

Scheme 18.1 Calculated dissociation energies for reaction transition-metal complexes of (a) Ru; (b) Pt, and (c) Pd. Source: Adapted from Weymuth et al. [53].

P. Chen and coworkers described various DFT methods like BP86, BP86-D3, B3LYP, B3LYP-D3, B97D, PBE, PBE0, TPSS, and TPSSh on the bases of dissociation of ligand energies of transition-metal complexes (Scheme 18.1) [53]. The dissociation energy of experimented gas-phase as mean absolute deviation (MAD), which obtained for PBE0 (MAD = 6.4 kcal mol^{-1}), whereas larger deviations observed for BP86-D3 (MAD = 10.7 kcal mol^{-1}) and B97D (MAD = 8.6 kcal mol^{-1}). For palladium-based reaction (Scheme 18.1c) BP86-D3 observed where, DFT methods highly system dependent.

18.3 Organometallic Catalysis Reactions

18.3.1 Palladium

Palladium catalysis offers various new transformation reactions with many structural species and is easily accessible. In a few cases, palladium was apparent and also had various pathways to consider. The mechanism of the reaction depends on reaction conditions, ligands, and additives [47]. Experimental chemistry had the necessary insight to interpret inconclusive results [15, 54]. In that case, computational chemistry had a great understanding of special chemical problems [55–57]. Computational chemistry also shown new insights, even though conducted before the experiment [15]. Here, we summarized the experimental and computational methods in different oxidation states of Pd [54] related to cross-coupling reaction mechanisms [58, 59] and C–H activation reactions [60–62] with Pd catalysis reactions.

18.3.1.1 C–H Bond Activation

The C—H bond used as functionalizing compounds of atoms. Thus, it motivates to develop catalytic reaction system [63]. There are various different TM-catalyzed C–H reactions as follows, (a) oxidative addition, (b) σ-bond metathesis, (c) electrophilic substitution, (d) 1,2-addition, (e) ambiphilic metal ligand activation (AMLA) via 4- (σ-complex-assisted metathesis, σ-CAM) and 6-membered transition states [60, 62]. Palladium had following ways of catalysis: Pd(II)/Pd(0), Pd(II)/Pd(IV), Pd(0)/Pd(II)/Pd(IV), and Pd(0)/Pd-(II) [64].

Alonso and coworkers described the experiments of Pd-catalyzed $C(sp^3)$–H activated and γ-arylation of N-covered amino acids described with both computation methods, at the CPCM (MeCN) M06/6-311+G- (2df,2p)//B3LYP/6-31G(d)(SDD) theory [65] and also described catalytic activity of complexes of monomeric Pd(II), but at ground state catalytic cycle were bimetallic. Reaction proceeds via Pd(II)/Pd(IV) catalytic cycle, which consists of initial activated C–H of the amino acid and subsequent aryl iodide added to the complexes of Pd(II). According to reaction mechanism, the aryl group transmitted from L2Pd(Ar)X complex to Pd–Alkyl complex. The catalytic cycle of Pd(II)/Pd(IV) showed with another $C(sp^3)$–H activated reaction of amino acid derivatives [66]. Here, they also described, unreported H-bonding, which had interaction with transition state of activated C–H and added two anions of bicarbonate in it (Scheme 18.2).

Ess and coworkers described the relation between C–H activation barrier and bond energy of Pd–C in transition state and for calculation used the M06/6-31G(d,p) (LANL2DZ) method, which influenced for prediction of regioselective activated C–H of heteroaromatic compounds [67]. Houk and Yu described the meta selective Pd and Ag catalyzed direct activation reaction of C–H used M06/6-311+G(d,p) (SDD)//B3LYP/6-31G(d)(LANL2DZ) method (Scheme 18.3a) [68]. And results were negative from MP2 and CCSD(T) calculations. The selectivity of Pd–Ag heterodimeric complex, coordinate with nitrile directing group, which placed an acetate anion after meta C–H position. Similarly Sunoj and Schaefer described bimetallic Pd–Ag complex in context of ortho C–H activation [70]. Zhang et al. observed and described the meta-selective C–H activation at the CPCM (benzene) B3LYP/6–311+G(d,p)(SDD)//B3LYP/6-31G(d)- (LANL2DZ) theory (Scheme 18.3b) [69]. They also explained steric repulsion between pyridine ligand and the reaction to C–H occurs at meta position. In some cases, electronic effects occurs para C–H activation.

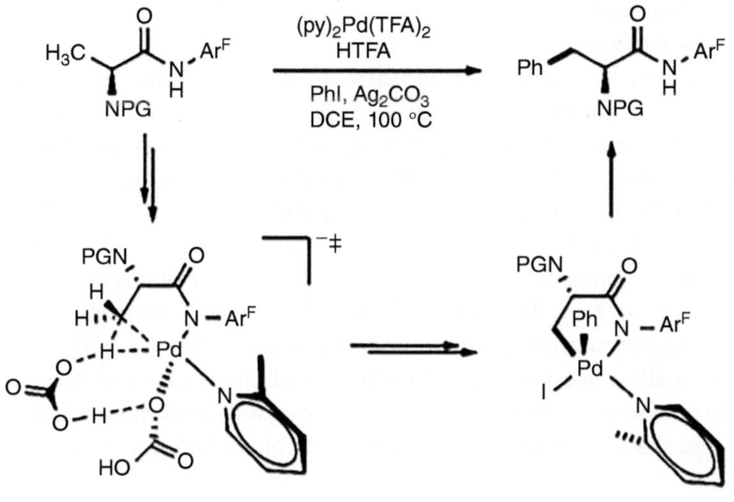

Scheme 18.2 C(sp^3)–H activated Pd-catalyzed thorough anionic bis(bicarbonate) and C–H activated transition state with Pd(IV) intermediate. Source: Adapted from Dang et al. [66].

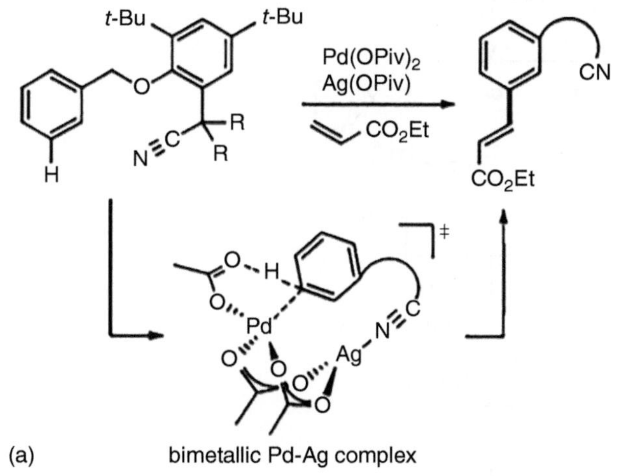

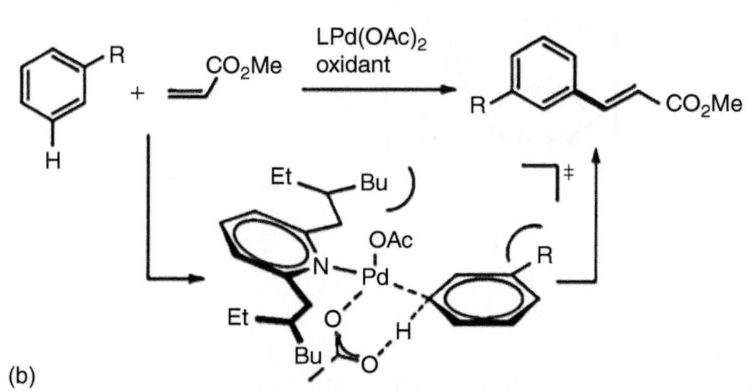

Scheme 18.3 Meta-selective C–H activation reactions: (a) nitrile directing group with Ag and Pd atoms. Source: Adapted from Yang et al. [68] (b) steric repulsion between ligand and R group directs Pd to meta position. Source: Zhang et al. Adapted from [69].

18.3.1.2 Coupling Reactions

18.3.1.2.1 Oxidative Addition

Oxidative addition was primitive step in various catalytic cycles with rate-determining step of reaction [59]. Thus, this step was prerequisite for transition-metal catalysis reaction calculations [71]. Harvey and Fey described calculation of oxidative addition of aryl halides (Cl, Br, I) to different PdL_n species with their results and kinetic data, where, L = $PtBu_3$ and PhBr [72, 73]. The IEF-PCM [74] (toluene) B3LYPD2/6-311+G(d) (aug-cc-pVTZ)//B3LYP/6-31G(d) (SDD) methods use for great applications.B3LYP method used for poor accounts the dispersion and failed to reproduce experimental observations. Depending on ligand properties, which played different and important pathway: bulky and electron-rich ligands allow access the low-coordinate and highly active species via dissociative pathways (Scheme 18.4). Whereas, smaller ligands preferred bisligated pathways. The dissociation of ligand was more relevant when catalytic cycle considered [74].

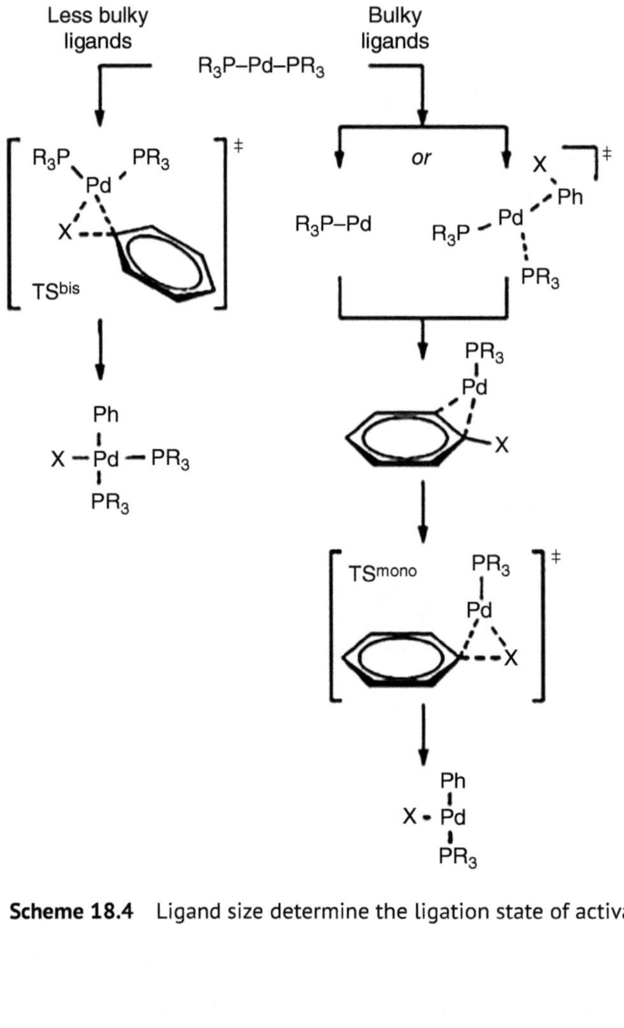

Scheme 18.4 Ligand size determine the ligation state of activated Pd [72, 73].

18.3.1.2.2 Reductive Elimination

The reductive elimination reaction, depend on hybridization state of carbon atoms, which involved in reaction and also bidentate ligands affects reaction [75, 76]. Reductive elimination from higher oxidation state (+III and +IV), Pd complexes also involved in various reactions [77, 78]. Ritter and Goddard described, reductive elimination of C–Cl and C–O from binuclear Pd(III) complexes [79].

18.3.1.2.3 Suzuki–Miyaura Coupling

Recently, Suzuki–Miyaura coupling reaction investigated via various computational studies [80, 81]. Kozuch and Martin described various ligands like PMe_3, PPh_3, and $PtBu_3$, which affect the catalytic cycle and frequency of Suzuki–Miyaura coupling reaction [82]. Different DFT methods compared with CCSD(T), which use to find out values of PCM (THF) PBE0-D3/def2-TZVPP//B97D/ def2-SV(P) method. The inefficiency of PMe_3computationally reproduced by formation of trans-PdL_2PhBr (Scheme 18.5a). For PPh_3, the high stability of PdL_3 species had turnover-limiting factor (Scheme 18.5a). For $PtBu_3$, the ligand displacement step had rate determining step (Scheme 18.5b). According to, monoligated mechanism showed, oxidative addition reaction diminished which lead to higher turnover frequency (TOF) [72, 73].

Maseras et al. described, in presence of water, trans-metalation step of Suzuki–Miyaura coupling reaction. It had following two mechanisms as follows:

1) Boronate mechanism, where R–$B(OH)_3$ – transfers the R group;
2) Hydroxo pathway through [LnPd(R′)(OH)] intermediate (Scheme 18.6) [83].

Previously, experimental and computational methods were preferred for the boronate mechanism [84–86]. Later, experimental data also supported the hydroxo mechanism [87]. Maseras and coworkers also described all pathways of water molecules coordinated with boronate, which lead for implicit solvation model. For calculations, SMD (THF)//M06/6-31G-(d,p) (SDD) level of theory used and concluded boronate mechanism was 7.4 kcal mol^{-1} over hydroxo mechanism. According to the SMD solvation model, in the optimization of species, energies changed by 1–3 kcal mol^{-1} and did not observed any change in energetic ways.

18.3.1.2.4 Heck Coupling

In recent years, various computational methods have been used for the Heck reaction with different Pd-catalysis [88, 89]. Norrby and coworkers described, the dependence of regioselectivity on tBu groups of ligand calculated via PBF (DMF) M06-D/LACVP*//B3LYP/LACVP* method for Pd-catalyzed Mizoroki–Heck coupling reaction in between aryl halide and ethyl vinyl ether (Figure 18.3) [90].

18.3.1.2.5 Allylic Substitution

Sunoj and Jindal described asymmetric multicatalytic Tsuji–Trost allylation of aldehydes with SMD (toluene) M06/6-31G(d,p)//B3LYP/6-31G(d,p)(LANL2DZ) method [91]. They also described achiral organ catalyst, which activate aldehyde.

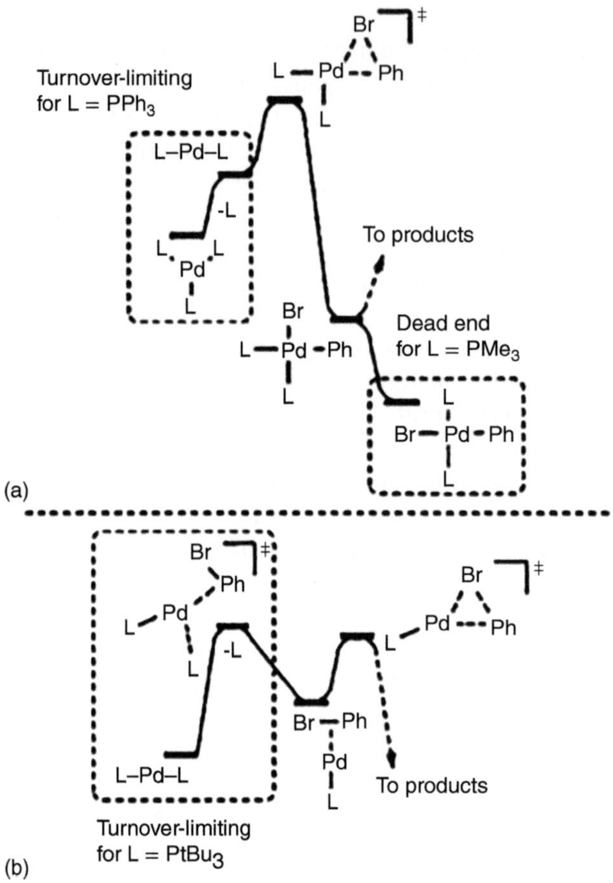

Scheme 18.5 (a) Bis-ligated mechanism of Suzuki–Miyaura coupling, where L = PPh_3 or PMe_3 with corresponding Intermediates; (b) Turnover-limiting Associative Displacement for L = $PtBu_3$ [72, 73].

Enantioselective induced through hydrogen bonding between enamine nitrogen and chiral BINOL–phosphate which acts counter ion than ligand in stereo controlling transition state.

According to Sunoj and Jindal, the formation of Pd-catalyzed, spirocyclic indene from indenyl cyclobutanol had the impotent role of ligand exchange [92]. According to calculations with SMD (toluene), M06/6-31G- (d,p)//B3LYP/6-31G(d,p) (LANL2DZ) level of theory, whereas phosphate ligands replaced with acetate anions on palladium, which substantially lower reaction activation barrier. According to Wacker mechanism, the allylic C–H activation pathway began with proton abstraction from cyclo-butanol via the outer-sphere of acetate and resulted in ring expansion. This catalytic cycle, related with ligand–ligand proton transfer and β-hydrogen abstraction steps. Thus, the optimization at SMD (toluene) M06/6-31G(d,p)(LANL2DZ) theory, proposes analogs energetic trends with aforementioned method.

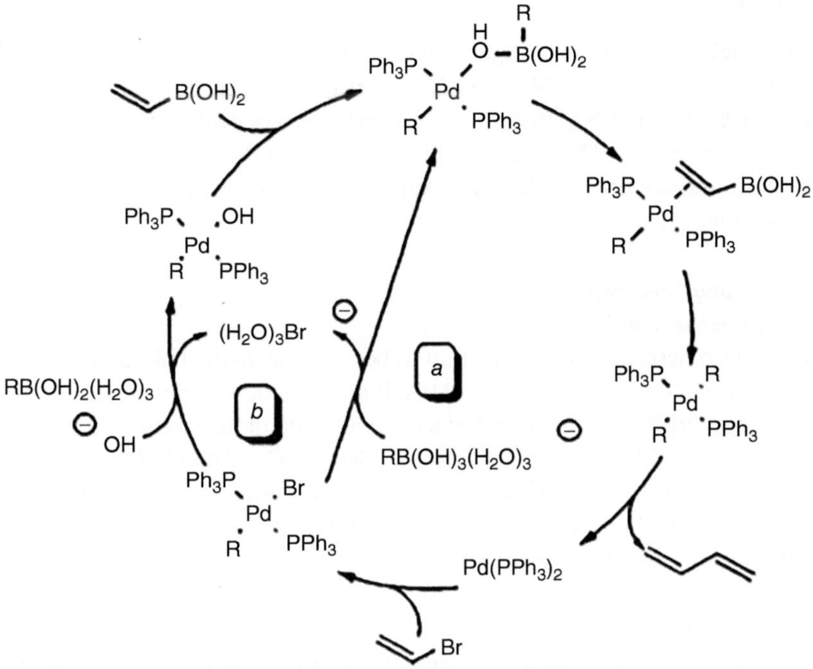

Scheme 18.6 Demonstration of (a) Boronate and (b) Hydroxo mechanisms of Suzuki–Miyaura coupling reaction. Source: Adapted from Ortuño et al. [83].

Figure 18.3 Regioselectivity of Mizoroki–Heck coupling, which depends upon *t*Bu groups on ligand. Source: Adapted from Nilsson Lill et al. [90].

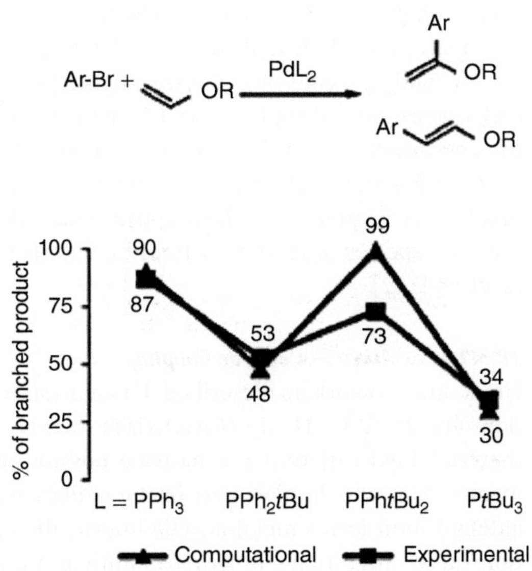

18.3.2 Nickel

These days, nickel is used in unique reactions [93, 94]. As like Pd chemistry, catalytic cycles of Ni followed Ni(0)/Ni(II) pathway. Thus, in comparison of Pd, odd oxidation states, like Ni(I)/Ni(III) cycles, which accessible with various Ni catalyzed transformations. The highest oxidation state of +IV, uncommon in nickel chemistry [95], and opposed to Pd where various catalytic cycles involved Pd(IV) intermediates [96, 97].

18.3.2.1 Coupling Reactions

18.3.2.1.1 Oxidative Addition

Allen and Locklin described about reaction in between aromatic halides and Ni(0) with B3LYP-D3/TZ2P(LANL2TZ(f)-LANL08d [98] method, which validate against CCSD(T) results [99]. They also described π complexation of haloarenes with Ni(0), which was an irreversible reaction and also took place on C—C bond of the aromatic ring. In oxidative addition, a series of ring-shaped reaction occur (Scheme 18.7). According to calculated pathways, 13CNMR kinetic isotope effect (KIE) with positive KIE values for C atoms away from C–X (X = Br, Cl).

18.3.2.1.2 Kumada Coupling

Hu and coworkers described both experimental and computational experiments with COSMO-RS(THF) PBE0-dDsC304/TZ2P//M06/def2-SVP on Ni-catalyzed, aryl–alkyl, Kumada coupling reaction [100]. They also described, aryl–alkyl coupling with Ni(I)/Ni(III) catalytic cycle, where, transmetalation reduced to turnover-determining step. The transmetalation involved bimetallic, in which two Ni(II) complexes with two electrons of alkyl halide.

Nakamura et al. described an accelerating effect of hydroxyphosphine ligands in the Ni-catalyzed coupling of unreactive aryl electrophiles, e.g. fluorides, chlorides, carbamates, and phosphates and Grignard reagents [101]. The effect and result of the coordination of ArMgX to the pendant hydroxyl group allowed a cooperative push–pull action (Scheme 18.8), which in turn lowered the barrier of C–X activation by 12 kcal mol^{-1}. Similarly coordination reaction of Grignard reagents with Ni centered species observed for fluoride anions in NiF2-catalyzed Kumada coupling reaction [102].

18.3.2.1.3 Alkyne–Aldehyde Coupling

Houk and coworkers described Ni-catalysis of alkyne–aldehyde coupling reaction via 3LYP/6-31G(d) (LANL2DZ) theory [103]. In qualitative comparison, they obtained different mechanistic possibilities with alkyne–aldehyde coupling mechanism, which takes part in the oxidative cyclization reaction of alkyne with aldehyde and forms metal-acyclic intermediate, which is followed by transmetalation (alkyl migration), β-hydride elimination, and reductive elimination reaction (Scheme 18.9).

The regioselective coupling of alkynes and aldehydes also observed with different sized carbene ligands [104]. And via M06/6–311+G(d,p) (SDD)//B3LYP/6-31G(d)-

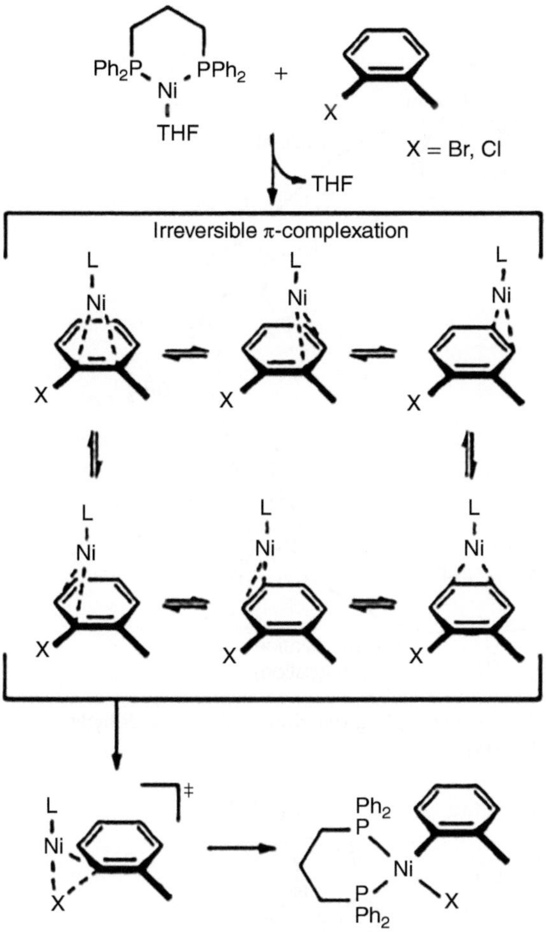

Scheme 18.7 Oxidative addition reaction of Ni with aryl bromides and chlorides (L = dppp). Source: Adapted from Sontag et al. [99].

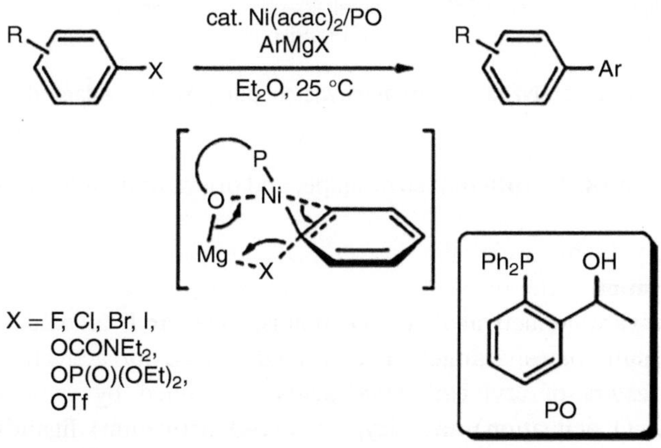

Scheme 18.8 Coordination reaction of Grignard reagent with Ni complex by hydroxyphosphine ligand. Source: Adapted from Yoshikai et al. [101].

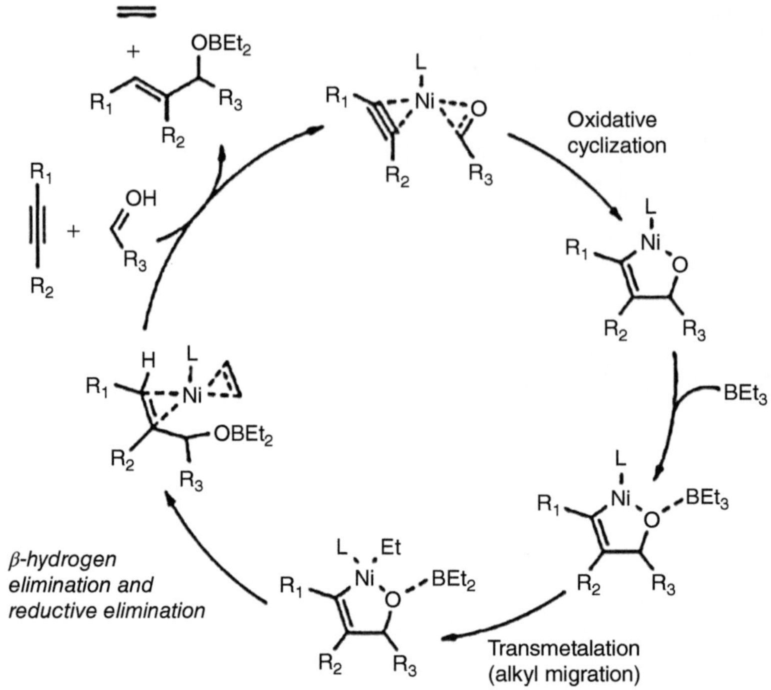

Scheme 18.9 Catalytic cycle of Ni-catalyzed coupling reaction of alkyne–aldehyde. Source: Adapted from McCarren et al. [103].

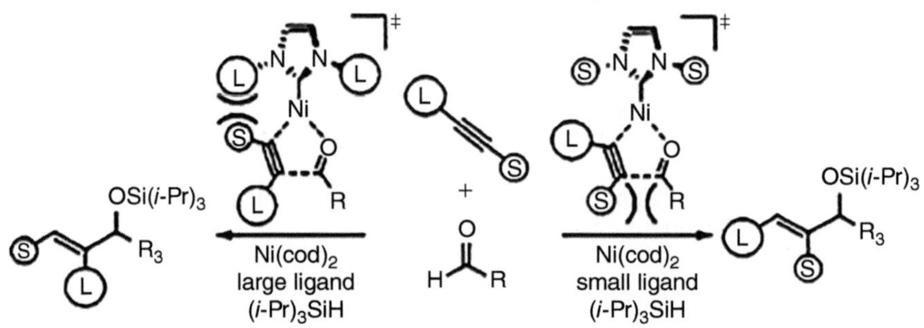

Scheme 18.10 Regioselective Ni-catalyzed alkyne–aldehyde coupling. Source: Adapted from Liu et al. [104].

(LANL2DZ) theory (Scheme 18.10), different size, shape, and orientation of ligands obtained.

18.3.2.2 C–O Bond Activation

Ni catalysts reactive toward unactivated C—O bonds, like aryl alkoxides, esters, carbamates, and sulfamates. Itamiet al. described, selectivity activation of C—O bond in aryl esters of aryl carboxylic acids, controlled by switching between PCy3 (Aryl–O activation) and dcype (Acyl–O activation) ligands (Scheme 18.11, [106]). According to Houk et al. calculations, SMD (dioxane)

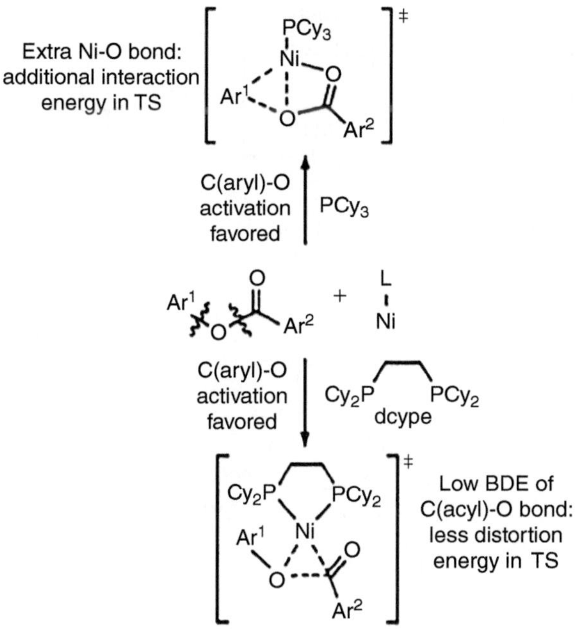

Scheme 18.11 Activation of Ni-catalyzed Ar—OBz and ArO—Bz Bond. Source: Adapted from Hong et al. [105].

M06/6-311+G(d,p)(SDD)//B3LYP/6-31G(d) (SDD) theory, differentiate between bonds through ligand from different transition state structures (Scheme 18.11, [105]). BDE of O—Acyl bond was lower and transition state associated with low distortion energy. With monoligated PCy3, five-centered transition state, which compensates for added distortion energy via interaction energy.

18.3.2.3 Cyclization Reactions

Various controlled cycloaddition reactions provided access of functionalized special ring systems [107, 108]. Morokuma, Uchiyama et al. described computational experiment at M06/6- 31+G(d)(LANL2DZ) theory of Ni-catalyzed [3+2+2] cycloaddition of cyclopropylidene acetate and two alkynes to determine reagents, which affects reaction (Scheme 18.12) [109]. Two pathways followed with various isomers of cycloheptadiene depended upon weight of alkyne substituents and their electronic state. All transition states occurs with the artificial force induced reaction (AFIR) method, which is used in multicomponent reactions [110, 111].

18.3.2.4 Olefin Functionalization

In the last few years, Ni-catalyzed olefin functionalization reactions have been investigated via computation [112]. Xie and Xiong described hydrosilylation of allenes, where specific reaction determined by metal (Pd and Ni) with ligand (1,3-bis(2,4,6-rimethylphenyl)imidazol-2-ylidene (IMes) or 1,3-bis(2,6-diisopropylphenyl)imidazol-2-ylidene (IPr)) [113]. The steric bulk of the ligand use to determine whether allene first is silylated (with bulky IPr) or hydrogenated (with IMes).

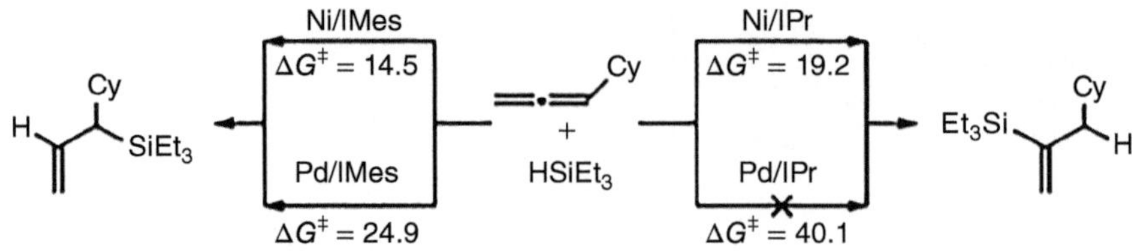

Scheme 18.12 Ni-catalyzed [3+2+2] cycloaddition of cyclopropylidene acetate with two terminal alkynes. Source: Adapted from Komagawa et al. [109].

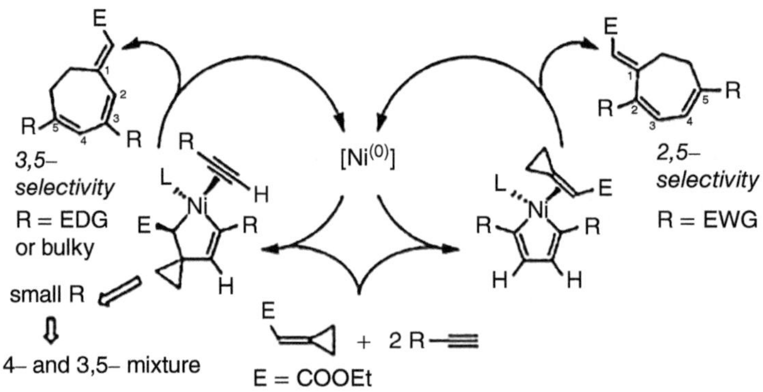

Scheme 18.13 Hydrosilylation of allenes with Ni/IPr or Pd/IMes Systems. Source: Adapted from Xie et al. [113].

Computational results get through CPCM (THF)//B3LYP/6-31G(d)(LANL2DZ+d,f) theory, which indicated Ni lowest barriers for all reactions (Scheme 18.13). The resultant product of Pd and IPr was vinylsilane, had high barrier (40 kcal mol^{-1}).

18.3.3 Rhodium

Rhodium complexes were highly versatile and also catalysts for industrial hydrogenations and hydroformylations [114]. A broad range of Rh-catalyzed reactions occurs with olefin functionalization, conjugate additions, C—H and other activated bond, as well as transformation with carbenes and cyclization reactions [115, 116]. Various experiments of Rh-catalysis reactions done by DFT as follows for example;

18.3.3.1 Hydrogenation

Hydrogenations were applied to polar double bonds like ketones and imines. Classical hydrogenation used dihydrogen (H_2), mainly employed for olefins [115]. Here, various process had hydrogen transfer reaction: (a) direct hydride transfer through Meerwein–Ponndorf–Verley (MPV) reduction; (b) indirect transfer through metal hydride species (Figure 18.4) [117]. Formation of hydride intermediate species considered by both mono and di-hydridic process. In mono-hydridic process, both inner-sphere mechanism, migratory insertion mechanism with coordinate substrate to metal center and outer-sphere mechanism, without direct coordinated

Figure 18.4 Demonstration of transition-metal catalyzed hydrogenations [117].

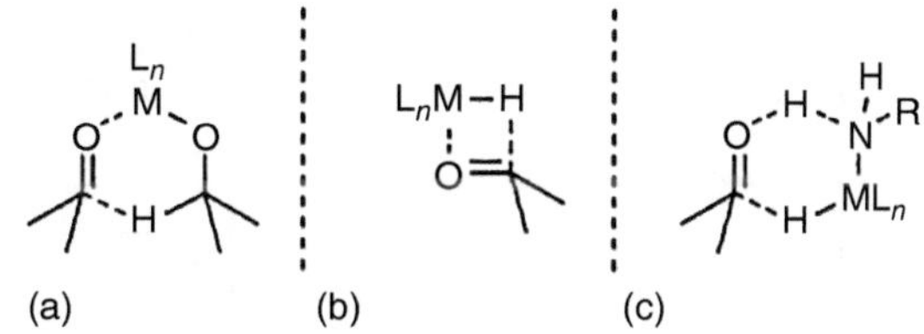

substrate to metal described [118, 119]. σ-bond metathesis considered in classical hydrogenation through H_2 [120]. Those proposal described with DFT calculations [118, 119].

18.3.3.2 Olefin Functionalization

18.3.3.2.1 Hydroformylation

It was the most important industrial method for functionalization of hydrocarbons, which was selective hydroformylation [121]. Carvajal and Shaik described mechanistic theoretical study based on CPCM (toluene) B3LYP/6-31G(d)(SDD) and hybrid ONIOM calculations [121]. It showed large xanthene ligands with η^2-coordination of terminal olefins compared with internal olefins, which increased the concentration of terminal alkene complexes and led to linear aldehyde products. Stability was increased and terminal olefin complexes were less reactive. Kumar and Jackson described, DFT calculations for investigating the olefin insertion step on Wilkinson catalytic cycle [122]. While dispersion-corrected methods like M06L and B3LYP-D3 had repeatable regioselectivities experimentally, the commonly used B3LYP functionally failed. Thus, it proposed for equatorial–equatorial arrangement of phosphine ligands, linear Rh–alkyl intermediates stabilized by noncovalent π–π (ligand–ligand) and π-CH (ligand–substrate) interactions.

18.3.3.3 Reactions of Rh Carbenoids and Nitrenoids

18.3.3.3.1 Rh Carbenoids

Transition-metal-catalyzed reactions of α-diazocarbonyl compounds used in various reactions including cyclopropanation and cyclopropenation [123, 124]. The intermolecular reactions were limited because of the elimination of β-hydride. Whereas, intramolecular Rh-catalyzed reactions of α-diazocarbonyl compounds with primary or secondary alkyl substituents were realized, intermolecular reactions proved to be more limited (Scheme 18.14). β-hydride elimination avoids the faster intramolecular reaction. Some chemo-selective reactions of α- diazocarbonyl compounds with β-tertiary alkyl substituents did not occur [125].

18.3.3.3.2 Rh Nitrenoids

Rh nitrenoids, as intermediates in C–H amination reactions [126]. Ess, Kürti, and Falck described synthesis of N–H and N–Me aziridines from olefins catalyzed by dinuclear Rh(I) [127]. According to calculation through CPCM (TFE) (U)M06/6–311+G(2d,2p)(LANL2TZ)//(U)M06/6-31G(d,p)(LANL2DZ) theory, a catalytic pathway which had Rh–nitrene species over Rh–amine and Rh–alkene catalytic cycles. Stereospecific aziridination occurs via formation of

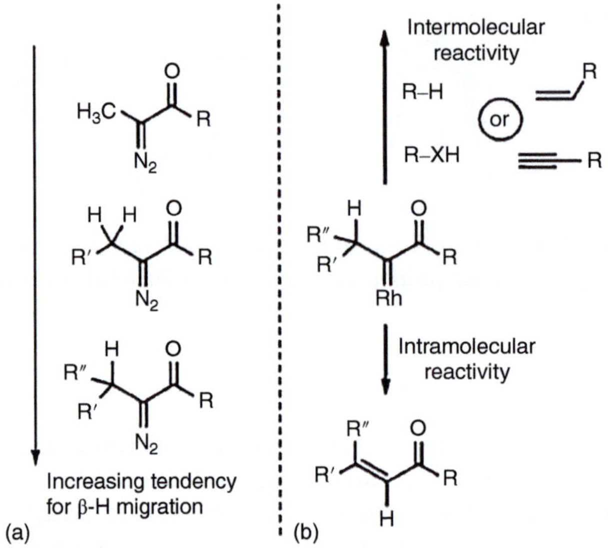

Scheme 18.14 Reactivity of α-diazocarbonyl compounds: (a) increasing the tendency of β-hydride migration and (b) demonstration of intramolecular with intermolecular reactivity. Source: Adapted from Taber et al. [125].

triplet Rh–nitrene, its reaction with olefin to form C—N bond through triplet transition state, which followed by spin interconversion to singlet diradical. The second C—N bond formed by thermoneutral coupling of singlet diradical intermediate.

18.3.3.3.3 Cyclization Reactions

[5+2] and [6+2] Cycloadditions Houk, Trost, Wender, and Z.-X. Yu described, seminal mechanistic reaction on vinylcyclopropane (VCP) cycloadditions through DFT calculations [128, 129]. That reaction pathway involves Rh–allyl complex, where, alkyne inserted into Rh—allyl bond, C—C bond formed with terminal alkenyl carbon of VCP, and resultant 7-membered ring was eliminated (Scheme 18.15). They also described a model which based on CPCM

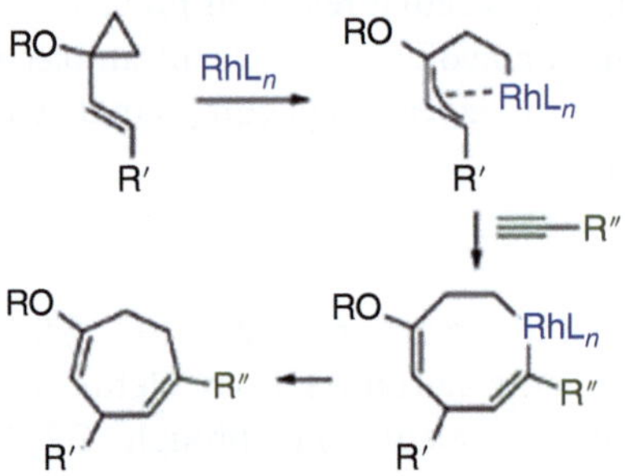

Scheme 18.15 Mechanism of Rh-catalyzed [5+2] cycloaddition of vinyl cyclopropanes (VCP) with alkynes [128, 129].

(DCE) B3LYP/6-31G(d)(SDD) calculations [130]. It showed, alkyne groups attached with distal orientation in alkyne insertion step, whereas, substitution of electron-withdrawing alkyne which decreased proximal orientation.

J. Wang et al. described, [6+2] cycloaddition of internal alkynes with cycloheptatriene [131]. On the basis of M06/6-31G(d)(LANL2DZ) calculations, reaction mechanism with oxidative coupling of cycloheptatriene and alkyne, intramolecular migration of Rh(III)–C(sp^3) bond, and reductive elimination were proposed.

Cabaleiro–Lago et al. reported hetero [5+2] cycloaddition reaction of cyclopropylimines and alkynes [132]. With DFT computations they elucidate the differences between Rh(I)-catalyzed reaction with noncatalytic variant. Rh-catalyzed transformation, two-step process with high activation barriers of 30.8 kcal mol^{-1} and multistep process possessing the lowest barriers (14.9, 15.9 kcal mol^{-1}).

[3+2] Cycloadditions Z.-X. Yu et al. reported the mechanism of Rh(I)-catalyzed intramolecular [3+2] cycloadditions of 1-ene- and 1-yne-VCPs by computations at CPCM (DCE) B3LYP/6-31G(d)- (LANL2DZ) theory [133]. This mechanism had cyclopropane cleavage, alkene or alkyne insertion, and reductive elimination, in which alkene and alkyne insertion rate were high and a stereoselective determining step (Scheme 18.16). Thus, insertion of alkynes proved more facile compared with alkenes, resulting in the observed higher reactivity of 1-yne-VCPs over 1-ene-VCPs. β-Hydride elimination is known as a possible side reaction. Thus, this reaction mechanism could inhibited completely by bidentate phosphine ligand.

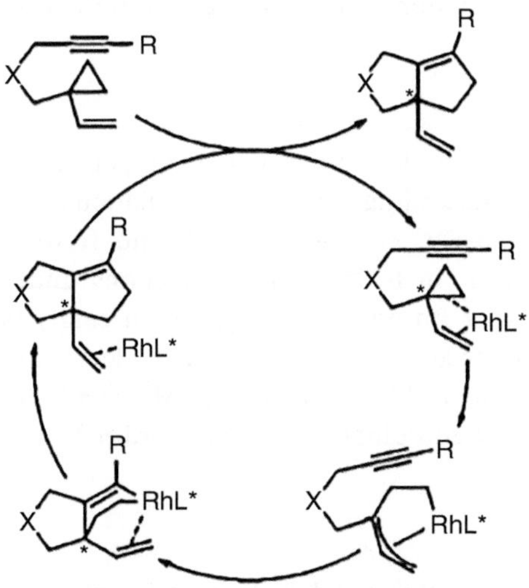

Scheme 18.16 Demonstration of mechanism of Rh-catalyzed intramolecular [3+2] cycloadditions of 1-ene and 1-yne-VCPs. Source: Adapted from Liao and Yu [134].

Z.-X. Yu et al. described their mechanistic study and also established an asymmetric species of intramolecular [3+2] cycloaddition of 1-yne-VCPs, which formed a chiral quaternary carbon center [135]. DFT studies on CPCM (DCE) M06/6-31G(d)//B3LYP/6-31G(d)(LANL2DZ) theory were used to determine the origin of enantioinduction and showed steric repulsion between alkyne moiety and BINAP ligand backbone, which responsible for cycloaddition product. They also elaborate the stereoselectivity intramolecular Rh(I)-catalyzed, electronically neutral Diels–lder reaction leading to cis-substituted bridged hydrogen atoms [134].

18.3.4 Iridium

In like rhodium, catalytic cycles occur with Rh(I)/Rh(III) manifold, which involved with Ir(V). Ir had numerous new possibilities and transformations that were inaccessible with Rh [136]. Thus, here described are selected computational reports on Ir-catalyzed.

18.3.4.1 Hydrogenation

18.3.4.1.1 Classical Hydrogenation

Ke, Zhao, and coworkers reported mechanical experiment of CO_2 hydrogenation, which provided new designed catalysts for reaction [137]. Their calculations were done by B3LYP/6–311+G(d,p)//B3LYP/6-31G(d,p)(CEP-121G [138, 139] theory. Their mechanism consist two steps as follows: First, heterolytic cleavage of H_2 which followed by hydride transfer. Later, it proceed through ligand-assisted outer-sphere reaction. The highest catalytic activity of half-sandwich Ir complexes was rationalized through analogs of Rh and Co. The higher activity of Ir is attributed to an increased ability of orbital back-donation in the rate-determination of heterolytic cleavage of H_2 (Scheme 18.17).

18.3.4.1.2 Transfer Hydrogenation

Lei et al. reported transition-metal catalysts for molecular hydrogen as a source of hydrogen in ketones [140]. According to DFT calculations via B3LYP/6–31+G(d,p)(LANL2DZ) theory, 16e–Cp* complexes of Rh and Ir were both efficient catalysts and also activated H_2 in hydrogenation reactions under acidic conditions [141, 142]. The inability of Rh and Ir hydrogenation catalysts used to activate H_2 and attributed to delocalized M—N π bonds which cleave and coordinate with H_2. Thus, added TfOH influenced the formation of M—N π bond in 16e – complexes and provided a provacant coordinating site for coordinate and activating H_2 (Scheme 18.18).

18.3.4.1.3 Dehydrogenation

Sunoj and coworkers described, Ir PCP pincer-catalyzed dehydrogenation reaction of n-hexane via DFT calculations through M06/6–311+G(d,p)-(SDD)//B3LYP/6-31G (d,p)(LANL2DZ) theory [143]. They also described six steps of catalytic cycle, where hydrogenation of tert-butylethylene used as acceptor, which formed intermediates of hexane, hexadiene, hexatriene, and electrocyclization of triene for cyclohexadiene. It was later converted to benzene via dehydroaromatization.

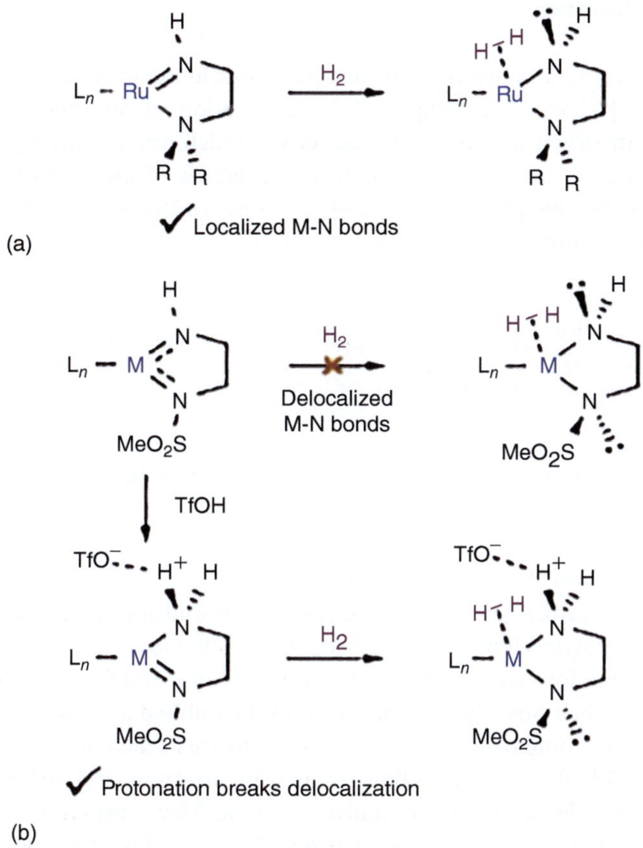

Scheme 18.17 (a) M–L weakens hydrogen bond and facilitates hydrogen cleavage by outer-sphere pathway; (b) Brønsted acid transfer to CO_2. Source: Adapted from Hou et al. [137].

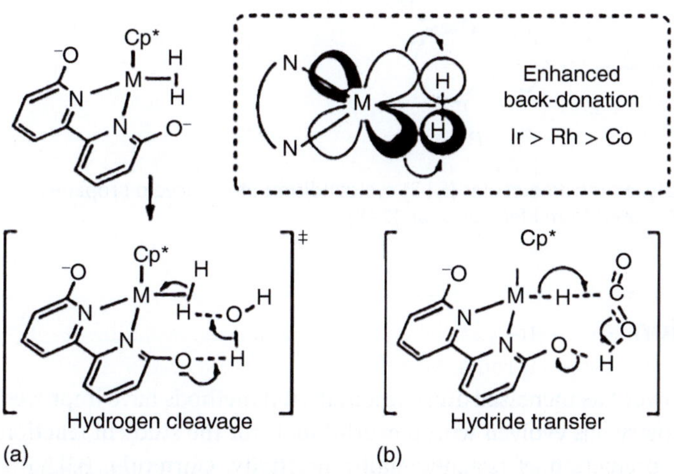

Scheme 18.18 (a) Ru-catalyzed hydrogenation [141, 142]; (b) transition-metal catalysts for transfer activated H_2 in hydrogenation reaction under acidic conditions. Source: Adapted from Lei et al. [140].

18.3.4.2 Other Bond Activations

18.3.4.2.1 C–F Activation

Macgregor et al. reported C–F activation process for hexafluorobenzene reaction with [IrMe-(PEt3)3] [144]. According to DFT calculation of benzene via BP86/6-31G(d,p)(SDD) theory. It was an ideal mechanism which proceed through four-centered transition state and involved nucleophilic attack of electron-rich Ir and trapped fluoride by the phosphine ligand (Scheme 18.19). After initial preparation of irida-phosphorane, ethyl transfer from P to Ir.

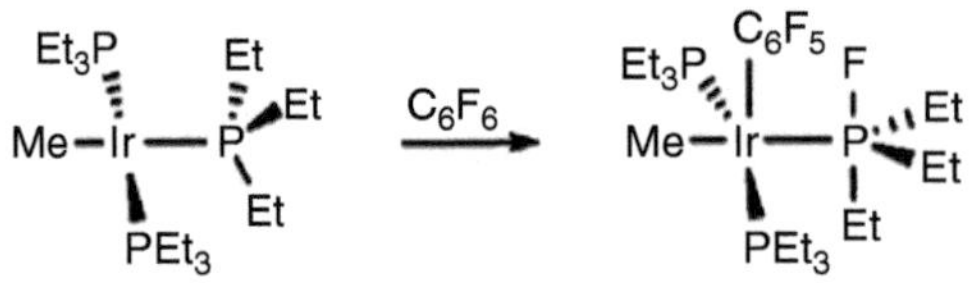

Scheme 18.19 Ir-catalyzed, activation of phosphine-assisted C–F. Source: Adapted from Erhardt and Macgregor [144].

18.3.4.2.2 Cyclization Reactions

Recently, Strand group, described the first Ir catalyzed inter- and intramolecular [5+2] cycloadditions of vinylcyclopropane and alkynes (Scheme 18.20) [145]. Previously, these transformations were carried out with Rh, Ni, Ru, and Fe catalysis [146–148]. They also described how the cationic Ir catalyst exhibited a substantial increase in reaction rate in comparison to Rh. According to DFT calculations for PBF(DCE) M06-2X/LACVP theory imply. Both migratory insertion and reductive elimination are highly facile in the Ir catalytic system. Those experiments showed increased functional group tolerance and quantitative yields of cationic Ir catalysis. It encouraged Ir catalysis and also improved catalytic processes with Rh catalysis.

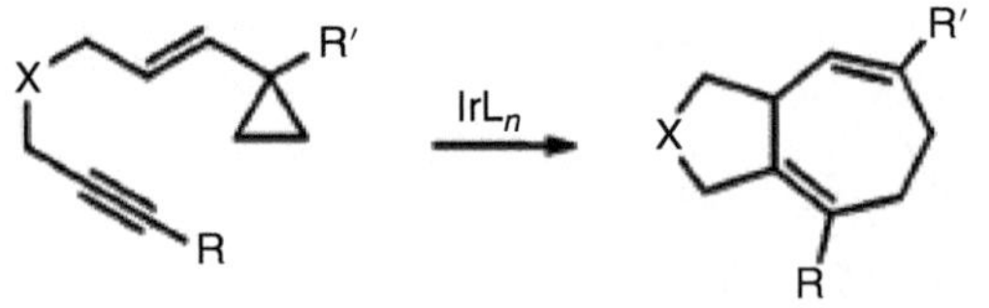

Scheme 18.20 Ir-catalyzed intramolecular [5+2] cycloaddition of vinylcyclo propanes with alkynes. Source: Adapted from Melcher et al. [145].

18.4 Conclusion

As computational power has increased and computational methods have improved, computational chemistry has evolved with powerful tools for the study of reaction, rationalization, and prediction of organometallic reactivity. Currently, B3LYP is primitively used for optimization of geometry but for energy calculations used for dispersion like M06 and DFT-D3, which are becoming highly useful because of their ability to provide a better quantitative description of reaction energies and activation barriers. Most of the reactions and calculations where Pd, Ni, Rh, and Ir complexes are involved, postexperimentally to rationalize the chemical reactivity observed. Here, we described the use of computational software or tools not only in studies but also in the development of reactions. Currently, the application of computational chemistry in the development of new synthetic concepts is limited. In the future, computational chemistry will stimulate more preexperimental applications with combined experimental-computational methods and also more prominent for organometallic reaction development.

References

1 Sameera, W. and Maseras, F. (2012). Transition metal catalysis by density functional theory and density functional theory/molecular mechanics. *Wiley Interdiscip. Rev.: Comput. Mol. Sci.* 2 (3): 375–385.

2 Foscato, M. and Jensen, V.R. (2020). Automated in silico design of homogeneous catalysts. *ACS Catal.* 10 (3): 2354–2377.

3 Burai Patrascu, M., Pottel, J., Pinus, S. et al. (2020). From desktop to benchtop with automated computational workflows for computer-aided design in asymmetric catalysis. *Nat. Catal.* 3 (7): 574–584.

4 Jacobson, L.D., Bochevarov, A.D., Watson, M.A. et al. (2017). Automated transition state search and its application to diverse types of organic reactions. *J. Chem. Theory Comput.* 13 (11): 5780–5797.

5 Soyemi, A. and Szilvási, T. (2021). Trends in computational molecular catalyst design. *Dalton Trans.* 50 (30): 10325–10339.

6 Young, T.A., Silcock, J.J., Sterling, A.J., and Duarte, F. (2021). Automated calculation of reaction energy profiles – application to organic and organometallic reactions. *Angew. Chem.* 133 (8): 4312–4320.

7 Tacey, S.A., Chen, B.W.J., Szilvási, T., and Mavrikakis, M. (2021). An automated cluster surface scanning method for exploring reaction paths on metal-cluster surfaces. *Comput. Mater. Sci.* 186: 110010.

8 Cassone, G., Giaquinta, P.V., Saija, F., and Saitta, A.M. JTJoPCB(2014). Proton conduction in water ices under an electric field. *J. Phys. Chem. B* 118 (16): 4419–4424.

9 Balcells, D., Nova, A., Clot, E. et al. (2008). Mechanism of homogeneous iridium-catalyzed alkylation of amines with alcohols from a DFT study. *Organometallics* 27 (11): 2529–2535.

10 Chamberlayne, C.F. and Zare, R.N. (2022). Microdroplets can act as electrochemical cells. *J. Chem. Phys.* 156 (5): 054705.

11 Chamberlayne, C.F. and Zare, R.N. (2022). What role does the electric double layer play in redox reactions at planar electrostatically charged insulating surfaces? *Top. Catal.* 65 (1): 228–233.

12 Xu, L., Xie, M., Yang, H. et al. (2022). In-silico screening the nitrogen reduction reaction on single-atom electrocatalysts anchored on MoS_2. *Top. Catal.* 65 (1): 234–241.

13 Barona, M., Johnson, S.I., Mbea, M. et al. (2022). Computational investigations of the reactivity of metalloporphyrins for ammonia oxidation. *Top. Catal.* 65 (1): 341–353.

14 Li, X. and Panetier, J.A. (2022). Mechanistic study of tungsten bipyridyl tetracarbonyl electrocatalysts for CO_2 fixation: exploring the roles of explicit proton sources and substituent effects. *Top. Catal.* 65 (1): 325–340.

15 Tsang, A.S.K., Sanhueza, I.A., and Schoenebeck, F. (2014). Combining experimental and computational studies to understand and predict reactivities of relevance to homogeneous catalysis. *Chem. Eur. J.* 20 (50): 16432–16441.

16 Hohenberg, P. and Kohn, W. (1964). Inhomogeneous electron gas. *Phys. Rev.* 136 (3B): B864.

17 Kohn, W. and Sham, L.J. (1965). Self-consistent equations including exchange and correlation effects. *Phys. Rev.* (4A): 140, A1133.

18 Brandenburg, J.G. and Grimme, S. (2013). Dispersion corrected Hartree–Fock and density functional theory for organic crystal structure prediction. In: *Prediction and Calculation of Crystal Structures*, 1–23. Springer.

19 Raghavachari, K., Trucks, G.W., Pople, J.A., and Head-Gordon, M. (1989). A fifth-order perturbation comparison of electron correlation theories. *Chem. Phys. Lett.* 157 (6): 479–483.

20 Deegan, M.J. and Knowles, P.J. (1994). Perturbative corrections to account for triple excitations in closed and open shell coupled cluster theories. *Chem. Phys. Lett.* 227 (3): 321–326.

21 Risthaus, T. and Grimme, S. (2013). Benchmarking of London dispersion-accounting density functional theory methods on very large molecular complexes. *J. Chem. Theory Comput.* 9 (3): 1580–1591.

22 Cramer, C.J. and Truhlar, D.G. (2009). Density functional theory for transition metals and transition metal chemistry. *Phys. Chem. Chem. Phys.* 11 (46): 10757–10816.

23 Tsipis, A.C. (2014). RETRACTED: DFT flavor of coordination chemistry. *Coord. Chem. Rev.* 272: 1–29.

24 Simón, L. and Goodman, J.M. (2011). How reliable are DFT transition structures? Comparison of GGA, hybrid-meta-GGA and meta-GGA functionals. *Org. Biomol. Chem.* 9 (3): 689–700.

25 Minenkov, Y., Singstad, Å., Occhipinti, G., and Jensen, V.R. (2012). The accuracy of DFT-optimized geometries of functional transition metal compounds: a validation study of catalysts for olefin metathesis and other reactions in the homogeneous phase. *Dalton Trans.* 41 (18): 5526–5541.

26 Bühl, M. and Kabrede, H. (2006). Geometries of transition-metal complexes from density-functional theory. *J. Chem. Theory Comput.* 2 (5): 1282–1290.

27 Niu, S. and Hall, M.B. (2000). Theoretical studies on reactions of transition-metal complexes. *Chem. Rev.* 100 (2): 353–406.

28 Schreckenbach, G., Wolff, S.K., and Ziegler, T. (2000). NMR shielding calculations across the periodic table: diamagnetic uranium compounds. 1. Methods and issues. *J. Phys. Chem. A* 104 (35): 8244–8255.

29 Chai, J.-D. and Head-Gordon, M. (2009). Long-range corrected double-hybrid density functionals. *J. Chem. Phys.* 131 (17): 174105.

30 Lee, C., Yang, W., and Parr, R.G. (1988). Development of the Colle-Salvetti correlation-energy formula into a functional of the electron density. *Phys. Rev. B* 37 (2): 785.

31 Perdew, J.P. and Yue, W. (1986). Accurate and simple density functional for the electronic exchange energy: generalized gradient approximation. *Phys. Rev. B* 33 (12): 8800.

32 Zhao, Y. and Truhlar, D.G. (2006). A new local density functional for main-group thermochemistry, transition metal bonding, thermochemical kinetics, and noncovalent interactions. *J. Chem. Phys.* 125 (19): 194101.

33 Zhao, Y. and Truhlar, D.G. (2008). The M06 suite of density functionals for main group thermochemistry, thermochemical kinetics, noncovalent interactions, excited states, and transition elements: two new functionals and systematic testing of four M06-class functionals and 12 other functionals. *Theor. Chem. Acc.* 120 (1): 215–241.

34 Grimme, S. (2006). Semiempirical GGA-type density functional constructed with a long-range dispersion correction. *J. Comput. Chem.* 27 (15): 1787–1799.

35 Grimme, S., Antony, J., Ehrlich, S., and Krieg, H. (2010). A consistent and accurate ab initio parametrization of density functional dispersion correction (DFT-D) for the 94 elements H-Pu. *J. Chem. Phys.* 132 (15): 154104.

36 Perdew, J.P., Burke, K., and Ernzerhof, M. JPrl(1996). Generalized gradient approximation made simple. *Phys. Rev. Lett.* 77 (18): 3865.

37 Tao, J., Perdew, J.P., Staroverov, V.N., and Scuseria, G.E. (2003). Climbing the density functional ladder: nonempirical meta–generalized gradient approximation designed for molecules and solids. *Phys. Rev. Lett.* 91 (14): 146401.

38 Quintal, M.M., Karton, A., Iron, M.A. et al. (2006). Benchmark study of DFT functionals for late-transition-metal reactions. *J. Phys. Chem. A* 110 (2): 709–716.

39 Holthausen, M.C. (2005). Benchmarking approximate density functional theory. I. s/d excitation energies in 3d transition metal cations. *J. Comput. Chem.* 26 (14): 1505–1518.

40 Waller, M.P., Braun, H., Hojdis, N., and Bühl, M. (2007). Geometries of second-row transition-metal complexes from density-functional theory. *J. Chem. Theory Comput.* 3 (6): 2234–2242.

41 Katcher, M.H., Norrby, P.-O., and Doyle, A.G. (2014). Mechanistic investigations of palladium-catalyzed allylic fluorination. *Organometallics* 33 (9): 2121–2133.

42 Pratt, L.M., Voit, S., Okeke, F.N., and Kambe, N. (2011). Nickel-catalyzed alkyl coupling reactions: evaluation of computational methods. *J. Phys. Chem. A* 115 (11): 2281–2290.

43 Yuan, J., Hughes, R.P., Golen, J.A., and Rheingold, A.L. (2010). Synthesis and structural and computational studies of a conformationally locked (η1-perfluoroalkylidene)(η2-alkene) transition metal complex: Ir (Cp*)($CFCF_3$) (C_2H_4). *Organometallics* 29 (8): 1942–1947.

44 Feng, X., Gu, J., Xie, Y. et al. JJoCT, Computation(2007). Homoleptic carbonyls of the second-row transition metals: evaluation of Hartree–Fock and density functional theory methods. *J. Chem. Theory Comput.* 3 (4): 1580–1587.

45 Buhl, M., Reimann, C., Pantazis, D.A. et al. (2008). Geometries of third-row transition-metal complexes from density-functional theory. *J. Chem. Theory Comput.* 4 (9): 1449–1459.

46 Schreiner, P.R., Chernish, L.V., Gunchenko, P.A. et al. (2011). Overcoming lability of extremely long alkane carbon–carbon bonds through dispersion forces. *Nature* 477 (7364): 308–311.

47 Schoenebeck, F. (2014). Ligand, additive, and solvent effects in palladium catalysis–mechanistic studies en route to catalyst design. In: *Understanding Organometallic Reaction Mechanisms Catalysis: Computational Experimental Tools* (ed. V.P. Ananikov), 69–92. Wiley.

48 Zhao, Y. and Truhlar, D.G. (2008). Density functionals with broad applicability in chemistry. *Acc. Chem. Res.* 41 (2): 157–167.

49 Hansen, A., Bannwarth, C., Grimme, S. et al. (2014). The thermochemistry of London dispersion-driven transition metal reactions: getting the 'right answer for the right reason'. *ChemistryOpen* 3 (5): 177–189.

50 Lyngvi, E., Sanhueza, I.A., and Schoenebeck, F. (2015). Dispersion makes the difference: Bisligated transition states found for the oxidative addition of Pd (Pt $Bu_3)_2$ to Ar-OSO2R and dispersion-controlled chemoselectivity in reactions with Pd [P(iPr) $(tBu_2)]_2$. *Organometallics* 34 (5): 805–812.

51 Becke, A.D. and Johnson, E.R. (2005). Exchange-hole dipole moment and the dispersion interaction. *J. Chem. Phys.* 122 (15): 154104.

52 Johnson, E.R. and Becke, A.D. (2006). A post-Hartree-Fock model of intermolecular interactions: inclusion of higher-order corrections. *J. Chem. Phys.* 124 (17): 174104.

53 Weymuth, T., Couzijn, E.P., Chen, P., and Reiher, M. (2014). New benchmark set of transition-metal coordination reactions for the assessment of density functionals. *J. Chem. Theory Comput.* 10 (8): 3092–3103.

54 Bonney, K.J. and Schoenebeck, F. (2014). Experiment and computation: a combined approach to study the reactivity of palladium complexes in oxidation states 0 to IV. *Chem. Soc. Rev.* 43 (18): 6609–6638.

55 Holder, J.C., Zou, L., Marziale, A.N. et al. (2013). Mechanism and enantioselectivity in palladium-catalyzed conjugate addition of arylboronic acids to β-substituted cyclic enones: insights from computation and experiment. *J. Am. Chem. Soc.* 135 (40): 14996–15007.

56 Li, Y. and Lin, Z. (2014). Understanding the reaction mechanisms of Pd-catalysed oxidation of alcohols and domino oxidation–arylation reactions using phenyl chloride as an oxidant. *Org. Chem. Front.* 1 (10): 1188–1196.

57 Sunesson, Y., Lime, E., Nilsson Lill, S.O. et al. (2014). Role of the base in Buchwald–Hartwig amination. *J. Org. Chem.* 79 (24): 11961–11969.

58 Xue, L. and Lin, Z. (2010). Theoretical aspects of palladium-catalysed carbon–carbon cross-coupling reactions. *Chem. Soc. Rev.* 39 (5): 1692–1705.

59 García-Melchor, M., Braga, A.A., Lledós, A. et al. (2013). Computational perspective on Pd-catalyzed C–C cross-coupling reaction mechanisms. *Acc. Chem. Res.* 46 (11): 2626–2634.

60 Balcells, D., Clot, E., and Eisenstein, O. (2010). C–H bond activation in transition metal species from a computational perspective. *Chem. Rev.* 110 (2): 749–823.

61 Musaev, D.G., Figg, T.M., and Kaledin, A.L. (2014). Versatile reactivity of Pd-catalysts: mechanistic features of the mono-N-protected amino acid ligand and cesium-halide base in Pd-catalyzed C–H bond functionalization. *Chem. Soc. Rev.* 43 (14): 5009–5031.

62 Balcells, D. and Eisenstein, O. (2013). Theoretical studies on the reaction mechanism of metal-assisted CH activation. In: *Comprehensive Inorganic Chemistry II*, From Elements to Applications, 2e (ed. J. Reedijk and K. Poeppelmeier), 695–726. Elsevier.

63 Chaumontet, M., Piccardi, R., Audic, N. et al. (2008). Synthesis of benzocyclobutenes by palladium-catalyzed C–H activation of methyl groups: method and mechanistic study. *J. Am. Chem. Soc.* 130 (45): 15157–15166.

64 Chen, X., Engle, K.M., Wang, D.H., and Yu, J.Q. (2009). Palladium (II)-catalyzed C–H activation/C–C cross-coupling reactions: versatility and practicality. *Angew. Chem. Int. Ed.* 48 (28): 5094–5115.

65 Poveda, A., Alonso, I., and Fernández-Ibáñez, M.Á. (2014). Experimental and computational studies on the mechanism of the Pd-catalyzed C (sp 3)–H γ-arylation of amino acid derivatives assisted by the 2-pyridylsulfonyl group. *Chem. Sci.* 5 (10): 3873–3882.

66 Dang, Y., Qu, S., Nelson, J.W. et al. (2015). The mechanism of a ligand-promoted C (sp 3)–H activation and arylation reaction via palladium catalysis: theoretical demonstration of a Pd (II)/Pd (IV) redox manifold. *J. Am. Chem. Soc.* 137 (5): 2006–2014.

67 Petit, A., Flygare, J., Miller, A.T. et al. (2012). Transition-state metal aryl bond stability determines regioselectivity in palladium acetate mediated C–H bond activation of heteroarenes. *Org. Lett.* 14 (14): 3680–3683.

68 Yang, Y.-F., Cheng, G.-J., Liu, P. et al. (2014). Palladium-catalyzed meta-selective C–H bond activation with a nitrile-containing template: computational study on mechanism and origins of selectivity. *J. Am. Chem. Soc.* 136 (1): 344–355.

69 Zhang, S., Shi, L., and Ding, Y. (2011). Theoretical analysis of the mechanism of palladium (II) acetate-catalyzed oxidative Heck coupling of electron-deficient arenes with alkenes: effects of the pyridine-type ancillary ligand and origins of the meta-regioselectivity. *J. Am. Chem. Soc.* 133 (50): 20218–20229.

70 Anand, M., Sunoj, R.B., and Schaefer, H.F. III, (2014). Non-innocent additives in a palladium (II)-catalyzed C–H bond activation reaction: insights into multimetallic active catalysts. *J. Am. Chem. Soc.* 136 (15): 5535–5538.

71 Algarra, A., Macgregor, S., and Panetier, J. (2013). Mechanistic Studies of CX Bond Activation at Transition-Metal Centers. In: *Comprehensive*

Inorganic Chemistry II, From Elements to Applications, 2e (ed. J. Reedijk and K. Poeppelmeier), 635–694. Elsevier.

72 McMullin, C.L., Fey, N., and Harvey, J.N. (2014). Computed ligand effects on the oxidative addition of phenyl halides to phosphine supported palladium (0) catalysts. *Dalton Trans.* 43 (36): 13545–13556.

73 McMullin, C.L., Jover, J., Harvey, J.N., and Fey, N. (2010). Accurate modelling of Pd (0)+PhX oxidative addition kinetics. *Dalton Trans.* 39 (45): 10833–10836.

74 Cances, E., Mennucci, B., and Tomasi, J. (1997). A new integral equation formalism for the polarizable continuum model: theoretical background and applications to isotropic and anisotropic dielectrics. *J. Chem. Phys.* 107 (8): 3032–3041.

75 Garcia-Melchor, M., Solans-Monfort, X., and Ujaque, G.-C.B. (2013). *Comprehensive Inorganic Chemistry II* (ed. J. Reedijk and K.R. Poeppelmeier). Amsterdam, The Netherlands: Elsevier.

76 Ananikov, V.P., Musaev, D.G., and Morokuma, K. (2005). Theoretical insight into the C–C coupling reactions of the vinyl, phenyl, ethynyl, and methyl complexes of palladium and platinum. *Organometallics* 24 (4): 715–723.

77 Powers, D.C., Geibel, M.A., Klein, J.E., and Ritter, T. (2009). Bimetallic palladium catalysis: direct observation of Pd (III)–Pd (III) intermediates. *J. Am. Chem. Soc.* 131 (47): 17050–17051.

78 Ball, N.D., Gary, J.B., Ye, Y., and Sanford, M.S. (2011). Mechanistic and computational studies of oxidatively-induced aryl–CF_3 bond-formation at Pd: rational design of room temperature aryl trifluoromethylation. *J. Am. Chem. Soc.* 133 (19): 7577–7584.

79 Powers, D.C. and Ritter, T. (2009). Bimetallic Pd(III) complexes in palladium-catalysed carbon–heteroatom bond formation. *Nat. Chem.* 1 (4): 302–309.

80 Zhang, T.-X. and Li, Z. (2013). A DFT study on Pd-catalyzed Suzuki cross-coupling polycondensation of aryl bromide monomers. *Comput. Theor. Chem.* 1016: 28–35.

81 Zhou, Y., Zhang, X., Liang, H. et al. (2014). Enantioselective synthesis of axially chiral biaryl monophosphine oxides via direct asymmetric Suzuki coupling and DFT investigations of the enantioselectivity. *ACS Catal.* 4 (5): 1390–1397.

82 Kozuch, S. and Martin, J.M. (2011). What makes for a bad catalytic cycle? A theoretical study on the Suzuki–Miyaura reaction within the energetic span model. *ACS Catal.* 1 (4): 246–253.

83 Ortuño, M.A., Lledós, A., Maseras, F., and Ujaque, G. (2014). The transmetalation process in Suzuki–Miyaura reactions: calculations indicate lower barrier via boronate intermediate. *ChemCatChem* 6 (11): 3132–3138.

84 Braga, A.A., Morgon, N.H., Ujaque, G., and Maseras, F. (2005). Computational characterization of the role of the base in the Suzuki–Miyaura cross-coupling reaction. *J. Am. Chem. Soc.* 127 (25): 9298–9307.

85 Matos, K. and Soderquist, J.A. (1998). Alkylboranes in the Suzuki–Miyaura coupling: stereochemical and mechanistic studies. *J. Org. Chem.* 63 (3): 461–470.

86 Miyaura, N. (2002). Cross-coupling reaction of organoboron compounds via base-assisted transmetalation to palladium (II) complexes. *J. Organomet. Chem.* 653 (1–2): 54–57.

87 Carrow, B.P. and Hartwig, J.F. (2011). Distinguishing between pathways for transmetalation in Suzuki–Miyaura reactions. *J. Am. Chem. Soc.* 133 (7): 2116–2119.

88 Blacque, O. and Frech, C.M. (2010). Pincer-type heck catalysts and mechanisms based on PdIV intermediates: a computational study. *Chemistry* 16 (5): 1521–1531.

89 Bäcktorp, C. and Norrby, P.-O. (2011). A DFT comparison of the neutral and cationic heck pathways. *Dalton Trans.* 40 (42): 11308–11314.

90 Nilsson Lill, S.O., Ryberg, P., Rein, T. et al. (2012). tBu or not tBu? *Chem. Eur. J.* 18 (6): 1640–1649.

91 Jindal, G. and Sunoj, R.B. (2014). Mechanistic insights on cooperative asymmetric multicatalysis using chiral counterions. *J. Org. Chem.* 79 (16): 7600–7606.

92 Jindal, G. and Sunoj, R.B. (2014). Importance of ligand exchanges in Pd(II)-Brønsted acid cooperative catalytic approach to spirocyclic rings. *J. Am. Chem. Soc.* 136 (45): 15998–16008.

93 Wilke, G. (1988). Contributions to organo-nickel chemistry. *Angew. Chem. Int. Ed. Engl.* 27 (1): 185–206.

94 Ananikov, V.P. (2015). Nickel: the "spirited horse" of transition metal catalysis. *ACS Catal.* ACS Publications 1964–1971.

95 Camasso, N.M. and Sanford, M.S. (2015). Design, synthesis, and carbon-heteroatom coupling reactions of organometallic nickel (IV) complexes. *Science* 347 (6227): 1218–1220.

96 Xu, L.-M., Li, B.-J., Yang, Z., and Shi, Z.-J. (2010). Organopalladium (IV) chemistry. *Chem. Soc. Rev.* 39 (2): 712–733.

97 Zhang, H. and Lei, A. (2011). Palladium (IV) chemistry supported by pincer type ligands. *Dalton Trans.* 40 (35): 8745–8754.

98 Roy, L.E., Hay, P.J., and Martin, R.L. (2008). Revised basis sets for the LANL effective core potentials. *J. Chem. Theory Comput.* 4 (7): 1029–1031.

99 Sontag, S.K., Bilbrey, J.A., Huddleston, N.E. et al. (2014). π-Complexation in nickel-catalyzed cross-coupling reactions. *J. Org. Chem.* 79 (4): 1836–1841.

100 Breitenfeld, J., Wodrich, M.D., and Hu, X. (2014). Bimetallic oxidative addition in nickel-catalyzed alkyl–aryl kumada coupling reactions. *Organometallics* 33 (20): 5708–5715.

101 Yoshikai, N., Matsuda, H., and Nakamura, E. (2009). Hydroxyphosphine ligand for nickel-catalyzed cross-coupling through nickel/magnesium bimetallic cooperation. *J. Am. Chem. Soc.* 131 (27): 9590–9599.

102 Hatakeyama, T., Hashimoto, S., Ishizuka, K., and Nakamura, M. (2009). Highly selective biaryl cross-coupling reactions between aryl halides and aryl Grignard reagents: a new catalyst combination of N-heterocyclic carbenes and iron, cobalt, and nickel fluorides. *J. Am. Chem. Soc.* 131 (33): 11949–11963.

103 McCarren, P.R., Liu, P., Cheong, P.H.-Y. et al. (2009). Mechanism and transition-state structures for nickel-catalyzed reductive alkyne–aldehyde coupling reactions. *J. Am. Chem. Soc.* 131 (19): 6654–6655.

104 Liu, P., Montgomery, J., and Houk, K. (2011). Ligand steric contours to understand the effects of N-heterocyclic carbene ligands on the reversal of regioselectivity in Ni-catalyzed reductive couplings of alkynes and aldehydes. *J. Am. Chem. Soc.* 133 (18): 6956–6959.

105 Hong, X., Liang, Y., and Houk, K. (2014). Mechanisms and origins of switchable chemoselectivity of Ni-catalyzed C (aryl)–O and C (acyl)–O activation of aryl esters with phosphine ligands. *J. Am. Chem. Soc.* 136 (5): 2017–2025.

106 Amaike, K., Muto, K., Yamaguchi, J., and Itami, K. (2012). Decarbonylative C–H coupling of azoles and aryl esters: unprecedented nickel catalysis and application to the synthesis of muscoride A. *J. Am. Chem. Soc.* 134 (33): 13573–13576.

107 Poater, A., Vummaleti, S.V.C., and Cavallo, L. (2013). Catalytic role of nickel in the decarbonylative addition of phthalimides to alkynes. *Organometallics* 32 (21): 6330–6336.

108 Hong, X., Liu, P., and Houk, K. (2013). Mechanism and origins of ligand-controlled selectivities in [Ni (NHC)]-catalyzed intramolecular (5+2) cycloadditions and homo-ene reactions: a theoretical study. *J. Am. Chem. Soc.* 135 (4): 1456–1462.

109 Komagawa, S., Wang, C., Morokuma, K. et al. (2013). Mechanistic origin of chemo-and regioselectivity of nickel-catalyzed [3+2+2] cyclization reaction. *J. Am. Chem. Soc.* 135 (39): 14508–14511.

110 Ohno, K. and Maeda, S. (2004). A scaled hypersphere search method for the topography of reaction pathways on the potential energy surface. *Chem. Phys. Lett.* 384 (4–6): 277–282.

111 Maeda, S., Ohno, K., and Morokuma, K. (2013). Systematic exploration of the mechanism of chemical reactions: the global reaction route mapping (GRRM) strategy using the ADDF and AFIR methods. *Phys. Chem. Chem. Phys.* 15 (11): 3683–3701.

112 Joseph, J., Rajan Babu, T.V., and Jemmis, E.D. (2009). A theoretical investigation of the Ni (II)-catalyzed hydrovinylation of styrene. *Organometallics* 28 (12): 3552–3566.

113 Xie, H., Zhao, L., Yang, L. et al. (2014). Mechanisms and origins of switchable regioselectivity of palladium-and nickel-catalyzed allene hydrosilylation with N-heterocyclic carbene ligands: a theoretical study. *J. Org. Chem.* 79 (10): 4517–4527.

114 Trzeciak, A.M. and Ziółkowski, J.J. (1999). Perspectives of rhodium organometallic catalysis. Fundamental and applied aspects of hydroformylation. *Coord. Chem. Rev.* 190: 883–900.

115 Tsuji, J. (2005). *Modern Rhodium-Catalyzed Organic Reactions*. Wiley.

116 Colby, D.A., Bergman, R.G., and Ellman JAJCr. (2010). Rhodium-catalyzed C–C bond formation via heteroatom-directed C–H bond activation. *Chem. Rev.* 110 (2): 624–655.

117 Perutz, R., Torres, O., and Vlček, A. Jr., (2013). *Comprehensive Inorganic Chemistry II* (ed. K. Poeppelmeier). Amsterdam: Elsevier.

118 Comas-Vives, A., Ujaque, G., and Lledós, A. (2009). Mechanistic evaluation of metal-catalyzed hydrogen-transfer processes: the Shvo catalyst as an example of computational unravelling. *J. Mol. Struct. THEOCHEM* 903 (1–3, 123): –32.

119 Sabater, S., Baya, M., and Mata, J.A. (2014). Highly active cp* Ir catalyst at low temperatures bearing an N-heterocyclic carbene ligand and a chelated primary benzylamine in transfer hydrogenation. *Organometallics* 33 (23): 6830–6839.

120 Polo, V., Al-Saadi, A.A., and Oro, L.A. (2014). Theoretical studies on the mechanism of iridium-catalyzed alkene hydrogenation by the cationic complex $[IrH_2(NCMe)_3(P^iPr_3)]^+$. *Organometallics* 33 (19): 5156–5163.

121 MAn, C., Kozuch, S., and Shaik, S. (2009). Factors controlling the selective hydroformylation of internal alkenes to linear aldehydes. 1. The isomerization step. *Organometallics* 28 (13): 3656–3665.

122 Kumar, M., Chaudhari, R.V., Subramaniam, B., and Jackson, T.A. (2014). Ligand effects on the regioselectivity of rhodium-catalyzed hydroformylation: density functional calculations illuminate the role of long-range noncovalent interactions. *Organometallics* 33 (16): 4183–4191.

123 Doyle, M.P. (1986). Catalytic methods for metal carbene transformations. *Chem. Rev.* 86 (5): 919–939.

124 Doyle, M.P., Duffy, R., Ratnikov, M., and Zhou, L. (2010). Catalytic carbene insertion into C–H bonds. *Chem. Rev.* 110 (2): 704–724.

125 Taber, D.F., Green, J.H., Zhang, W., and Song, R. (2000). Preparation of a cis-isoprostane synthon. *J. Org. Chem.* 65 (17): 5436–5439.

126 Roizen, J.L., Harvey, M.E., and Du Bois, J. (2012). Metal-catalyzed nitrogen-atom transfer methods for the oxidation of aliphatic C–H bonds. *Acc. Chem. Res.* 45 (6): 911–922.

127 Jat, J.L., Paudyal, M.P., Gao, H. et al. (2014). Direct stereospecific synthesis of unprotected NH and N-Me aziridines from olefins. *Science* 343 (6166): 61–65.

128 Yu, Z.-X., Wender, P.A., and Houk, K. (2004). On the mechanism of $[Rh(CO)_2Cl]_2$-catalyzed intermolecular (5+2) reactions between vinylcyclopropanes and alkynes. *J. Am. Chem. Soc.* 126 (30): 9154–9155.

129 Xu, X., Liu, P., Lesser, A. et al. (2012). Ligand effects on rates and regioselectivities of Rh (I)-catalyzed (5+2) cycloadditions: a computational study of cyclooctadiene and dinaphthocyclooctatetraene as ligands. *J. Am. Chem. Soc.* 134 (26): 11012–11025.

130 Liu, P., Sirois, L.E., Cheong, P.H.-Y. et al. (2010). Electronic and steric control of regioselectivities in Rh (I)-catalyzed (5+2) cycloadditions: experiment and theory. *J. Am. Chem. Soc.* 132 (29): 10127–10135.

131 Zhang, X., Wang, J., Zhao, H. et al. (2013). Rhodium-catalyzed [6+2] cycloaddition of internal alkynes with cycloheptatriene: catalytic study and DFT calculations of the reaction mechanism. *Organometallics* 32 (12): 3529–3536.

132 Montero-Campillo, M.M., Cabaleiro-Lago, E.M., and Rodriguez-Otero, J. (2008). A density functional theory study of rhodium-catalyzed hetero-[5+2]-cycloaddition of cyclopropyl imine derivatives and alkynes. *J. Phys. Chem. A* 112 (38): 9068–9074.

133 Jiao, L., Lin, M., and Yu, Z.-X. (2011). Density functional theory study of the mechanisms and stereochemistry of the Rh (I)-catalyzed intramolecular [3+2] cycloadditions of 1-ene-and 1-yne-vinylcyclopropanes. *J. Am. Chem. Soc.* 133 (3): 447–461.

134 Liao, W. and Yu, Z.-X. (2014). DFT study of the mechanism and stereochemistry of the Rh (I)-catalyzed Diels–Alder reactions between electronically neutral dienes and dienophiles. *J. Org. Chem.* 79 (24): 11949–11960.

135 Lin, M., Kang, G.-Y., Guo, Y.-A., and Yu, Z.-X. (2012). Asymmetric Rh (I)-catalyzed intramolecular [3+2] cycloaddition of 1-Yne-vinylcyclopropanes for bicyclo [3.3.0] compounds with a chiral quaternary carbon stereocenter and density functional theory study of the origins of enantioselectivity. *J. Am. Chem. Soc.* 134 (1): 398–405.

136 Takeuchi, R. and Kezuka, S. (2006). Iridium-catalyzed formation of carbon-carbon and carbon-heteroatom bonds. *Synthesis* 2006 (20): 3349–3366.

137 Hou, C., Jiang, J., Zhang, S. et al. (2014). Hydrogenation of carbon dioxide using half-sandwich cobalt, rhodium, and iridium complexes: DFT study on the mechanism and metal effect. *ACS Catal.* 4 (9): 2990–2997.

138 Stevens, W.J., Basch, H., and Krauss, M. (1984). Compact effective potentials and efficient shared-exponent basis sets for the first-and second-row atoms. *J. Chem. Phys.* 81 (12): 6026–6033.

139 Stevens, W.J., Krauss, M., Basch, H., and Jasien, P.G. (1992). Relativistic compact effective potentials and efficient, shared-exponent basis sets for the third-, fourth-, and fifth-row atoms. *Can. J. Chem.* 70 (2): 612–630.

140 Lei, M., Zhang, W., Chen, Y., and Tang, Y. (2010). Preference of H_2 as hydrogen source in hydrogenation of ketones catalyzed by late transition metal complexes. A DFT Study. *Organometallics* 29 (3): 543–548.

141 Sandoval, C.A., Ohkuma, T., Muñiz, K., and Noyori, R. (2003). Mechanism of asymmetric hydrogenation of ketones catalyzed by BINAP/1, 2-diamine–ruthenium (II) complexes. *J. Am. Chem. Soc.* 125 (44): 13490–13503.

142 Noyori, R., Yamakawa, M., and Hashiguchi, S. (2001). Metal–ligand bifunctional catalysis: a nonclassical mechanism for asymmetric hydrogen transfer between alcohols and carbonyl compounds. *J. Org. Chem.* 24 (66): 7931–7944.

143 Thawani, A., Rajeev, R., and Sunoj, R.B. (2013). On the mechanism of the dehydroaromatization of hexane to benzene by an iridium pincer catalyst. *Chemistry* 19 (12): 4069–4077.

144 Erhardt, S. and Macgregor, S.A. (2008). Computational study of the reaction of C6F6 with [IrMe (PEt3) 3]: identification of a phosphine-assisted C–F activation pathway via a metallophosphorane intermediate. *J. Am. Chem. Soc.* 130 (46): 15490–15498.

145 Melcher, M.C., von Wachenfeldt, H., Sundin, A., and Strand, D. (2015). Iridium catalyzed carbocyclizations: efficient (5+2) cycloadditions of vinylcyclopropanes and alkynes. *Chemistry* 21 (2): 531–535.

146 Fürstner, A., Majima, K., Martin, R. et al. (2008). A cheap metal for a "noble" task: preparative and mechanistic aspects of cycloisomerization and cycloaddition reactions catalyzed by low-valent iron complexes. *J. Am. Chem. Soc.* 130 (6): 1992–2004.

147 Trost, B.M., Toste, F.D., and Shen, H. (2000). Ruthenium-catalyzed intramolecular [5+2] cycloadditions. *J. Am. Chem. Soc.* 122 (10): 2379–2380.

148 Zuo, G. and Louie, J. (2005). Selectivity in nickel-catalyzed rearrangements of cyclopropylen-ynes. *J. Am. Chem. Soc.* 127 (16): 5798–5799.

Index

Organometallic Compounds: Synthesis, Reactions, and Applications, First Edition.
Edited by Dakeshwar Kumar Verma and Jeenat Aslam.

C

d

p